Plant Derived Bioactive Compounds in Human Health and Disease

Edited by
Surajit Pathak and Antara Banerjee

CRC Press
Taylor & Francis Group
Boca Raton London New York

CRC Press is an imprint of the
Taylor & Francis Group, an **informa** business

Designed cover image: Shutterstock

First edition published 2025
by CRC Press
2385 NW Executive Center Drive, Suite 320, Boca Raton FL 33431

and by CRC Press
4 Park Square, Milton Park, Abingdon, Oxon, OX14 4RN

CRC Press is an imprint of Taylor & Francis Group, LLC

Library of Congress Cataloging-in-Publication Data
Names: Pathak, Surajit, editor. | Banerjee, Antara (Associate professor of
stem cell biology), editor.
Title: Plant derived bioactive compounds in human health and disease /
edited by Surajit Pathak and Antara Banerjee.
Description: First edition. | Boca Raton : CRC Press, 2025. | Includes
bibliographical references and index. | Identifiers: LCCN 2024023198 (print) |
LCCN 2024023199 (ebook) | ISBN 9781032780764 (hbk) |
ISBN 9781032780801 (pbk) | ISBN 9781003486237 (ebk)
Subjects: LCSH: Plant bioactive compounds. | Phytochemicals. |
Plant bioactive compounds--Physiological effect. |
Phytochemicals--Physiological effect. |
Plant bioactive compounds--Health aspects. |
Phytochemicals--Health aspects. | Plant bioactive compounds--Therapeutic use. |
Phytochemicals--Therapeutic use.
Classification: LCC QP801.P636 P53 2025 (print) |
LCC QP801.P636 (ebook)
| DDC 572/.2--dc23/eng/20241001
LC record available at https://lccn.loc.gov/2024023198
LC ebook record available at https://lccn.loc.gov/2024023199

ISBN: 978-1-032-78076-4 (hbk)
ISBN: 978-1-032-78080-1 (pbk)
ISBN: 978-1-003-48623-7 (ebk)

DOI: 10.1201/9781003486237

Typeset in Times
by KnowledgeWorks Global Ltd.

Contents

About the Editors

Prof. Surajit Pathak, PhD, is currently working as Professor at Chettinad Academy of Research and Education, Kelambakkam, Chennai (since 2015). He has received his PhD degree from India in 2007 and completed his post-doctoral training (2007–2015) at the University of Alabama, United States; University of Padova, Italy; and University of Linkoping, Sweden.

He currently focuses his research on role of microRNAs for early detection of colorectal cancer. He has published over 140 research and review articles in peer-reviewed international journals of repute with 3800 citations and 34 h-index. He serves as an editorial board member for several international high-impact journals. Dr. Pathak has edited several books of Springer Nature and Elsevier publishers. He is a member of several professional research organizations, including the Indian Association of Cancer Research (IACR); Indian Science Congress Association; Society of Toxicology (SOT), United States; the Aesthetic Multispecialty Society, Europe (AMS); and the ReGenera Research Group in Italy and the founding member of the Indian Extracellular Vesicles Society (IEVS).

Dr. Antara Banerjee is presently working as Associate Professor at Chettinad Academy of Research and Education, Chennai (since 2016). Dr. Banerjee completed her Post-Doctoral research on Cancer Biology and Regenerative Medicine at the University of Padova, Italy and Linkoping University, Sweden. This research fostered a deep interest in the role of the mesenchymal niche and extracellular vesicles in tissue regeneration, as well as the signaling pathways that govern stem cell fate. Dr. Banerjee's laboratory focuses on analyzing the tumor-specific microenvironment in colon cancer development, with a particular emphasis on epigenetic signals as a potential target for novel therapeutic strategies.

She has published over 125 research and review articles in peer-reviewed international and national journals with citation index-2387 and h-index-24 and has active collaboration with research laboratories in India and abroad. She has been granted with two patents on her research works. She is a lifetime member and affiliated with professional research bodies including Indian Association of Cancer Research, Indian Science Congress, and Indian Society of Extracellular Vesicles.

Contributors

Suman Adhikari
Department of Chemistry
Government Degree College
Dharmanagar, Tripura(N), India

Anushka Agrawal
Division of Molecular Biology
School of Life Sciences
JSS Academy of Higher Education & Research
Mysuru, Karnataka, India

Mahadesh Prasad AJ
Department of Studies in Biochemistry
School of Life Science
SBRR Mahajana First Grade College (Autonomous)
Mysore, Karnataka, India

Anjani
Department of Biotechnology
SBRR Mahajana First Grade College (Autonomous)
Mysore, Karnataka, India

Malarvizhi Arthanari
Department of Microbiology
Vivekanandha College of Arts and Sciences for Women (Autonomous)
Namakkal, Tamil Nadu, India

Raksa Arun
Department of Genetic Engineering
School of Bioengineering
SRM Institute of Science and Technology
Kattankulathur, Tamil Nadu, India

Dhanu AS
Division of Molecular Biology
School of Life Sciences
JSS Academy of Higher Education & Research
Mysuru, Karnataka, India

Kishore Kumar B
Department of Studies in Biotechnology
Davangere University
Davangere, Karnataka, India

Usha Rani Balu
Endocrine Disruption and Reproductive Toxicology Laboratory
Department of Biotechnology
SRM Institute of Science & Technology
Chennai, Tamil Nadu, India

Antara Banerjee
Faculty of Allied Health Sciences
Chettinad Academy of Research and Education (CARE)
Chettinad Hospital and Research Institute (CHRI)
Chennai, Tamil Nadu, India

Kanthesh M Basalingappa
Department of Microbiology
Vivekanandha College of Arts and Sciences for Women (Autonomous)
Namakkal, Tamil Nadu, India

and

Division of Molecular Biology
School of Life Sciences, JSS AHER
Mysore, Karnataka, India

Indrani Biswas
MGMARI, SBV, Pillaiyarkuppam
Pillaiyarkuppam, Puducherry, India

Koushik BR
Department of Studies in Biotechnology
Davangere University
Davangere, Karnataka, India

Akshaya Chandrakanth
Endocrine Disruption and Reproductive Toxicology Laboratory
Department of Biotechnology
SRM Institute of Science & Technology
Chennai, Tamil Nadu, India

Akansha Chakraborty
Integrative Biology Research Unit
Department of Life Sciences
Presidency University
Kolkata, West Bengal, India

Nishtha Chauhan
Centre for Medical Biotechnology
Amity Institute of Biotechnology
Amity University
Noida, Uttar Pradesh, India

Tanishqua Chopra
Centre for Medical Biotechnology, Amity Institute of Biotechnology
Amity University
Noida, Uttar Pradesh, India

Sudipta Choudhury
Division of Molecular Biology
School of Life Sciences, JSS AHER
Mysore, Karnataka, India

Diptimayee Das
Faculty of Allied Health Sciences
Chettinad Academy of Research and Education (CARE)
Chettinad Hospital and Research Institute (CHRI)
Chennai, Tamil Nadu, India

Jayanta Kumar Das
Florida Memorial University
Miami Gardens, FL, United States

Madhumita Das
Miami Dade College
Miami, FL, United States

and

Palm Beach State College
Lake Worth, FL, United States

Vishal Kumar Deb
School of Health Sciences and Technology
UPES
Dehradun, Uttarakhand, India

Pradip Debnath
IGNOU Regional Centre
Siliguri, West Bengal, India

Arikketh Devi
Cancer and Stem Cell Biology Laboratory
Department of Genetic Engineering, School of Bioengineering
SRM Institute of Science and Technology
Kattankulathur, Tamil Nadu, India

Asim K Duttaroy
Department of Nutrition
Institute of Medical Sciences, Faculty of Medicine
University of Oslo
Oslo, Norway

Haema G
Pharmaceutical Chemistry
School of Pharmacy, A Unit of Sri Balaji Vidyapeeth (Deemed to be University)
Pillaiyarkuppam, Puducherry, India

Ramakrishnan Ganapathy
MGMARI, SBV
Pillaiyarkuppam, Puducherry, India

Harin N Ganesh
Department of Genetic Engineering
School of Bioengineering
SRM Institute of Science and Technology
Kattankulathur, Tamil Nadu, India

Agnishwar Girigoswami
Medical Bionanotechnology
Faculty of Allied Health SciencesChettinad Hospital and Research Institute
Chettinad Academy of Research and Education, Chettinad Health City
Kelambakkam, Tamil Nadu, India

Koyeli Girigoswami
Medical Bionanotechnology
Faculty of Allied Health Sciences
Chettinad Hospital and Research Institute
Chettinad Academy of Research and Education, Chettinad Health City
Kelambakkam, Tamil Nadu, India

Anupam Guha
Michael Madhusudan Dutta College
Sabroom, Tripura, India

Hemarangan J
Department of Genetic Engineering
School of Bioengineering
SRM Institute of Science and Technology
Kattankulathur, Tamil Nadu, India

Shambhavi Jha
Department of Genetic Engineering
School of Bioengineering, SRM Institute of Science and Technology
Kattankulathur, Tamil Nadu, India

Mangalasamy Jothika
Department of Microbiology
Dr. ALM PG IBMS, University of Madras
Chennai, Tamil Nadu, India

Kumaran K
Department of Genetic Engineering
School of Bioengineering
SRM Institute of Science and Technology
Kattankulathur, Tamil Nadu, India

Yashoda K
Department of Genetic Engineering
School of Bioengineering
SRM Institute of Science and Technology
Kattankulathur, Tamil Nadu, India

Srisri Satish Karthik
Department of Genetic Engineering
School of Bioengineering
SRM Institute of Science and Technology
Kattankulathur, Tamil Nadu, India

Lohapriya KB
Cancer and Stem Cell Biology Laboratory
Department of Genetic Engineering, School of Bioengineering
SRM Institute of Science and Technology
Kattankulathur, Tamil Nadu, India

Arul Jothi KN
Department of Genetic Engineering
School of Bioengineering
SRM Institute of Science and Technology
Kattankulathur, Tamil Nadu, India

Gobianand Kuppannan
Department of Microbiology
Vivekanandha College of Arts and Sciences for Women (Autonomous)
Namakkal, Tamil Nadu, India

Nirmala Mahendran
Department of Microbiology
Vivekanandha College of Arts and Sciences for Women (Autonomous)
Namakkal, Tamil Nadu, India

Sayani Mondal
Integrative Biology Research Unit
Department of Life Sciences
Presidency University
Kolkata, West Bengal, India

Manohar Murugan
Department of Microbiology
Vivekanandha College of Arts and Sciences for Women (Autonomous)
Namakkal, Tamil Nadu, India

Shanmuga Priya Murugan
Department of Microbiology
Vivekanandha College of Arts and Sciences for Women (Autonomous)
Namakkal, Tamil Nadu, India

Kavitha Murugesan
Department of Microbiology
Vivekanandha College of Arts and Sciences for Women (Autonomous)
Namakkal, Tamil Nadu, India

Yuvarani Muthusamy
Department of Microbiology
Vivekanandha College of Arts and Sciences for Women (Autonomous)
Namakkal, Tamil Nadu, India

Keerthi N
Faculty of Allied Health Sciences
Chettinad Academy of Research and Education (CARE)
Chettinad Hospital and Research Institute (CHRI)
Chennai, Tamil Nadu, India

Sourav Nath
Department of Chemistry, Government Degree College
Dharmanagar, Tripura(N), India

Shangavy P
Medical Bionanotechnology
Faculty of Allied Health Sciences, Chettinad Hospital and Research Institute
Chettinad Academy of Research and Education, Chettinad Health City
Kelambakkam, Tamil Nadu, India

Shriya P
Department of Genetic Engineering
School of Bioengineering
SRM Institute of Science and Technology
Kattankulathur, Tamil Nadu, India

Kesavaram Padmavathy
Department of Microbiology
Sree Balaji Dental College & Hospital
Bharath Institute of Higher Education & Research
Chennai, Tamil Nadu, India

Varsha Virendra Palol
MGMARI, SBV
Pillaiyarkuppam, Puducherry, India

Ankita Panigrahi
Division of Molecular Biology
School of Life Sciences
JSS Academy of Higher Education & Research
Mysuru, Karnataka, India

Seema J Patel
Department of Studies in Biotechnology
Davangere University
Davangere, Karnataka, India

Surajit Pathak
Faculty of Allied Health Sciences
Chettinad Academy of Research and Education (CARE)
Chettinad Hospital and Research Institute (CHRI)
Chennai, Tamil Nadu, India

Agiesh Kumar Balakrishna Pillai
MGMARI, SBV
Pillaiyarkuppam, Puducherry, India

Jayadevan Pooja
Cancer and Stem Cell Biology Laboratory
Department of Genetic Engineering, School of Bioengineering
SRM Institute of Science and Technology
Kattankulathur, Tamil Nadu, India

Dhandapani Prabu
Department of Microbiology
Dr. ALM PG IBMS, University of Madras
Chennai, Tamil Nadu, India

Kousik Pramanick
Integrative Biology Research Unit
Department of Life Sciences, Presidency University
Kolkata, West Bengal, India

J Jashwanth Raj
Medical Bionanotechnology
Faculty of Allied Health Sciences, Chettinad Hospital and Research Institute
Chettinad Academy of Research and Education, Chettinad Health City
Kelambakkam, Tamil Nadu, India

Priya Rajendran
Department of Microbiology
Vivekanandha College of Arts and Sciences for Women (Autonomous)
Namakkal, Tamil Nadu, India

Abinaya Ramachandran
Department of Microbiology
Dr. ALM PG IBMS, University of Madras
Chennai, Tamil Nadu, India

Satish Ramalingam
Department of Genetic Engineering
School of Bioengineering
SRM Institute of Science and Technology
Kattankulathur, Tamil Nadu, India

Rachana R Rao
Division of Molecular Biology
School of Life Sciences, JSS AHER
Mysore, Karnataka, India

Vijay Murali RM
Department of Genetic Engineering
School of Bioengineering
SRM Institute of Science and Technology
Kattankulathur, Tamil Nadu, India

Logesh S
Department of Biotechnology and Bioinformatics
School of Life Sciences, JSS AHER
Mysore, Karnataka, India

Winkins Santosh
PG & Research Department of Advanced Zoology & Biotechnology
Government Arts College for Men
Chennai, Tamil Nadu, India

Barathi Seetharaman
Endocrine Disruption and Reproductive Toxicology Laboratory
Department of Biotechnology
SRM Institute of Science & Technology
Chennai, Tamil Nadu, India

Girish Sharma
Centre for Medical Biotechnology, Amity Institute of Biotechnology
Amity University
Noida, Uttar Pradesh, India

and

Amity Centre for Cancer Epidemiology & Cancer Research
Amity University
Noida, Uttar Pradesh, India

Manikandan Sivaprakasam
MGM Advanced Research Institute (MGMARI)
Sri Balaji Vidyapeeth (SBV) (Deemed to be University)
Pillaiyarkuppam, Puducherry, India

Akshatha SJ
Department of Studies in Microbiology, School of Life Science
SBRR Mahajana First Grade College (Autonomous)
Mysore, Karnataka, India

Anusha SN
Department of Studies in Microbiology
School of Life Science
SBRR Mahajana First Grade College (Autonomous)
Mysore, Karnataka, India

Rose Mary Stiffin
Florida Memorial University
Miami Gardens, FL, United States

Veni Subramanyam
MGM Advanced Research Institute (MGMARI)
Sri Balaji Vidyapeeth (SBV) (Deemed to be University)
Pillaiyarkuppam, Puducherry, India

Ganesh G Tilve
Department of Biotechnology
GM Institute of Technology
Davangere, Karnataka, India

Vasanth Kanth TL
Department of Genetic Engineering
School of Bioengineering
SRM Institute of Science & Technology
Kattankulathur, Tamil Nadu, India

Gopenath TS
Department of Biotechnology and Bioinformatics
School of Life Sciences
JSS AHER
Mysore, Karnataka, India

Saranya Udayakumar
Medical Bionanotechnology
Faculty of Allied Health Sciences, Chettinad Hospital and Research Institute
Chettinad Academy of Research and Education
Chettinad Health City
Kelambakkam, Tamil Nadu, India

Murugesan Amirtha Varshini
Cancer and Stem Cell Biology Laboratory
Department of Genetic Engineering, School of Bioengineering
SRM Institute of Science and Technology
Chennai, Tamil Nadu, India

Ramasamy Vasantharekha
Endocrine Disruption and Reproductive Toxicology Laboratory
Department of Biotechnology
SRM Institute of Science & Technology
Chennai, Tamil Nadu, India

1 Harnessing Nature's Richness

Delving into the Impact of Phytochemicals from Fruit Peels on Health and Wellness

Koushik BR[1], *Seema J Patel*[1], *Kishore Kumar B*[1], *and Ganesh G Tilve*[2]

[1]Department of Studies in Biotechnology, Davangere University Davangere, India

[2]Department of Biotechnology, GM Institute of Technology Davangere, India

1.1 INTRODUCTION

Plants produce bioactive compounds known as phytochemicals that help in their protection. They are produced in various parts of plants like whole grains, vegetables, fruits, nuts, herbs, flowers, bark, and roots and present in waste parts like fruit peel and middle lamella. From ages, thousands of phytochemicals are discovered. Carotenoids, isoprenoids, polyphenols, saponins, phytosteroids, dietary fibres, and certain polysaccharides are a few examples of important phytochemicals that are known to exist. Many of the phytochemicals possess antioxidant properties and exhibit many properties like antimicrobial, antiviral, antidiarrhoeal, antispasmodic, antiallergic, and anthelmintic properties (Kumar *et al.*, 2023). In addition, they support immunity, increase gap junction communication, control gene transcription, and guard against prostate and lung malignancies (Rowles & Erdman, 2020).

1.2 USES OF FRUIT PEELS IN AGES

Many fruit peels are not edible but they are used in the preparation of medicine in tradition because they contain the extremely useful bioactive compounds. Peels from fruits are utilized in folk medicine in the treatment of various health conditions like cough, cold, digestive problems, infections, pains, and inflammations (Hussain *et al.*, 2022). In trade name "Chimpi" in Japan, the peels of *Citrus reticulata* and *Citrus unshiu* are used as crude drugs (Li Han *et al.*, 2020). In Moroccan traditional medicine, the fruit peel infused water of fresh or dried loquat (*Eriobotrya japonica* (Thunb.) Lindl.) taken for 45 days is helpful in the treatment of hypercholesterolaemia, hyperglycaemia, and cardiovascular diseases.

1.3 NUTRITIONAL PROFILE OF FRUIT PEELS

General nutrient content of fruit peels is like fibres, vitamins and minerals, antioxidants, and more. The presence of high fibre content in fruit and vegetable peels helps in promoting gut health and digestion. Most of the fruit peels contain vitamin A, vitamin C, and vitamin K. Most of the citrus family fruits contain vitamin C in abundance. However, the majority of fruits have significant in vitro antioxidant capability and biochemical indices but poor profiles of amino acids and some minerals (Hussain *et al.*, 2023) (Table 1.1).

DOI: 10.1201/9781003486237-1

TABLE 1.1
Major Phytochemicals in Different Fruit Peels and Their Activities

Fruit	Phytochemicals	Activity
***Punica granatum* L. (pomegranate)**	Ellagic acid Punicalagins Anthocyanins Quercetin Catechins Gallic acid Beta-carotene Tannins	Antimicrobial Antiallergic Anti-inflammatory Antidiabetic
***Citrus sinensis* (orange)**	Limonene Naringin Hesperidin Quercetin Polyphenols Carotenoids Pectin	Antimicrobial Anticarcinogenic Antioxidant Anti-inflammatory
***Musa paradisiaca* (banana)**	Dopamine Catechins Lutein Zeaxanthin Polyphenols Tannins Carotenoids	Antimicrobial Antioxidant
***Citrus limon* (lemon)**	Limonene Citral Naringin Hesperidin Quercetin Polyphenols Pectin	Antimicrobial Anti-inflammatory Anticarcinogenic Antioxidant
***Annona squamosa* (sugar apple)**	Annonaceous acetogenins Polyphenols Flavonoids Tannins Alkaloids Terpenoids Carotenoids	Antimicrobial Antioxidant Antitumour
***Artocarpus heterophyllus* (jackfruit)**	Tannins Saponins Alkaloids Flavonoids Carotenoids Phenolic compounds Terpenoids	Antioxidant Anti-inflammatory
***Mangifera indica* (mango)**	Quercetin Mangiferin Catechins Beta-carotene Polyphenols Tannins Anthocyanins (in some varieties) Lupeol Beta-cryptoxanthin Gallic acid	Antioxidant Antiproliferative Antidiabetic

TABLE 1.1 ***(Continued)***
Major Phytochemicals in Different Fruit Peels and Their Activities

Fruit	Phytochemicals	Activity
Vitis vinifera (grape)	Resveratrol Quercetin Anthocyanins Proanthocyanidins Catechins Flavonoids Ellagic acid Stilbenes Carotenoids Polyphenols	Antioxidant Antiproliferative Antidiabetic

1.4 PHYTOCHEMICAL EXTRACTION FROM FRUIT PEELS

Phytochemical extraction from fruit peels employs various methods such as cold extraction, plant tissue homogenization, and serial exhaustive extraction. Soxhlet extraction is notable for its efficiency in extracting non-polar phytochemicals like essential oils, while maceration and decoction are favoured for their simplicity and effectiveness in extracting polar compounds. Other techniques like infusion, digestion, and percolation offer unique approaches to extracting bioactive compounds, each with its advantages. Ultrasonic and microwave-assisted extraction methods have gained prominence for their rapidity and efficiency in extracting a diverse range of phytochemicals while reducing solvent use. These methods utilize mechanical or microwave energy to enhance extraction yields while preserving delicate compounds. Overall, a variety of extraction techniques are employed, each tailored to extract specific classes of phytochemicals efficiently from fruit peels (Nortjie *et al.*, 2022; Patel *et al.*, 2019; Shikov *et al.*, 2022).

1.5 QUALITATIVE PHYTOCHEMICAL ANALYSIS

The preliminary qualitative analysis of phytochemicals can be conducted according to the accepted procedures outlined by Trease and Evans in 1989. Trease and Evans' methods offer a valuable starting point for exploring the phytochemical landscape of plants. Their simplicity and accessibility make them suitable for educational purposes and preliminary analysis in resource-limited settings. However, it is crucial to interpret the results with caution. Common tests conducted are as follows:

Detection of alkaloids: Dragendorff's test and Mayer's test.
Detection of flavonoids: Shinoda test, ferric chloride test, lead acetate test, and sulphuric acid test.
Detection of steroids: Liebermann-Burchard test.
Detection of terpenoids: Salkowski's test.
Detection of phenols: Ferric chloride test and lead acetate test.
Detection of anthraquinones: Bontrager's test.
Detection of tannins: Ferric chloride test and gelatin test.

1.6 THERAPEUTIC BENEFITS OF SOME FRUIT PEELS

1.6.1 *Punica granatum* L. (Pomegranate)

Belongs to the *Punicaceae* family, the pomegranate (*Punica granatum*) peels make up approximately 26–30% of its overall fruit weight, forming an intricate network of membranes. Notably, these peels are abundant in phenolic compounds, flavonoids, and hydrolysable tannins. These bioactive compounds play a pivotal role in conferring antioxidant properties to both the fruit and its peels. Pomegranate is the most common researched fruit in recent years. The anti-inflammatory, antibacterial, antiallergic, and antioxidant activities of the peel are studied and published. Besides, anticancer, wound healing, and cardioprotective properties are also studied (Singh *et al.*, 2023).

1.6.2 *Citrus sinensis* (Orange)

After the development of advanced flavonoid isolation methods, the plethora of biological properties of citrus peels have got importance in health and disease management. Citrus peels are rich in various compounds, including polymethoxylated flavones (PMFs), flavanones, limonoids, and coumarins. PMFs, such as nobiletin and tangeretin, are abundant in citrus peels and have been shown to have neuroprotective effects in various neurodegenerative disease models. Flavanones, such as hesperidin, naringin, and narirutin, are also present in citrus peels and have been linked to improved cognitive performance and decreased chance of illness in clinical and epidemiological studies. Limonoids and coumarins are other classes of compounds found in citrus peels that have been associated with various health benefits, including anti-inflammatory and anticancer effects (Matsuzaki *et al.*, 2022).

1.6.3 *Musa paradisiaca* (Banana)

Belongs to family *Musaceae*, banana is a tropical fruit grown in most of the countries worldwide. Banana peels are rich in a variety of phytochemicals and biologically active compounds, including phenolic compounds such as gallic acid, carotenoids like α-carotene and β-carotene, flavonoids such as catechins and quercetin, as well as triterpenes, sterols, and alkaloids like dopamine and serotonin. These compounds have been found to possess antioxidant, anti-inflammatory, mood-enhancing, and cholesterol-lowering properties. The antioxidant and anti-inflammatory properties make these compounds valuable for protecting against chronic diseases, while the presence of carotenoids and lutein suggests potential uses in skincare and UV protection. Additionally, the alkaloids present in banana peels may have mood-enhancing effects. Overall, the diverse array of compounds found in banana peels presents a promising avenue for various health and wellness applications (Hikal *et al.*, 2022).

1.6.4 *Citrus limon* (Lemon)

Belongs to family *Rutaceae*, lemon is a common culinary fruit in kitchens for flavouring. Numerous bioactive substances, including vitamin C, citric acid, phenolic compounds, dietary fibre, and essential oil, are found in *Citrus limon* peels. High concentrations of polyphenolic chemicals, including narirutin, hesperidin, diosmin, and eriocitrin, have been found in the peel and have been linked to a number of biological benefits, including anti-inflammatory, anticancer, and antioxidant activities. *C. limon* peel powder's health advantages include lowering intestinal barrier deficiencies and inflammation in a colitic mouse experimental model, mainly to these components. Furthermore, the dietary fibres in the peel have the power to affect the makeup and activity of intestinal microbiota, which is intimately linked to human health. The microbial breakdown of dietary fibres generates organic acids that control the intestinal barrier and inflammation, which helps to lessen the symptoms of colitis (Thanh *et al.*, 2021).

1.6.5 *Annona squamosa* (Sugar Apple)

Annona squamosa belongs to family *Annonaceae* and a native fruit of Central America, usually referred to as sugar apples. *A. squamosa* peel contains a variety of particular compounds, such as tannins, triterpenoids, reducing sugars, anthraquinones, and saponins. These substances, which may also be found in other *Annonaceae* fruits, exhibit a number of biological properties, including anti-inflammatory, cytotoxic, antioxidant, and antibacterial properties. The substances found in *A. squamosa* peel have demonstrated advantages and possible applications in a number of fields. Their pharmacological properties, such as their antiviral, anti-inflammatory, and antineoplastic properties, have been investigated. Additionally, it has been discovered that *A. squamosa* peel can alter blood glucose levels, suggesting that it may have antihyperglycaemic and antidiabetic properties. Furthermore, the peel is an excellent source of bioactive substances with antioxidant qualities, such as total phenols. These substances may be able to scavenge free radicals and provide protection against oxidative stress. All things considered, the chemicals found in *A. squamosa* peel show promise as possible treatments for a range of illnesses (Ponce-Sánchez *et al.*, 2022).

1.6.6 *Artocarpus heterophyllus* (Jackfruit)

The majority of jackfruit is produced in South India and a lot of the peel is thrown away as agricultural trash. The fibrous outer peel of the ripe fruit makes up over half of its weight and is sticky, rich in calcium and pectin. The existence of alkaloids, flavonoids, glycosides, phenols, saponins, sterols, and terpenoids has been verified by a few earlier investigations. Research has also indicated that the peel extract has anti-inflammatory properties.

1.6.7 *Mangifera indica* (Mango)

The most popular tropical fruit consumed globally is the mango. The majority of the mango tree's components are proven to have therapeutic qualities. A common byproduct of the pulp production business is mango peel. Vitamin C, E, and carotenoids, as well as polyphenols, are all present in mango peel. Antioxidant activity, anti-inflammatory properties, immune system support, better digestion, and possible defence against chronic illnesses are some of the health advantages of *Mangifera indica* components (Kim *et al.*, 2021).

1.6.8 *Vitis vinifera* (Grape)

Significant levels of dietary fibre and a variety of phytochemicals, including non-flavonoids like hydroxycinnamic acids and flavonoids like anthocyanins, proanthocyanidins, and flavan-3-ols, with strong antioxidant properties may be found in grape peel. These substances may have positive effects on gut and skin health as well as the prevention and treatment of illness. For instance, human keratinocytes treated with grape peel extract enriched with resveratrol and other antioxidants demonstrated cytoprotective activity against oxidative stress, and sustained long-term administration of grape pomace extract stimulated microbial gut homeostasis and produced significant anti-aging benefits (Averilla *et al.*, 2019).

1.7 GENERAL PHYTOCHEMICAL HEALTH BENEFITS

Strong antioxidant qualities are exhibited by a variety of phytochemicals, including flavonoids, phenolic acids, and carotenoids. By scavenging free radicals, they protect cells and tissues from oxidative damage, which is linked to ageing, inflammation, and chronic illnesses including cancer, heart disease, and neurological problems.

Anti-inflammatory effects: A few types of phytochemicals have the ability to reduce inflammation in the body. They lessen the symptoms of inflammatory diseases such arthritis, inflammatory bowel disease, and asthma by inhibiting pro-inflammatory enzymes and cytokines.

Support for the immune system: By increasing immune cell activity and encouraging the generation of cytokines and antibodies, phytochemicals can influence immunological function. By strengthening the body's defences against illnesses and infections, this immune-boosting impact enhances general immune function.

Cardioprotective effects of several phytochemicals: Fruits, vegetables, and whole grains are rich sources of flavonoids and polyphenols, which are particularly beneficial for the heart. They lower blood pressure, enhance blood lipid profiles, lessen vascular inflammation, and stop blood clots from forming, all of which minimize the risk of heart disease and stroke.

Cancer prevention: By halting the proliferation of cancer cells, causing apoptosis (cell death), and preventing the development of tumours, many phytochemicals have anticancer qualities. They also lower the chance of developing different forms of cancer by protecting against DNA damage and neutralizing toxins.

Support for detoxification: The body's natural detoxification processes are aided by some phytochemicals, such as glucosinolates, which are present in cruciferous vegetables like broccoli and cabbage. They activate the enzymes that help detoxify toxic chemicals, facilitating the removal of toxins and promoting overall health.

Effects on neuroprotection: Flavonoids and polyphenols are two examples of phytochemicals with neuroprotective qualities that shield brain cells from inflammation and oxidative stress. They could lessen the chance of developing neurodegenerative illnesses like Alzheimer's and Parkinson's, as well as aid with memory improvement and cognitive performance.

Bone health: Studies have linked some phytochemicals, such as the lignans in flaxseeds and the isoflavones in soybeans, to better bone health. They could aid in promoting bone development and mineralization, lowering the risk of osteoporosis, and increasing bone density.

Skin health: By preventing UV ray damage, encouraging the production of collagen, and enhancing skin suppleness and moisture, some phytochemicals, such as flavonoids, vitamins, and carotenoids, contribute to the health of the skin. They could lessen wrinkles, improve the look of the skin overall, and stop premature ageing.

(Rai, 2021)

1.8 CONCLUSION

The examination of the impact of phytochemicals derived from fruit peels on health and wellness underscores the incredible potential inherent in nature's abundance. By reaching into the diverse range of phytochemicals within fruit peels, particularly those extracted from fruits like pomegranate, banana, orange, jackfruit, and more, wealth of compounds, including phenolic compounds, flavonoids, and hydrolysable tannins, is revealed. These bioactive substances showcase significant antioxidant properties, offering a promising avenue for health promotion and the prevention of various ailments. The utilization of these natural components symbolizes a harmonious alliance between scientific advancements and the inherent qualities of the natural world, presenting exciting opportunities for the development of functional foods and nutraceuticals that contribute to overall well-being. This exploration not only highlights the importance of fruit peels in nutritional science but also instils a deeper appreciation for a holistic approach to health that embraces the abundant offerings of nature.

REFERENCES

Averilla, Janice N., Jisun Oh, Hyo Jung Kim, Jae Sik Kim, and Jong-Sang Kim. 2019. "Potential Health Benefits of Phenolic Compounds in Grape Processing By-Products." Food Science and Biotechnology 28 (6): 1607–15. https://doi.org/10.1007/s10068-019-00628-2

Hikal, Wafaa M., Hussein A. H. Said-Al Ahl, Amra Bratovcic, Kirill G. Tkachenko, Javad Sharifi-Rad, Miroslava Kačániová, Mohammed Elhourri, and Maria Atanassova. 2022. "Banana Peels: A Waste Treasure for Human Being." Edited by Marinella de Leo. Evidence-Based Complementary and Alternative Medicine 2022 (May): 1–9. https://doi.org/10.1155/2022/7616452

Hussain, Hidayat, Nilufar Z. Mamadalieva, Amjad Hussain, Uzma Hassan, Aisha Rabnawaz, Ishtiaq Ahmed, and Ivan R. Green. 2022. "Fruit Peels: Food Waste as a Valuable Source of Bioactive Natural Products for Drug Discovery." Current Issues in Molecular Biology 44 (5): 1960–94. https://doi.org/10.3390/cimb44050134

Hussain, Tarique, Dildar Hussain Kalhoro, and Yulong Yin. 2023. "Identification of Nutritional Composition and Antioxidant Activities of Fruit Peels as a Potential Source of Nutraceuticals." Frontiers in Nutrition 9 (February). https://doi.org/10.3389/fnut.2022.1065698

Kim, Hyemee, Maria Joselyn Castellon-Chicas, Shirley Arbizu, Stephen T. Talcott, Nicholas L. Drury, Shayna Smith, and Susanne U. Mertens-Talcott. 2021. "Mango (*Mangifera indica* L.) Polyphenols: Anti-Inflammatory Intestinal Microbial Health Benefits, and Associated Mechanisms of Actions." Molecules 26 (9): 2732. https://doi.org/10.3390/molecules26092732

Kumar, Ashwani, P. Nirmal, Mukul Kumar, Anina Jose, Vidisha Tomer, Emel Oz, Charalampos Proestos, et al. 2023. "Major Phytochemicals: Recent Advances in Health Benefits and Extraction Method." Molecules 28 (2): 887. https://doi.org/10.3390/molecules28020887

Matsuzaki, Kentaro, Akira Nakajima, Yuanqiang Guo, and Yasushi Ohizumi. 2022. "A Narrative Review of the Effects of Citrus Peels and Extracts on Human Brain Health and Metabolism." Nutrients 14 (9): 1847. https://doi.org/10.3390/nu14091847

Nortjie, Elvino, Moses Basitere, Doice Moyo, and Pardon Nyamukamba. 2022. "Extraction Methods, Quantitative and Qualitative Phytochemical Screening of Medicinal Plants for Antimicrobial Textiles: A Review." Plants 11 (15): 2011. https://doi.org/10.3390/plants11152011

Patel, Komal, Namrata Panchal, and Pradnya Ingle. 2019. "Review of Extraction Techniques Extraction Methods: Microwave, Ultrasonic, Pressurized Fluid, Soxhlet Extraction, Etc." International Journal of Advanced Research in Chemical Science 6 (3). https://doi.org/10.20431/2349-0403.0603002

Ponce-Sánchez, Claudia, Víctor Hugo Oidor-Chan, Erika Lorena Álvarez-Ramírez, Rocío Gómez-Cansino, Addy Leticia Zarza-García, José Luís Gómez-Olivares, Fernando Díaz de-León-Sánchez, and José Alberto Mendoza-Espinoza. 2022. "Chemical Profile and Study of the Antidiabetic Effect of *Annona Squamosa* L. Peel." Research Square (Research Square), September. https://doi.org/10.21203/rs.3.rs-1951602/v1

Rai, Dilip K. 2021. "Phytochemicals in Food and Health." Foods 10 (4): 901. https://doi.org/10.3390/foods10040901

Rowles, Joe L., and John W. Erdman. 2020. "Carotenoids and Their Role in Cancer Prevention." Biochimica et Biophysica Acta. Molecular and Cell Biology of Lipids 1865 (11): 158613. https://doi.org/10.1016/j.bbalip.2020.158613

Saini, R. K., Ranjit, A., Sharma, K., Prasad, P., Shang, X., Gowda, K. G. M., & Keum, Y. S. (2022). Bioactive Compounds of Citrus Fruits: A Review of Composition and Health Benefits of Carotenoids, Flavonoids, Limonoids, and Terpenes. Antioxidants (Basel, Switzerland), 11(2): 239. https://doi.org/10.1016/j.bbalip.2020.158613

Shikov, Alexander N., Irina Yu. Mikhailovskaya, И. А. Наркевич, E. V. Flisyuk, and Olga N. Pozharitskaya. 2022. "Methods of Extraction of Medicinal Plants." Elsevier EBooks (January), 771–96. https://doi.org/10.1016/b978-0-323-85542-6.00029-9

Singh, Jyoti, Harpreet Kaur, Anjali Verma, A. S. Chahal, Kaushal Jajoria, Prasad Rasane, Sawinder Kaur, et al. 2023. "Pomegranate Peel Phytochemistry, Pharmacological Properties, Methods of Extraction, and Its Application: A Comprehensive Review." ACS Omega 8 (39): 35452–69. https://doi.org/10.1021/acsomega.3c02586

Thanh, Thi, Gertrude Cynthia Sitolo, Yoshinari Yamamoto, and Takuya Suzuki. 2021. "*Citrus limon* Peel Powder Reduces Intestinal Barrier Defects and Inflammation in a Colitic Murine Experimental Model." Foods 10 (2): 240–40. https://doi.org/10.3390/foods10020240

2 The Role of Bioactive Principles from Traditional Medicinal Plants in Metabolic Disorder

Shanmuga Priya Murugan[1], *Kavitha Murugesan*[1], *Yuvarani Muthusamy*[1], *Kanthesh M Basalingappa*[2], *Manohar Murugan*[3], *Nirmala Mahendran*[1], *Priya Rajendran*[1], *Gobianand Kuppannan*[1], *and Malarvizhi Arthanari*[1]

[1]Department of Microbiology, Vivekanandha College of Arts and Sciences for Women (Autonomous) Elayampalayam, Tiruchengode, Namakkal, India

[2]Course Coordinator & Associate Professor, Division of Molecular Biology at FLS, JSS Academy of Higher Education & Research Mysore, India

[3]Department of Microbiology, Sadakathullah Appa College (Autonomous) Tirunelveli, India

2.1 INTRODUCTION

Metabolism refers to the series of biochemical reactions occurring within the cells of the body that transform food into the necessary energy for various activities. This energy is vital for tasks ranging from basic movements to daily functions. Enzymes are used by the digestive system to break down protein into amino acids, lipids into triglycerides, and carbohydrates into simple sugars to aid in metabolism. The body uses the carbohydrates, amino acids, and triglycerides that are produced as energy sources by absorbing them into the bloodstream and delivering them to the cells. Once within the cells, other enzymes help to speed up or control the chemical processes that these substances undergo during their metabolism. The energy produced by these molecules can either be released into the human body during various metabolic processes or is retained in particular tissues such as the muscles, adipose tissue, and liver [1].

There are two main types of metabolism:

1. The production and storage of materials to enable the development of new cells, preserve bodily tissues, and store energy for later use are at the center of anabolism, which is defined as the constructive digestion process phase. Smaller molecules are changed into larger and more complex forms, such as proteins, lipids, and carbohydrates, by this process.
2. On the other hand, a destructive metabolism phase of catabolism is in charge of producing the energy needed for every cellular function. Cells break down big molecules in this phase to release energy, which serves as the fuel for anabolic mechanisms. This release of energy helps in muscular contractions and general bodily movement in addition to warming the body.

DOI: 10.1201/9781003486237-2

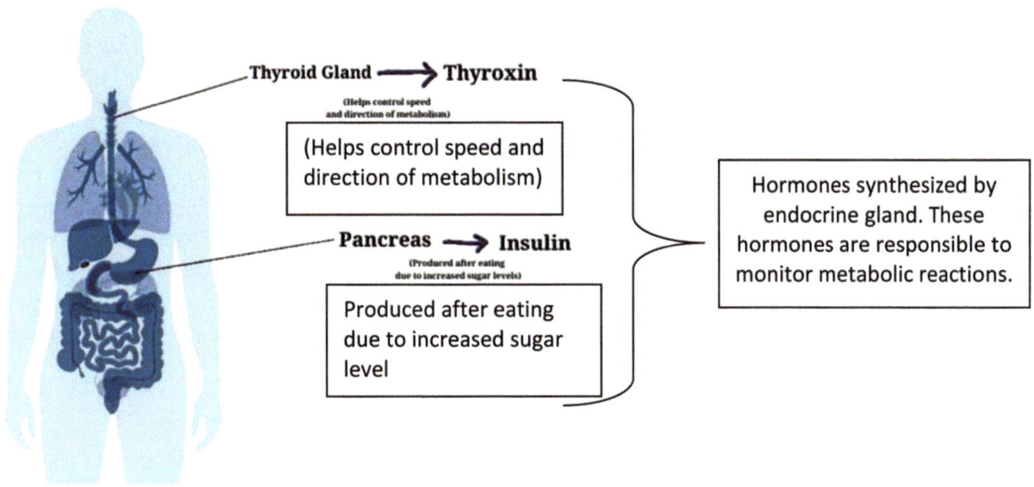

FIGURE 2.1 Regulation of Metabolism in human body by thyroid gland and pancreas.

2.2 REGULATION OF METABOLISM

The endocrine system utilizes several hormones to regulate the rate and orientation of metabolism. The thyroid gland produces and releases thyroxin, which influences the speed of metabolic chemical reactions within the body. Meanwhile, the pancreas secretes hormones that determine whether the body's primary metabolic activity is anabolic or catabolic at any given time. For instance, after a meal, there is usually a greater anabolic activity. This is because eating raises blood sugar levels in the body, which the pancreas detects and responds to by releasing insulin, a hormone that instructs cells to intensify their anabolic processes (Figure 2.1).

2.3 METABOLIC DISORDERS

Metabolic disorders encompass a group of conditions marked by disruptions in the usual metabolic processes occurring within the cells of the body. These disorders can be either genetic or acquired and affect various aspects of metabolism, breakdown of nutrients for energy, synthesis of essential molecules, and elimination of waste products [2]. They may involve deficiencies or abnormalities in enzymes, hormones, or other molecules essential for metabolic processes. Common metabolic disorders are:

- The hereditary disorder **phenylketonuria** (PKU) is defined by the body's incapacity to properly metabolize the amino acid phenylalanine, which causes an accumulation of the amino acid in the blood.
- A metabolic disease called **diabetes mellitus** causes blood sugar levels to remain elevated for an extended period.
- **Galactosemia** is a genetic disorder that impairs the body's capacity to metabolize the sugar galactose, which is present in milk and can cause damage to the kidneys and liver.
- **Maple syrup urine disease (MSUD)** is an uncommon hereditary condition that causes the brain to accumulate potentially hazardous compounds by hindering the digestion of particular amino acids, including valine, isoleucine, and leucine.
- A hereditary condition known as **Wilson's disease** results in an excess of copper buildup in the liver, brain, and other organs.
- **Glycogen storage diseases:** These disorders are characterized by improper breakdown or storage of glycogen, which affects the release and storage of energy. The errors occur in enzymes that are essential for the metabolism of glycogen.

- **Lysosomal storage disorders:** A wide range of illnesses, including Gaucher's disease, Fabry disease, and Pompe disease, are caused by defects in certain enzymes that affect how certain compounds are broken down within lysosomes.
- **Mitochondrial disorders:** These conditions impact the function of the mitochondria, which produce energy for cells. Leigh syndrome and mitochondrial myopathy are two examples.

Metabolic disorders can present with a wide range of symptoms such as development and energy metabolism irregularities, cognitive impairment, developmental delays, and organ malfunction. Medications, supportive care, enzyme replacement therapy, and dietary modifications are frequently used in treatment. For families with a track record of metabolic diseases, genetic counseling may be advised. Early identification and management are essential to treating the effects of these conditions [3].

2.4 PHYTOCHEMICALS OR BIOACTIVE PRINCIPLES

- Plants develop bioactive molecules known as phytochemicals to defend themselves. There are numerous sources of phytochemicals, notably whole grains, fruits, vegetables, nuts, and herbs. To date, over a thousand phytochemicals have been detected.
- Due to their numerous biological effects, which include anti-inflammatory, anti-carcinogenic, anti-allergenic, anti-mutagenic, and antimicrobial properties, plant bioactive compounds are crucial for maintaining human health. These properties can also help with a range of non-communicable diseases, including autoimmune, inflammatory, cardiovascular, cancer, metabolic, and neurodegenerative conditions [4, 5].

2.4.1 Medicinal Plants

Medicinal plants are known for their traditional or scientifically established therapeutic properties in treating various medical ailments. These plants contain active compounds such as alkaloids and flavonoids, which have pharmacological effects on the human body. Across different global cultures and traditional systems, including Ayurveda and Chinese Medicine, plants have been utilized for their medicinal properties, prepared in various forms such as herbal teas, tinctures, powders, essential oils, and decoctions. Examples of medicinal plants include aloe vera, chamomile, ginseng, and turmeric. While many of these medicinal plants have been demonstrated with more health benefits, they cannot be prescribed in modern pharmaceuticals, caution and consultation with healthcare professionals are needed, particularly for individuals with medical conditions or those taking regular medications, as some plants may interact in due course with the metabolism or may cause adverse effects [6].

2.5 ADRENOLEUKODYSTROPHY: ADDISON'S DISEASE

A peroxisomal condition called adrenoleukodystrophy is brought on by an uneven metabolism of very long-chain fatty acids (VLCFA). Based on the organs associated, age of onset, clinical presentation, and path of inheritance, it is divided into several subgroups. The majority of affected patients will die or experience neurological disorders, and the prognosis is severe. The pathophysiology, assessment, therapy, and prognosis of the various kinds of adrenoleukodystrophy are reviewed in this exercise, which also highlights recent advancements in the diagnosis and available or researched treatments for this illness.

2.5.1 Ashwagandha (*Withania somnifera*)

Ashwagandha, also known as *Withania somnifera*, is a unique plant that is native to areas of Africa, the Middle East, and India that has long been utilized in traditional medicine, particularly in Ayurveda. Ashwagandha's name, which means the "smell of a horse" in Sanskrit, derives from the peculiar smell of its roots. This plant is regarded as an adaptogen, meaning it is thought to enhance general well-being and assist the body in managing stress. *W. somnifera* is a well-known medicinal herb that possesses a wide variety of phytochemicals that contribute to its healing qualities.

- Withanolides are a category of naturally occurring steroids that possess the ability to adapt.
- Alkaloids: The herb includes tangerine, somniferous, and someone, among other alkaloids. Steroid lactones: It is thought that ashwagandha's antioxidant and anti-inflammatory qualities stem from these substances, which also include withaferins A and withanone. They have demonstrated promise in the study of cancer treatments and modulate the immune system.
- Saponins: Saponins are glycosides possessing soap-like qualities found in ashwagandha, and they help the plant's adaptogenic and anti-stress actions. They might also have immunomodulatory qualities.
- Flavonoids: The body uses these substances to counteract dangerous free radicals. Tannins: The astringent qualities of ashwagandha's tannins are well-known, and they may have an impact on wound healing as well as anti-inflammatory benefits.

***Withania somnifera*—Ashwagandha**

It thrives in arid regions of subtropical weather. Madhya Pradesh, Rajasthan, Punjab, Haryana, Uttar Pradesh, Gujarat, and Maharashtra.

The Chemicals found in ashwagandha may aid in lowering blood pressure, shrinking edema, relaxing the brain, and changing the immune system.

The synergistic interaction of these phytochemicals is thought to be in charge of the diverse therapeutic effects of ashwagandha, including its adaptogenic, anti-stress, anti-inflammatory, and antioxidant properties. While research on ashwagandha's phytochemicals is ongoing, current findings suggest a promising array of bioactive compounds that contribute to its traditional use in promoting overall well-being and managing various health conditions.

2.5.2 Tulsi (*"Ocimum sanctum"*)

- Also called holy basil or tulsi, is *Ocimum sanctum*. The therapeutic effects of this plant are attributed to its complex phytochemical composition.
- **Olianolic acid and ursolic acid:** These triterpenoids have anti-inflammatory, antioxidant, and anti-cancer characteristics. They enhance the plant's ability to fend off inflammation and oxidative damage.
- **Flavanoids:** Holy basil is a good source of flavonoids such as vicenin and orientin. The potential health benefits of flavonoids are well-known, and they include anti-inflammatory and antioxidant properties.
- **Eugenol:** This substance gives tulsi its distinct scent and possesses antibacterial and anti-inflammatory qualities. Other aromatic herbs, such as cloves, also contain eugenol.

- **Rosmarinic acid:** Possessing antioxidant and anti-inflammatory properties, rosmarinic acid contributes to tulsi's ability to combat oxidative stress and inflammation in the body.
- **Triterpenes:** Holy basil contains triterpenes such as ursolic acid, which may contribute to its potential anti-cancer properties and immune-modulating effects.
- **Essential oils:** The essential oils of *Ocimum sanctum*, including eugenol and linalool contribute to its aromatic properties and have been studied for their antimicrobial and stress-relieving effects.
- **Camphor:** Present in smaller amounts, camphor contributes to the overall phytochemical profile of tulsi and may have mild analgesic and anti-inflammatory properties.

***Ocimum sanctum*—Tulsi**

In India's Sasangir National Park and Gir Wildlife Sanctuary, the tulsi is an abundant plant. In humid regions and tropical areas, tulsi grows wild.

2.5.3 Bhumi Amla (*Phyllanthus niruri*)

Phyllanthus niruri, commonly known as Bhumi Amla or Chanca Piedra, is a medicinal plant containing various phytochemicals that contribute to its therapeutic properties. *P. niruri* is a plant with medicinal properties that is enriched with numerous phytochemicals. These include alkaloids, flavonoids, lignans (such as phyllanthin and hypophyllanthin), tannins, polyphenols, and terpenoids. The alkaloids from plant sources may have diverse pharmacological effects while flavonoids and polyphenols contribute to antioxidant and anti-inflammatory activities. Lignans specifically phyllanthin and hypophyllanthinare unique to *Phyllanthus* species and have been studied for their hepatoprotective effects. Tannins provide astringent properties and terpenoids have varied biological activities. All the above-mentioned phytochemicals form the basis of *P. niruri*'s traditional uses including the treatment of kidney stones, liver disorders, and antimicrobial agent. Ongoing research explores the specific mechanisms and potential applications of these compounds.

***Phyllanthus niruri* (Bhumiamla)**

It is found that Bhumi Amla grows well as a weed in various parts of the entire country, notably Punjab, Uttar Pradesh, Tamil Nadu, Maharashtra, and Sikkim.

Because of its hepatoprotective, antioxidant, and antiviral qualities, Bhumi Amla aids in the management of liver diseases and restores any liver damage that may have arisen.

2.5.4 Curcumin (*Curcuma longa*)

Curcuma longa, commonly known as turmeric, is rich in phytochemicals, with the primary active compound being curcumin, a polyphenol renowned for its antioxidant, anti-inflammatory, and potential anticancer properties. Within turmeric, curcuminoids, including curcumin, contribute to its vibrant color and numerous health benefits. Turmericones, essential oils, curcumin analogs, and polysaccharides are additional compounds found in turmeric, collectively offering a spectrum of potential therapeutic effects. The antioxidant activity of curcumin helps to combat oxidative stress, potentially benefiting overall cellular health. Also, some studies suggest a role for curcumin in cancer prevention, inhibiting tumor growth, but the bioavailability of curcumin can be enhanced when consumed with black pepper.

Curcuma longa (Curcumin)

Turmeric is one of the world's leading manufacturers and traders from India. Among the most significant states that harvest are Andhra Pradesh, Tamil Nadu, Orissa, Karnataka, West Bengal, Gujarat, Meghalaya, Maharashtra, and Assam.

Multiple medical advantages of curcumin have been scientifically demonstrated, including the potential to prevent cancer, and Alzheimer's disorder and to boost heart health. It has powerful antioxidant and anti-inflammatory properties. Likewise, it might lessen the signs of arthritis and sadness.

2.5.5 Fenugreek (*Trigonellafoenum graecum*)

This herbaceous plant contains high concentrations of plant compounds that are effective in reinstating overall health. It contains phosphates, lecithin, trigonelline, choline, triethylamine, and iron. It is also rich in sterols and therefore may be effective for the treatment of hypocortisolism.

Trigonella foenum graecum (Fenugreek)

It is found in Asia (India and China), parts of Europe, Africa, Australia, and North and South America. Fenugreek was traditionally used for diabetes and to increase milk supply in women who were breastfeeding.

It has medicinal properties such as carminative, gastric stimulant, antidiabetic, and galactogogue (lactation-inducer) effects; newer research has identified hypocholesterolemic, antilipidemia, antioxidant, hepatoprotective, anti-inflammatory, antibacterial, antifungal, antiulcer, antilithigenic, etc. galactagogue (lactation-inducer) actions.

2.5.6 Licorice

It is the root extract of a shrub called *Glycyrrhiza glabra*. Frequent consumption of licorice might be hazardous for healthy individuals but for Addison's disease, it may be effective if taken under supervision.

Licorice

The name *Glycyrrhiza* is derived from the Greek terms "glykos", meaning sweet, and "rhiza", meaning root. It is native to the Mediterranean region and grown in India, China, Spain, Iran, Russia, and Italy.

Licorice root is promoted as a dietary supplement for conditions such as digestive problems, menopausal symptoms, cough, and bacterial and viral infections.

2.6 DIABETES MELLITUS

High blood glucose levels are a hallmark of this set of metabolic illnesses related to the metabolism of carbohydrates. Type 1 diabetes is often caused by insufficient production of the hormone insulin, while type 2 diabetes is caused by poor cell response to insulin. Insulin, which is secreted by the pancreas, is necessary to carry blood glucose into cells. Diabetes is a major cause of adult blindness and a significant cardiovascular disease risk factor. Additional chronic issues comprise poor circulation leading to lower limb amputation, nerve damage, and renal failure.

2.6.1 Type 1 Diabetes

Once known as juvenile-onset or insulin-dependent diabetes, it can strike anyone at any age, but it usually manifests in late childhood as a result of insufficient insulin secretion by the pancreas. Even though there is a strong genetic component, the majority of cases are caused by an autoimmune disease that may be brought on by a virus, foreign protein, or poison in the environment. While increased blood sugar or carbohydrates in the diet do not directly cause the illness. Insulin injections, in addition to tiny, regularly scheduled meals and snacks that distribute glucose intake throughout the day and reduce blood glucose swings, are used to manage it.

2.6.1.1 *Acacia senegal*

The small, prickly deciduous tree *Senegalia senegal* (sometimes called *Acacia senegal*) belongs to the genus *Senegalia*, which is also known by several other common names, including gum arabic tree, gum acacia, Sudan gum, and Sudan gum arabic. The scientific term for *A. senegal* gum is gumarabic. Gum arabic is made up of branching oligosaccharides, a lengthy chain of galactosyl units, and glucuronic acid residues. The widespread consensus is that gum Arabic is safe to use directly as a food ingredient.

On day 16 following the injection, *S. senegal* dramatically reduced the levels of blood glucose, serum total cholesterol (TC), serum triglycerides (TTG), serum low-density lipoprotein (LDL), serum urea, and creatinine while raising the serum high-density lipoprotein (HDL) level in alloxan-induced diabetic albino rats [3]. Also, this plant extract were challenged.

> ***Acacia senegal (Senegalia senegal)***
>
> Native to the arid regions of India, such as southeast Punjab, some regions of Gujarat, and the stony Aravallis highlands, is *A. senegal.*
>
> *Acacia* is utilized as a film-forming agent in peel-off skin masks and as a pharmaceutical ingredient in medications for stomach and throat discomfort. This gum is frequently used to treat small cuts and scrapes on the skin that is afflicted. Moreover, it has a demulcent effect, which calms mucous membranes.

2.6.1.2 *Aloe sinkatana*

The leaves are thick and fleshy, green to grey-green, with some varieties showing white flecks on their upper and lower stem surfaces. The margin of the leaf is serrated and has small white teeth.

The effects of aqueous extracts of *A. sinkatana* leaves on blood glucose and lipid profile in type 2 diabetic patients were evaluated [6]. The volunteers of the experimental group received the aqueous extract (500 g/l) at a dose of 5 mL/day. A significant reduction in levels of fasting blood sugar, TTG, TC, and LDL and a significant increase in HDL levels was observed.

> ***Aloe sinkatana***
>
> *Aloe sinkatana* blooms make long-lasting cut flowers. The leaf juice is valued locally to treat a variety of ailments including skin diseases and digestion problems, fevers, diabetes, tonsillitis, and inflammatory colon.
>
> Among these single species, Aloe vera (Linnaeus) Burman f. is present in India and distributed in the extremely dry parts of Rajasthan and Gujarat state. In other states of India, it is grown as a medicinal plant.

2.6.2 Onion (*Allium cepa*)

An **onion**, also known as the **bulb onion** or **common onion**, is a vegetable that is the most widely cultivated species of the genus *Allium*. Its close relatives include garlic, scallion, leek, and chive. Onion has been known to possess various medicinal properties from ancient times. Among various activities of *Allium cepa*, regulation of hypoglycemic activity is considered one of its important effects in DM. Sulfur compounds including S-methylcysteine and flavonoids such as quercetin are mainly responsible for the hypoglycemic activity of *A. cepa*. S-methyl cysteine and flavonoids help to decrease the levels of blood glucose, serum lipids, oxidative stress, and lipid peroxidation, as well as increase antioxidant enzyme activity and insulin secretion [7].

Allium cepa

Grown for its edible bulb, the onion (*A. cepa*) is a herbaceous biennial plant in the *Amaryllis* family (*Amaryllidaceae*). Though it is currently grown all throughout the world, mostly in temperate regions, onions are most likely native to southwest Asia.

The herb has long been used to treat a variety of illnesses, including bronchitis-related cough, asthma, inflammatory conditions, dysentery, ulcer wounds, scars, keloids, and pain and swelling following wasp or bee stings [8].

2.6.3 Garlic (*Allium sativum*)

One species of bulbous blooming plant in the *Allium* genus is garlic (*Allium sativum*). The onion, shallot, leek, chive, Welsh onion, and Chinese onion are some of its near relatives. It is indigenous to northeastern Iran, Central Asia, and South Asia. The perennial flowering plant *A. sativum* emerges from a bulb. It has an up to 1 m (3 ft) tall, erect blooming stem. The leaf blade has an acute tip and is flat, linear, solid, and measures around 1.25–2.5 cm (0.5–1.0 in) in width. When rats with diabetes were given an aqueous extract of *A. sativum*, blood glucose, total serum lipids, and cholesterol levels all decreased [9].

Allium sativum

It is indigenous to northeastern Iran, Central Asia, and South Asia.

Garlic reduces blood pressure, cholesterol, and sugar levels, inhibits and eliminates bacteria, fungus, and parasites, and guards the liver by preventing blood clotting. It has anticancer properties as well [10, 11].

2.6.3.1 *Ambrosia maritima*

Ragweed is extremely drought tolerant and will grow on poor soils when exposed to full light. In late summer and early fall, the male flowers discharge large volumes of pollen, which the wind disperses. When albino rats were given water, 50% ethanolic, ether, or petroleum ether extracts of the entire *A. maritima* plant, blood glucose levels were dramatically lowered after 1.5 and 2 hours, but insulin levels were not significantly affected [12]. The majority of biochemical parameters, such as serum insulin, total proteins, albumin, globulin, HDL, aspartate aminotransferase (AST), alanine aminotransferase (ALT), urea, creatinine, uric acid, serum TC, TTG, and LDL, significantly improved in alloxan-induced diabetic albino rats administered with an aqueous extract of *A. maritima* aerial parts twice daily [13].

Ambrosia maritima

Ambrosia maritima L. is commonly used in North African traditional medicine to treat infections, inflammatory illnesses, gastrointestinal and urinary tract abnormalities, rheumatic pain, and respiratory issues. It is traditionally known as Hachich el Aouinet in Algeria and as Damsissa in Egypt and Sudan.

It is indigenous to North America and can be found all over the continent as well as in Canada.

2.6.3.2 *Ammi visnaga*

Visnaga daucoides is a species of flowering plant in the carrot family that goes by a variety of common names, such as bishop's weed, *Bisnaga*, toothpickweed, and occasionally even toothpick-plant.

After repeated oral administration for nine days, an aqueous extract of *A. visnaga* at a concentration of 20 mg/kg b.w. effectively lowered blood glucose in rats with diabetes caused by streptozotocin (Jouadet al., 2002).

Ammi visnaga

Although it is an invasive species that is found all over the world, its native regions include Europe, Asia, and North Africa.

It is a plant-based treatment for kidney stones and other conditions.

2.6.3.2.1 Obesity

An enduring (chronic) medical disease that gets worse with time is obesity. Excess adipose tissue (body fat) that may be harmful to one's health is called obesity. The body mass index, also known as the BMI, is a metric used to assess body size that accounts for both height and weight. Usually, doctors utilize it as a tool for obesity screening. As per the Centers for the Control of Disease and Prevention (CDC), adults are considered obese if their BMI is 30 or more. It increases a larger chance of major ailments such as heart disease and cancer.

2.6.4 Rhizoma Coptidis (Huang Lian)

The roots of *Coptis chinensis* Franch., *Coptis deltoidea* C. Y. Cheng et Hsiao, or *Coptis teeta* Wall are the source of rhizomacoptidis (RC) [14]. Its principal constituents consist of lignans and alkaloids. RC's primary active ingredient is berberine, one of the alkaloids [15]. According to the research, RC can prevent adipogenesis [16], decrease lipids [17], and decrease lipid synthesis [18]. RC (200 mg/kg) and berberine (200 mg/kg) significantly decreased blood glucose and lipid levels, slowed the breakdown of dietary polysaccharides in high-fat diets, and decreased body and visceral adipose weight, according to Xie et al. [19].

2.6.5 *Panax Ginseng* C. A. Mey (Ren Shen)

The dried roots and rhizome of *Panax ginseng* C. A. Mey (Araliaceae) are the source of Ren Shen [14]. *P. ginseng* C. A. Mey's primary active ingredients are ginseng saponins and polysaccharides [20]. Structure allows for the division of ginseng saponins into Rb1, Rb2, Rc, Rd, Re, and Rl [21, 22]. *P. ginseng* C. A. Mey may improve insulin resistance, lower body weight [23], attenuate fat formation [24], decrease lipid accumulation, and produce less reactive oxygen species (ROS) [25].

2.6.6 *Radix Lithospermi* (Zicao)

The root of *Arnebia euchroma* (Royle) Johnst., *Lithospermum erythrorhizon* Sieb. Et Zucc, or *Arnebia guttata* Bunge is the source of Radix Lithospermi [14]. According to studies, Radix Lithospermi can control lipid metabolism, cause lipolysis, prevent lipid buildup, and help people lose weight. Shikonin and acetylshikonin are the primary active components of Radix Lithospermi [26–29]. Su et al. [30] discovered that giving obese rats intragastric injections of 100, 300, or 900 mg/kg of acetylshikonin extract for six weeks dramatically reduced the animals' weight, serum levels of TG, and serum free fatty acid (FFA). By inhibiting the expression of proteins specific to adipocytes and transcription factors involved in adipogenic development, as well as by boosting the activity of cAMP-dependent protein kinase (PKA) and phosphorylating hormone-sensitive lipase, acetylshikonin is an effective treatment for obesity.

2.6.7 *Ephedra sinica* Stapf. (Ma Huang)

Dried *Ephedra sinica* Stapf rhizome is the primary component of Ma Huang, a herbal remedy [30]. Obesity has been treated with it in recent years [31]. Ephedra intermedia Schrenk et C. A. Mey are among the other species that are employed. Ephedra equisetinaBge is another [14]. Stapf's *E. sinica* can improve glucose intolerance, lower weight, and modify gut microbiome. Oral gavage of 5% ephedra and 0.5% acarbose for six weeks was found to be effective in improving lipid profiles and glucose intolerance, as well as reducing weight gain and epididymal fat accumulation in mice fed a high-fat diet [32]. The primary method by which *E. sinica* lowers obesity and hyperglycemia is by raising the activity of diponectin and peroxisome proliferator-activated receptor alpha (PPAR-α), while also decreasing tumor necrosis factor-alpha.

2.6.8 Tea (Lvcha)

Tea polyphenols, catechins, caffeine, and amino acids are found in green tea, which is one of the most popular teas in China and is often used as a weight loss aid [33]. Many mechanisms, including activation of the nuclear factor erythroid-2-related-factor-2 (Nrf2) pathway [34], upregulation of neprilysin [35], prevention of gut dysbiosis [36], regulation of the body's metabolic balance, inhibition of fat accumulation and cholesterol synthesis, and reduction of abdominal fat, are involved in the weight loss effects of green tea. In diet-induced obesity (DIO) mice, Choi et al.'s research [37] showed that giving an HFD with 0.25% (w/w) green tea extract for a 12-week period improved the mice's insulin resistance, hepatic steatosis, and obesity. Through transcriptional responses to the metabolism of fat, glucose, and amino acids, green tea extract helped to regulate systemic metabolic homeostasis.

2.7 PANCREATIC CANCER

Pancreatic cancer is a type of cancer that begins with an increase in pancreatic cells. The pancreas is situated behind the lower part of the stomach. It produces digestive enzymes and hormones that help control blood sugar. Among pancreatic cancers, pancreatic ductal adenocarcinoma is the most common type. This type begins in the cells that line the ducts that carry the pancreatic digesting enzymes. Pancreatic cancer is rarely detected when there is the greatest possibility of a cure. This is due to the fact that symptoms frequently don't show up until the illness has affected other organs. Your healthcare team considers the stage of your pancreatic cancer when formulating a treatment plan for you. There may be options for radiation therapy, chemotherapy, surgery, or other treatments [38].

2.7.1 *Viscum album* Extract

Climbing on the stems of other trees, *Viscum album* is a hemiparasitic shrub that is native to Europe, Western Asia, and Southern Asia. The ingredients that give this product its chemical characteristics include phenolics, polysaccharides, several proteins, and other substances [39]. Apoptotic, cytotoxic, and immune-stimulating effects can be found in extracts from the *V. album*. Viscose residues on the cell surface of glycoproteins are bound by the cytotoxic protein of *V. album*, viscumin, which inactivates the 60S ribosomal subunit and prevents the synthesis of new proteins. When *V. album* extract was given three times a week at increasing doses (0.01–10 mg) to 220 patients with locally advanced or metastatic PC without concurrently administered antineoplastic therapy, Tröger et al. observed that the median overall survival for *V. album* patients was 4.8 months, while that of the control group was 2.7 months.

2.7.2 Mistletoe

The primary components of mistletoe, an obligatory hemiparasitic plant, include lectins, flavonoids, sterols, alkaloids, and iscotoxins [40]. Some adverse effects, such as diarrhea, vomiting, irregular heartbeats, and cardiac arrest, are brought on by pyramine and phorotoxin. It has been used as an adjuvant therapy for cancer patients in addition to other treatments. The FDA in the USA has not, however, approved the use of this plant as a cancer treatment. Because of its cytotoxic and immunomodulatory qualities, which include inducing apoptosis and trapping chemotherapeutic medications within cancer cells via P-glycoprotein's inhibitory capability, mistletoe leaves are highly valued in Europe. In terms of clinical data, an observational pharmaco-epidemiological study including PC patients was carried out by Matthes et al. Gemcitabine was administered to patients while mistletoe was used as support [41].

2.7.3 Crocetinic Acid (*Gardenia jasminoides*)

Gardenia jasminoides has several chemical constituents that have been identified and isolated, including glucosides, triterpenoids, and iridoids. It has been possible to isolate three main bioactive compounds: gardenoside, geniposide, and genipin. Additionally, crocetin, a naturally occurring apocarotenoid dicarboxylic acid, is present in *G. jasminoides* [42]. This compound is a carotenoid component of saffron that has demonstrated antitumor activity in animal studies. According to Rangarajan et al., crocetinic acid caused apoptosis and reduced PC cell line growth in a dose- and time-dependent manner.

2.7.4 Hispidin (*Phellinus linteus*)

Using reversed-phase high-performance liquid chromatography, Lee et al. isolated six compounds from the *Phellinus linteus* mushroom. They also used spectroscopic methods to identify the compounds' structures: caffeic acid, inotilone, 4-(3,4-dihydroxyphenyl)-3-buten-2-one, phellilane H, (2E,4E)-(+)-4′-hydroxy-γ-ionylideneacetic acid, and (2E,4E)-γ-ionylideneacetic acid; the first three showing the strongest antioxidant activity [43]. Hispidin, a polyphenolic molecule with demonstrated anti-tumor effects, is also present in this mushroom. The cytotoxic effects of hispidin on PC cells (BxPC-3 and AsPC-1) were assessed by Chandimali *et al.* Additionally, they looked at the potential synergistic effects of hispidin and gemcitabine on PC stem cells in vitro. They discovered that hispidin had anticancer effects on PC cells as well as stem cells. Furthermore, gemcitabine's therapeutic efficiency was enhanced by hispidin's sensitization of PC stem cells to the drug [43].

2.8 HEREDITARY HEMOCHROMATOSIS

A blood condition called hereditary hemochromatosis makes your body absorb excessive amounts of iron from your diet. A common mineral in diet is iron. Iron overload, however, is harmful to the body. Your body stores extra iron in its tissues and organs. Your body accumulates iron over time (iron overload) [44].

2.8.1 Manjistha

Manjistha (*Rubia cordifolia*) is the major herb that is used to prepare this wonderful formulation for hemochromatosis. According to Ayurveda, this herb is used as the best blood purifier that regulates both lymph and blood circulation in the whole body. In this disorder both rakta dhatu and pitta dosha are imbalanced and these capsules help in the process of detoxification and protect our liver from the attack of free radicals and toxins that cause major problems in the body.

2.8.2 Gandhak Rasayan

Gandhak Rasayan is a very effective antioxidant, anti-viral, and anti-microbial ayurvedic mineral preparation that is used to treat skin disorders and to remove internal impurities that cause serious disorders in our body internally. It is an effective herbal supplement for hemochromatosis because it is used to purify blood and protect our liver from the toxicity of harmful toxins.

2.8.3 Green Essentials Capsules

Green essential is a very effective supplement provided at house of planet, ayurveda for patients of hemochromatosis. Grape seeds (*Vitis vinifera*), wheat grass (*Triticulum aestivum*), Spirulina (*Spirulina platensis*), Green tea (*Camellia sinensis*), and Amla (*Emblica officinalis*) are the major ingredients that are used to prepare this herbal and effective remedy.

2.8.4 Liver Detox

Liver detox formula is again a very effective formulation to maintain proper health of liver. It regulates production of hormones in the body. According to Ayurveda liver is also the house of pitta dosha and these wonderful supplements help to pacify pitta dosha.

2.8.5 Navkarshik Churna

Manjistha (*Rubia cordifolia*), Katuki (*Picrorrhizakurrao*), Giloy (*Tinospora cordifolia*), Daruhaldi (*Berberis aristata*), Neem (*Azadirachta indica*), etc., are the various ingredients that are used to prepare this wonderful churna for liver disorders. It is a wonderful supplement for fatty liver and liver cirrhosis. Neem is a natural blood purifier that is used to remove toxins from blood and improve the circulatory system in our body. It is a natural antioxidant and helps to boost immune system. This herbal formulation is hepato-protective in nature. All these herbs are anti-inflammatory, antioxidants, hypoglycemic, carminative, hypotensive, stomachic, and astringent in nature.

2.9 LESCH–NYHAN SYNDROME

An uncommon congenital (present at birth) condition that affects the brain of children and behavior is called Lesch–Nyhan syndrome (LNS). Self-harming behaviors that are out of control, such as head banging or biting one's lips or fingers, are a major sign. Uric acid is a naturally occurring waste product that accumulates in the body as a result of the condition. Scientists think LNS might have an impact on dopamine levels as well. A neurotransmitter essential to normal brain activity is dopamine. Gout, a very painful and severe form of arthritis, is experienced by children with LNS. They also suffer from dystonia, weak muscle control, and mental impairment [45].

2.10 MAPLE SYRUP URINE DISEASE

A genetic condition known as "maple syrup urine disease" occurs when the body is unable to appropriately metabolize specific amino acids, which are the building blocks of proteins. The exceptionally sweet-smelling pee of affected infants gives rise to the condition's moniker. Along with poor feeding and vomiting, lethargy, odd motions, and delayed development are further characteristics. Seizures, a coma, and even death may result from untreated maple syrup urine illness. The pattern of signs and symptoms of maple syrup urine disease is typically used to categorize it. The classic kind manifests early after birth and is the most common and severe form of the disease. Though usually milder, variant types of the condition manifest later in infancy or childhood and still cause delayed development [46–48].

2.11 MENKES SYNDROME

A hereditary condition known as Menkes disease causes the body to have trouble absorbing and distributing copper to the body's organs. The illness has an impact on both physical and mental development. In a newborn or older infant, hypotonia is frequently an indication of abnormality and can point to the existence of muscle diseases, genetic disorders, or problems with the central nervous system. Infants with normal tone often have flexed elbows and knees when they sleep, whereas hypotonic infants have their elbows and knees loosely stretched. The floppy infant may have poor head control, with the head falling forward, backward, or to the side. Menkes syndrome typically runs in families since it is hereditary. Since the gene is located on the X chromosome, if a mother has the faulty [49].

2.12 NIEMANN–PICK DISEASE

A class of hereditary metabolic diseases known as Niemann–Pick disease is characterized by aberrant lipid accumulation in the brain, spleen, liver, lungs, and bone marrow. Lipids are fatty substances that include waxes, oils, and cholesterol. Enzymes that are defective or lacking in quantity cannot break down lipids into smaller parts so that the body can produce energy from them. An uncommon genetic condition called Niemann–Pick impairs the body's capacity to digest fat (lipids and cholesterol) inside of cells. These cells eventually break down and perish. The brain, nerves, liver, spleen, bone marrow, and, in extreme situations, the lungs can all be impacted by Niemann–Pick disease [50, 51].

2.13 PRADER–WILLI SYNDROME

The symptoms of Prader–Willi syndrome (PWS) include short stature if growth hormone (GH) treatment is not received, childhood-onset obesity if excessive eating is not curbed, behavioral findings, neonatal hypotonia with poor suck and poor weight gain without nutritional support, developmental delay, mild cognitive impairment, and typically a characteristic facial appearance. Reduced fetal movements, small hands and/or feet, hypopigmentation in comparison to family members, skin picking, strabismus, abnormal visual acuity, sleep disturbance (including daytime sleepiness and occasionally sleep apnea), thick, viscous saliva, and articulation abnormalities are less common but common features. Depending on the molecular etiology of PWS, clinical manifestations can vary somewhat [52, 53].

2.14 REFSUM DISEASE

An uncommon genetic condition known as adult Refsum disease (ARD) results in peripheral neuropathy, or paralysis or numbness in the hands and feet. People with ARD illness are deficient in the enzyme in peroxisomes that breaks down phytanic acid, a type of fat present in some diets, due to a genetic mutation. As a result, the blood, brain, and other tissues accumulate hazardous amounts of phytanic acid. Due to retinal degeneration, the condition typically manifests as increasing night blindness in late childhood or early adulthood (retinitis pigmentosa). If the illness worsens, further signs and symptoms could be as follows:

- Deafness
- anosmia, or loss of smell
- Issues related to coordination and balance (ataxia)
- Skin that is scaly and dry (ichthyosis)
- Abnormal heartbeats (cardiac arrhythmias)

2.15 TANGIER DISEASE

HDL in the bloodstream is severely reduced or absent in Tangier disease, which leads to tissue buildup of cholesteryl esters throughout the body, especially in the reticuloendothelial system. Hepatosplenomegaly, hyperplastic yellow-orange tonsils, and peripheral neuropathy—which can be either chronically progressive or relapsing-remitting—are the main clinical indicators of Tangier illness. Less common side effects could be corneal opacities, which usually don't impair vision, early atherosclerotic coronary artery disease that develops in the sixth and seventh decades of life (usually not before the age of 40), and mild hematologic symptoms like hemolytic anemia, reticulocytosis, stomatocytosis, or mild thrombocytopenia. Tangier illness can manifest clinically in a variety of ways; in certain cases, infected individuals only exhibit biochemical abnormalities [54, 55].

2.16 TAY–SACHS DISEASE

Parents can pass on a rare genetic illness called Tay–Sachs disease to their offspring. The lack of an enzyme that aids in the breakdown of fatty compounds is the cause. In the brain and spinal cord, these fat molecules, known as gangliosides, accumulate to toxic concentrations and impair nerve cell activity. About three to six months of age is when signs and symptoms of Tay–Sachs disease's most prevalent and severe form first appear. Muscle weakness starts when the condition worsens and development slows down. This eventually results in seizures, loss of hearing and vision, paralysis, and other serious problems. The typical lifespan of a child with this type of Tay–Sachs illness is a few years. The juvenile version of Tay–Sachs disease affects some children less frequently.

2.17 PHENYLKETONURIA

A hereditary disorder termed phenylketonuria (PKU) results in the accumulation of high amounts of phenylalanine in the body. The body needs phenylalanine for regular metabolic processes, but when levels stay consistently higher than normal, issues can emerge. One type of amino acid is phenylalanine. Proteins are composed of chemicals called amino acids. Phenylalanine is present in a lot of meals that you eat that are high in protein and artificial sweeteners aspartame. Phenylalanine accumulation in the body can lead to symptoms such as difficulties developing cognitively (intellectual impairment) if the problem is left untreated.

- Wood betony: This plant's main chemical constituents include betulinic acid, rutin, urosolic acid, stachydrine, rutin, oleonilic acid, and glycosides.
- Nettle: The main chemical constituents of this plant are vitamins, acetylcholine, serotonin, histamine, and formic acid.
- Plantago ovate: Xylose, arabinose, alanine, valine, glutamic acid, glycine, cysteine, lysine, leucine, tyrosine, and xylose are the main chemical components of this plant.
- Turmeric: Curcumin and camphene are the two main chemicals found in this plant.
- Dandelion: Taraxacin, laevulin, resin, and inulin are the main chemical components of this plant.

Heart-leaved moonseed, or
Tinospora cordifolia-giloy

The heart-leaved moonseed can be discovered in tropical regions up to 1200 meters high, ranging from Kumaon to Assam and northward through West Bengal. helps to boost immunity and regulate the amount of sugar in the blood.

Phyllanthus emblica

Phyllanthus emblica frequently referred to as amla or Indian gooseberry. It is indigenous to southern Asia and the tropics. Grown commercially in the Indian states of Uttar Pradesh, Tamil Nadu, and Rajasthan. Encourage heart health, have anti-aging properties, boost immunity, lessen the intensity of heartburn, and lower the risk of cancer.

Terminalia chebula

Terminalia chebula, a variety of Terminalia, usually referred to as black- or chebulic myrobalan, is indigenous to South Asia, extending from Pakistan, India, and Nepal eastward to southwest China. It helps in the healthy secretion of mucosal membranes in the respiratory, digestive, urinary, and reproductive systems.

Centella asiatica

Common names for *Centella asiatica* include Asian and Indian pennywort. It is indigenous to the western Pacific Ocean and tropical parts of Africa, Asia, and Australia. In addition to wound healing, the herb is suggested for the treatment of a number of skin disorders, including diarrhea, psoriasis, eczema, leprosy, and varicose ulcers.

Asparagus racemosus

Asparagus racemosus—Shatavari found at low altitudes throughout India. The dried roots of the plant are used as drug, native from Africa through southern Asia, including the Indian subcontinent, to northern Australia. Shatavari attempts to treat conditions related to hormone imbalance such as polycystic ovarian syndrome (PCOS) and infertility.

Aegle marmelos

Aegle marmelos—Indian bael also known as golden apple, Japanese bitter orange, stone apple. It is grown in various parts of Southeast Asia including India, Sri Lanka, Pakistan, Burma, Bangladesh, Thailand, etc. Help to reduce swelling (inflammation). This might help treat asthma, diarrhea, and other conditions.

Piper longum

Long pepper—In India's tropical woods, it grows untamed, grown in tiny areas of Eastern Uttar Pradesh, Madhya Pradesh, Maharashtra, Kerala, Karnataka, and Tamil Nadu; the Khasi highlands. Treat cholera, diarrhea, intestinal gas, heartburn, stomachaches, and indigestion in addition to enhancing appetite and digestion.

Andrographis paniculata

Andrographis paniculata—Green chiretta. Native to India and Sri Lanka, it is widely cultivated in Southern and Southeastern Asia, where it has been believed to be a treatment for bacterial infections and some diseases.

Acorus calamus

Acorus calamus—Sweet flag. India, Central Asia, southern Russia, Siberia, Europe, and North America are all home to sweet flag plants. It has anti-spasmodic properties that aid in the relief of pain and cramping. Additionally, it aids in the prevention of several digestive problems like diarrhea, bloating, and constipation. Additionally, it exhibits gas-reducing and colic-relieving qualities.

Terminalia arjuna

Terminalia arjuna—Arjun tree from Uttar Pradesh, Bihar, Maharashtra, Madhya Pradesh, West Bengal, Odisha and south and central India. It has been used as a cardiotonic in heart failure, ischemic, cardiomyopathy, atherosclerosis, and myocardium necrosis.

Zingiber officinale

Zingiber officinale—Ginger. It is also cultivated in Jamaica, Nigeria, Sierra Leone, Brazil, China, Japan, and Indonesia. The major ginger-growing states in India are Kerala, West Bengal, and North Eastern Region. It is used to treat colds, nausea, arthritis, migraines, and hypertension.

Cassia fistula

Cassia fistula—Golden shower tree, Indian Laburnum. It grows in parts of the Himalayas, and other northern parts of India, Asia, and Australia. Golden Shower Tree is known as a "disease killer" and it pacifies the three doshas of vaat, pitta, and kapha, as well as cardiac conditions and stomach problems such as acid reflux.

Flame lily, Gloriosa

Originating from tropical and southern Africa, they have spread to Asia and Australia, as well as the Pacific. Gout, infertility, open wounds, snakebite, ulcers, arthritis, cholera, colic, renal issues, typhus, itching, leprosy, bruises, sprains, and hemorrhoids have all been treated with it.

Momordica charanti

The bitter melon, *Momordica charanti*, from Africa is the home of bitter melon. It is a common ingredient in Southeast Asian, South Asian, and East Asian cuisines. It aids in the treatment of diabetes-related illnesses. Recent research has validated the fruit's function in regulating blood sugar levels.

Saracaasoca

The Ashoka tree, or *Saracaasoca* from India is home to the Ashoka tree, which is particularly prevalent in Bengal, Kerala, the Himalayas, and the southern areas. It combats cancer. It can be applied to depression, arthritis, ulcers, and bacterial infections.

***Alpinia galangal*—Greater Galangal**

It is native to Indonesia. It is used as an antimicrobial, antioxidant (acts against free radicals), anti-inflammatory properties, helps to enhance immunity, and protects the liver.

2.18 CONCLUSION

The intricate relationship between traditional medicinal plants and metabolic disorders has been thoroughly explored in this comprehensive overview. Metabolic disorders, arising from disruptions in the body's biochemical processes, present a significant challenge to human health. However, nature, in its abundance, offers a plethora of bioactive compounds within medicinal plants that have demonstrated potential in managing these disorders.

From the adaptogenic properties of Ashwagandha (Withania somnifera) to the antioxidant prowess of Tulsi (Ocimum sanctum) and the hepatoprotective effects of Bhumi Amla (Phyllanthus niruri), the diverse array of phytochemicals present in these plants offers a multifaceted approach to addressing metabolic dysfunction. We've seen how curcumin from turmeric and compounds in fenugreek and licorice may aid in managing diabetes, while others may mitigate the effects of obesity, pancreatic cancer, and even rare genetic disorders.

However, the path to fully integrating these natural remedies into mainstream healthcare necessitates further rigorous scientific investigation. Understanding the precise mechanisms of action, potential interactions, and optimal dosages remains crucial to ensuring their safe and effective utilization. The synergistic effects of various phytochemicals, as often observed in traditional formulations, also warrant in-depth exploration.

While modern medicine continues to advance, the wisdom embedded within traditional practices and the therapeutic potential of medicinal plants offer a complementary avenue for managing metabolic disorders. The journey towards harnessing these natural resources calls for a harmonious blend of ancient knowledge and contemporary scientific inquiry, ultimately contributing to a more holistic and effective approach to human health and well-being.

REFERENCES

1. A. Sliwinska, M. A. Kasinska, and J. Drzewoski, "MicroRNAs and metabolic disorders - Where are we heading?," *Archives of Medical Science*, vol. 13, no. 4, pp. 885–896, 2017 Jun, doi: 10.5114/aoms.2017.65229. Epub 2017 Jan 19. PMID: 28721157; PMCID: PMC5507111.
2. R. H. Eckel, K. G. Alberti, S. M. Grundy, and P. Z. Zimmet, "The metabolic syndrome," *Lancet*, vol. 375, no. 9710, pp. 181–183, 2010 Jan 16, doi: 10.1016/S0140-6736(09)61794-3. PMID: 20109902.
3. P. Batra, S. K. Das, T. Salinardi, L. Robinson, E. Saltzman, T. Scott, A. G. Pittas, and S. B. Roberts, "Relationship of cravings with weight loss and hunger. Results from a 6 month worksite weight loss intervention," *Appetite*, vol. 69, no. 1–7, 2013 Oct, doi: 10.1016/j.appet.2013.05.002. Epub 2013 May 16. PMID: 23684901.
4. M. Pal, et al., "Structural basis for phosphorylation-dependent recruitment of Tel2 to Hsp90 by Pih1," *Structure*, vol. 22, no. 6, pp. 805–818.
5. A. M. Abdelhady, Z. Desta, F. Jiang, C. W. Yeo, J. G. Shin, and B. R. Overholser, "Population pharmacogenetic-based pharmacokinetic modeling of efavirenz, 7-hydroxy- and 8-hydroxyefavirenz," *The Journal of Clinical Pharmacology*, vol. 54, no. 1, pp. 87–96, 2014 Jan, doi: 10.1002/jcph.208. Epub 2013 Nov 19. PMID: 24142869; PMCID: PMC4412845.
6. M. Gaber, H. Selim, and T. El-Nahas, "Prospective study evaluating the radiosensitizing effect of reduced doses of temozolomide in the treatment of Egyptian patients with glioblastoma multiforme," *Cancer Management and Research*, vol. 5, pp. 349–356, 2013, doi: 10.2147/CMAR.S52147
7. M. S. Akash, K. Rehman, and S. Chen, "Spice plant Allium cepa: Dietary supplement for treatment of type 2 diabetes mellitus," *Nutrition*, vol. 30, no. 10, pp. 1128–1137, 2014 Oct, doi: 10.1016/j.nut.2014.02.011. Epub 2014 Mar 2. PMID: 25194613.
8. R. Thomson, "The qualitative longitudinal case history: Practical, methodological and ethical reflections," *Social Policy and Society*, vol. 6, no. 4, pp. 571–582, 2007, doi:10.1017/S1474746407003909
9. J. C. Ozougwu, and J. E. Eyo, "Studies on the anti-diabetic activity of Allium sativum (garlic) aqueous extracts on alloxan-induced diabetic albino rat," *Pharmacologyonline*, vol. 2, pp. 1079–1088, 2010.
10. S. L. Badole, S. M. Chaudhari, P. P. Bagul, S. P. Mahamuni, R. D. Khose, A. C. Joshi, C. G. Raut, and A. A. Zanwar, "Effect of concomitant administration of L-glutamine and cycloart-23-ene-3β, 25-diol (B2) with sitagliptin in GLP-1 (7–36) amide secretion, biochemical and oxidative stress in streptozotocin - Nicotinamide induced diabetic Sprague Dawley rats," *PLOS ONE*, vol. 8, no. 8, p. e72817, 2013 Aug 30, doi: 10.1371/journal.pone.0072817. PMID: 24023648; PMCID: PMC3758338.
11. N. W. Fraulini, G. M. Hancock, A. R. Neigel, V. L. Claypoole, and J. L. Szalma, "A critical examination of the research and theoretical underpinnings discussed in Thomson, Besner, and Smilek (2016)," *Psychological Review*, vol. 124, no. 4, pp. 525–531, 2017 Jul, doi: 10.1037/rev0000066. PMID: 28639818.
12. N. M. Ammar, Y. A. O. Sahar, and H. B. Ibrahim, "The hypoglycemic effect of different extracts of *Ambrosia maritima, L.* Compositae," *Journal of Islamic Academy of Sciences*, vol. 6, no. 4, pp. 298–301, 1993.
13. A. Helal, et al., "Bioaccessibility of polyphenols and cinnamaldehyde in cinnamon beverages subjected to in vitro gastro-pancreatic digestion," *Journal of Functional Foods*, vol. 7, pp. 506–516, 2014.
14. *Chinese Pharmacopoeia Commission, Pharmacopoeia of the People's Republic of China*, Chinese Medical Science Press, Beijing, China, 2010.
15. W.-J. Ni, H.-H. Ding, and L.-Q. Tang, "Berberine as a promising anti-diabetic nephropathy drug: An analysis of its effects and mechanisms," *European Journal of Pharmacology*, vol. 760, pp. 103–112, 2015.
16. J. S. Choi, J.-H. Kim, M. Y. Ali, B.-S. Min, G.-D. Kim, and H. A. Jung, "*Coptis chinensis* alkaloids exert anti-adipogenic activity on 3T3-L1 adipocytes by downregulating C/EBP-α and PPAR-γ," *Fitoterapia*, vol. 98, pp. 199–208, 2014.

17. Z.-Y. Zou, Y.-R. Hu, H. Ma, et al., "Coptisine attenuates obesity-related inflammation through LPS/TLR-4-mediated signaling pathway in Syrian golden hamsters," *Fitoterapia*, vol. 105, pp. 139–146, 2015.
18. W.-L. Zhang, L. Zhu, and J.-G. Jiang, "Active ingredients from natural botanicals in the treatment of obesity," *Obesity Reviews*, vol. 15, no. 12, pp. 957–967, 2014.
19. W. Xie, D. Gu, J. Li, K. Cui, and Y. Zhang, "Effects and action mechanisms of berberine and rhizomacoptidis on gut microbes and obesity in high-fat diet-fed C57BL/6J mice," *PLOS ONE, vol.* 6, no. 9, p. e24520, 2011.
20. Y. Li, T. J. Zhang, S. X. Liu, R. R. Wang, and J. Luo, "Research on chemical compositions and pharmacology role of Ginseng," *Chinese Traditional and Herbal Drugs*, vol. 40, no. 1, pp. 164–166, 2009.
21. L. P. Christensen, "Chapter 1 ginsenosides: Chemistry, biosynthesis, analysis, and potential health effects," *Advances in Food and Nutrition Research*, vol. 55, pp. 1–99, 2008.
22. S. A. Palaniyandi, J. W. Suh, and S. H. Yang, "Preparation of ginseng extract with enhanced levels of ginsenosides Rgl and Rbl using high hydrostatic pressure and polysaccharide hydrolases," *Pharmacognosy Magazine*, vol. 13, no. 49, pp. 142–147, 2017.
23. M. C. Kho, Y. J. Lee, J. H. Park, et al., "Fermented red ginseng potentiates improvement of metabolic dysfunction in metabolic syndrome rat models," *Nutrients*, vol. 8, no. 6, p. 369, 2016.
24. N. Karu, R. Reifen, and Z. Kerem, "Weight gain reduction in mice fed *Panax ginseng* saponin, a pancreatic lipase inhibitor," *Journal of Agricultural and Food Chemistry*, vol. 55, no. 8, pp. 2824–2828, 2007.
25. Lee, K., Seo, Y. J., Song, J. H., Chei, S., & Lee, B. Y. (2019). Ginsenoside Rgl promotes browning by inducing UCP1 expression and mitochondrial activity in 3T3-L1 and subcutaneous white adipocytes. Journal of ginseng research, 43(4), 589–599. https://doi.org/10.1016/j.jgr.2018.07.005..
26. Zhang, X., Cui, J. H., Meng, Q. Q., Li, S. S., Zhou, W., & Xiao, S. (2018). Advance in Anti-tumor Mechanisms of Shikonin, Alkannin and their Derivatives. Mini reviews in medicinal chemistry, 18(2), 164–172. https://doi.org/10.2174/1389557517666170228114809.
27. S.-H. Park, N. M. Phuc, J. Lee, et al., "Identification of acetyl shikonin as the novel CYP2J2 inhibitor with anti-cancer activity in HepG2 cells," *Phytomedicine*, vol. 24, pp. 134–140, 2017.
28. Y. He, Q. Li, M. Su, W. Huang, and B. Zhu, "Acetylshikonin from Zicao exerts antifertility effects at high dose in rats by suppressing the secretion of GTH," *Biochemical and Biophysical Research Communications*, vol. 476, no. 4, pp. 560–565, 2016.
29. R. Gul, S. U. Jan, S. Faridullah, S. Sherani, and N. Jahan, "Preliminary phytochemical screening, quantitative analysis of alkaloids, and antioxidant activity of crude plant extracts from Ephedra intermedia indigenous to Balochistan," *Scientific World Journal, vol.* 2017, p. 5873648, 2017.
30. E. A. Abourashed, A. T. El-Alfy, I. A. Khan, and L. Walker, "Ephedra in perspective—A current review," *Phytotherapy Research*, vol. 17, no. 7, pp. 703–712, 2003.
31. M.-K. Song, J.-Y. Um, H.-J. Jang, and B.-C. Lee, "Beneficial effect of dietary *Ephedra sinica* on obesity and glucose intolerance in high-fat diet-fed mice," *Experimental and Therapeutic Medicine*, vol. 3, no. 4, pp. 707–712, 2012.
32. D. Türközü, and N. A. Tek, "A minireview of effects of green tea on energy expenditure," *Critical Reviews in Food Science and Nutrition*, vol. 57, no. 2, pp. 254–258, 2017.
33. C. Sampath, M. R. Rashid, S. Sang, and M. Ahmedna, "Green tea epigallocatechin 3-gallate alleviates hyperglycemia and reduces advanced glycation end products via nrf2 pathway in mice with high fat diet-induced obesity," *Biomedicine and Pharmacotherapy*, vol. 87, pp. 73–81, 2017.
34. M. Muenzner, N. Tappenbeck, F. Gembardt, et al., "Green tea reduces body fat via upregulation of neprilysin," *International Journal of Obesity*, vol. 40, no. 12, pp. 1850–1855, 2016.
35. M. Cheng, X. Zhang, Y. Miao, J. Cao, Z. Wu, and P. Weng, "The modulatory effect of (-)-epigallocatechin 3-o-(3-o-methyl) gallate (EGCG3"Me) on intestinal microbiota of high fat diet-induced obesity mice model," *Food Research International*, vol. 92, pp. 9–16, 2017.
36. J.-Y. Choi, Y. J. Kim, R. Ryu, S.-J. Cho, E.-Y. Kwon, and M.-S. Choi, "Effect of green tea extract on systemic metabolic homeostasis in diet-induced obese mice determined via RNA-seq transcriptome profiles," *Nutrients*, vol. 8, no. 10, p. 640, 2016.
37. Chee-Chee Stucky, M.D. Pancreatic cancer. https://www.mayoclinic.org/diseases-conditions/pancreatic-cancer/symptoms-causes/syc-20355421
38. K. S. Zänker, and S. V. Kaveri editors, *Mistletoe: From Mythology to Evidence-Based Medicine*, Karger Medical and Scientific Publishers, Basel, Switzerland, 2015, vol. 4, pp. 11–23.
39. C. N. Ishiwu, J. E. Obiegbuna, and N. M. Aniagolu, "Evaluation of chemical properties of mistletoe leaves from three different trees (avocado, African oil bean and kola)," *The Nigerian Food Journal*, vol. 31, pp. 1–7, 2013, doi: 10.1016/S0189-7241(15)30070-9

40. H. Matthes, W. E. Friedel, P. R. Bock, and K. S. Zänker, "Molecular mistletoe therapy: Friend or foe in established anti-tumor protocols? A multicenter, controlled, retrospective pharmaco-epidemiological study in pancreas cancer," *Current Molecular Medicine*, vol. 10, pp. 430–439, 2010, doi: 10.2174/156652410791317057
41. W. Xiao, S. Li, S. Wang, and C. T. Ho, "Chemistry and bioactivity of *Gardenia jasminoides*," *Journal of Food and Drug Analysis*, vol. 25, pp. 43–61, 2017, doi: 10.1016/j.jfda.2016.11.005
42. M. S. Lee, B. S. Hwang, I. K. Lee, G. S. Seo, and B. S. Yun, "Chemical constituents of the culture broth of *Phellinus linteus* and their antioxidant activity," *Mycobiology, vol.* 43, pp. 43–48, 2015, doi: 10.5941/MYCO.2015.43.1.43
43. J. L. Babitt, and H. Y. Lin, "The molecular pathogenesis of hereditary hemochromatosis," *Seminars in Liver Disease*, vol. 31, no. 3, pp. 280–292, 2011 Aug, doi: 10.1055/s-0031-1286059. Epub 2011 Sep 7.
44. J. C. Harris, "Lesch-Nyhan syndrome and its variants: Examining the behavioral and neurocognitive phenotype," *Current Opinion in Psychiatry*, vol. 31, no. 2, pp. 96–102, 2018, https://pubmed.ncbi.nlm.nih.gov/29227296/. Accessed 7/17/2022.
45. National Library of Medicine (U.S.). Maple Syrup Urine Disease. http://ghr.nlm.nih.gov/condition/maple-syrup-urine-disease. Last updated 7/1/2017. Accessed 9/25/2023.
46. S. A. Hassan, and V. Gupta, "Maple Syrup Urine Disease," in *StatPearls* [Internet], StatPearls Publishing, Treasure Island (FL), 2022 Sep 5, https://www.ncbi.nlm.nih.gov/books/NBK557773/#:~:text=Maple%20syrup%20urine%20disease%20is,the%20severity%20of%20these%20complications. Accessed 9/25/2023.
47. I. Bertini, and A. Rosato, "Menkes disease," *Cellular and Molecular Life Sciences*, vol. 65, no. 1, pp. 89–91, 2008 Jan, doi: 10.1007/s00018-007-7439-6
48. P. de Bie, P. Muller, C. Wijmenga, and L. W. Klomp, "Molecular pathogenesis of Wilson and Menkes disease: Correlation of mutations with molecular defects and disease phenotypes," *Journal of Medical Genetics*, vol. 44, no. 11, pp. 673–688, 2007 Nov, doi: 10.1136/jmg.2007.052746. Epub 2007 Aug 23.
49. N. X. Cawley, C. Sojka, A. Cougnoux, A. T. Lyons, E. R. Nicoli, C. A. Wassif, and F. D. Porter, "Abnormal LAMP1 glycosylation may play a role in Niemann-Pick disease, type C pathology," *PLOS ONE*, vol. 15, no. 1, p. e0227829, 2020.
50. N. Kresojević, G. Mandić-Stojmenović, V. Dobričić, I. Petrović, L. Brajković, E. Stefanova, M. Svetel, and V. Kostić, "Very late-onset Niemann pick type C disease: Example of progressive supranuclear palsy look-alike disorder," *Movement Disorders Clinical Practice*, vol. 7, no. 2, pp. 211–214, 2020 Feb.
51. D. C. Bittel, and M. G. Butler, "Prader-Willi syndrome: Clinical genetics, cytogenetics and molecular biology," *Expert Reviews in Molecular Medicine*, vol. 7, no. 14, pp. 1–20, 2005 Jul 25, doi: 10.1017/S1462399405009531
52. S. B. Cassidy, and D. J. Driscoll, "Prader-Willi syndrome," *European Journal of Human Genetics*, vol. 17, no. 1, pp. 3–13, 2009 Jan, doi: 10.1038/ejhg.2008.165. Epub 2008 Sep 10.
53. M. Dean, K. Moitra, R. Allikmets, "The human ATP-binding cassette (ABC) transporter superfamily," *Human Mutation*, vol. 43, no. 9, pp. 1162–1182, 2022 Sep, doi: 10.1002/humu.24418. Epub 2022 Jun 22.
54. G. D. Kolovou, D. P. Mikhailidis, K. K. Anagnostopoulou, S. S. Daskalopoulou, D. V. Cokkinos, "Tangier disease four decades of research: A reflection of the importance of HDL," *Current Medicinal Chemistry*, vol. 13, no. 7, pp. 771–782, 2006, doi: 10.2174/092986706776055580

3 Phytochemicals as a Potential Alternative for Combating Multidrug-Resistant Bacterial Infections

Abinaya Ramachandran[1], *Mangalasamy Jothika*[1], *Kesavaram Padmavathy*[2], *and Dhandapani Prabu*[1]

[1]Department of Microbiology, Dr. ALM PG IBMS, University of Madras Taramani, Chennai, India

[2]Department of Microbiology, Sree Balaji Dental College & Hospital, Bharath Institute of Higher Education & Research Chennai, India

3.1 INTRODUCTION

Phytochemical, plant-derived compounds have proven to exhibit bioactive properties. Phytochemicals are detectable in all plant foods, such as vegetables, nuts, fruits and legumes. Plants possess an inherent capacity to synthesize/secrete phytochemicals which serve as a crucial defence strategy against harsh environmental conditions (UV, climate), insufficient nutrient supply and attack by pests/pathogens-insects, bacteria, fungi and viruses. Owing to their enormous therapeutic potential, antioxidant, neuroprotective, anti-carcinogenic and anti-inflammatory, phytocompounds are currently being re-explored in both medicinal and nutritional science.

Bright-coloured vegetables and fruits contain a high concentration of phytochemicals that have long been used to treat cardiovascular diseases and cancer. Phytochemicals, carotenoids and polyphenols are known to possess phytonutrient properties. Also, phytocompounds classified as phenolic derivatives (quercetin, apigenin and curcumin), organosulphur compounds (allicin), terpenes (thymol, eugenol) and aldehydes (cinnamaldehyde and vanillin) exhibit antimicrobial properties (antibacterial, antiviral, antifungal and antiprotozoal) antioxidant and immunomodulatory properties.

The last few decades have witnessed an enormous increase in use of phytochemical compounds as dietary supplements. In the Western countries (USA), the use of plant-derived supplements is reported to be as high as 44% in males, 53% in females (Bailey *et al.*, 2011). The first advancement in drug research is Morphine, a medicinal alkaloid derived from *Papaver somniferum*. Despite an increased use of phytocompounds as dietary supplements and alternative therapeutic strategies, their potential toxicity is poorly understood. Though phytocompounds have gained popularity and are increasingly used, data on the safety and efficacy of phytochemicals for the prevention and treatment of chronic illnesses like cancer is very limited (Correa-velaz *et al.*, 2005; Fratamico PM *et al.*, 2014; Funk *et al.*, 2014; Gaigé *et al.*, 2014). Previous studies have documented that phytochemicals like epigallocatechin gallate (EGCG) (Stanker *et al.*, 2013 ; Funk *et al.*, 2014) or resveratrol (Lorenz *et al.*, 2013) are potential anti-cancer compounds of plant origin.

The golden era of antibiotics (1940–1965) marks the period when several new antibiotics were brought into modern medicine, revolutionizing the treatment of bacterial illness and more than half of the antibiotics used till date were identified (Prestinaci *et al.*, 2015). Antibiotics remain the

DOI: 10.1201/9781003486237-3

cornerstone of modern medicine. Antibiotics are and will remain an incredible part of the treatment and control of infectious diseases. Nevertheless, inappropriate antibiotic usage in human healthcare and non-therapeutic use in animal husbandry have led to the development, persistence and dissemination of AMR (Prestinaci *et al.*, 2015). Multidrug resistance is defined as non-susceptibility to two or more classes of antimicrobials and those strains are classified as multidrug-resistant (MDR) bacterial pathogens (Yu *et al.*, 2020).

Antibiotic/MDR organisms have evolved primarily in response to the antibiotic selection pressure by unpredicted genetic jugglery. Currently, combating MDR strains is a cause of concern. Research and development of new antimicrobials is of global priority, as there is an antibiotic pipeline that is dry as well there exists an antibiotic access crisis. As per the WHO's latest annual review on antibacterial agents that are in pipeline, 27 antibiotics that focus on the bacterial priority pathogens are in the clinical development phase, nevertheless, only six were considered as novel (Gigante *et al.*, 2022).

3.2 MAJOR DETERMINANTS OF ANTIBIOTIC RESISTANCE

Microbes have evolved various self-defense mechanisms to counteract antibiotics. These strategies primarily include biochemical defenses and genetic alterations, such as mutations and the horizontal transfer of mobile genetic elements.

- Efflux based – Upregulation, downregulation of efflux pumps.
- Enzyme-based – Enzymatic deactivation.
- Location-based – Intracellular localization, sequestration.
- Target approach – Modification, bypass, repair, amplification of the target.
- Biofilm approach – Biofilm production – molecular cross-talk, quorum sensing.

The main strategies that have been targeted by researchers to combat bacterial resistance to antibiotics include

- Cell membrane inhibitors.
- Cell wall synthesis inhibitors.
- Drug efflux pump inhibitors.
- Enzyme inhibitors.
- Attenuating bacterial virulence.
- Targeting bacterial biofilm and quorum sensing.

3.2.1 Cell Membrane Inhibitors

The cell membrane of bacteria acts as a permeability barrier for the entry of antimicrobials. Cellular inflow of antibiotics is proposed to be monitored by modifying the fatty acid, proteins in the bacterial cell membrane that alters its permeability (Yu *et al.*, 2020). Nonetheless, phytoconstituents which are hydrophobic in nature are known to interact with membrane lipids, thereby disturbing the structure, increasing the permeability of the cell membrane, may lead to and thus avert the bacterial cells from monitoring the flow of cellular molecules (Scazzocchio *et al.*, 2017).

3.2.2 Inhibitors of Cell Wall Synthesis

Peptidoglycan that is composed of alternating units of NAG (N-acetylglucosamine) and NAM (N-acetylmuramic acid) contributes to rigidity and strength of the cell wall and also acts as a mechanism of defence. Several phytochemicals have been proven to inhibit bacterial cell wall synthesis (Upadhyay *et al.*, 2014). When the phytochemicals attach to the bacterial cell wall protein that was eventually disrupts the cell membrane penetrability. The profound ability of flavonoids to combine with the bacterial cell wall contributes to their strong antibacterial potential in combating infectious diseases (AlSheikh *et al.*, 2020).

3.2.3 Inhibitors of Efflux Pump in Bacteria

Efflux pumps elaborated by bacteria are broadly classified based on their energy source as:

1. Primary efflux pumps/primary transporters derive energy from ATP hydrolysis.
2. Secondary transporters derive energy from electrochemical gradients of protons (H+) or ions, such as sodium (Na+).

Certain phytochemicals are reported to reverse antibiotic resistance by acting as EPI (efflux pump inhibitors). Also, a few phytochemicals exhibit antibacterial activity by interfering with FtsZ Z-ring formation and thereby subsequently inhibiting cytokinesis in bacteria (Kelley *et al.*, 2012).

3.2.4 Enzyme Inhibitors

Phytocompounds are also known to inhibit nucleic acid synthesis by inhibiting the DNA gyrase, an enzyme that is crucial for DNA replication (Wu *et al.*, 2013). Certain flavonoids are known to intrude with helicase activity (DnaB and RecBCD) thus preventing replication of bacterial DNA (Xu *et al.*, 2001).

3.2.5 Tormenting Biofilm Synthesis and Anti-Quorum Sensing

Recent approaches to combat bacterial resistance rely on inhibiting the synthesis of biofilms and quorum sensing (QS). Most phytochemicals also exhibit anti-biofilm and anti-QS activity. The QS inhibitory actions of the phytocompounds are largely responsible for transcription inhibition of genes encoding biofilm formation (Packiavathy *et al.*, 2014).

3.2.6 Attenuation of Bacterial Virulence

Bacterial capsular polysaccharides are important virulence factors that significantly contribute to pathogenicity (Taylor and Roberts, 2005). Those bacterial strains that are exposed to plant-derived phytochemicals produce fewer capsular polysaccharides (Derakhshan *et al.*, 2008). As well owe the capacity to modulate the gene expression of the regulators of capsule synthesis.

3.3 MECHANISM OF PHYTOCHEMICALS

3.3.1 Regulation and Control of Metabolism

Phytochemicals also increase the metabolism by regulating the expression of metabolism-related proteins and anti-oxidant stress-associated proteins. In fructose-fed rats, cinnamon extract (bark) is reported to enhance metabolism of glucose and alter the lipid profile (Kannappan *et al.*, 2006). As a result, phytochemicals play a crucial role in the regulation and control of proteins or genes that are related to energy metabolism, formation of hepatocytes, transportation of fatty acid and the oxidation process.

3.3.2 Antioxidant Activity

Plant-based antioxidants are highly promising molecules and are extensively studied for potent antioxidant activity by scavenging reactive oxygen species (ROS). Prolonged inflammation, a characteristic feature of chronic microbial infections, is often associated with persistent upregulation of pro-inflammatory signals that eventually result in the release of lysosomal enzymes and free radicals – ROS. Metabolic syndrome and intestinal dysfunction are associated with ROS overproduction and decreased antioxidant activity. Oxidative stress is documented to play a prominent role in various metabolic and pathological disorders *viz.*, diabetes, obesity, dyslipidemia, insulin resistance, chronic inflammation and cardiovascular diseases (Čolak and Pap, 2021; Forrester *et al.*, 2018; Padgett *et al.*, 2013). Excess ROS that results in oxidative stress is proven to induce oxidation of biological substances, cellular damage, DNA damage and peroxidize lipids (Juan *et al.*, 2021; Wani *et al.*, 2021). Polyphenols present in the olives (used in Mediterranean diet) have the potency to reduce oxidative stress and thereby lower the risk of metabolic syndrome.

3.3.3 Antimicrobial Property

Phytochemicals are potential antimicrobial agents of natural origin. Anti-inflammatory and antimicrobial properties of α-amyrin and lupeol – nonpolar compounds – present in herbal extracts potentially damage the cell membrane and internal structures, which improves the flow of propidium iodide into bacterial cells that are treated (Famuyide *et al.*, 2020). β-amyrin is confirmed to be a potent antimicrobial agent. β-amyrin is reported to induce the production of ROS, which results in apoptosis-like death in *E. coli*. The antimicrobial activity of essential oils is confined not only to their chemical constituents, and also to the intermolecular interactions between the constituents, which can result in synergistic or antagonistic activities (Wani *et al.*, 2021). Essential oils are derivatives of terpenes and terpenoids that are aromatic and aliphatic acid esters and phenolic compounds. The unique property exhibited by the plant essential oils i.e. hydrophobicity, enables them to permeate into the bacterial cell by partitioning with the lipids in the bacterial cell membrane as well as the mitochondria. Essential oils disrupt cell membrane integrity eventually leading to bacterial cell death owing to the leakage of critical molecules and ions from the bacterial cell. Also, certain compounds (e.g. geraniol) in essential oils are capable of modulating drug resistance in Gram-negative bacteria by targeting efflux mechanisms (Lorenzi *et al.*, 2009). Recent reports have confirmed the antibacterial potential of nanoparticle (NP) functionalized with essential oils against MDR pathogens. This could be attributed to enhanced chemical stability and solubility, reduced evaporation and minimal destruction of active principles present in the essential oil. Of note, controlled and sustained release of the essential oil components from the encapsulated formulations has substantially increased their bioavailability and antibacterial efficacy against MDR strains (Chouhan *et al.*, 2017).

3.3.4 Immunomodulatory Activity

Phytochemicals showing immunostimulant properties, providing a limited alternative to chemotherapeutics. In a host immune system phytochemicals can stimulated either by directly boosting the innate immune system or inhibiting commensal bacteria, substances and pathogens (López Nadal *et al.*, 2020; Firmino *et al.*, 2021). Phytochemicals when consumed as dietary supplements can regulate the innate immune system, especially effector cells – leucocytes. i.e. "leukocyte activation", "neutrophil degranulation" and "neutrophil activation" (Firmino *et al.*, 2021). Plants with immunostimulatory properties improve the activity of lymphocytes, macrophages and natural killer cells, showing increased phagocytosis and stimulating the interferon production (Kuralkar and Kuralkar, 2021). The dietary supplementation of phytochemicals can substantially increase the WBC count and other immunological markers which possess strong anti-inflammatory effects (Roome *et al.*, 2008).

Natural immunomodulators are bio-molecules that can modulate adaptive or innate immunity and can reduce inflammatory responses through oxidation or alkylation. In clinical practice, immunomodulators have been broadly classified as immunosuppressants, immunostimulants and immunoadjuvants. The mechanism by which medicinal plant products primarily exert immunomodulatory effects is by inducing the production of cytokines, stimulating phagocyte cells and inhibiting the synthesis of iNOS, PGE and COX-2. Natural plant products with immunomodulatory activity include low molecular weight bioactive compounds like alkaloids, terpenoids, phenolics and high molecular weight bioactive compounds like lectins and polysaccharides. The common plant-derived immunostimulants include curcumin (polyphenol), genistein (phytoestrogen), immunoadjuvants include, pectic arabinogalactan (polysaccharide), immunosuppressants include resveratrol and epigallocatechin-3-gallate (polyphenols), colchicine, capsaicin (alkaloids) and andrographolide (diterpenoid) (Zebeaman *et al.*, 2023).

3.3.5 Mycotoxin Detoxification

Mycotoxins are low molecular weight, toxic secondary metabolites that are produced by filamentous fungi. Nearly 400 compounds have been identified as mycotoxins, of which 30 compounds have received great medical attention, as they pose a great threat to human or animal lives. The medically important mycotoxins include aflatoxins (mainly aflatoxin B1), ochratoxins (mainly ochratoxin A), fumonisins (mainly fumonisins B1, B2 and B3), trichothecenes (type A mainly HT-2 toxin and T-2 toxin and type B deoxynivalenol), zearalenone (Agriopoulou *et al.*, 2020). Consequences of intoxication of mycotoxins include immunological disturbances, intestinal barrier dysfunction, cytotoxicity and subsequently cell death (Marin *et al.*, 2013). Essential oils (EOs) and their main bioactive compounds have been documented to inhibit the mycotoxin production and anti-mycotoxigenic activity (Perczak *et al.*, 2019; Piemontese *et al.*, 2018), and they were demonstrated. The oil of Clove oil (especially eugenol) and turmeric essential oil are reported to inhibit the growth and production of AFB1 by *Aspergillus*. Also, clove buds are known to inhibit in vitro cultivation and production of mycotoxins by *A. flavus* and *P. citrinum* (Luo et al., 2018).

3.3.6 Phytochemicals can Negate Bacterial Drug Resistance

Due to the increasing emergence and dissemination of antibiotic-resistant pathogens, it is critical to identify/develop novel therapeutic alternatives to manage drug-resistant infections. As a result, bioactive phytocompounds have been considered as a potential, effective and safe alternative to the existing antibiotics to treat infections caused by MDR pathogens (Khare *et al.*, 2021). Plants are rich in phytochemicals, which are increasingly being explored for the identification and synthesis of novel drugs against AMR pathogens.

Also, phytocompounds have been shown to inhibit the major resistance mechanisms exhibited by MDR bacterial strains viz., efflux pump, replication machinery, membrane permeability and integrity, cell microstructure as well as other virulence mechanisms such as QS and biofilm formation. In addition to bactericidal activities, several plant-derived compounds have been recently discovered that potentiate/enhance the action of antibiotics i.e. serve as antibiotic adjuvants that re-sensitize or revert antibiotic resistance capacity.

These phytochemicals have been documented to modify/alter the cell membrane and cell wall of bacteria, thereby increasing the membrane permeability and cellular leakage. These alterations eventually result in the reduction of ATP, decreased protein synthesis, intracytoplasmic damage, Ph alteration, DNA fragmentation, downregulation of gene expression, binding of ions, inactivation of DNA gyrase, increase in ROS, upregulation of efflux pumps and mobile genetic elements (Bazzaz *et al.*, 2010; Yu *et al.*, 2020).

3.3.7 Combating MDR Bacteria

Antibiotics are chemotherapeutic drugs that are either, bactericidal or bacteriostatic and have been used to control bacterial infections. As per the Global Antimicrobial Resistance and Use Surveillance System (GLASS) report (2022), there alarming increase in the incidence of antibiotic resistance among bacterial pathogens. Nearly 42% of *E. coli* is reported to exhibit resistance to cephalosporins and 35% of *Staphylococcus aureus* are found to be methicillin-resistant. In 2020, one in five cases of urinary tract infections (UTI) caused by *E. coli*, exhibited decreased susceptibility to the conventional antibiotics. Thereby it is harder to effectively treat common infections. Examples include beta-lactam antibiotics, tetracyclines, macrolide, aminoglycosides and quinolones (Gilbert and McBain, 2003). Infections by MDR bacterial pathogens, MRSA, VRE and Carbapenem-resistant *Klebsiella pneumoniae* are on the rise and are associated with significant increase in the incidence of difficult-to-treat nosocomial infections (Golkar *et al.*, 2014; Luyt *et al.*, 2014). According to the projected scores by the Organization for Economic Cooperation and Development (OECD), it is predicted that the resistance rates to the last-resort antibiotics would increase by twofold by the year 2035.

The medicinal value and the therapeutic potential of plants, crude extracts/pure active phytomolecules have long been used in traditional/alternative medicine (Perumal Samy and Gopalakrishnakone, 2010). This first use of plant derivates in alleviating pain dates back to 1803 with the identification and extraction of morphine from *Papaver somniferum* (Schmitz and Friedrich, 1985). Crude extracts or pure active phytomolecules derived from plants are known to possess high therapeutic potential (Perumal Samy and Gopalakrishnakone, 2010). Extraction of morphine from *Papaver somniferum* dates back to 1803 (Schmitz and Friedrich, 1985). Till date the Food and Drug Administration (FDA) has approved only a few phytochemicals, including codeine, reserpine and colchicines (Kongkham *et al.*, 2020). Herbal extracts of medicinal plants, including *Zingiber officinale*, *Curcuma longa* and *Tinospora cordifolia*, shown to be bactericidal against both Gram-positive (*S. aureus*) and Gram-negative pathogens (*E. coli*, *P. aeruginosa* and *K. pneumoniae*) (Chakraborty *et al.*, 2014). Some pure substances have demonstrated their activity, either alone or in combination with conventional antibiotics. The antibacterial activity of berberine, an alkaloid isolated from *Berberis* spp. (root/stem), can be attributed to its potential to intercalate DNA, inhibit enzymes such as RNA polymerase, gyrase/topoisomerase, as well as inhibit cell division (Iwasa *et al.*, 2001; Yi *et al.*, 2007; Domadia *et al.*, 2008).

3.4 MAJOR PHYTOCHEMICALS THAT ARE INVOLVED IN COMBATING MDR STRAINS

3.4.1 Alkaloids

Alkaloids are heterocyclic molecules that contain organic nitrogen which have a wide range of chemical structures. Alkaloids such as morphine, nicotine and caffeine are water-soluble salts produced from amino acids (Compean and Ynalvez, 2014). Alkaloids have shown broad-spectrum antimicrobial activities. Alkaloid-rich biorational plant extracts, including Amaryllidaceae, have demonstrated antimycobacterial activity, *E. coli* and *S. aureus* (Thawabteh *et al.*, 2019; Omar *et al.*, 2021).

Piperine, an alkaloid isolated from *Piper nigrum* that inhibits mutant strains of *S.aureus* and also, piperine when co-administered with ciprofloxacin, significantly reduces the MIC against *S. aureus* (Khan *et al.*, 2006). The use of piperine as an EP inhibitor suggests that this alkaloid inhibits the MRSA and *S. aureus* NorA EPs. Berberine, alkaloid found in root, stem and bark of Berberis exhibits antibacterial activity by intercalating DNA, targeting RNA polymerase and gyrase (Yi *et al.*, 2007; Kang *et al.*, 2015; Peng *et al.*, 2015).

Quinoline alkaloids (dictamnine, kokusagine, masculine) extracted from *Tecleaafzelii* (stem bark) exhibit excellent antibacterial activity (Kuete *et al.*, 2008). Quinolone alkaloids prevent DNA replication by inhibiting type II topoisomerase (Heeb *et al.*, 2011). Alkaloids act as potent inhibitors of efflux pump, reserpine (an indole alkaloid from *Rauwolfia serpentina)* in *Streptococcus* spp., *Staphylococcus* spp., (Abdelfatah and Efferth, 2015; Sridevi *et al.*, 2017), Conessine (from Holarrhena antidysenterica) in *P. aeruginosa*, and Lysergol (from *Ipomoea muricata*) in *E. coli*. (Siriyong *et al.*, 2017; Maurya *et al.*, 2013). Tomatidine, a steroidal alkaloid compound isolated from solanaceous plant species such as *S. melongena* (aubergine), *S. tuberosum* (potato) and *S. lycopersicum* (tomato) exhibits significant anti-staphylococcal activity either alone or in synergy with aminoglycosides (Jiang *et al.*, 2016). Also,

Tomatidine is a potent ATP synthase inhibitor of *Staphylococcus* spp., *Bacillus* spp. and *Listeria* (Lamontagne Boulet *et al*., 2017; Guay *et al*., 2018). 8-epidiosbulbin E-acetate (from Dioscorea bulbifera) is capable of curing plasmids in *E. coli*, *P. aeruginosa*, *Shigella sonnei* and *Enterococcus faecalis* (Shriram *et al*., 2008).

3.4.2 Phenolic Compounds

Owing to their broad-spectrum antimicrobial activity and extraordinary pharmacological properties, plant phenolic compounds are considered remarkable natural bioactive chemicals. The molecular structure of phenolic compounds of plant origin contains aromatic ring with ≥1 OH group. Based on their structural properties, plant phenolics have been categorized into many classes, flavonoids, non-flavonoids and phenolic acids (de Souza *et al*., 2019). Plant phenolic compounds are used in the treatment of many chronic illnesses viz., diabetes, cancer, cardiovascular disease and chronic bacterial infections (Del Rio *et al*., 2013; Lim *et al*., 2016; Zacchino *et al*., 2017). Plant phenolics represent prospective natural antimicrobial weapons as they exhibit antibacterial activity against both susceptible and MDR strains (Miklasińska-Majdanik *et al*., 2018). Baicalei extracted from *Scutellaria baicalensis* (Chan *et al*., 2011) and Kaempferol from *Alpinia calcarata* (Randhawa *et al*., 2016) are potent inhibitors of efflux pump in MRSA. Also, resveratrol from *Vitis vinifera* is proven to be an effective EPI in *Campylobacter jejuni* (Klančnik *et al*., 2017).

Phenols like pyrogallol and catechol exhibit antibacterial activity against both Gram-positive and Gram-negative bacteria. Caffeic acid, a phenolic acid elaborates is effective against *S. aureus* and *E. coli*. However, MIC is much higher (1600 μg/ml) compared to ampicillin (0.1 and 3.2 μg/ml) (Lim *et al*., 2016). Of note, gallic acid exhibits potent antibacterial potential against *Streptococcus pneumoniae*, *S. aureus*, *Streptococcus agalactiae* strains *Enterococcus faecalis*, *Moraxella catarrhalis*, *E. coli* and *P. aeruginosa* (Aires *et al*., 2013; Cueva *et al*., 2012). Gallic acid is bactericidal to *Campylobacter jejuni* and *C. coli* strains with MICs ranging from 15.63 to 250 μg/ml (Sarjit *et al*., 2015). Taxifolin derived from *Allium cepa* acts as Beta-Ketoacyl acyl carrier protein synthase inhibitor in *E. faecalis* (Jeong *et al*., 2009).

3.4.3 Flavonoids

Plant flavonoids contain 2-phenyl-benzo-γ-pyrane nucleus and belong to the category of phenolic compounds that exhibit antimicrobial-potentiator properties (Górniak *et al*., 2019) had documented that different classes of flavonoids, *viz*., flavanones, isoflavonoids, flavonols, chalcones and dihydrochalcones act as allelochemicals that inhibit microbial growth. Also, the derivatives of chalcone-ferrocene/steroidal/fluorinated/chlorinated chalcones are reported to exhibit effective antimicrobial action against MDR pathogens (Xu *et al*., 2019). Catechin mediates cell membrane disruption, potassium leakage and the subsequent loss of membrane integrity in MRSA (Cushnie *et al*., 2008).

Flavonoids are potent inhibitors of biofilm synthesis and molecular crosstalk-QS. In *P. aeruginosa* PAO1, quercetin had been proven to inhibit biofilm formation with substantially low-level expression of genes associated with QS – *lasR*, *lasI*, *rhlR* and *rhlI* and reduced virulence *viz*., pyocyanin, protease and elastase (Ouyang *et al*., 2016). The synergistic potential of pinostrobin-A with ciprofloxacin on MDR strains of *P. aeruginosa* and *E. coli* is essentially by pinostrobin inhibited NorA efflux pumps (Christena *et al*., 2015).

3.4.4 Terpenes

Terpenes are hydrocarbons (chemical formula (C5H8)n) that are varying numbers of isoprene units composed of five carbon atoms (Yang *et al*., 2020). They are naturally synthesized in the cytoplasm of the vegetative cells initially from acetyl CoA via mevalonic acid pathway. Terpenes confer flavour and taste to the plant parts. The important terpene being, cannabis, recognized for its rich smell and aroma. Further, they play a substantial role in plant growth and development, physiological processes and environmental responses. Monoterpenes like menthol, sabinene, limonene and carvone exhibit strong antibacterial properties against *S. aureus* (Trombetta *et al.*, 2005; Yang *et al.*, 2020). Sesquiterpene (Patchouli alcohol), Diterpene (Artemisinin and Andrographolide) and Triterpene (Oleanolic acid) also show antibacterial properties against bacterial strains (Appalasamy *et al.*, 2014; Xu *et al.*, 2017; Yang *et al.*, 2020; Zhao *et al.*, 2013).

3.4.5 Terpenoids

Terpenoids are terpenes with additional oxygen molecules or that have lost/translocated methyl groups. The common terpenoids include carvacrol, menthol, thymol, linalyl acetate, linalool, citronellal, piperitone and geraniol. The incredible antimicrobial activity of terpenoids is largely linked to a triad of factors, presence of functional groups, presence of OH group of the phenolic terpenoids and presence of delocalized electrons. Carvacrol is more effective than other EOs, such as *p*-cymene (Ultee *et al.*, 2002; Ben Arfa *et al.*, 2006). The antimicrobial activity and hydrophobicity of carvacrol can be affected by the exchange of OH group and methyl ether. The antimicrobial activity of thymol against *B. cereus*, *P. aeruginosa* and *S. aureus* is equivalent to that of carvacrol (Ben Arfa *et al.*, 2006; Lambert *et al.*, 2001). Both carvacrol and thymol possess prominent OM disintegrating properties. *p*-Cymene, a monoterpene, precursor of carvacrol can enhance the antimicrobial action of carvacrol.

Carvacrol is one of the highly investigated compounds for antibacterial activity. Carvacrol is reported to be to cause:

1. Disintegration of the outer membrane of bacteria
2. Release of lipopolysaccharide
3. Alters the transport of ions across the cytoplasmic membrane to alter
4. Alters the fatty acid composition of cell membranes thereby elevating the fluidity and permeability
5. Affects the insertion/folding of OMPs
6. Affects protein folding (In *E. coli* cells that are exposed to sub-lethal concentration of carvacrol)
7. Inhibited the synthesis of flagellin (gives rise to bacterial cells that exhibit decreased motility)
8. Diminishes flagellar movement by reducing proton motive force (Nazzaro *et al.*, 2013; Burt *et al.*, 2007)

3.4.6 Essential Oils

Essential oils (EOs) are low-molecular-weight phytochemicals that possess antibacterial activity and are frequently employed in medicine (Yu *et al.*, 2020). EOs act as potent inhibitors of MDR bacterial pathogens by targeting and/or disrupting the most common mechanisms of drug resistance viz., altered permeability of the cell membrane/cell wall, drug efflux pumps, mobile genetic elements, QS in biofilms (Yu *et al.*, 2020).

It is well documented that Gram-negative organisms are relatively resistant to EOs compared to Gram-positive bacteria. EOs being hydrophobic are difficult to diffuse into the complex cell wall of Gram-negative bacteria as it has a lipopolysaccharide (LPS) layer and a rigid outer membrane. Nevertheless, Gram-positive bacteria are devoid of outer membrane and LPS. Though it is enclosed by a thick murein layer, it cannot prevent the entry of the low molecular weight EOs. Being small antimicrobial molecules, EOs can gain access to the cell membrane. Moreover, lipoteichoic acids present in the Gram-positive bacterial cell membrane facilitate the infiltration of the hydrophobic compounds of Eos (Azhdarzadeh and Hojjati, 2016; Huang *et al.*, 2014; Cox *et al.*, 2000). Being lipophilic, EOs can influence the unsaturated fatty acids (quantity and structure) present in the cell membrane of Gram-positive bacteria (Nazzaro *et al.*, 2013).

M. alternifolia EOs is proven to enhance the permeability of cell membranes (Cox *et al.*, 2000). EOs of *Dodartia oriental* disrupt bacterial cells and prevents the formation of biofilm in *S. aureus*, *E. coli* and *Salmonella enteritidis* (Wang *et al.*, 2017). EO of *Mentha longifolia* is documented to damage the cell wall of *E. coli*, *Micrococcus luteus*, *Salmonella typhimurium*. EOs of *Allium sativum* and *Piper nigrum* are reported to induce leakage of intracellular material in *E. coli* (Hyldgaard *et al.*, 2012; Zhang *et al.*, 2017). EO is present in *Litsea cubeba* to destroy both the inner and outer membranes of *E. coli* (Li *et al.*, 2014). EO of *Foeniculum vulgare* potentiates loss of membrane integrity in *Shigella dysenteriae* (Diao *et al.*, 2014). EO of *Ocimum gratissimum* and *Origanum vulgare* (Hyldgaard *et al.*, 2012; Lambert *et al.*, 2001) are known to permeabilize the cell membrane of Gram-negative pathogenic bacteria – *E. coli*, *Pseudomonas aeruginosa* and the ubiquitous Gram-positive pathogenic cocci – *Staphylococcus aureus*. Also, EO derived from *Dipterocarpus gracilis* and Cinnamon are known to disrupt the cell membrane of *Proteus mirabilis*, *Bacillus cereus*, *S. aureus* and *E. coli*, respectively (Zhang *et al.*, 2015; Zhang *et al.*, 2016). *Cuminum cyminum* is reported to induce alterations in the cytoplasm of *Bacillus* species, *B. cereus* and *B. subtilis* (Hyldgaard *et al.*, 2012). *Coriandrum sativum* EOs are documented to alter cell membrane permeability, impair cellular integrity and hence are a potent inhibitor of uropathogenic *E. coli*- MDR strains (Scazzocchio *et al.*, 2017). Essential oils (EOs) such as cinnamaldehyde, eugenol, and farnesol are terpenes that induce cell membrane disruption in pathogenic bacterial species like E. coli and S. aureus. Eugenol, a specific type of terpene, is documented to disrupt the cell membrane of MRSA (Togashi *et al.*, 2010; Yadav *et al.*, 2015; Zhang *et al.*, 2015).

3.5 REMOVAL OF MOBILE GENETIC ELEMENTS – PLASMID CURING

The well-established mechanisms of horizontal gene transfer (HGT) include conjugation transduction, transformation and transposition. Bacterial mobile genetic elements are the DNA pieces that are capable of being shuttled between bacterial cells and are largely represented by plasmids, transposons and phages. These mobile genetic elements, especially plasmids play a crucial role in the acquisition and dissemination of antibiotic resistance genes by HGT. EOs are documented to possess both bactericidal as well as resistance reversal activity, primarily plasmid curing ability (especially R-plasmids). Schelz *et al.* (2006) found that EOs from several plants, pure turpentine oil, is a compound with plasmid-curing activity against *E. coli* and *S. epidermidis*. Oregano EOs are potent antimicrobial molecules against MDR, ESBL *E. coli* strains (Si *et al.*, 2008). Also, EOs mixed with antibiotics such as polymyxin, amoxicillin and lincomycin exhibit synergistic plasmid-curing activity.

3.6 ATTENUATION OF BACTERIAL VIRULENCE TO COMBAT BACTERIAL PATHOGENS

Attenuation of bacterial virulence plays a key role in combating antimicrobial resistance by the phytochemicals. Bioactive phytocompounds with antimicrobial potential include alkaloids, polyphenols, flavonoids, tannins, terpenoids, glycosides and steroids. Of these bioactive compounds, polyphenols exhibit broad-spectrum antibacterial activity against a wide variety of pathogens. Notably, phenolic acids and flavanol (polyphenol) exhibit antibacterial action by attenuating the bacterial virulence factors-toxins/enzymes, diminishing extracellular polysaccharide (EPS) production, and also serving as EPS inhibitors. Isolation and phytochemical profiling of the bioactive compounds is of utmost importance and is the crucial part in the discovery and development of novel, natural plant-derived antimicrobials of clinical importance. These potentially safer phytocompounds would revolutionize the current era by mitigating the problem of AMR with potentially impossible resistance development in the future (Khameneh *et al.*, 2015).

3.7 TRANSLATIONAL RESEARCH ON PHYTOCOMPOUDS

The translation of *in vitro* studies and *in vivo* studies into clinical trials is a major barrier in the development of newer compounds of plant origin. Phytochemicals derived from medicinal plants exhibit antibacterial effects are potential alternatives for conventional therapy. Introduction of a new drug or antibiotic in the market is both time-consuming and expensive. Identification of chemical compounds from natural sources like plants is important to combat infections by MDR strains (Mandal *et al.*, 2014). Stringent measures need to be adopted by way of animal and human clinical trials to ensure safety and efficacy of phytochemical compound formulations. Compounds that are derived from herbs need to be validated and preclinically tested against a variety of infectious pathogens before being licensed. It could be a single chemical or several bioactive constituents being co-administered.

3.7.1 Hurdles in the Development of Phytomedicines

The major hurdle in the discovery and development of newer phytocompounds is the transition of *in vitro* and *in vivo* studies into clinical trials. Preclinical and clinical studies for phytochemical compounds need to be meticulously performed to be approved by the FDA and for licensed human use. Recommendation of a phytocompound/herbal/polyherbal formulation in routine clinical use necessitates ensuring consistency in drug formulation, advocating efficient methods of evaluation of efficacy, and safety with minimal or no side effects. The phytocompounds need a multilevel screening for efficacy and safety before being licensed for human administration, be it a single bioactive compound or a combination of bioactive constituents being co-administered.

3.8 DRAWBACKS

1. Phytochemicals have low bioavailability and absorption, high clearance, resistance and toxicity.
2. Certain phytochemicals, like alkaloids, can be harmful in large doses.
3. Though phytochemicals/herbal medicines/supplements are generally regarded as safe and well-tolerated, there often exist under-regulated quality, efficacy and safety issues.
4. Phytomedicine might cause modest to severe adverse effects such as mild gastrointestinal problems, hemolytic anemia and increased risk of hepatotoxicity.
5. Phytochemical profiling may not reveal all the compounds present.
6. Phytotoxins that could be rarely present in the phytoformulations may be hazardous to humans when consumed.

3.9 FUTURE PERSPECTIVE

With the alarming increase in the incidence of difficult-to-treat infections caused by antibiotic-resistant MDR strains, the development of novel alternative therapeutic approaches is the need of the hour. Plant-derived bioactive substances (phytochemicals) have proven to possess bactericidal, bacteriostatic, antibiofilm, virulence attenuating and antibiotic-reversal properties. Furthermore, the combined/synergistic action of phytochemicals with available antibiotics needs to be explored against MDR pathogens. Nevertheless, translational research is required to obtain conversion success, licensing and marketing as commercial phytocompounds with medicinal applications.

REFERENCES

Abdelfatah SA, Efferth T. Cytotoxicity of the indole alkaloid reserpine from *Rauwolfia serpentina* against drug-resistant tumor cells. Phytomedicine. 2015 Feb 15;22(2):308–18. https://doi.org/10.1016/j.phymed.2015.01.002. Epub 2015 Jan 19. PMID: 25765838.

Agriopoulou S, Stamatelopoulou E, Varzakas T. Advances in occurrence, importance, and mycotoxin control strategies: Prevention and detoxification in foods. Foods. 2020 Jan 28;9(2):137. https://doi.org/10.3390/foods9020137. PMID: 32012820; PMCID: PMC7074356.

Aires A, Marques E, Carvalho R, Rosa EA, Saavedra MJ. Evaluation of biological value and appraisal of polyphenols and glucosinolates from organic baby-leaf salads as antioxidants and antimicrobials against important human pathogenic bacteria. Molecules. 2013 Apr 19;18(4):4651–68. https://doi.org/10.3390/molecules18044651. PMID: 23603948; PMCID: PMC6269804.

AlSheikh HMA, Sultan I, Kumar V, Rather IA, Al-Sheikh H, Tasleem Jan A, Haq QMR. Plant-based phytochemicals as possible alternative to antibiotics in combating bacterial drug resistance. Antibiotics (Basel). 2020;9(8):480. https://doi.org/10.3390/antibiotics9080480. PMID: 32759771; PMCID: PMC7460449.

Appalasamy S, Lo KY, Ch'ng SJ, Nornadia K, Othman AS, Chan LK. Antimicrobial activity of artemisinin and precursor derived from in vitro plantlets of *Artemisia annua* L. Biomed Res Int. 2014;2014:215872. https://doi.org/10.1155/2014/215872. Epub 2014 Jan 19. PMID: 24575401; PMCID: PMC3915762.

Azhdarzadeh F, Hojjati M. Chemical composition and antimicrobial activity of leaf, ripe andunripe peel of bitter orange (*Citrus aurantium*) essential oils. Nutr Food Sci Res. 2016;3:43–50.

Bailey RL, Gahche JJ, Lentino CV, Dwyer JT, Engel JS, Thomas PR, Betz JM, Sempos CT, Picciano MF. Dietary supplement use in the United States, 2003–2006. J Nutr. 2011;141(2):261–6. https://doi.org/10.3945/jn.110.133025. Epub 2010 Dec 22. PMID: 21178089; PMCID: PMC3021445.

Bazzaz BS, Memariani Z, Khashiarmanesh Z, Iranshahi M, Naderinasab M. Effect of galbanic acid, a sesquiterpene coumarin from *Ferula szowitsiana*, as an inhibitor of efflux mechanism in resistant clinical isolates of Staphylococcus aureus. Braz J Microbiol. 2010;41(3):574–80. https://doi.org/10.1590/S1517-83822010000300006. Epub 2010 Sep 1. PMID: 24031531; PMCID: PMC3768656.

Ben Arfa A, Combes S, Preziosi-Belloy L, Gontard N, Chalier P. Antimicrobial activity of carvacrol related to its chemical structure. Lett Appl Microbiol. 2006;43(2):149–54. https://doi.org/10.1111/j.1472-765X.2006.01938.x. PMID: 16869897.

Burt SA, van der Zee R, Koets AP, de Graaff AM, van Knapen F, Gaastra W, Haagsman HP, Veldhuizen EJ. Carvacrol induces heat shock protein 60 and inhibits synthesis of flagellin in *Escherichia coli* O157:H7. Appl Environ Microbiol. 2007;73(14):4484–90. https://doi.org/10.1128/AEM.00340-07. Epub 2007 May 25. PMID: 17526792; PMCID: PMC1932834.

Chakraborty B, Nath A, Saikia H, Sengupta M. Bactericidal activity of selected medicinal plants against multidrug resistant bacterial strains from clinical isolates. Asian Pac J Trop Med. 2014 Sep;7S1:S435–41. https://doi.org/10.1016/S1995-7645(14)60271-6. PMID: 25312164.

Chan BC, Ip M, Lau CB, Lui SL, Jolivalt C, Ganem-Elbaz C, Litaudon M, Reiner NE, Gong H, See RH, Fung KP, Leung PC. Synergistic effects of baicalein with ciprofloxacin against NorA over-expressed methicillin-resistant *Staphylococcus aureus* (MRSA) and inhibition of MRSA pyruvate kinase. J Ethnopharmacol. 2011;137(1):767–73. https://doi.org/10.1016/j.jep.2011.06.039. Epub 2011 Jul 1. PMID: 21782012.

Christena LR, Subramaniam S, Vidhyalakshmi M, Mahadevan V, Sivasubramanian A, Nagarajan S. (2015). Dual role of pinostrobin-a flavonoid nutraceutical as an efflux pump inhibitor and antibiofilm agent to mitigate food Borne pathogens. RSC Adv. 5, 61881–61887. https://doi.org/10.1039/C5RA07165H

Chouhan S, Sharma K, Guleria S. Antimicrobial activity of some essential oils-present status and future perspectives. Medicines (Basel). 2017;4(3):58. https://doi.org/10.3390/medicines4030058. PMID: 28930272; PMCID: PMC5622393.

Čolak E, Pap D. The role of oxidative stress in the development of obesity and obesity-related metabolic disorders. J Med Biochem. 2021 Jan 26;40(1):1–9. https://doi.org/10.5937/jomb0-24652. PMID: 33584134; PMCID: PMC7857849.

Compean KL, Ynalvez RA (2014). Antimicrobial activity of plant secondary metabolites: A review. Res J Med Plant 8, 204–213. https://doi.org/10.3923/rjmp.2014.204.213

Correa-Velez I, Clavarino A, Eastwood H. Surviving, relieving, repairing, and boosting up: Reasons for using complementary/alternative medicine among patients with advanced cancer: A thematic analysis. J Palliat Med. 2005;8(5):953–61. https://doi.org/10.1089/jpm.2005.8.953. PMID: 16238508

Cox SD, Mann CM, Markham JL, Bell HC, Gustafson JE, Warmington JR, Wyllie SG. The mode of antimicrobial action of the essential oil of *Melaleuca alternifolia* (tea tree oil). J Appl Microbiol. 2000;88(1):170–5. https://doi.org/10.1046/j.1365-2672.2000.00943.x. PMID: 10735256.

Cueva C, Mingo S, Muñoz-González I, Bustos I, Requena T, del Campo R, Martín-Álvarez PJ, Bartolomé B, Moreno-Arribas MV. Antibacterial activity of wine phenolic compounds and oenological extracts against potential respiratory pathogens. Lett Appl Microbiol. 2012;54(6):557–63. https://doi.org/10.1111/j.1472-765X.2012.03248.x. Epub 2012 Apr 23. PMID: 22449241.

Cushnie TP, Taylor PW, Nagaoka Y, Uesato S, Hara Y, Lamb AJ. Investigation of the antibacterial activity of 3-o-octanoyl-(-)-epicatechin. J Appl Microbiol. 2008;105(5):1461–9. https://doi.org/10.1111/j.1365-2672.2008.03881.x. Epub 2008 Sep 13. PMID: 18795977.

de Souza EL, de Albuquerque TMR, Dos Santos AS, Massa NML, de Brito Alves JL. Potential interactions among phenolic compounds and probiotics for mutual boosting of their health-promoting properties and food functionalities - a review. Crit Rev Food Sci Nutr. 2019;59(10):1645–1659. https://doi.org/10.1080/10408398.2018.1425285. Epub 2018 Jan 29. PMID: 29377718.

Del Rio D, Rodriguez-Mateos A, Spencer JP, Tognolini M, Borges G, Crozier A. Dietary (poly)phenolics in human health: Structures, bioavailability, and evidence of protective effects against chronic diseases. Antioxid Redox Signal. 2013 May 10;18(14):1818–92. https://doi.org/10.1089/ars.2012.4581. Epub 2012 Aug 27. PMID: 22794138; PMCID: PMC3619154.

Derakhshan S, Sattari M, Bigdeli M. Effect of subinhibitory concentrations of cumin (*Cuminum cyminum* L.) Seed essential oil and alcoholic extract on the morphology, capsule expression and urease activity of klebsiella pneumoniae. Int J Antimicrob Agents. 2008;32(5):432–6. https://doi.org/10.1016/j.ijantimicag. 2008.05.009. Epub 2008 Aug 19. PMID: 18715764.

Diao WR, Hua QP, Zhang H, Xu JG. Chemical composition, antibacterial activity and mechanism of action of essential oil from seeds of fennel (*Foeniculum vulgare* Mill). Food Control. 2014;35:109–116.

Domadia PN, Bhunia A, Sivaraman J, Swarup S, Dasgupta D. Berberine targets assembly of *Escherichia coli* cell division protein FtsZ. Biochemistry. 2008 Mar 11;47(10):3225–34. https://doi.org/10.1021/bi7018546. Epub 2008 Feb 15. PMID: 18275156.

Famuyide IM, Fasina FO, Eloff JN, McGaw LJ. The ultrastructural damage caused by *Eugenia zeyheri* and *Syzygium legatii* acetone leaf extracts on pathogenic Escherichia coli. BMC Vet Res. 2020;16(1):326. https://doi.org/10.1186/s12917-020-02547-5. PMID: 32887606; PMCID: PMC7472583.

Firmino JP, Vallejos-Vidal E, Balebona MC, Ramayo-Caldas Y, Cerezo IM, Salomón R, Tort L, Estevez A, Moriñigo M, Reyes-López FE, Gisbert E. Diet, immunity, and Microbiota interactions: An integrative analysis of the intestine transcriptional response and Microbiota modulation in gilthead seabream (*Sparus aurata*) fed an essential oils-based functional diet. Front Immunol. 2021 Mar 4;12:625297. https://doi.org/10.3389/fimmu.2021.625297. PMID: 33746962; PMCID: PMC7969985.

Fratamico PM, Wasilenko JL, Garman B, Demarco DR, Varkey S, Jensen M, Rhoden K, Tice G. Evaluation of a multiplex real-time PCR method for detecting Shiga toxin-producing Escherichia coli in beef and comparison to the U.S. Department of agriculture food safety and inspection service microbiology laboratory guidebook method. J Food Prot. 2014;77(2):180–8. https://doi.org/10.4315/0362-028X.JFP-13-248. PMID: 24490911.

Forrester SJ, Kikuchi DS, Hernandes MS, Xu Q, Griendling KK. Reactive oxygen species in metabolic and inflammatory signaling. Circ Res. 2018 Mar 16;122(6):877–902. https://doi.org/10.1161/CIRCRESAHA.117.311401. PMID: 29700084; PMCID: PMC5926825.

Funk J, Stoeber H, Hauser E, Schmidt H. Molecular analysis of subtilase cytotoxin genes of food-borne shiga toxin-producing *Escherichia coli* reveals a new allelic subAB variant. BMC Microbiol. 2013 Oct 15;13:230. https://doi.org/10.1186/1471-2180-13-230. Erratum in: BMC Microbiol. 2014;14:32. PMID: 24128013; PMCID: PMC3852854.

Gaigé S, Djelloul M, Tardivel C, Airault C, Félix B, Jean A, Lebrun B, Troadec JD, Dallaporta M. Modification of energy balance induced by the food contaminant t-2 toxin: A multimodal gut-to-brain connection. Brain Behav Immun. 2014 Mar;37:54–72. https://doi.org/10.1016/j.bbi.2013.12.008. Epub 2013 Dec 16. PMID: 24355099.

Gigante V, Sati H, Beyer P. Recent advances and challenges in antibacterial drug development. ADMET DMPK. 2022;10(2):147–151. https://doi.org/10.5599/admet.1271. PMID: 35350117; PMCID: PMC8957241.

Gilbert P, McBain AJ. Potential impact of increased use of biocides in consumer products on prevalence of antibiotic resistance. Clin Microbiol Rev. 2003;16(2):189–208. https://doi.org/10.1128/CMR.16.2.189-208.2003. PMID: 12692093; PMCID: PMC153147.

Golkar Z, Bagasra O, Pace DG. Bacteriophage therapy: A potential solution for the antibiotic resistance crisis. J Infect Dev Ctries. 2014 Feb 13;8(2):129–36. https://doi.org/10.3855/jidc.3573. PMID: 24518621.

Górniak I, Bartoszewski R, Króliczewski J (2019). Comprehensive review of antimicrobial activities of plant flavonoids. Phytochem Rev. 18, 241–272. https://doi.org/10.1007/s11101-018-9591-z

Guay I, Boulanger S, Isabelle C, Brouillette E, Chagnon F, Bouarab K, Marsault E, Malouin F. Tomatidine and analog FC04-100 possess bactericidal activities against listeria, *Bacillus* and *Staphylococcus* spp. BMC Pharmacol Toxicol. 2018 Feb 13;19(1):7. https://doi.org/10.1186/s40360-018-0197-2. PMID: 29439722; PMCID: PMC5812199.

Heeb S, Fletcher MP, Chhabra SR, Diggle SP, Williams P, Cámara M. Quinolones: From antibiotics to autoinducers. FEMS Microbiol Rev. 2011;35(2):247–74. https://doi.org/10.1111/j.1574-6976.2010.00247.x. PMID: 20738404; PMCID: PMC3053476.

Huang DF, Xu JG, Liu JX, Zhang H, Hu QP. Chemical constituents, antibacterial activity and mechanism of action of the essential oil from *Cinnamomum cassia* bark against four food related bacteria. Microbiology. 2014;83:357–365.

Hyldgaard M, Mygind T, Meyer RL. Essential oils in food preservation: Mode of action, synergies, and interactions with food matrix components. Front Microbiol. 2012 Jan 25;3:12. https://doi.org/10.3389/fmicb.2012.00012. PMID: 22291693; PMCID: PMC3265747.

Iwasa K, Moriyasu M, Yamori T, Turuo T, Lee DU, Wiegrebe W. In vitro cytotoxicity of the protoberberine-type alkaloids. J Nat Prod. 2001;64(7):896–8. https://doi.org/10.1021/np000554f. PMID: 11473418.

Jeong KW, Lee JY, Kang DI, Lee JU, Shin SY, Kim Y. Screening of flavonoids as candidate antibiotics against *Enterococcus faecalis*. J Nat Prod. 2009;72(4):719–24. https://doi.org/10.1021/np800698d. PMID: 19236029.

Jiang QW, Chen MW, Cheng KJ, Yu PZ, Wei X, Shi Z. Therapeutic potential of steroidal alkaloids in cancer and other diseases. Med Res Rev. 2016;36(1):119–43. https://doi.org/10.1002/med.21346. Epub 2015 Mar 27. PMID: 25820039.

Juan CA, Pérez de la Lastra JM, Plou FJ, Pérez-Lebeña E. The chemistry of reactive oxygen species (ROS) revisited: Outlining their role in biological macromolecules (DNA, lipids and proteins) and induced pathologies. Int J Mol Sci. 2021 Apr 28;22(9):4642. https://doi.org/10.3390/ijms22094642. PMID: 33924958; PMCID: PMC8125527.

Kang S, Li Z, Yin Z, Jia R, Song X, Li L, Chen Z, Peng L, Qu J, Hu Z, Lai X, Wang G, Liang X, He C, Yin L. The antibacterial mechanism of berberine against *Actinobacillus pleuropneumoniae*. Nat Prod Res. 2015;29(23):2203–6. https://doi.org/10.1080/14786419.2014.1001388. Epub 2015 Jan 23. PMID: 25613587.

Kannappan S, Jayaraman T, Rajasekar P, Ravichandran MK, Anuradha CV. Cinnamon bark extract improves glucose metabolism and lipid profile in the fructose-fed rat. Singapore Med J. 2006;47(10):858–63. PMID: 16990960.

Kelley C, Zhang Y, Parhi A, Kaul M, Pilch DS, LaVoie EJ. 3-Phenyl substituted 6,7-dimethoxyisoquinoline derivatives as FtsZ-targeting antibacterial agents. Bioorg Med Chem. 2012 Dec 15;20(24):7012–29. https://doi.org/10.1016/j.bmc.2012.10.009. Epub 2012 Oct 17. PMID: 23127490; PMCID: PMC3947851.

Khameneh B, Iranshahy M, Ghandadi M, Ghoochi Atashbeyk D, Fazly Bazzaz BS, Iranshahi M. Investigation of the antibacterial activity and efflux pump inhibitory effect of co-loaded piperine and gentamicin nanoliposomes in methicillin-resistant *Staphylococcus aureus*. Drug Dev Ind Pharm. 2015;41(6):989–94. https://doi.org/10.3109/03639045.2014.920025. Epub 2014 May 20. PMID: 24842547.

Khan IA, Mirza ZM, Kumar A, Verma V, Qazi GN. Piperine, a phytochemical potentiator of ciprofloxacin against *Staphylococcus aureus*. Antimicrob Agents Chemother. 2006;50(2):810–2. https://doi.org/10.1128/AAC.50.2.810-812.2006. PMID: 16436753; PMCID: PMC1366922.

Khare T, Anand U, Dey A, Assaraf YG, Chen ZS, Liu Z, Kumar V. Exploring phytochemicals for combating antibiotic resistance in microbial pathogens. Front Pharmacol. 2021 Jul 21;12:720726. https://doi.org/10.3389/fphar.2021.720726. PMID: 34366872; PMCID: PMC8334005.

Klančnik A, Šikić Pogačar M, Trošt K, Tušek Žnidarič M, Mozetič Vodopivec B, Smole Možina S. Anti-campylobacter activity of resveratrol and an extract from waste pinot noir grape skins and seeds, and resistance of Camp. Jejuni planktonic and biofilm cells, mediated via the CmeABC efflux pump. J Appl Microbiol. 2017;122(1):65–77. https://doi.org/10.1111/jam.13315. Epub 2016 Nov 15. PMID: 27709726.

Kongkham B, Prabakaran D, Puttaswamy H. Opportunities and challenges in managing antibiotic resistance in bacteria using plant secondary metabolites. Fitoterapia. 2020 Nov;147:104762. https://doi.org/10.1016/j.fitote.2020.104762. Epub 2020 Oct 16. PMID: 33069839.

Kuete V, Wansi JD, Mbaveng AT, Kana Sop MM, Beng VP, Etoa FX, et al. (2008). Antimicrobial activity of the methanolic extract and compounds from *Teclea afzelii* (Rutaceae). South Afr J Bot. 74, 572–576. https://doi.org/10.1016/j.sajb.2008.02.004

Kuralkar P, Kuralkar SV. Role of herbal products in animal production - An updated review. J Ethnopharmacol. 2021 Oct 5;278:114246. https://doi.org/10.1016/j.jep.2021.114246. Epub 2021 May 27. PMID: 34052352.

Lambert RJ, Skandamis PN, Coote PJ, Nychas GJ. A study of the minimum inhibitory concentration and mode of action of oregano essential oil, thymol and carvacrol. J Appl Microbiol. 2001;91(3):453–62. https://doi.org/10.1046/j.1365-2672.2001.01428.x. PMID: 11556910.

Lamontagne Boulet M, Isabelle C, Guay I, Brouillette E, Langlois JP, Jacques P, Rodrigue S, Brzezinski R, Beauregard PB, Bouarab K, Boyapelly K, Boudreault PL, Marsault É, Malouin F. Tomatidine is a lead antibiotic molecule that targets *Staphylococcus aureus* ATP synthase subunit c. Antimicrob Agents Chemother. 2018 May 25;62(6):e02197–17. https://doi.org/10.1128/AAC.02197-17. PMID: 29610201; PMCID: PMC5971568.

Li W-R, Shi Q-S, Liang Q, Xie X-B, Huang X-M, Chen YB. Antibacterial activity and kinetics of litseacubeba oil on *Escherichia coli*. PLoS ONE. 2014;9:e110983.

Lim A, Subhan N, Jazayeri JA, John G, Vanniasinkam T, Obied HK. Plant phenols as antibiotic boosters: In vitro interaction of olive leaf phenols with ampicillin. Phytother Res. 2016;30(3):503–9. https://doi.org/10.1002/ptr.5562. Epub 2016 Jan 11. PMID: 26931616.

López Nadal A, Ikeda-Ohtsubo W, Sipkema D, Peggs D, McGurk C, Forlenza M, Wiegertjes GF, Brugman S. Feed, Microbiota, and gut immunity: Using the zebrafish model to understand fish health. Front Immunol. 2020 Feb 5;11:114. https://doi.org/10.3389/fimmu.2020.00114. PMID: 32117265; PMCID: PMC7014991.

Lorenz SC, Son I, Maounounen-Laasri A, Lin A, Fischer M, Kase JA. Prevalence of hemolysin genes and comparison of ehxA subtype patterns in Shiga toxin-producing *Escherichia coli* (STEC) and non-STEC strains from clinical, food, and animal sources. Appl Environ Microbiol. 2013;79(20):6301–11. https://doi.org/10.1128/AEM.02200-13. Epub 2013 Aug 9. PMID: 23934487; PMCID: PMC3811216.

Lorenzi V, Muselli A, Bernardini AF, Berti L, Pagès JM, Amaral L, Bolla JM. Geraniol restores antibiotic activities against multidrug-resistant isolates from gram-negative species. Antimicrob Agents Chemother. 2009;53(5):2209–11. https://doi.org/10.1128/AAC.00919-08. Epub 2009 Mar 2. PMID: 19258278; PMCID: PMC2681508.

Luo Y, Liu X, Li J. Updating techniques on controlling mycotoxins—A review. Food Control. 2018;89:123–132.

Luyt CE, Bréchot N, Trouillet JL, Chastre J. Antibiotic stewardship in the intensive care unit. Crit Care. 2014 Aug 13;18(5):480. https://doi.org/10.1186/s13054-014-0480-6. PMID: 25405992; PMCID: PMC4281952.

Mandal SM, Roy A, Ghosh AK, Hazra TK, Basak A, Franco OL. Challenges and future prospects of antibiotic therapy: From peptides to phages utilization. Front Pharmacol. 2014 May 13;5:105. https://doi.org/10.3389/fphar.2014.00105. PMID: 24860506; PMCID: PMC4027024.

Marin S, Ramos AJ, Cano-Sancho G, Sanchis V. Mycotoxins: Occurrence, toxicology, and exposure assessment. Food Chem Toxicol. 2013 Oct;60:218–37. https://doi.org/10.1016/j.fct.2013.07.047. Epub 2013 Jul 29. PMID: 23907020.

Maurya A, Dwivedi GR, Darokar MP, Srivastava SK. Antibacterial and synergy of clavine alkaloid lysergol and its derivatives against nalidixic acid-resistant *Escherichia coli*. Chem Biol Drug Des. 2013;81(4): 484–90. https://doi.org/10.1111/cbdd.12103. Epub 2013 Mar 21. PMID: 23290001.

Miklasińska-Majdanik M, Kępa M, Wojtyczka RD, Idzik D, Wąsik TJ. Phenolic compounds diminish antibiotic resistance of *Staphylococcus aureus* clinical strains. Int J Environ Res Public Health. 2018 Oct 22;15(10):2321. https://doi.org/10.3390/ijerph15102321. PMID: 30360435; PMCID: PMC6211117.

Nazzaro F, Fratianni F, De Martino L, Coppola R, De Feo V. Effect of essential oils on pathogenic bacteria. Pharmaceuticals (Basel). 2013 Nov 25;6(12):1451–74. https://doi.org/10.3390/ph6121451. PMID: 24287491; PMCID: PMC3873673.

Omar F, Tareq AM, Alqahtani AM, Dhama K, Sayeed MA, Emran TB, Simal-Gandara J. Plant-based indole alkaloids: A comprehensive overview from a pharmacological perspective. Molecules. 2021 Apr 15; 26(8):2297. https://doi.org/10.3390/molecules26082297. PMID: 33921093; PMCID: PMC8071433.

Ouyang J, Sun F, Feng W, Sun Y, Qiu X, Xiong L, Liu Y, Chen Y. Quercetin is an effective inhibitor of quorum sensing, biofilm formation and virulence factors in *Pseudomonas aeruginosa*. J Appl Microbiol. 2016;120(4):966–74. https://doi.org/10.1111/jam.13073. Epub 2016 Mar 7. PMID: 26808465.

Packiavathy IA, Priya S, Pandian SK, Ravi AV. Inhibition of biofilm development of uropathogens by curcumin - An anti-quorum sensing agent from *Curcuma longa*. Food Chem. 2014 Apr 1;148:453–60. https://doi.org/10.1016/j.foodchem.2012.08.002. Epub 2012 Aug 10. PMID: 24262582.

Padgett LE, Broniowska KA, Hansen PA, Corbett JA, Tse HM. The role of reactive oxygen species and proinflammatory cytokines in type 1 diabetes pathogenesis. Ann N Y Acad Sci. 2013;1281(1):16–35. https://doi.org/10.1111/j.1749-6632.2012.06826.x. Epub 2013 Jan 16. PMID: 23323860; PMCID: PMC3715103.

Peng L, Kang S, Yin Z, Jia R, Song X, Li L, Li Z, Zou Y, Liang X, Li L, He C, Ye G, Yin L, Shi F, Lv C, Jing B. Antibacterial activity and mechanism of berberine against *Streptococcus agalactiae*. Int J Clin Exp Pathol. 2015;8(5):5217–23. PMID: 26191220; PMCID: PMC4503092.

Perczak A, Juś K, Gwiazdowska D, Marchwińska K, Waśkiewicz A. The efficiency of deoxynivalenol degradation by essential oils under in vitro conditions. Foods. 2019 Sep 11;8(9):403. https://doi.org/10.3390/foods8090403. PMID: 31514336; PMCID: PMC6769570.

Perumal Samy R, Gopalakrishnakone P. Therapeutic potential of plants as anti-microbials for drug discovery. Evid Based Complement Alternat Med. 2010;7(3):283–94. https://doi.org/10.1093/ecam/nen036. Epub 2008 Jun 24. PMID: 18955349; PMCID: PMC2887332.

Piemontese L, Messia MC, Marconi E, Falasca L, Zivoli R, Gambacorta L, Perrone G, Solfrizzo M. Effect of gaseous ozone treatments on DON, microbial contaminants and technological parameters of wheat and semolina. Food Addit Contam Part A Chem Anal Control Expo Risk Assess. 2018;35(4):760–771. https://doi.org/10.1080/19440049.2017.1419285. Epub 2018 Jan 19. PMID: 29279049.

Prestinaci F, Pezzotti P, Pantosti A. Antimicrobial resistance: A global multifaceted phenomenon. Pathog Glob Health. 2015;109(7):309–18. https://doi.org/10.1179/2047773215Y.0000000030. Epub 2015 Sep 7. PMID: 26343252; PMCID: PMC4768623.

Randhawa HK, Hundal KK, Ahirrao PN, Jachak SM, Nandanwar HS. (2016). Efflux pump inhibitory activity of flavonoids isolated from *Alpinia calcarata* against methicillin-resistant *Staphylococcus aureus*. Biologia 71, 484–493. https://doi.org/10.1515/biolog-2016-0073

Roome T, Dar A, Naqvi S, Ali S, Choudhary MI. *Aegiceras corniculatum* extract suppresses initial and late phases of inflammation in rat paw and attenuates the production of eicosanoids in rat neutrophils and human platelets. J Ethnopharmacol. 2008 Nov 20;120(2):248–54. https://doi.org/10.1016/j.jep.2008.08.025. Epub 2008 Sep 2. PMID: 18809484.

Sarjit A, Wang Y, Dykes GA. Antimicrobial activity of gallic acid against thermophilic campylobacter is strain specific and associated with a loss of calcium ions. Food Microbiol. 2015 Apr;46:227–233. https://doi.org/10.1016/j.fm.2014.08.002. Epub 2014 Aug 27. PMID: 25475290.

Scazzocchio F, Mondì L, Ammendolia MG, Goldoni P, Comanducci A, Marazzato M, Conte MP, Rinaldi F, Crestoni ME, Fraschetti C, Longhi C. Coriander (*Coriandrum sativum*) essential oil: Effect on multidrug resistant uropathogenic *Escherichia coli*. Nat Prod Commun. 2017;12(4):623–626. PMID: 30520610.

Schelz Z, Molnar J, Hohmann J. Antimicrobial and antiplasmid activities of essential oils. Fitoterapia. 2006;77(4):279–85. https://doi.org/10.1016/j.fitote.2006.03.013. Epub 2006 May 11. PMID: 16690225.

Schmitz R. Friedrich Wilhelm Sertürner and the discovery of morphine. Pharm Hist. 1985;27(2):61–74. PMID: 11611724.

Shriram V, Jahagirdar S, Latha C, Kumar V, Puranik V, Rojatkar S, Dhakephalkar PK, Shitole MG. A potential plasmid-curing agent, 8-epidiosbulbin E acetate, from *Dioscorea bulbifera* L. Against multidrug-resistant bacteria. Int J Antimicrob Agents. 2008;32(5):405–10. https://doi.org/10.1016/j.ijantimicag.2008.05.013. Epub 2008 Aug 20. PMID: 18718743.

Si H, Hu J, Liu Z, Zeng ZL. Antibacterial effect of oregano essential oil alone and in combination with antibiotics against extended-spectrum beta-lactamase-producing *Escherichia coli*. FEMS Immunol Med Microbiol. 2008;53(2):190–4. https://doi.org/10.1111/j.1574-695X.2008.00414.x. Epub 2008 Apr 23. PMID: 18435748.

Siriyong T, Srimanote P, Chusri S, Yingyongnarongkul BE, Suaisom C, Tipmanee V, Voravuthikunchai SP. Conessine as a novel inhibitor of multidrug efflux pump systems in *Pseudomonas aeruginosa*. BMC Complement Altern Med. 2017Aug 14;17(1):405. https://doi.org/10.1186/s12906-017-1913-y. PMID: 28806947; PMCID: PMC5557310.

Sridevi D, Shankar C, Prakash P, Park J, Thamaraiselvi K. Inhibitory effects of reserpine against efflux pump activity of antibiotic resistance bacteria. Chem Biol. 2017;Lett:469–72.

Stanker LH, Scotcher MC, Cheng L, Ching K, McGarvey J, Hodge D, Hnasko R. A monoclonal antibody based capture ELISA for botulinum neurotoxin serotype b: Toxin detection in food. Toxins (Basel). 2013;5(11):2212–26. https://doi.org/10.3390/toxins5112212. PMID: 24253240; PMCID: PMC3847722.

Taylor CM, Roberts IS. Capsular polysaccharides and their role in virulence. Contrib Microbiol. 2005;12: 55–66. https://doi.org/10.1159/000081689. PMID: 15496776.

Thawabteh A, Juma S, Bader M, Karaman D, Scrano L, Bufo SA, Karaman R. The biological activity of natural alkaloids against herbivores, cancerous cells and pathogens. Toxins (Basel). 2019 Nov 11;11(11):656. https://doi.org/10.3390/toxins11110656. PMID: 31717922; PMCID: PMC6891610.

Togashi N, Hamashima H, Shiraishi A, Inoue Y, Takano A. Antibacterial activities against *Staphylococcus aureus* of terpene alcohols with aliphatic carbon chains. J Essent Oil Res. 2010;22:263–269.

Trombetta D, Castelli F, Sarpietro MG, Venuti V, Cristani M, Daniele C, et al. Mechanisms of antibacterial action of three monoterpenes. Antimicrob Agents Chemother. 2005;49(6):2474–2478.

Ultee A, Bennik MH, Moezelaar R. The phenolic hydroxyl group of carvacrol is essential for action against the food-borne pathogen *Bacillus cereus*. Appl Environ Microbiol. 2002;68(4):1561–8. https://doi.org/10.1128/AEM.68.4.1561-1568.2002. PMID: 11916669; PMCID: PMC123826.

Upadhyay A, Upadhyaya I, Kollanoor-Johny A, Venkitanarayanan K. Combating pathogenic microorganisms using plant-derived antimicrobials: A minireview of the mechanistic basis. Biomed Res Int. 2014;2014:761741. https://doi.org/10.1155/2014/761741. Epub 2014 Sep 14. PMID: 25298964; PMCID: PMC4178913.

Wang J, Yang T, Chen H, Xu YN, Yu LF, Liu T, Tang J, Yi Z, Yang CG, Xue W, Yang F. The synthesis and antistaphylococcal activity of 9, 13-disubstituted berberine derivatives. Eur J Med Chem. 2017 Feb 15;127:424–433. https://doi.org/10.1016/j.ejmech.2017.01.012. Epub 2017 Jan 9. PMID: 28092858.

Wani AR, Yadav K, Khursheed A, Rather MA. An updated and comprehensive review of the antiviral potential of essential oils and their chemical constituents with special focus on their mechanism of action against various influenza and coronaviruses. Microb Pathog. 2021 Mar;152:104620. https://doi.org/10.1016/j.micpath.2020.104620. Epub 2020 Nov 16. PMID: 33212200; PMCID: PMC9159739.

Wu T, Zang X, He M, Pan S, Xu X. Structure-activity relationship of flavonoids on their anti-*Escherichia coli* activity and inhibition of DNA gyrase. J Agric Food Chem. 2013Aug 28;61(34):8185–90. https://doi.org/10.1021/jf402222v. Epub 2013 Aug 20. PMID: 23926942.

Xu H, Ziegelin G, Schröder W, Frank J, Ayora S, Alonso JC, Lanka E, Saenger W. Flavones inhibit the hexameric replicative helicase RepA. Nucleic Acids Res. 2001 Dec 15;29(24):5058–66. https://doi.org/10.1093/nar/29.24.5058. PMID: 11812837; PMCID: PMC97556.

Xu M, Wu P, Shen F, Ji J, Rakesh KP. Chalcone derivatives and their antibacterial activities: Current development. Bioorg Chem. 2019 Oct;91:103133. https://doi.org/10.1016/j.bioorg.2019.103133. Epub 2019 Jul 19. PMID: 31374524.

Xu YF, Lian DW, Chen YQ, Cai YF, Zheng YF, Fan PL, Ren WK, Fu LJ, Li YC, Xie JH, Cao HY, Tan B, Su ZR, Huang P. In vitro and in vivo antibacterial activities of patchouli alcohol, a naturally occurring tricyclic sesquiterpene, against helicobacter pylori infection. Antimicrob Agents Chemother. 2017 May 24;61(6):e00122–17. https://doi.org/10.1128/AAC.00122-17. PMID: 28320722; PMCID: PMC5444145.

Yadav MK, Chae SW, Im GJ, Chung JW, Song JJ. Eugenol: A phyto-compound effective against methicillin-resistant and methicillin-sensitive *Staphylococcus aureus* clinical strain biofilms. PLoS One. 2015 Mar 17;10(3):e0119564. https://doi.org/10.1371/journal.pone.0119564. PMID: 25781975; PMCID: PMC4364371.

Yang W, Chen X, Li Y, Guo S, Wang Z, Yu X. Advances in pharmacological activities of terpenoids. Nat Prod Commun. 2020;15.

Yi ZB, Yu Y, Liang YZ, Zeng B. Evaluation of the antimicrobial mode of berberine by LC/ESI-MS combined with principal component analysis. J Pharm Biomed Anal. 2007;44(1):301–4. https://doi.org/10.1016/j.jpba.2007.02.018. Epub 2007 Feb 23. PMID: 17383137.

Yu Z, Tang J, Khare T, Kumar V. The alarming antimicrobial resistance in ESKAPEE pathogens: Can essential oils come to the rescue? Fitoterapia. 2020 Jan;140:104433. https://doi.org/10.1016/j.fitote.2019.104433. Epub 2019 Nov 21. PMID: 31760066.

Zacchino SA, Butassi E, Liberto MD, Raimondi M, Postigo A, Sortino M. Plant phenolics and terpenoids as adjuvants of antibacterial and antifungal drugs. Phytomedicine. 2017 Dec 15;37:27–48. https://doi.org/10.1016/j.phymed.2017.10.018. Epub 2017 Oct 31. PMID: 29174958.

Zebeaman M, Tadesse MG, Bachheti RK, Bachheti A, Gebeyhu R, Chaubey KK. Plants and plant-derived molecules as natural immunomodulators. Biomed Res Int. 2023 Jun 5;2023:7711297. https://doi.org/10.1155/2023/7711297. PMID: 37313550; PMCID: PMC10260316.

Zhang J, Ye KP, Zhang X, Pan DD, Sun YY, Cao JX. Antibacterial activity and mechanism of action of black pepper essential oil on meat-Borne *Escherichia coli*. Front Microbiol. 2017 Jan 4;7:2094. https://doi.org/10.3389/fmicb.2016.02094. PMID: 28101081; PMCID: PMC5209337.

Zhang Y, Liu X, Wang Y, Jiang P, Quek SY. Antibacterial activity and mechanism of cinnamon essential oil against *Escherichia coli* and *Staphylococcus aureus*. Food Control. 2016;59:282–289.

Zhang YB, Liu XY, Jiang PP, Li WD, Wang YF. Mechanism and antibacterial activity of cinnamaldehyde against *Escherichia coli* and *Staphylococcus aureus*. Mod Food Sci Technol. 2015;31:31–35.

Zhao LX, Wang ZW, Zheng CJ, et al. Synthesis, characterization and antibacterial activity of oleanolic acid c-28 modification. J Liaoning Normal Univ. 2013;36:529–533.

4 Mechanism of Phytochemicals in Modulating Signalling Pathways Involved in Inflammation and Cancer

Hemarangan J, Yashoda K, Srisri Satish Karthik, Harin N Ganesh, Vasanth Kanth TL, Vijay Murali RM, Shriya P, Shambhavi Jha, Kumaran K, Raksa Arun, Arul Jothi KN, and Satish Ramalingam
Department of Genetic Engineering
College of Engineering and Technology
SRM Institute of Science and Technology
Kattankulathur, Chengalpattu, India

4.1 INTRODUCTION

Cancer is the second most prevalent disease and the leading cause of death worldwide. Globally, the prevalence of cancer is rising, and despite the availability of numerous treatments, the number of deaths from cancer is also rising. According to estimates on the global cancer burden, there were 19.3 million cases in 2020 and 28.4 million cases predicted by 2040. The three most common cancer types are colorectal (10.0%), lung (11.4), and breast (11.7%). Potential diagnostic/prognostic markers that can identify cancer at an early stage are needed by the medical community to prevent the sharp rise in the death rate. Researchers discovered several substances that can be used as chemotherapeutic medications to treat cancer and prolong patient lives. Because of treatment failure and tumour recurrence, the rise in fatalities is uncontrollable even with these chemotherapeutic medications. Cancer cells can continue to spread and harm a patient's quality of life even after surgery, chemotherapy, and radiation treatment. They might also impact nearby or distant organs. Additionally, cancer cells can develop treatment resistance and evade immune monitoring. Changes in the gene regulators governing different signalling pathways give cancer cells the ability to elude treatment and immune response.

Using natural herbs for treating various diseases and illnesses was started ten decades ago by our ancestors. All the parts of plants such as roots, leaves, stems, and bark have been used both as medicine and also to enhance the taste of the food. Animals like chimps, baboons, and elephants also drink the fermented extracts from marula fruit, since it contains various antioxidants and essential fatty acids. Rather than plant extracts, our ancestors used vegetable and animal extracts for various diseases. Furthermore, the trend of repurposing the use of those extracts for chronic diseases such as cancer, cardiovascular, and skin-related disorders to improve the patient's survival. Even though we developed various chemotherapeutic drugs, the cancer cells undergo drug resistance and multiple side effects that end up with a decrease in the survival rate of the patients. The idea of replacing the natural components as a drug to replace artificial chemotherapeutic drugs is the strategy to overcome the multiple side effects associated with any cancer treatments. The extracts from the plant parts are called phytochemicals or phytonutrients, in this chapter, we will discuss various types of cancer and phytochemicals, their advantage, and mechanisms involved in cancer prevention. Some of the major types of cancer include

DOI: 10.1201/9781003486237-4

breast cancer, prostate cancer, gastrointestinal cancer, lung cancer, gynaecological cancer, colorectal cancer, stomach cancer, pancreatic cancer, skin cancer, oesophageal cancer, euro endocrine, multiple myeloma, leukaemia, head and neck cancer (HNC), and osteosarcoma [1–3].

4.2 PHYTOCHEMICALS

No specific classification of phytochemicals has been identified yet. The knots in the mechanism of phytochemicals in cancer cell proliferation and metastasis are currently removed through the knowledge of various research. So far, the phytochemicals can be grouped as primary (sugars, amino acids, and lipids) and secondary metabolites (alkaloids, terpenes, and steroids). Together, these compounds provide essential nutrients, colour, and defence mechanisms for plants. Phytochemicals regulate various signalling molecules by modulating the redox system and controlling the immune system by synthesising inflammatory cytokines.

4.2.1 Curcumin

Curcumin is a yellow colour pigment extracted from turmeric and it belongs to the ginger family. Curcumin has various therapeutic benefits starting from wound healing, anti-coagulant, anti-bacterial, and antioxidant activity. In PubMed over 100 clinical trial studies have been published, which shows that curcumin has a significant therapeutic index in disorders such as rheumatoid arthritis, neurological disorders, and cancers. The mechanism of curcumin in cancer cell progression and metastasis is not well known, based on the research it is known that it has antioxidant activity through inhibiting the lipid peroxide formation, acid phosphatases, and cathepsin D. In another study, it was demonstrated that it increases superoxide dismutase and glutathione peroxidase activity in serum. It also has the property of scavenging oxidative free radicals and suppressing proteoglycans, thus it can be used as an alternative to non-steroidal anti-inflammatory drugs (NSAIDs). Curcumin potentially regulates the cellular immune response by altering the function of M1 macrophage, B and T lymphocytes, and TLR4 binding capacity.

4.2.2 Ellagic Acid (EA)

It is a polyphenolic compound present mostly in fruits such as pomegranate and berries and nuts like walnuts and almonds. It exists either in free form or in combined form called ellagitannins. EA has antioxidants and anti-cancer activity. It inhibits cancer cell growth through the regulation of various signalling pathways and increases cancer cell apoptosis. It also involves cell cycle arrest in the G0/G1 phase and activates the tumour suppressor genes. In MCF-7 cells, the EA treatment shows that there was an increased expression of CDK inhibitors and decreased expression of cyclins A_2 and E_2. It also regulates SMAD3/TGF-β, thus it has an influential role in the cell growth and repair mechanism. EA is weekly soluble in water and has poor bioavailability, a very minimal amount of this compound is only absorbed by the small intestine and converted to active metabolites such as glucuronides and various forms of methyl esters. It is reported that EA can inhibit vascular endothelial growth factor (VEGF) and VEGF receptors, thus decreasing angiogenesis, which is an important hallmark of cancer. However, due to its poor biocompatibility, the researchers used a different strategy to develop a nano-particle to enhance drug delivery. It is shown that the chitosan, tween 80 coated nanoparticle has a higher efficacy rate in killing the cancer cells than the native EA. One more study's finding showed that the combination of 10mM EA and radiotherapy targets the cancer cells and increases DNA damage. Also, commercially available drugs such as paclitaxel cause cell shrinkage, DNA damage, and fragmentation of the nucleus [4, 5].

4.2.3 Resveratrol

Resveratrol is a phytoalexin, which is produced by plants to withstand environmental stress and protect them from any foreign substance entering into the plants. The topical application of this compound on skin cancer-induced mice shows a significant decrease in tumorigenesis, this property of resveratrol was highlighted by the researcher to develop them into a novel anticancer agent. This compound is abundantly present in grape wine and peanuts and exists in different isomers such as cis- and trans-isomers. The trans-isomers are more stable and have higher anticancer capacity than the cis-isomers. These compounds are sensitive to PH and temperature and have poor pharmaco-kinetics and low potency, hence the bioavailability of resveratrol has to be increased by synthesising nano-particles.

4.2.4 Cyclopamine

Cyclopamine is the natural steroid present in the corn lily, veratrum californium. It can bind with SMO (the smoothened receptor) protein which is involved in Hedgehog signalling pathways. The SMO protein has an important role in embryonic development and tissue homeostasis. A mutation in this protein causes birth defects and cancer. Cyclopamine inhibits the SMO proto-oncogenic effect. Cyclopamine inhibits the invasion of cells, hence it blocks the movement of the cell to adjacent or distant organs. The soluble derivatives of cyclopamine are stable and were developed into prodrugs and tested in glioblastoma cells, and these prodrugs are under clinical trials with minimal adverse effects and decreased metastasis. Cyclopamine also inhibits the formation of neurospheres, and the cyclopamine injected into the athymic cells shows blockage in the shh signal pathway and there were no more tumours formed [1, 6].

4.2.5 Sulforaphane

Sulforaphane is present mainly in broccoli, it is an isothiocyanate that exists in an inactive form called glucoraphanin, upon mechanical damage, the enzyme myrosinase hydrolysed the inactive form glucoraphanin and converted into active sulforaphane. This phytochemical has prevention towards cancer and anti-cancer properties. Sulforaphane can prevent cancer by decreasing DNA damage, and accumulation of DNA adducts and regulates the signalling pathways such as nuclear factor erythroid related factor 2 (Nrf2)-Keapl and nuclear factor Kappa B. It also prevents cancer initiation and progression by modulating histone deacetylases, an epigenetic factor.

4.2.6 Apigenin

Similar to all other phytochemicals, apigenin also has anti-cancer properties and is extracted from fruits, alcoholic beverages, nuts, and vegetables. Apigenin has a potential role in the mitochondrial membrane and increases the cytochrome C level in the cytoplasm.

Some of the other important phytochemicals include rosmarinic acid, EGCG, ferulic acid, fistein, quercetin, caffeic acid, geneistin, astaxanthin, lycopene, etc. [7–9] (Table 4.1).

4.3 INFLAMMATION AND CANCER

The body's immune system triggers the intricate process of inflammation in response to any pathogen invasion or other form of irritation. It manifests as redness or swelling where the attack or injury occurred. As part of the normal healing process, the body sends out inflammatory cells to kill these infections after first recognizing the signals supplied by the cells. It's a complicated process wherein all immune cells receive signals, surround the location, and secrete particular chemicals.

4.3.1 Types of Inflammation

There are typically two forms of inflammation: acute and chronic. When the body continues sending out inflammatory cells in the absence of pathogens, it leads to chronic inflammation, for instance, inflammatory cells and chemicals target joint tissues in rheumatoid arthritis, resulting in a cyclical inflammation that can seriously harm joints, causing pain and deformity. It is a slow and long-lasting

TABLE 4.1
Role of Phytochemicals in Modulating Major Signalling Pathways in Inflammation and Cancer

Signalling Pathway	Phytochemical	Target Response
MAPK	Sulforaphane Quercetin Caffeic acid EGCG Resveratrol Silibinin Ellagic acid Curcumin	Apoptosis [10]
JAK-STAT	Quercetin Silibinin Ellagic acid Gallic acid Resveratrol Curcumin Ursolic acid Anthocyanin Genistein	Suppressing effect on p-STAT1/3
PI3K/Akt	Quercetin EGCG Curcumin Kaempferol Silibinin Resveratrol Ellagic acid Caffeic acid Ursolic acid Indole-3-carbinol Anthocyanin	Inhibiting cell proliferation by suppressing PI3K/Akt/mTOR [11–14]

effect that causes adverse effects on the affected tissues. On the other hand, acute inflammation is the reaction to an unexpected injury to the body, such as a cut finger. Your body sends inflammatory cells to the cut to aid in healing. These cells initiate the process of healing [15, 16].

4.3.2 Inflammation and Cancer

There is strong evidence linking inflammation to the formation, progression, and overall effectiveness of anti-cancer treatments across the entire spectrum for the vast majority of cancer types. Inflammatory responses are also induced by anti-cancer agents/therapies. Acute inflammation has a role in the death of cancer cells by triggering an immune response that attacks tumours. On the other hand, prolonged inflammation caused by therapy increases the likelihood of resistance to treatment and the advancement of cancer.

Under normal circumstances, inflammation doesn't lead to any adverse conditions or effects but in case of prolonged/untreated chronic inflammation cancer results due to DNA damage. In certain organs like the liver, the liver cells/immune cells work towards repairing the inflammation of the liver, but these efforts can also cause the organ's cells to divide at a much faster rate compared to normalcy. If these cells possess damaged DNA, then it passes along to the other cells as they divide.

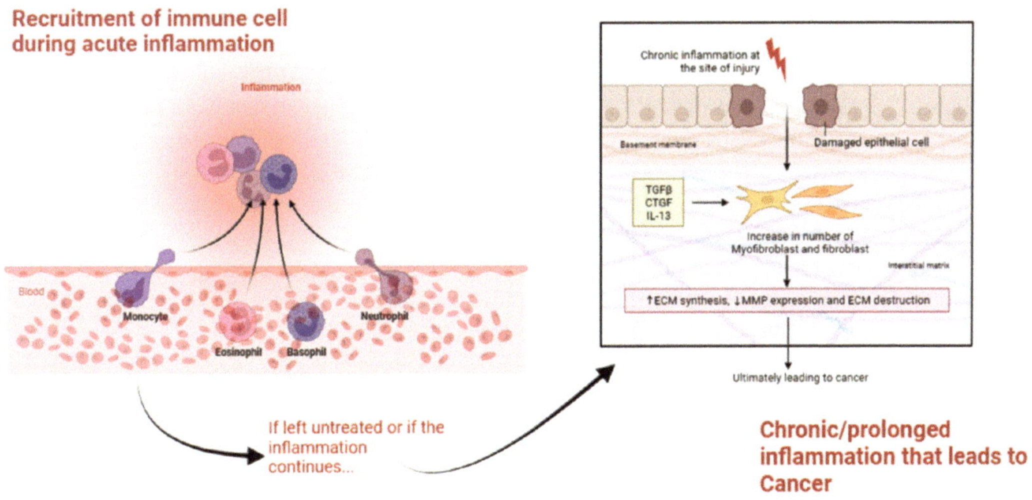

FIGURE 4.1 The progression of a simple injury to chronic inflammation and finally to cancer by the increase in numbers of fibroblasts and recruitment of chemokines like interleukins and tumour growth factors [17].

Cellular healing processes and tissue surveillance usually eliminate damaged cells or restore them without any detectable side effects. On the other hand, cancer may develop when aberrant cells take over and prevent the organ from recovering (Figure 4.1).

Numerous types of cancers are likely to be developed due to inflammation or in an inflammatory environment. For example, an increased risk of cancer known as Barrett's oesophagus can develop as a result of a reflux disease, which causes chronic inflammation at the stomach-oesophageal junction; colorectal cancer risk factors include inflammatory bowel illnesses like Crohn's disease or ulcerative colitis; the chance of developing lung cancer is increased when smokers inflict chronic inflammation and damage on their lungs [17, 18].

Sethi et al. describe that the risk of cancer varies from person to person including whether or not a person has a genetic risk for cancer. Inflammation promotes tumour progression at every stage and makes people more likely to get cancer. The cancer cells and the surrounding tumour cells form an inflammatory tumour microenvironment (TME) wherein multiple interactions that lead to the growth of the tumour take place. Inflammation has a greater influence over the components of the TME and the plasticity of both stromal and tumour cells. Few articles believe that the immune system plays a major role in tumorigenesis as it can act as both pro-tumorigenic and anti-tumorigenic agents.

The positive evolutionary pressure on the capacity to heal wounds and regenerate tissues is far greater than that on the ability to prevent tumour growth, which typically occurs after reproductive age. There seems to be a well-established link between tumorigenesis and inflammation, chronic inflammation in particular. However, if acute inflammation is considered it plays a role in enhancing anti-tumour immunity by boosting the dendritic cell (DC) development and effector T cell initiation. As it is evident from the above-mentioned points, the inflammation is almost non-resolving during the cancer development stages. Let us understand the major signalling pathways that are modulated by various phytochemicals in cancer and inflammation in the upcoming section [15, 19].

4.4 MAJOR SIGNALLING PATHWAYS MODULATED BY PHYTOCHEMICALS IN CANCER AND INFLAMMATION

4.4.1 MAPK Pathway

The Mitogen-Activated Protein Kinase (MAPK) pathway is a complex system of pathways responsible for a variety of cell functions like proliferation, response to cell stress, apoptosis, etc. This pathway functions due to the phosphorylation of these proteins by three kinases: MAPK kinase kinase (MAP3K), MAPK kinase (MAP2K), and MAP kinase (MAPK).

There have been four identified mammalian categories to the MAPK pathway system, namely the ERK1/2, JNKs, P38, and the ERK5 systems. The JNK and the p38 pathways are the ones that are responsive to cellular stress and inflammatory responses. The basic structure of a three-tier MAPK cascade involves the upstream activator, MAP3K, MAP2K, and MAPK. However, each pathway has a different protein to induce responses including the modification of the Ras-Raf-MEK-ERK pathway instead of the MAP3K-MAP2K-MAPK pathway [20, 21].

ERK is a serine/threonine protein kinase found in an inactive state in the cytoplasm that functions as a signal transduction protein. On activation by mitogen signals, these proteins enter the nucleus and regulate gene expression and transcriptional activity.

Ras is an upstream protein that has a GDP-binding shape when it's not in use, but it changes into a GTP-binding shape when it's activated by stimulating receptor binding. Binding to Ras activates Raf, which then continues downstream to deliver differentiation and proliferation signals into the nucleus. This ensures a change in the transcription or expression of a target gene [22–24].

Phosphorylated serine residues of Raf activate MEK, a dual-specificity kinase. MEK, in turn, phosphorylates the Tyr and the regulatory sites in order to activate ERK. In an inactive state, ERK is localised to the cytoplasm of inactive cells. The phosphorylation and dimerization of ERK allow it to get activated. The formation of such a homologous dimer is essential for the nuclear translocation of ERK [11]. Activated ERK enters the nucleus and phosphorylates the target proteins, transcription factors, and other kinases. This is done to transform cell proliferation, apoptosis, and transcription due to changes in the expression of certain genes [25–30].

Mutations that take place in the Ras or Raf genes, as well as the kinase genes, are implicated in the process of cancer induction. The notion that apigenin can suppress the MAPK/ERK signalling pathway is supported by the fact that it can modulate genes such as Akt, ERK, DR4/DR5, and NF-kB. Apigenin's ability to inhibit ERK1/2 and FAK makes it more likely that cells will be unable to generate integrins, which ultimately results in the death of the cells by apoptosis. Figure 4.2 illustrates the modulation of MAPK pathway by apigenin [31, 32].

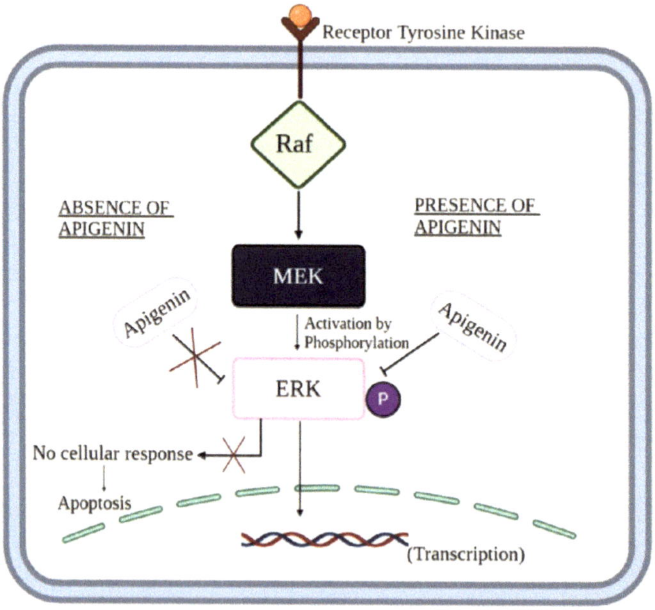

FIGURE 4.2 Modulation of MAPK pathway by apigenin, apigenin presence leads to apoptosis while its absence leads to ERK phosphorylation and transcription.

Apigenin, also known as 40,5,7-trihydroxyflavone, is a member of the flavonoids compound family, which can be found in a broad variety of plant species *Artemisia*, *Achillea*, and *Matricaria* are examples of plant species that have been shown to possess higher quantities of this apigenin compound. It can also be found in other types of plants, like Sideritis and Teucrium. In three plants, the compound can be found as an aglycone, an O-methyl ester, or one of its acetylated forms. Apigenin is a bioactive chemical that is produced as a result of the phenylpropanoid pathway. It has also been investigated that apigenin can be obtained through the phenylalanine and tyrosine amino acids, both of which are classified as aromatic amino acids. It has been demonstrated that apigenin, which is one of the flavonoid compounds that is most well-known, possesses a wide variety of health qualities in addition to its nutritional value, which contributes to the efficiency of the compound. Apigenin has also been shown to protect neurons by participating in Ca2+ signalling and lowering the amount of reactive oxygen species (ROS) [33, 34].

In addition to the fact that apigenin has the potential to be used as a medicine in the majority of cases, it has also been demonstrated that it possesses anticancer action to some degree. Apigenin's anticancer activity in head and neck cancer was examined using HN-30 cells in a dose- and time-dependent manner. From the study, it was revealed that there was a reduction in the amounts of NANOG, CD44, and CD105 mRNAs that were observed as a result of the impact of apigenin on cancer cells. In addition to this, it was discovered that apigenin has the ability to infect the population of cancer stem cells, which is an additional factor that contributes to its usefulness as an anticancer agent. With the help of apigenin, cancer prevention can be accomplished through the process of apoptosis in a wide variety of cancer cell lines and animal models that have received testing. Apoptosis is induced in these cells when apigenin is involved in the modulation of the pathways that are involved in signalling between cancer cells. Apigenin is implicated in each of these pathways, and one such action is through the regulation of the MAPK pathway. Apigenin was found to cause apoptosis in mice ANA-1 cells, as was discovered in. This was accomplished by lowering the levels of phospho-JNK and phospho-ERK expression while simultaneously raising the levels of phospho-p38 expression, which ultimately resulted in the death of the cells. The downregulation of Bcl-2 was a consequence of this process of cell death, which was further accompanied by increased amounts of ROS [31, 35, 36].

4.4.2 Notch Pathway

The Notch pathway, which functions as a regulatory mechanism for the cells and ultimately leads to carcinogenesis, is responsible for determining the fate of the cell in cases of osteosarcoma. This includes the cell's ability to differentiate, proliferate, and survive. To this day, researchers have found four distinct types of Notch receptors in mammals. These receptors are referred to as Notch 1, 2, 3, and 4. These Notch receptors are specific to their ligand molecules, which are referred to as jagged 1, jagged 2, delta-like 1, or delta-like 4. The binding of these molecules to the Notch receptors results in the cleavage and nuclear translocation of the Notch intracellular domain (NICD), which in turn leads to the downstream signalling cascade. The two-stage proteolytic cleavage of the domain that results from the contact between the ligand and the receptor precedes the intracellular translocation, which then transports the domain to the cell's nucleus. Once the intracellular domain is translocated into the nucleus, it then forms a protein complex with a highly conserved transcription factor domain called CSL (CBF1/Suppressor of Hairless Lag1) and transcriptional coactivators of the mastermind-like (MAML) family. The formation of this protein complex leads to the transcription of the target gene, thereby forming the protein product that includes Hes and the Hes-related YRPW motif (Hey). The protein product Hes-1 is proven to play a pivotal role in the progression of osteosarcoma via the Notch pathway, and thereby modulation of the Notch pathway paves the way for the detrimental effects of osteosarcoma. The effect of the Notch pathway post-curcumin treatment revealed that the proteins Hes-1, cyclin D1, MMP-2 and MMP-9 were downregulated. This inhibitory action occurs when curcumin reacts with the NICD-1 complex that

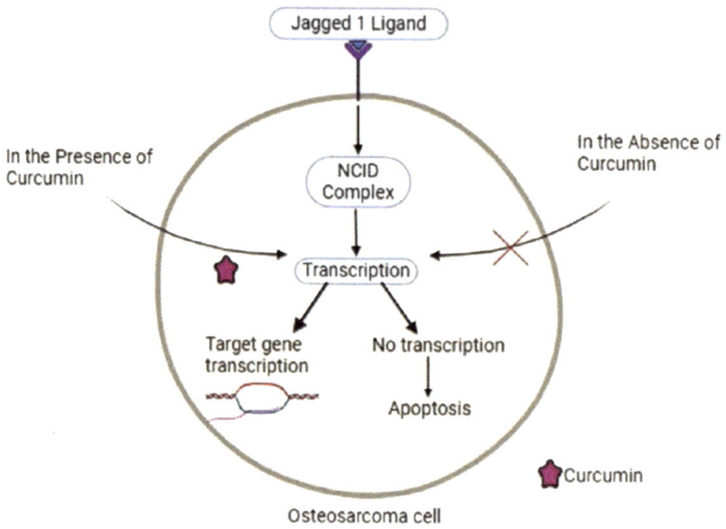

FIGURE 4.3 Modulation of notch pathway by curcumin in the case of osteosarcoma, NCID complex which leads to carcinogenesis is blocked in the absence of curcumin, while its presence activates transcription.

is involved in the transcription of the target genes that code for the specific protein product. In the absence of NICD-1, the MAML protein is unable to transcribe the target protein, thereby leading to the downregulation of the pathway. Added on to this, curcumin inhibited cell growth and differentiation via siRNA modulation, which ultimately resulted in cell death in osteoblastoma cell lines. Figure 4.3 illustrates the modulation of Notch pathway by curcumin [37, 38].

Another name for curcumin is 1,7-bis(4-hydroxy-3-methoxyphenyl)-1,6-heptadiene-3,5-dione. It is a polyphenol that is mostly found in the Curcuma longa plant, which is what people call turmeric. The chemicals that are known as curcuminoids make up to 10% of turmeric powder when it is in its dry state. These compounds are among the many different compounds that have been identified in the turmeric plant. Curcumin, dimethoxy-curcumin, and bisdemethoxy-curcumin are the chemicals that make up the majority of the compounds that are classified as belonging to this category. When it comes to turmeric, which is known to possess a variety of medical characteristics, including antibacterial, antimicrobial, and anti-inflammatory properties, as well as treatment for Alzheimer's disease, the principal component that plays a role in this process is curcumin, which is responsible for the wide range of applications that turmeric may carry out. The efficacy of this molecule has not only been demonstrated to be useful in the situations described above, but it has also been demonstrated to be rather effective as an anticancer drug. Studies have shown that curcumin's anticancer effect can be demonstrated in two different ways: first, it can inhibit angiogenesis, which has been shown to be a root cause of cancer; second, it can stop the proliferation of cancer cells by inhibiting metastasis. In addition, the anticancer properties of curcumin have been investigated in relation to osteosarcoma. One study discovered that curcumin-induced apoptosis on osteosarcoma cell lines such as MG63, U2OS, and HOS by modulating the signalling pathways that are involved in the progression of cancer [24, 39–41].

4.4.3 P13K/Akt Pathway

The PI3K/Akt pathway, also known as the PI3K/mTOR pathway, is a crucial intracellular signalling cascade of protein kinases involved in diverse cellular processes, including cell growth, survival, metabolism, and inflammation. It involves phosphoinositide 3-kinase (PI3K) activation, leading to the phosphorylation and subsequent activation of Akt (also known as protein kinase B). Activation of this pathway is triggered by various extracellular stimuli such as growth factors, cytokines, and hormones binding to their respective receptors, initiating a signalling cascade through PI3K.

Once PI3K is turned on, it turns on other proteins that work with it, like Akt and mTOR (Mammalian Target of Rapamycin). These proteins help make target proteins and lipids, which in turn causes cells to grow and differentiate. When there is an abnormal expression of Akt, the cells tend to evade the apoptotic process. This evasion mechanism causes uncontrolled cell proliferation and differentiation, which leads to the formation of cancer. This is accomplished through a variety of target molecules, including proteins from the Bcl-2 (B-cell lymphoma type 2) family. Curcumin has been shown in studies to be an effective anticancer drug by inhibiting cancer by targeting the PAM (PI3K/Akt/mTOR) pathway A study revealed that, in MDA-MB-231 cancer cell lines, curcumin was effective in downregulating Akt in a dose-dependent fashion, leading to PAM pathway suppression and thereby inhibiting further cell growth and differentiation [42–45].

The mechanism of PI3K inhibition by curcumin is initiated by the binding of curcumin to the RTKs present on the cell surface. Upon binding of the growth receptor to the RTK, it leads to the phosphorylation of the RTK, which further activates PI3K and other proteins involved in the signalling cascade. When curcumin binds to the RTKs, the phosphorylation of RTKs is inhibited, thereby ceasing the activation of other proteins. Curcumin is found to inhibit a wide range of proteins such as IKKβ, AKt, GSK3β, and HER2 expression, which is proven to facilitate the processes of cell growth, invasion, and metastasis [44, 46–48].

Quercetin (3,5,7,30,40-pentahydroxyflavone) is the dominant representative of the flavonoid family and belongs to the subclass of flavanols that is quite often found in various fruits, vegetables and other dietary sources of food. The compound is primarily acquired from *quercetum* (oak forest) and is also studied to function as a polar auxin transport inhibitor. With the presence of quercetin present in abundance in natural food sources, it is regarded as having a wide variety of applications, such as anti-inflammatory and anti-carcinogenic in human as well as animal models. The anticancer activity of quercetin has been studied on various types of cancer, like breast, colorectal, head and neck, etc., where it was found that quercetin is capable of inducing apoptosis, suppressing tumour growth and inhibiting the process of angiogenesis [49, 50].

The PI3K/Akt pathway can be changed by quercetin because it stops IGF-1 from joining. This is necessary for the phosphorylation of a number of proteins, including PI3K, Akt, and Glycogen synthase kinase-3 beta (GSKβ), which are essential for further downstream signalling mechanisms. The activation of these mechanisms leads to the expression of the target gene. The mechanism of action of quercetin on the pathway can be of various types: Quercetin can inhibit the Ras protein, which is involved in the activation process of PI3K. Inactivation of Ras leads to further inhibition of PI3K inactivation, and thereby the cellular process is halted. Quercetin is also capable of directly inactivating PI3K, which leads to the same fate for the cell. Quercetin can bind to proteins such as IRS1 (Insulin Receptor Substrate-1) bound to the RTK, thereby inactivating them. Inactivation of IRS1 leads to the shutting down of the pathway, thereby inhibiting further signalling. The protein PDK1 (Protein Dependent Kinase 1) is involved in the activation of Akt in the signalling process; upon the action of quercetin on PDK1, further activation of Akt is inhibited. The process of PI3K/Akt inactivation by quercetin can also be achieved by the activation of PTEN (Phosphatase and Tensin homologue) protein, which further inactivates PIP3 (Phosphatidylinositol-3,4,5-triphosphate), which is involved in the activation of Akt. Figures 4.4 and 4.5 illustrate the mechanism by which P13/AKt pathway is modulated by curcumin and quercetin in the presence and absence of the phytochemicals [51, 52].

4.4.4 JAK-STAT Pathway

The Janus kinase (JAK)/signal transducers and activators of transcription (STATs) is a signalling pathway that deals with the activation and communication of cellular responses to the binding of cytokines. This pathway is involved in a variety of processes, not limited to inflammation, haematopoiesis,

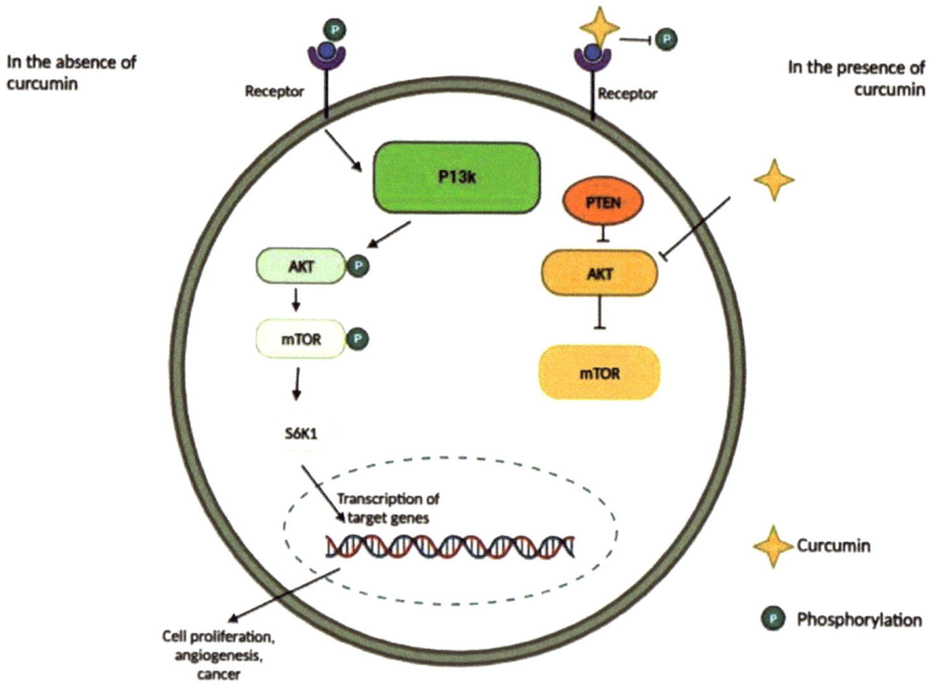

FIGURE 4.4 Modulation of P13/AKt pathway in the presence and absence of curcumin.

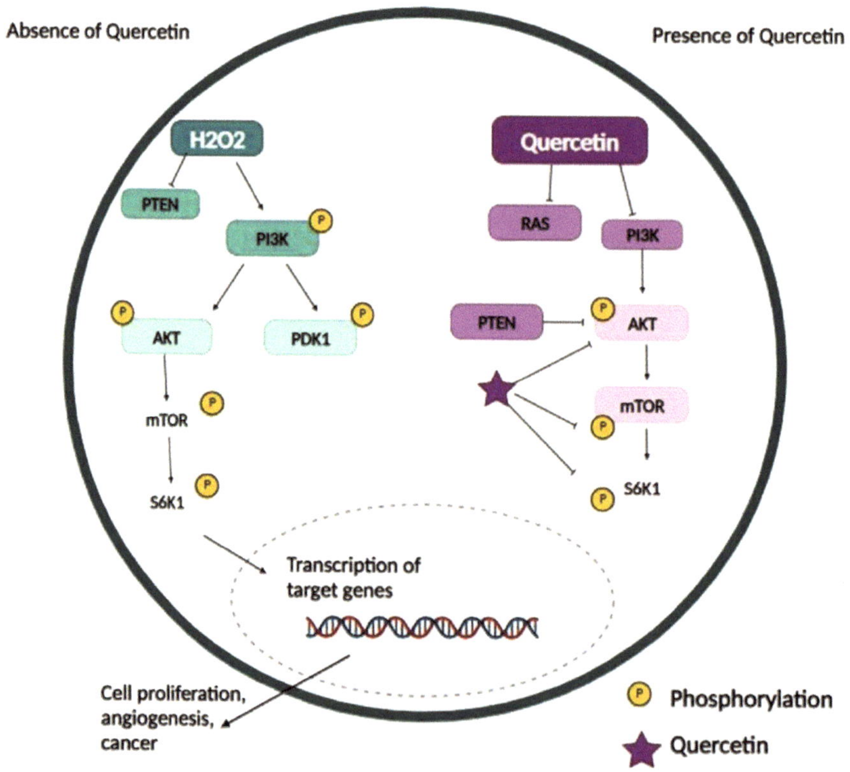

FIGURE 4.5 Modulation of P13/AKt pathway in the presence and absence of quercetin.

tumour formation, apoptosis, etc. This pathway has two main components: the JAKs and the STAT proteins, which are highly conserved [13, 44, 53].

The four members of the JAK family are JAK1-3 and TYK2 (Tyrosine Kinase-2). JAKs have seven homology regions or domains from 1–7 (JH1–JH7). JH1 encodes for a kinase that is involved in the phosphorylation of a substrate. JH2 is a pseudokinase domain that can work to inhibit Tyr kinase activity. JH3 and JH4 are part of the Src-homology 2 domain, while JH4, JH5, JH6, and JH7 make up the FERM domain. Both of these domains come together to work as mediators for cytokine receptors [54, 55].

A family of proteins called STATs controls transcription by phosphorylating JAKs. These proteins are capable of shuttling in between the nucleus and the cytoplasm in order to regulate transcription. They are usually present in the form of homodimers in the cytoplasm when they're inactive and resting. They become active on ligand binding. They have conserved SH2 and transcriptional activation domains (TAD) regions. When TADs are present, they make it possible for coactivators and other chromatin-modifying enzymes to bind to promoters [56, 57].

The binding of a cytokine to a receptor triggers dimerization of the receptors. This triggers the phosphorylation of the JAK1 and TYK2 enzymes, which act as docking sites for the phosphorylation of STATs. The STAT dimerizes in an antiparallel sense to enable the translocation into the nucleus. Here, STAT regulates transcription by binding to target gene sequences. Negative regulation of the JAK-STAT pathway is carried out in three different ways, which involve the inhibition of JAK kinase activity and STAT binding to the phosphorylated JAK or its receptor blocking the binding of DNA to the STAT, or dephosphorylated STAT to inhibit signal transduction of the JAK-STAT pathway [21, 29].

Quercetin has potential as a JAK-STAT pathway inhibitor, indicating possible uses in cancer prevention and treatment, as well as in the treatment of neurological and autoimmune illnesses. Quercetin has been extensively studied and found to possess anti-inflammatory, antiproliferative, and anti-angiogenic characteristics. As a result, it is considered a reliable and secure inhibitor of the JAK/STAT pathway. Quercetin effectively inhibits the activation of STAT1 and STAT3 produced by cytokines in several cancer cells, including cholangiocarcinoma (CCA) cells. The interference of this substance with the process of STAT3 phosphorylation reduces its movement into the nucleus and its ability to activate transcription. As a result, the expression of genes targeted by STAT3, which are involved in cell growth, migration, and invasion, is decreased. Quercetin demonstrates the capacity to restrain tumour growth and hinder metastasis in animal models, thereby showing its promise as an anti-cancer drug [58, 59].

Quercetin has been found to enhance the activation of the JAK-STAT signalling pathway generated by type-I interferon (IFN) by inhibiting SHP2. This suggests that there may be a synergistic effect when combining quercetin with specific anticancer treatments. Quercetin has been shown to play a role in preventing and treating glioblastoma, a type of neurodegenerative disease. It achieves this by suppressing the IL-6-induced JAK-STAT3 signalling pathway in glioblastoma cell lines. Quercetin treatment suppresses IL-6-induced activation of JAK–STAT3 signalling, phosphorylation of Rb protein, production of cyclin D1, and release of MMP-2 in glioblastoma cell lines.

Although quercetin exhibits potential, addressing its restricted bioavailability, low water solubility, and significant hepatic metabolism is necessary for optimal clinical utilisation. Additional investigation is important to fully comprehend the inhibitory effects of quercetin on the JAK-STAT pathway in different illnesses.

The effect of curcumin on the JAK-STAT pathway has been highly researched upon and it has been found that curcumin is generally inhibitive towards the JAK-STAT pathway. Because it inhibits the phosphorylation of STAT3 and other STAT molecules, curcumin has the potential to be an anti-inflammatory agent. This is because it leads to a reduction in the levels of cytokines that are typically associated with inflammation. This impact has been demonstrated in a variety of models, such as BV2 microglia among others.

There was a discernible reduction in cell proliferation, migration, and invasion in cancer models such as MG-63 OS cells, Rb cells, and HEp-2 cells, which is indicative of the possibility that the substance possesses anti-cancer capabilities. Not only was there a notable reduction in the expression of STAT3, but there was also an increase in apoptosis.

It has been shown that this compound, which possesses antioxidant characteristics, has the ability to reduce the death of cells and offer protection against oxidative stress in a variety of experimental models. These models include those that involve intestinal inflammation and spinal cord injury in rats.

The influence of curcumin on JAK-STAT is often more obvious at larger doses; however, the appropriate dosage ranges can vary based on the particular cell line and the outcome that is desired. The suppression of STAT appears to be an important mechanism; however, it is important to note that curcumin may also interact with other molecules that are implicated in this pathway, such as SOCS proteins. A wide variety of cell lines and animal models have been used to conduct significant research on the effects of curcumin, which indicates that it has the potential to be utilised in a wide variety of therapeutic applications [13, 60–62].

4.4.5 NF-kB Pathway

The NF-kB pathway is an important signalling chain that plays a part in many cellular processes, such as inflammation, immune response, cell proliferation, and cell survival. Its complex network of proteins regulates gene expression in response to various stimuli, such as pathogens, stress, and cytokines. The NF-kB pathway comprises a family and a team of five key proteins: p50, p52, RelA (p65), RelB, and c-Rel. These proteins function as dimers, each with distinct roles and preferences for partners. Normally sequestered in the cytoplasm bound to inhibitory proteins (IκBs), they await the right cues to unleash their transcriptional prowess.

Upon activation by stimuli like TNF-alpha or microbial products, the IKK (IkB kinase) complex comprising IKKα, IKKβ, and IKKγ (NEMO) is activated. This complex phosphorylates IkB proteins, marking them for ubiquitination and subsequent proteasomal degradation. Liberation of NF-kB dimers, primarily p50/p65 heterodimers, allows their translocation into the nucleus, where they modulate gene transcription or perform transcriptional regulation of genes involved in inflammation and immunity [63–65].

The NF-kB pathway is very important for controlling the activity of genes that cause inflammation. These genes include cytokines (like IL-6 and TNF-alpha), chemokines, adhesion molecules (like ICAM-1 and VCAM-1), and enzymes (like COX-2) and inducible nitric oxide synthase (iNOS). Its dysregulation is implicated in chronic inflammatory diseases, autoimmune disorders, and cancer. The potent influence of the NF-κB pathway demands meticulous regulation. This symphony relies on both positive and negative modulators to maintain harmonious cellular function. Positive regulators, like IKKs and TRAF proteins, amplify the signal. Negative regulators, such as IκBα and FLIP, act as brakes, inhibiting IKKs, promoting IκB synthesis, and targeting activated NF-κB for degradation. This intricate equilibrium ensures appropriate responses to external cues and prevents pathological activation [66, 67].

During inflammation, NF-kB activation is a central mediator, responding to various stimuli like pathogens, tissue injury, or stress signals. In response, it orchestrates the expression of inflammatory mediators, recruits immune cells to the affected site, and promotes tissue repair. Inhibiting the NF-kB pathway involves targeting various steps in its activation cascade. Some strategies are stopping the degradation of IkB, stopping the movement of NF-kB, or stopping the upstream kinases that phosphorylate IkB.

The drop in IkBα gene expression induced by quercetin is particularly noteworthy since it indicates a reduction in the activation and movement of NF-kB into the cell nucleus. The decrease in IL-1β levels further strengthens the anti-inflammatory effects of quercetin by affecting many stages of signal transduction. In addition to its effects on CAD, quercetin has been studied for its influence on the NF-kB pathway in different cellular settings. Quercetin has demonstrated the ability to counteract the increase in adhesion molecules, such as ICAM-1, VCAM-1, and E-selectin, by preventing the activation of NF-kB and AP-1. This demonstrates the extensive impact of quercetin in reducing inflammatory reactions [68–70].

Quercetin has demonstrated significant effects, specifically in regulating the nuclear factor-kappa B (NF-kB) signalling system. Quercetin treatment in patients with coronary artery disease (CAD) has demonstrated favourable results by affecting crucial elements of the NF-kB pathway, resulting in a decrease in pro-inflammatory indicators. Quercetin administration in CAD patients led to a significant reduction in the levels of inflammatory cytokines, including interleukin-1 beta (IL-1β) and tumour necrosis factor-alpha (TNF-α). Significantly, there was an inclination to reduce levels of interleukin-10 (IL-10) as well. The anti-inflammatory impact was accompanied by a significant drop in the expression of the inhibitor of the kappa B alpha (IkBα) gene, suggesting a reduction in the transcriptional activity of NF-kB. The results strongly indicate that quercetin demonstrates anti-inflammatory characteristics in the context of CAD.

Furthermore, research conducted on many cell types and under diverse experimental situations has demonstrated that quercetin regulates the NF-kB pathway by decreasing the expression of nitric oxide (NO) and inducible nitric oxide synthase (iNOS), so effectively controlling inflammation. Furthermore, quercetin improves the inflammatory response caused by lipopolysaccharide (LPS) in gingival fibroblasts by activating peroxisome proliferator-activated receptor-gamma (PPAR-γ) and inhibiting NF-kB activation. In addition, quercetin hinders oxidative stress by decreasing the expression of NADPH oxidase 2 (NOX2) and preventing the movement of NF-kB into the nucleus, which leads to the suppression of proinflammatory cytokines including TNF-α and IL-1β.

Quercetin's impact on NF-kB subunits has been studied, uncovering a complex adjustment. It elevates the levels of pro-inflammatory NFkB p65 in the nucleus while decreasing them in the cytoplasm. Significantly, when esculetin concentrations are high, a derivative of quercetin, there is a simultaneous impact: a reduction in NFkB p65 levels within the nucleus and an increase in anti-inflammatory NFkB p50 levels. This indicates a possible development of inhibitory p50 homodimers, which may be associated with an anti-inflammatory reaction [71, 72].

Quercetin has been shown to reduce the activation of NF-kB in allergic inflammation by decreasing the degradation of IkBa and the movement of p65 NF-kB into the cell nucleus. This leads to a decrease in the expression of pro-inflammatory cytokines. The combined results highlight the diverse effects of quercetin on the NF-kB pathway, indicating that it has a strong ability to regulate inflammation and could be used as a potential treatment in different situations. Similarly, some other major phytochemicals that interact with the NF-KB pathway are curcumin which inhibits NF-kB activation by blocking IkB degradation and interfering with NF-kB DNA binding, resveratrol inhibits NF-kB by targeting multiple points in its activation cascade. Epigallocatechin Gallate stops NF-kB from activating by stopping IkB from becoming phosphorylated and breaking down. Genistein stops NF-kB from activating by stopping IkB from being phosphorylated and breaking down. Sulforaphane stops NF-kB from activating by stopping IkB from becoming phosphorylated and then breaking down. Finally, luteolin stops NF-kB from activating by stopping IkB degradation and NF-kB nuclear translocation.

These phytochemicals' ability to modulate NF-kB signalling highlights their potential as natural compounds for therapeutic interventions in various inflammatory conditions and diseases associated with NF-kB dysregulation [73].

4.5 CONCLUSION

The role of natural compounds in modulating various signalling pathways in the case of cancer and inflammation is vast. Some of those important signalling pathways include the Notch pathway, MAPK pathway, JAK-STAT pathway, NF-KB pathway, and P13/AKt pathway. While there is no single classification for phytochemicals, we generally refer to them as primary and secondary.

Modulation of the redox system and the production of inflammatory cytokines are some of the common mechanisms by which phytochemicals modulate signalling pathways in general. Apoptosis was found to be the major cellular mechanism by which cancer cell death is usually witnessed in the case of phytochemicals such as apigenin, sulforaphane, curcumin and resveratrol, which modulate the MAPK and Notch pathways. The P13/AKt pathway was found to be modulated by the induction of apoptosis through suppression of important membrane proteins and the release and modulation of inflammatory cytokines. JAK-STAT and NF-KB were found to be associated with various inflammatory cytokines, which get downregulated and modified by various phytochemicals such as quercetin, caffeic acid, curcumin, resveratrol, apigenin, sulforaphane, etc. Along with these major phytochemicals, there are numerous other naturally important compounds that play a major role in initiating modifications of signalling pathways. Understanding the role of these phytochemicals has made us draw useful conclusions that create new insights in research. While unravelling the complex genetic mechanisms that are involved in these pathways, we can understand the role of phytochemicals in targeting these pathways and thus develop new strategies to develop therapeutics. The usage of natural compounds along with nanoparticles as drug conjugates is gaining more popularity due to its advantage in increasing stability and enhancement of drug delivery. While compared to artificial compounds these phytochemicals cause fewer side effects and are more efficient due to their synergistic aspect. Thus, the development of novel therapeutic methods based upon natural compounds can be achieved through proper understanding and usage of these phytochemicals to bring newer prospectives in the field of anti-cancer and anti-inflammatory medications.

REFERENCES

1. Kumar, A.; P, N.; Kumar, M.; Jose, A.; Tomer, V.; Oz, E.; Proestos, C.; Zeng, M.; Elobeid, T.; K, S.; et al. Major Phytochemicals: Recent Advances in Health Benefits and Extraction Method. *Molecules* 2023, *28*, 887, https://doi.org/10.3390/molecules28020887
2. Rudzińska, A.; Juchaniuk, P.; Oberda, J.; Wiśniewska, J.; Wojdan, W.; Szklener, K.; Mańdziuk, S. Phytochemicals in Cancer Treatment and Cancer Prevention—Review on Epidemiological Data and Clinical Trials. *Nutrients* 2023, *15*, 1896. https://doi.org/10.3390/nu15081896
3. Jang, J.-H.; Lee, T.-J. Mechanisms of Phytochemicals in Anti-inflammatory and Anti-cancer. *Int J Mol Sci* 2023, *24*, 7863. https://doi.org/10.3390/ijms24097863
4. Li, Y.; Zhang, J.; Ma, D.; Zhang, L.; Si, M.; Yin, H.; Li, J. Curcumin Inhibits Proliferation and Invasion of Osteosarcoma Cells through Inactivation of Notch-1 Signaling. *FEBS J* 2012, *279*, 2247–2259. https://doi.org/10.1111/j.1742-4658.2012.08607.x
5. El-Saadony, M.T.; Yang, T.; Korma, S.A.; Sitohy, M.; Abd El-Mageed, T.A.; Selim, S.; Al Jaouni, S.K.; Salem, H.M.; Mahmmod, Y.; Soliman, S.M.; et al. Impacts of Turmeric and Its Principal Bioactive Curcumin on Human Health: Pharmaceutical, Medicinal, and Food Applications: A Comprehensive Review. *Front Nutr* 2023, *9*, https://doi.org/10.3389/fnut.2022.1040259
6. Ornano, L.; Venditti, A.; Donno, Y.; Sanna, C.; Ballero, M.; Bianco, A. Phytochemical Analysis of Non-volatile Fraction of *Artemisia Caerulescens* Subsp. *Densiflora* (Viv.) (Asteraceae), an Endemic Species of La Maddalena Archipelago (Sardinia – Italy). *Nat Prod Res* 2016, *30*, 920–925. https://doi.org/10.1080/14786419.2015.1079189
7. Bitwell, C.; Indra, S.S.; Luke, C.; Kakoma, M.K. A Review of Modern and Conventional Extraction Techniques and Their Applications for Extracting Phytochemicals from Plants. *Sci Afr* 2023, *19*, e01585. https://doi.org/10.1016/j.sciaf.2023.e01585
8. Khan, A.; Lee, B. DeepGene Transformer: Transformer for the Gene Expression-Based Classification of Cancer Subtypes. *Expert Syst Appl* 2023, *226*, 120047. https://doi.org/10.1016/j.eswa.2023.120047
9. Anand David, A.; Arulmoli, R.; Parasuraman, S. Overviews of Biological Importance of Quercetin: A Bioactive Flavonoid. *Pharmacogn Rev* 2016, *10*, 84. https://doi.org/10.4103/0973-7847.194044
10. Pham, D.-C.; Shibu, M.A.; Mahalakshmi, B.; Velmurugan, B.K. Effects of Phytochemicals on Cellular Signaling: Reviewing Their Recent Usage Approaches. *Crit Rev Food Sci Nutr* 2020, *60*, 3522–3546. https://doi.org/10.1080/10408398.2019.1699014
11. Moon, S.Y.; Kim, K.D.; Yoo, J.; Lee, J.-H.; Hwangbo, C. Phytochemicals Targeting JAK–STAT Pathways in Inflammatory Bowel Disease: Insights from Animal Models. *Molecules* 2021, *26*, 2824. https://doi.org/10.3390/molecules26092824

12. Patra, S.; Mishra, S.R.; Behera, B.P.; Mahapatra, K.K.; Panigrahi, D.P.; Bhol, C.S.; Praharaj, P.P.; Sethi, G.; Patra, S.K.; Bhutia, S.K. Autophagy-Modulating Phytochemicals in Cancer Therapeutics: Current Evidences and Future Perspectives. *Semin Cancer Biol* 2022, *80*, 205–217. https://doi.org/10.1016/j.semcancer.2020.05.008
13. Ashrafizadeh, M.; Rafiei, H.; Mohammadinejad, R.; Afshar, E.G.; Farkhondeh, T.; Samarghandian, S. Potential Therapeutic Effects of Curcumin Mediated by JAK/STAT Signaling Pathway: A Review. *Phytother Res* 2020, *34*, 1745–1760. https://doi.org/10.1002/ptr.6642
14. Chauhan, A.; Islam, A.U.; Prakash, H.; Singh, S. Phytochemicals Targeting NF-KB Signaling: Potential Anti-cancer Interventions. *J Pharm Anal* 2022, *12*, 394–405. https://doi.org/10.1016/j.jpha.2021.07.002
15. Medzhitov, R. Origin and Physiological Roles of Inflammation. *Nature* 2008, *454*, 428–435. https://doi.org/10.1038/nature07201
16. Sandberg, M.; Patil, J.; D'Angelo, B.; Weber, S.G.; Mallard, C. NRF2-Regulation in Brain Health and Disease: Implication of Cerebral Inflammation. *Neuropharmacology* 2014, *79*, 298–306. https://doi.org/10.1016/j.neuropharm.2013.11.004
17. Shin, S.A.; Joo, B.J.; Lee, J.S.; Ryu, G.; Han, M.; Kim, W.Y.; Park, H.H.; Lee, J.H.; Lee, C.S. Phytochemicals as Anti-inflammatory Agents in Animal Models of Prevalent Inflammatory Diseases. *Molecules* 2020, *25*, 5932. https://doi.org/10.3390/molecules25245932
18. Ghafouri-Fard, S.; Shoorei, H.; Khanbabapour Sasi, A.; Taheri, M.; Ayatollahi, S.A. The Impact of the Phytotherapeutic Agent Quercetin on Expression of Genes and Activity of Signaling Pathways. *Biomedicine & Pharmacotherapy* 2021, *141*, 111847. https://doi.org/10.1016/j.biopha.2021.111847
19. Akira, S.; Takeda, K. Toll-Like Receptor Signalling. *Nat Rev Immunol* 2004, *4*, 499–511. https://doi.org/10.1038/nri1391
20. Plotnikov, A.; Zehorai, E.; Procaccia, S.; Seger, R. The MAPK Cascades: Signaling Components, Nuclear Roles and Mechanisms of Nuclear Translocation. *Biochim Biophys Acta Mol Cell Res* 2011, *1813*, 1619–1633. https://doi.org/10.1016/j.bbamcr.2010.12.012
21. Davis, R.J. Signal Transduction by the JNK Group of MAP Kinases. *Cell* 2000, *103*, 239–252. https://doi.org/10.1016/S0092-8674(00)00116-1
22. Rahmani, A.H.; Alsahli, M.A.; Almatroudi, A.; Almogbel, M.A.; Khan, A.A.; Anwar, S.; Almatroodi, S.A. The Potential Role of Apigenin in Cancer Prevention and Treatment. *Molecules* 2022, *27*, 6051. https://doi.org/10.3390/molecules27186051
23. Ahmed, K.; Zaidi, S. F.; Cui, Z. G.; Zhou, D.; Saeed, S. A.; & Inadera, H. (2019). Potential proapoptotic phytochemical agents for the treatment and prevention of colorectal cancer. Oncology letters, 18(1), 487–498. https://doi.org/10.3892/ol.2019.10349
24. Vandamme, D.; Herrero, A.; Al-Mulla, F.; Kolch, W. Regulation of the MAPK Pathway by Raf Kinase Inhibitory Protein. *Crit Rev Oncog* 2014, *19*, 405–415. https://doi.org/10.1615/CritRevOncog.2014011922
25. Simanshu, D.K.; Nissley, D.V.; McCormick, F. RAS Proteins and Their Regulators in Human Disease. *Cell* 2017, *170*, 17–33. https://doi.org/10.1016/j.cell.2017.06.009
26. Muñoz-Maldonado, C.; Zimmer, Y.; Medová, M. A Comparative Analysis of Individual RAS Mutations in Cancer Biology. *Front Oncol* 2019, *9*, https://doi.org/10.3389/fonc.2019.01088
27. O'Neill, E.; Kolch, W. Conferring Specificity on the Ubiquitous Raf/MEK Signalling Pathway. *Br J Cancer* 2004, *90*, 283–288. https://doi.org/10.1038/sj.bjc.6601488
28. Nagai, H.; Noguchi, T.; Takeda, K.; Ichijo, H. Pathophysiological Roles of ASK1-MAP Kinase Signaling Pathways. *BMB Rep* 2007, *40*, 1–6. https://doi.org/10.5483/BMBRep.2007.40.1.001
29. Torres, V.E.; Harris, P.C. Mechanisms of Disease: Autosomal Dominant and Recessive Polycystic Kidney Diseases. *Nat Clin Pract Nephrol* 2006, *2*, 40–55. https://doi.org/10.1038/ncpneph0070
30. Kiesel, V.A.; Stan, S.D. Modulation of Notch Signaling Pathway by Bioactive Dietary Agents. *Int J Mol Sci* 2022, *23*, 3532. https://doi.org/10.3390/ijms23073532
31. Javed, Z.; Sadia, H.; Iqbal, M.J.; Shamas, S.; Malik, K.; Ahmed, R.; Raza, S.; Butnariu, M.; Cruz-Martins, N.; Sharifi-Rad, J. Apigenin Role as Cell-Signaling Pathways Modulator: Implications in Cancer Prevention and Treatment. *Cancer Cell Int* 2021, *21*, 189, https://doi.org/10.1186/s12935-021-01888-x
32. Woo, J. S., Choo, G. S., Yoo, E. S., Kim, S. H., Lee, J. H., Han, S. H., Kim, H. J., Jung, S. H., Park, Y. S., Kim, B. S., Kim, S. K., Park, B. K., Cho, S. D., Nam, J. S., Choi, C. S., Che, J. H., & Jung, J. Y. (2020). Apigenin induces apoptosis by regulating Akt and MAPK pathways in human melanoma cell A375SM. Molecular medicine reports, 22(6), 4877–4889. https://doi.org/10.3892/mmr.2020.11572
33. Klegeris, A.; Pelech, S.; Giasson, B.I.; Maguire, J.; Zhang, H.; McGeer, E.G.; McGeer, P.L. α-Synuclein Activates Stress Signaling Protein Kinases in THP-1 Cells and Microglia. *Neurobiol Aging* 2008, *29*, 739–752. https://doi.org/10.1016/j.neurobiolaging.2006.11.013

34. Boulton, T.G.; Nye, S.H.; Robbins, D.J.; Ip, N.Y.; Radzlejewska, E.; Morgenbesser, S.D.; DePinho, R.A.; Panayotatos, N.; Cobb, M.H.; Yancopoulos, G.D. ERKs: A Family of Protein-Serine/Threonine Kinases That Are Activated and Tyrosine Phosphorylated in Response to Insulin and NGF. *Cell* 1991, *65*, 663–675. https://doi.org/10.1016/0092-8674(91)90098-J
35. Liao, Y.; Shen, W.; Kong, G.; Lv, H.; Tao, W.; Bo, P. Apigenin Induces the Apoptosis and Regulates MAPK Signaling Pathways in Mouse Macrophage ANA-1 Cells. *PLoS One* 2014, *9*, e92007. https://doi.org/10.1371/journal.pone.0092007
36. Bijani, S.; Dizaji, R.; Sharafi, A.; Hosseini, M.-J. Neuroprotective Effect of Apigenin on Depressive-Like Behavior: Mechanistic Approach. *Neurochem Res* 2022, *47*, 644–655. https://doi.org/10.1007/s11064-021-03473-0
37. Phytochemical Composition and Pharmacological Activities of Teucrium Polium L. Collected from Eastern Turkey. *Turk J Chem* 2021, https://doi.org/10.3906/kim-2107-13
38. Żyżelewicz, D.; Kulbat-Warycha, K.; Oracz, J.; Żyżelewicz, K. Polyphenols and Other Bioactive Compounds of Sideritis Plants and Their Potential Biological Activity. *Molecules* 2020, *25*, 3763. https://doi.org/10.3390/molecules25163763
39. Majrashi, T.A.; Alshehri, S.A.; Alsayari, A.; Muhsinah, A.B.; Alrouji, M.; Alshahrani, A.M.; Shamsi, A.; Atiya, A. Insight into the Biological Roles and Mechanisms of Phytochemicals in Different Types of Cancer: Targeting Cancer Therapeutics. *Nutrients* 2023, *15*, 1704. https://doi.org/10.3390/nu15071704
40. Ma, J.-Q.; Li, Z.; Xie, W.-R.; Liu, C.-M.; Liu, S.-S. Quercetin Protects Mouse Liver Against CCl4-Induced Inflammation by the TLR2/4 and MAPK/NF-KB Pathway. *Int Immunopharmacol* 2015, *28*, 531–539. https://doi.org/10.1016/j.intimp.2015.06.036
41. Maik-Rachline, G.; Hacohen-Lev-Ran, A.; Seger, R. Nuclear ERK: Mechanism of Translocation, Substrates, and Role in Cancer. *Int J Mol Sci* 2019, *20*, 1194. https://doi.org/10.3390/ijms20051194
42. Wang, Z., Zhang, K., Zhu, Y., Wang, D., Shao, Y., & Zhang, J. (2017). Curcumin inhibits hypoxia-induced proliferation and invasion of MG-63 osteosarcoma cells via downregulating Notch1. Molecular medicine reports, 15(4), 1747–1752. https://doi.org/10.3892/mmr.2017.6159
43. Walters, D.K.; Muff, R.; Langsam, B.; Born, W.; Fuchs, B. Cytotoxic Effects of Curcumin on Osteosarcoma Cell Lines. *Invest New Drugs* 2008, *26*, 289–297. https://doi.org/10.1007/s10637-007-9099-7
44. Ahmad, I.; Hoque, M.; Alam, S.S.M.; Zughaibi, T.A.; Tabrez, S. Curcumin and Plumbagin Synergistically Target the PI3K/Akt/MTOR Pathway: A Prospective Role in Cancer Treatment. *Int J Mol Sci* 2023, *24*, 6651. https://doi.org/10.3390/ijms24076651
45. Qiao, Q.; Jiang, Y.; Li, G. Inhibition of the PI3K/AKT-NF-KB Pathway With Curcumin Enhanced Radiation-Induced Apoptosis in Human Burkitt's Lymphoma. *J Pharmacol Sci* 2013, *121*, 247–256. https://doi.org/10.1254/jphs.12149FP
46. Farghadani, R.; Naidu, R. Curcumin: Modulator of Key Molecular Signaling Pathways in Hormone-Independent Breast Cancer. *Cancers (Basel)* 2021, *13*, 3427. https://doi.org/10.3390/cancers13143427
47. Retraction: Curcumin Suppresses Proliferation and Migration of MDA-MB-231 Breast Cancer Cells Through Autophagy-Dependent Akt Degradation. *PLoS One* 2023, *18*, e0283354. https://doi.org/10.1371/journal.pone.0283354
48. Nelson, K.M.; Dahlin, J.L.; Bisson, J.; Graham, J.; Pauli, G.F.; Walters, M.A. The Essential Medicinal Chemistry of Curcumin. *J Med Chem* 2017, *60*, 1620–1637. https://doi.org/10.1021/acs.jmedchem.6b00975
49. Ganapathy, A.; Ezekiel, U. Phytochemical Modulation of MiRNAs in Colorectal Cancer. *Medicines* 2019, *6*, 48. https://doi.org/10.3390/medicines6020048
50. Angulo, P.; Kaushik, G.; Subramaniam, D.; Dandawate, P.; Neville, K.; Chastain, K.; Anant, S. Natural Compounds Targeting Major Cell Signaling Pathways: A Novel Paradigm for Osteosarcoma Therapy. *J Hematol Oncol* 2017, *10*, 10. https://doi.org/10.1186/s13045-016-0373-z
51. Sun, Y.; Liu, L.; Wang, Y.; He, A.; Hu, H.; Zhang, J.; Han, M.; Huang, Y. Curcumin Inhibits the Proliferation and Invasion of MG-63 Cells through Inactivation of the p-JAK2/p-STAT3 Pathway</P>. *Onco Targets Ther* 2019, *12*, 2011–2021. https://doi.org/10.2147/OTT.S172909
52. Porro, C.; Cianciulli, A.; Trotta, T.; Lofrumento, D.D.; Panaro, M.A. Curcumin Regulates Anti-Inflammatory Responses by JAK/STAT/SOCS Signaling Pathway in BV-2 Microglial Cells. *Biology (Basel)* 2019, *8*, 51. https://doi.org/10.3390/biology8030051
53. Bielack, S.S.; Kempf-Bielack, B.; Delling, G.; Exner, G.U.; Flege, S.; Helmke, K.; Kotz, R.; Salzer-Kuntschik, M.; Werner, M.; Winkelmann, W.; et al. Prognostic Factors in High-Grade Osteosarcoma of the Extremities or Trunk: An Analysis of 1,702 Patients Treated on Neoadjuvant Cooperative Osteosarcoma Study Group Protocols. *Journal of Clinical Oncology* 2002, *20*, 776–790, https://doi.org/10.1200/JCO.2002.20.3.776

54. Owen, K.L.; Brockwell, N.K.; Parker, B.S. JAK-STAT Signaling: A Double-Edged Sword of Immune Regulation and Cancer Progression. *Cancers (Basel)* 2019, *11*, 2002. https://doi.org/10.3390/cancers11122002
55. Banerjee, S.; Biehl, A.; Gadina, M.; Hasni, S.; Schwartz, D.M. JAK–STAT Signaling as a Target for Inflammatory and Autoimmune Diseases: Current and Future Prospects. *Drugs* 2017, *77*, 521–546. https://doi.org/10.1007/s40265-017-0701-9
56. Katiyar, S. Emerging Phytochemicals for the Prevention and Treatment of Head and Neck Cancer. *Molecules* 2016, *21*, 1610. https://doi.org/10.3390/molecules21121610
57. Muta, Y.; Matsuda, M.; Imajo, M. Divergent Dynamics and Functions of ERK MAP Kinase Signaling in Development, Homeostasis and Cancer: Lessons from Fluorescent Bioimaging. *Cancers (Basel)* 2019, *11*, 513. https://doi.org/10.3390/cancers11040513
58. Cuenda, A.; Rousseau, S. P38 MAP-Kinases Pathway Regulation, Function and Role in Human Diseases. *Biochim Biophys Acta Mol Cell Res* 2007, *1773*, 1358–1375. https://doi.org/10.1016/j.bbamcr.2007.03.010
59. Kaushik, M.; Tiku, A.B. Molecular Pathways Modulated by Phytochemicals in Head and Neck Cancer. *J Cell Commun Signal* 2023, *17*, 469–483. https://doi.org/10.1007/s12079-022-00711-0
60. Sharifi-Rad, M.; Nazaruk, J.; Polito, L.; Morais-Braga, M.F.B.; Rocha, J.E.; Coutinho, H.D.M.; Salehi, B.; Tabanelli, G.; Montanari, C.; del Mar Contreras, M.; et al. Matricaria Genus as a Source of Antimicrobial Agents: From Farm to Pharmacy and Food Applications. *Microbiol Res* 2018, *215*, 76–88, https://doi.org/10.1016/j.micres.2018.06.010
61. Venditti, A.; Guarcini, L.; Bianco, A.; Rosselli, S.; Bruno, M.; Senatore, F. Phytochemical Analysis of *Achillea Ligustica* All. from Lipari Island (Aeolian Islands). *Nat Prod Res* 2016, *30*, 912–919. https://doi.org/10.1080/14786419.2015.1079188
62. Venditti, A.; Maggi, F.; Vittori, S.; Papa, F.; Serrilli, A.M.; Di Cecco, M.; Ciaschetti, G.; Mandrone, M.; Poli, F.; Bianco, A. Antioxidant and α -Glucosidase Inhibitory Activities of *Achillea Tenorii*. *Pharm Biol* 2015, *53*, 1505–1510. https://doi.org/10.3109/13880209.2014.991833
63. ten Hoeve, J.; de Jesus Ibarra-Sanchez, M.; Fu, Y.; Zhu, W.; Tremblay, M.; David, M.; Shuai, K. Identification of a Nuclear Stat1 Protein Tyrosine Phosphatase. *Mol Cell Biol* 2002, *22*, 5662–5668. https://doi.org/10.1128/MCB.22.16.5662-5668.2002
64. Yu, H.; Lin, L.; Zhang, Z.; Zhang, H.; Hu, H. Targeting NF-KB Pathway for the Therapy of Diseases: Mechanism and Clinical Study. *Signal Transduct Target Ther* 2020, *5*, 209. https://doi.org/10.1038/s41392-020-00312-6
65. Yamamoto, Y.; Gaynor, R.B. Therapeutic Potential of Inhibition of the NF-KB Pathway in the Treatment of Inflammation and Cancer. *Journal of Clinical Investigation* 2001, *107*, 135–142. https://doi.org/10.1172/JCI11914
66. Liu, T.; Zhang, L.; Joo, D.; Sun, S.-C. NF-KB Signaling in Inflammation. *Signal Transduct Target Ther* 2017, *2*, 17023. https://doi.org/10.1038/sigtrans.2017.23
67. Xia, Y.; Shen, S.; Verma, I.M. NF-KB, an Active Player in Human Cancers. *Cancer Immunol Res* 2014, *2*, 823–830. https://doi.org/10.1158/2326-6066.CIR-14-0112
68. Lawrence, T. The Nuclear Factor NF-B Pathway in Inflammation. *Cold Spring Harb Perspect Biol* 2009, *1*, a001651–a001651. https://doi.org/10.1101/cshperspect.a001651
69. Hayden, M.S.; Ghosh, S. NF-KB, the First Quarter-Century: Remarkable Progress and Outstanding Questions. *Genes Dev* 2012, *26*, 203–234. https://doi.org/10.1101/gad.183434.111
70. Hashemzaei, M., Delarami Far, A., Yari, A., Heravi, R. E., Tabrizian, K., Taghdisi, S. M., Sadegh, S. E., Tsarouhas, K., Kouretas, D., Tzanakakis, G., Nikitovic, D., Anisimov, N. Y., Spandidos, D. A., Tsatsakis, A. M., & Rezaee, R. (2017). Anticancer and apoptosis-inducing effects of quercetin in vitro and in vivo. Oncology reports, 38(2), 819–828. https://doi.org/10.3892/or.2017.5766
71. Kubina, R.; Iriti, M.; Kabała-Dzik, A. Anticancer Potential of Selected Flavonols: Fisetin, Kaempferol, and Quercetin on Head and Neck Cancers. *Nutrients* 2021, *13*, 845. https://doi.org/10.3390/nu13030845
72. Reyes-Farias, M.; Carrasco-Pozo, C. The Anti-cancer Effect of Quercetin: Molecular Implications in Cancer Metabolism. *Int J Mol Sci* 2019, *20*, 3177. https://doi.org/10.3390/ijms20133177
73. Krajka-Kuźniak, V.; Baer-Dubowska, W. Modulation of Nrf2 and NF-KB Signaling Pathways by Naturally Occurring Compounds in Relation to Cancer Prevention and Therapy. Are Combinations Better Than Single Compounds? *Int J Mol Sci* 2021, *22*, 8223. https://doi.org/10.3390/ijms22158223

5 Structure-Activity Relationship of Plant-Derived Bioactive Compounds in Anti-Cancer Therapy

Pradip Debnath[1,2], *Vishal Kumar Deb*[3], *Sourav Nath*[4], *Anupam Guha*[5], *and Suman Adhikari*[4]

[1]IGNOU Regional Centre
Siliuri, India

[2]Maharaja Bir Bikram College
Agartala, West Tripura, India

[3]School of Health Sciences and Technology
UPES, Dehradun, India

[4]Department of Chemistry, Govt. Degree College
Dharmanagar, Tripura(N), India

[5]Michael Madhusudhan Dutta College
Sabroom, South Tripura, India

5.1 INTRODUCTION

Cancer is one of the biggest threats to worldwide health, contributing significantly to morbidity and mortality globally [1, 2]. Cancer is a multifaceted illness that starts with unchecked tissue cell growth and proliferation, which leads to localized cell aggregation (tumor) and can disseminate systemically to other nearby tissues or an entire organ (metastasis). Oncogenes regulating cell death or proliferation may exhibit changes in their genetic or epigenetic makeup, leading to uncontrollable cell behavior [3]. The search for safe and effective anti-cancer drugs continues despite advancements in therapeutic approaches (radiation therapy, chemotherapy, immunotherapy, etc.) [4–8]. Multidrug resistance, drug-induced toxicities, treatment failure, and recurrence reduce the effectiveness of this therapy [9]. Additionally, lesser-income populations cannot afford the cost of these treatments. Due to rapid population expansion and rising life expectancy, the prevalence of cancers is projected to rise by 75% by 2030 [10–12]. Thus, developing novel approaches to cancer therapy is critical to minimizing patient suffering and cutting expenses associated with today's costly treatment [13, 14].

Plants are a gift from nature, and people from all over the world have accepted a variety of plant-based medications and health products to enhance the quality of life, prevent illness, and treat geriatric and chronic illnesses [15]. Secondary metabolites that originate in plants are called phytochemicals. About 10% of the roughly 250,000 plant species that comprise the plant kingdom have been examined for potential medical applications [16, 17]. Plant parts such as flowers, sprouts, pericarp, fruits, roots, seeds, stems, rhizomes, leaves, bark, embryos, etc., comprise phytochemicals and their derived analogs, which have various beneficial uses. Phytochemicals show an inordinate deal of structural diversity, tend to take on the ideal conformation and require steric complexity to perform various functions in biological test systems [18]. Phytochemicals generally have more chiral centers, fewer heteroatoms, fewer heavy atoms, and more diverse ring systems than synthesized organic compounds [19].

DOI: 10.1201/9781003486237-5

Generally, phytochemicals are thought to have "biologically friendly" molecular features and "drug-like" pharmacological effects [20]. Natural products, especially bioactive chemicals originating from plants, have drew a lot of consideration in the last decades due to their distinct chemical structures and strong biological effects [21]. The effect of bioactive plant ingredients, or phytochemicals, has been shown to reduce the risk of several chronic illnesses, such as cardiovascular disease, neurological, diabetes, and cancer, based on data from epidemiological and clinical trials [22]. Plants are a rich source of structurally distinct compounds that have the can block a variety of oncogenic pathways, making them an attractive target for the design and synthesis of anti-cancer drugs. Historically, plant-based compounds have produced crucial anti-cancer drugs like vincristine, camptothecin, paclitaxel, etc., which have completely changed the way that cancer is treated [23]. Most natural anti-neoplastic medications have anti-tumor properties. However, they often work by controlling human immune response instead of directly killing tumor cells. Several crucial cell cycle processes include division and duplication, and disruption of these processes may contribute to the emergence of cancer [24]. As a matter of fact, natural products account for 52% of all approved compounds between 1981 and 2014 [25]. Remarkably, of the 131 anti-cancer compounds made available during that time, 49% were from natural products and their derivatives [26]. These triumphs highlight the possibility of using plant-based bioactive compounds as a basis for anti-cancer drugs.

The vast majority of chemical compounds or chemicals that are natural products come from nature. A molecule's solubility, stability, and potential for therapeutic use are all attributed to its structural properties and functional groups [27–29]. Understanding the relationship between a substance's chemical structure and its biological or pharmacological effects allows us to predict its target biological impact in an organism [30, 31]. The intricate relationship between the bioactive compounds' structure and their anti-cancer property, commonly referred to as the structure-activity relationship (SAR), serves as a pivotal framework for rational drug design and optimization [32]. The identification of important structural motifs accountable for bioactivity is made easier by SAR, which clarifies the molecular interactions between ligands and their biological targets [33]. Understanding SAR is necessary for the development of innovative compounds and for comprehension of the molecular mechanisms underlying the therapeutic efficacy of plant-derived anti-cancer drugs [34].

Recent advancements in computational biology, structural biology, and analytical techniques have spurred efforts to comprehensively decipher SAR of plant-derived bioactive compounds [35]. A synergistic combination of molecular modeling, structure-based drug design, and high-throughput screening assays has sped up the validation and design of compounds with improved anti-cancer potency and selectivity [36]. The most well-known and significant plant-derived anti-cancer drugs target the cell microtubule cytoskeleton system [37]. Moreover, identifying novel phytochemicals from traditionally revered medicinal plants has been made easier by phytochemical profiling and metabolomic investigations, which has helped close the knowledge gap between ethnopharmacology and contemporary drug discovery [38]. Phytochemicals or plant-derived compounds should be demonstrated therapeutically and clinically safe throughout several stages of clinical trials before being developed into anti-cancer medicinal products. Data on SAR analysis and mechanism of action are crucial components of laboratory research safety and efficacy, speeding up the process of moving that study toward clinical trials [34]. Concerning the SAR of plant-based compounds in anti-cancer therapy, this chapter seeks to clarify current understanding. We want to identify the structural factors influencing anti-cancer activity, highlight new developments in the field, and highlight possibilities and difficulties for further research by combining knowledge from various research works. Deciphering the SAR of plant-derived bioactive compounds will ultimately accelerate the development of novel anti-cancer therapeutics and increase awareness of the unrealized potency of natural products in the fight against cancer.

5.2 APPROVED DRUGS FOR CANCER TREATMENT

Cancer remains a foremost global health issue, requiring ongoing research and development as well as therapeutic strategy modification [1]. Substantial developments in the identification and approval of anti-cancer medications have been made in the last several decades, improving patient outcomes and survival rates [5, 8, 14]. To advance cancer therapy and meet unmet medical needs, it is essential to overcome resistance mechanisms, optimize therapeutic regimens, and develop novel targeted therapies. In addition to its target specificity, the advancement of cancer therapy with conventional chemotherapy drugs has become promising. The molecular features of the tumor have been taken into consideration when developing this therapy approach. However, the tumor's heterogeneity also leads to issues with the way the body reacts to the drugs [39]. Gene mutations in ALK (anaplastic lymphoma kinase), VEGF (vascular endothelial growth factor), EGFR (epidermal growth factor receptor), and other related genes reveal the heterogenicity characteristic. To target certain genetic changes, several small molecule-based inhibitors and monoclonal antibodies (mAb) were developed.

The majority of NSCLC (non-small cell lung cancer) victims with EGFR mutations of the gene are female East Asian patients or those who never smoke [40]. The most frequent EGFR mutations studied include the removal of exon 19 and point mutations in exon 21 L858R [41]. As first-generation, EGFR-TKIs (tyrosine kinase inhibitors), gefitinib, and erlotinib were designed with this in consideration, specifically focusing on the deletion of exon 19 in treating NSCLC [42]. Moreover, clinical trials have suggested using second-generation medications, such as dacomitinib and afatinib, for patients who no longer respond to gefitinib or erlotinib due to the T790M mutation [43]. However, osimertinib, a third-generation EGFR-TKI, has also been developed for the T790 M-resistant mutation.

BRAF mutations have also been discovered in NSCLC patients; these mutations are specifically seen in patients with adenocarcinoma. In females as well as in smokers who are currently or have previously smoked, the predominant BRAF mutation is the substitution of valine for glutamate at codons 600 [44]. FDA-approved trametinib and dabrafenib are applied for treating NSCLC patients with the BRAF mutation (V600E). Additionally, one of the most common cancers in the world is colorectal cancer. VEGF is crucial in this disease because it encourages cell invasion and proliferation. Many drugs have been designed to target metastatic colorectal cancer specifically. In one study, for example, three groups were divided up and given different treatments: the first group received irinotecan and fluoropyrimidine along with oxaliplatin, leucovorin, fluorouracil, and bevacizumab; the second group received leucovorin only with other drugs; and the third group obtained bevacizumab along with other treatments. Consequently, the combination of leucovorin, bevacizumab, and fluoropyrimidine/irinotecan demonstrated a prolonged patient survival rate [45].

One tyrosine kinase receptor, EGFR, is accountable for angiogenesis, cell migration, and proliferation in colorectal cancer. As a result, when treating colorectal cancer, EGFR serves as a therapeutic marker. To further explore this, cetuximab was developed as a treatment for colorectal cancer that is positive for EGFR. According to the anti-cancer study, cetuximab can be used to treat colorectal cancer by preventing chemotherapy-induced resistance mediated by the EGFR pathway [46]. Crucially, metastatic colorectal cancer with EGFR (+) can also be treated with cetuximab in conjunction with FOLFIRI [47]. Furthermore, depending on the stage of the tumor, transforming growth factor (TGF)-β has been shown to play a dual part in encouraging and preventing tumor growth [48]. Interestingly, this marker has been linked to the progression of tumors in colorectal cancer, making it a desirable target for mAb that target the TGF receptor's kinase activity [49]. A small chemical compound called galunisertib was developed to specifically target the kinase domain of the TβRI (TGFβ-receptor I). This small chemical compound demonstrated anti-cancer activity against lung, breast, and liver malignancies [50].

The enzyme CYP17, essential in transforming progesterone into androstenedione and pregnenolone into androgen, aids in the production of testosterone from cholesterol. It has been documented that androgen is also produced by prostate cancerous cell. Abiraterone acetate, an inhibitor of the CYP17 enzyme, prevents the CYP17 from becoming active [51]. Patients with prostate cancer had a higher survival rate in the abiraterone clinical trial.

Furthermore, the FDA has permitted several drugs (Table 5.1) to block signaling pathways linked to cancer, including the EGFR, ALK, and other biomarker-associated pathways (Figure 5.1).

TABLE 5.1
FDA-Approved Therapeutics for Different Cancer Treatment

S. No	FDA-Approved Drugs	Targeted Protein	Types of Cancers
1.	Crizotinib	ALK	Lung cancer
2.	Brigatinib	ALK	Lung cancer
3.	Lorlatinib	ALK	Lung cancer
4.	Gefitinib	EGFR	Lung cancer
5.	Panitumumab	EGFR	Colorectal cancer
6.	Dabrafenib	BRAF	Lung cancer
7.	Encorafenib	BRAF	Colorectal cancer
8.	Bevacizumab	VEGF	Colorectal cancer
9.	Ramucirumab	VEGF	Colorectal cancer
10.	Galunisertib (under clinical trial)	TGFβRI	Colorectal cancer
11.	Apalutamide	Androgen receptor	Prostate cancer
12.	Abiraterone acetate	CYP17	Prostate cancer
13.	Rucaparib	PARP	Prostate cancer
14.	Talazoparib	PARP	Prostate cancer

5.2.1 Limitations of Approved Cancer Treatment Drugs

Regarding approved anti-cancer drugs, there are several limitations. Drug classes differ in toxicity profiles, which might lead to problems with patients' clinical efficacy. Furthermore, varying toxicity profiles present a greater challenge regarding administration routes. These problems could have a greater impact on drugs as a whole. Furthermore, these drugs have less available safety data after going on the market than those that have already received approval. Furthermore, the mechanisms of action of the therapeutics that were developed to inhibit tyrosine kinase are uncertain. Nonetheless, the advancement of drugs to combat cancer is ongoing, but the cost is so enormous that the general public cannot afford them [52].

5.3 NOVEL APPROACHES IN ANTI-CANCER THERAPY BY PLANT-DERIVED COMPOUNDS

Long-term monotherapy with chemotherapy drugs has also brought up several problems, including the development of drug resistance, lack of efficacy, and adverse reactions. Using phytochemicals, numerous novel methods have been anticipated to get around these restrictions. Here are some examples of them.

5.3.1 A Combination of Plant-Derived Compounds with Chemotherapeutic Drugs

Because of their ability to modulate the anti-cancer effect, plant-derived compounds are currently the focus of interest in cancer treatment. These natural compounds significantly, inhibited the P-glycoprotein (p-gp) expression, demonstrating anti-cancer properties [53]. These compounds also modulate several pathways linked to cancer and have negligible harmful effects. Consequently, the use of plant-based compounds in conjunction with anti-cancer medications represents a unique delivery strategy at the site of cancer. This novel approach also helps to advance the effectiveness of cancer therapy and lessen the side effects of chemotherapy medications (Figure 5.2; Table 5.2).

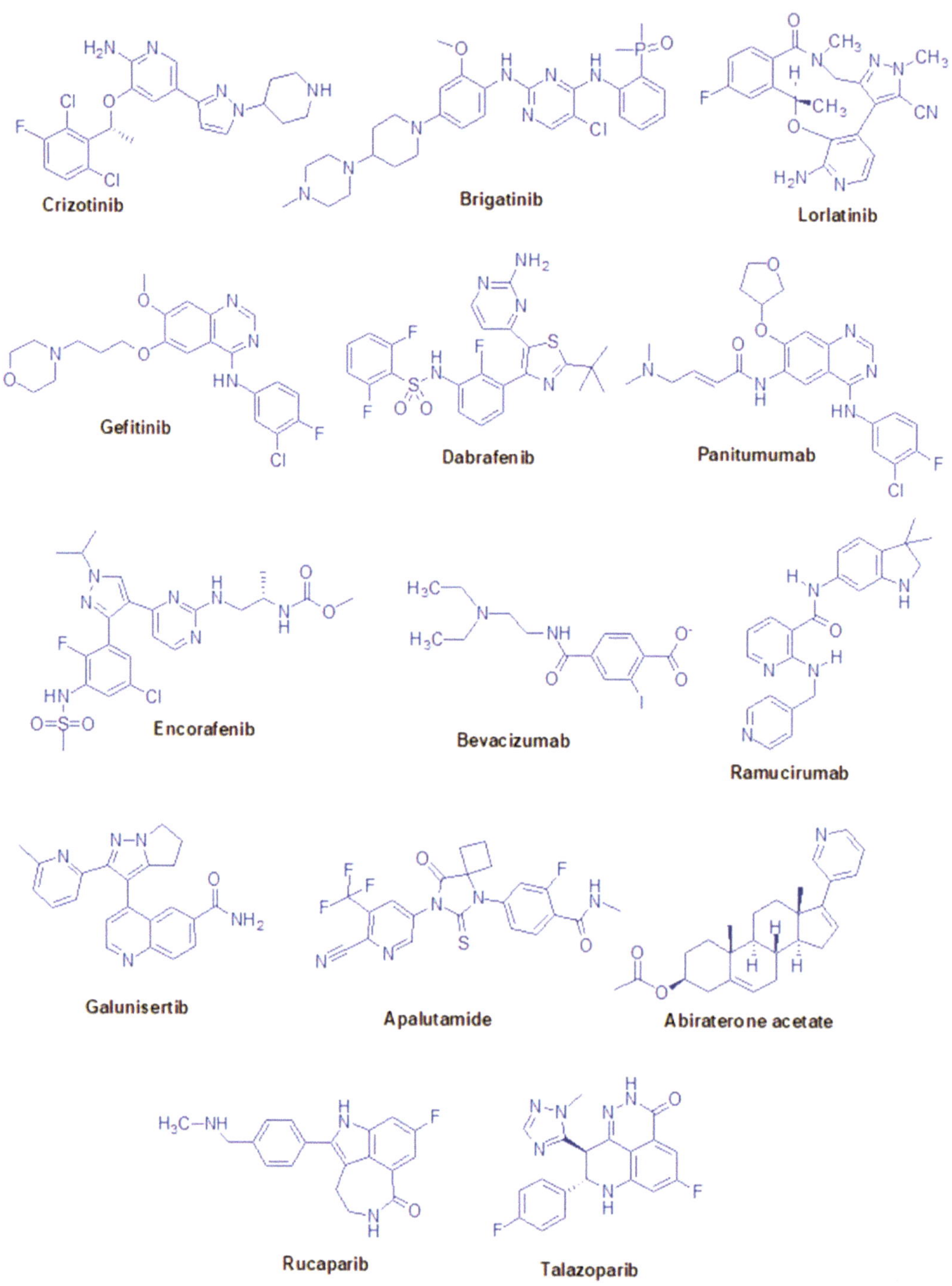

FIGURE 5.1 Chemical structure of FDA-approved drugs for different cancer therapies.

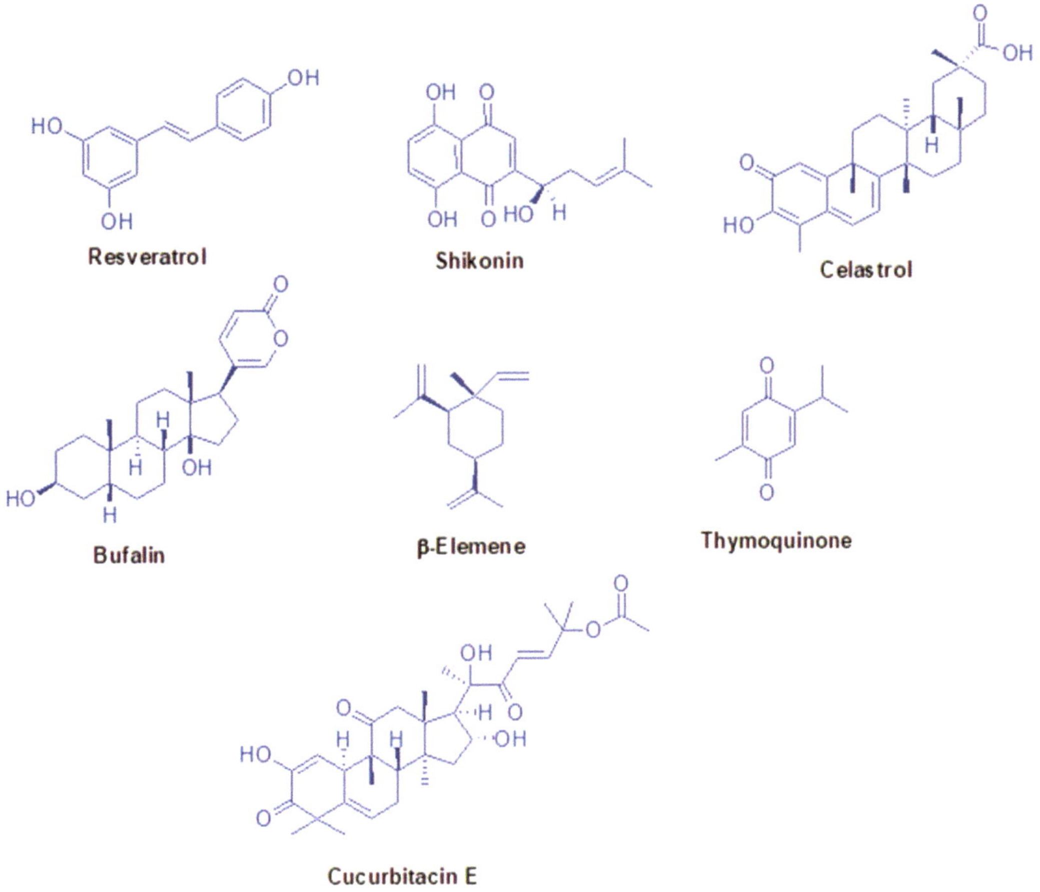

FIGURE 5.2 Chemical structure of plant-based compounds for cancer therapy.

TABLE 5.2
Effects of Combination Therapy by Plant-Derived Compounds and Chemotherapeutic Drugs in Different Cancers

Plant-Derived Compounds	Drugs	Types of Cancers	Effects	References
Resveratrol	Cisplatin	Lung cancer	Increase Bax expression (apoptotic protein) Decrease Bcl-2 expression (anti-apoptotic protein)	[54]
Shikonin	Gefitinib	Lung cancer	Inhibit expression of STAT3 and cyclin D1	[55]
Celastrol	Lapatinib	Liver cancer	NR	[56]
Bufalin	Sorafenib	Liver cancer	ERK pathway down-regulated	[57]
Curcumin	5-FU	Stomach cancer	Impair the expression of COX-2 protein, and NF-kB pathway	[58]
Cucurbitacin E	Doxorubicin	Stomach cancer	Inhibit Akt protein expression	[59]
β-Elemene	Cetuximab	Colorectal cancer	Reduce the EMT	[60]
Thymoquinone	Gemcitabine	Pancreatic cancer	Inhibit the PI3K expression	[61]

TABLE 5.3
Nanomaterials Used for the Nano-Formulation to Treat Different Types of Cancers

Types of Nanomaterials	Plant-Derived Compounds	Types of Cancers	IC_{50}	References
Polymeric nanoparticles	Betulinic acid	Breast cancer (Triple negative)	IC_{50} 2 µg/mL	[62]
Magnetic nanoparticles	Silibinin	Lung cancer	IC_{50} 34.24 µM	[63]
Liposomes	Resveratrol	Liver cancer	IC_{50} 34.2 µg/mL	[64]
Solid lipid nanoparticles	Ferulic acid and resveratrol	Colon cancer	IC_{50}10 µg/mL	[65]
Dendrimer	Celastrol	Colorectal cancer	IC_{50} 1.106 µM	[66]

5.3.2 Nanomaterials as a Novel Approach to Co-Deliver Plant-Derived Compounds and Anti-Cancer Drugs

Multidrug resistance, which is increased as an outcome of the overexpression of certain specialized proteins following repeated use of chemotherapeutic drugs, is another drawback of cancer treatment. Drugs' ability to pass through resistant cells' membranes is hindered by the overexpression of these specialized proteins, known as biomarkers. Moreover, in many cancer types, the production of multidrug resistance proteins reduces the effectiveness of the drugs. Nanotechnology is playing a critical role as a novel approach as an advanced carrier system to transport those drugs to therapeutic sites, helping to minimize this issue. Employing nanoparticles as a delivery carrier system for chemotherapy drugs to expand stability and shield them from enzymatic activity or destruction (Table 5.3).

5.3.3 Targeting Cancer Stem Cells and Signaling Pathways by Plant-Derived Compounds

A population of cells recognized as cancer stem cells can regenerate, hence promoting the formation of cancerous cells. These cells also can become resistant to chemotherapy drugs [67]. Though there are a number of drugs that are said to target cancer, naturally occurring plant-derived compounds, have recently attracted attention due to their safety profiles and capacity to target many cell types in a tumor microenvironment. For instance, by comprehending their role in signaling pathways control, numerous phytochemicals, including retinoic acid, sulforaphane, celastrol, 3,3-di-indolylmethane, and resveratrol, were researched for their anti-cancer potential toward breast cancer stem cells [67]. These plant-derived compounds may target the signaling pathway linked to cancer stem cells, which could be a unique strategy as a therapeutic agent.

5.3.3.1 Rutin-Mediated Targeting of a Cell Signaling Gene Jab1 in Cervical Cancer

Rutin is a bioflavonoid that has many biological effects against tumor cells. Additionally, this molecule has been shown to have the ability to suppress the oncogene Jabl in cervical cancer cell lines (SiHa). This oncogene can inhibit the action of tumor suppressor proteins like p27 and p53. To explore this issue further, it has been described that rutin can cause both apoptosis and cell cycle arrest in cancer cells. In cervical tumor cells, it also produces reactive oxygen species (ROS), besides apoptosis [68].

5.3.3.2 Combination of Plant-Derived Compounds and a Histone Demethylase Inhibitor in Prostate Cancer

Another plant-derived compound with anti-cancer potential is hesperetin, which inhibits the TGFβ-mediated invasion and transformation of cancer cells from the epithelium to the mesenchymal state. For example, low levels of H3K27me3 expression are detected in prostate cancer

TABLE 5.4
Plant-Derived Compounds Used for Targeting Immune Checkpoints to Treat Different Cancers

Plant-Derived Compounds	Types of Cancers	Targeted Immune Checkpoints	Effects	References
Cyanidin-3-O glucoside	Colorectal cancer	PD-1 and PD-L1	Down-regulated	[71]
Luteolin	Breast cancer	PD-L1	Down-regulated	[72]
Hesperidin	Breast cancer (triple negative)	PD-L1	Down-regulated	[73]
Resveratrol	Breast cancer	PD-L1 dysfunction	Up-regulated	[74]
Gallic acid	Lung cancer	PD-L1	Down-regulated	[75]

because demethylase expression is elevated [69]. To solve this, a new approach to treating cancer may involve the introduction of demethylase inhibitors in conjunction with anti-cancer phytochemicals. According to a recent study, hesperetin and the JMJD3/KDM6B (H3K27 demethylase) inhibitor GSK-J4 together can cause cytotoxicity and prevent the migration of prostate cancer cells driven by TGFs [69].

5.3.4 Plant-Derived Compounds in Cancer Immune Checkpoint Inhibitor Therapy

Recently, immunological checkpoint proteins have been found to play a part in preventing T cells from identifying tumor cells and diminishing the immune responses against cancer. Immune checkpoint inhibitors demonstrated a promising therapeutic potential against cancer cells in this particular domain. Nevertheless, immunogenic adverse effects have limited the use of immune checkpoint-mediated anti-cancer therapy. Numerous plant-based drugs have shown immune checkpoint regulatory actions to address this issue in various types of malignancies [70] (Table 5.4).

5.4 MECHANISM OF ANTI-CANCER ACTIVITY OF PLANT-DERIVED COMPOUNDS OR PHYTOCHEMICALS IN CANCER CELLS

Cancer is one of the deadliest illnesses in the world, and chronic inflammation can be another reason [76]. For instance, the hepatitis B virus is also a responsible virus that may cause chronic inflammation in liver cells and convert it into hepatocellular carcinoma [77]. Nonetheless, several clinical applications are currently undergoing trials to lower the cancer's death rate. Another use for cancer research is the treatment of cancer with chemotherapy-based pharmaceuticals or nanotechnology-based drug delivery systems, including organic, inorganic, and hybrid nanomaterials. Anti-microbial-based cancer therapy is also becoming attractive in the field of cancer research. Numerous mechanisms for treating cancer cells based on the phytochemicals have been described. The following highlights these mechanisms.

5.4.1 Cell Cycle Arrest

Currently, many small compounds, including those found in nature, are used to either stimulate or stop the cell cycle. Transcription and translation are two of the cell's dormant processes once the cell cycle is stopped. Many proteins function as targeted proteins in cancer therapeutic targets where they are involved in cell cycle regulation [78]. In addition, the cell cycle comprises four primary stages: the Gl phase, which is the replication of DNA, the S phase, which is the creation of identical strands from a single DNA template, the G2 phase, which is the cell's division, and the M stage, that is the division of the cell into two identical daughter cells. These stages also serve as targets for cancer treatment [78]. A current study has investigated that a phytochemical named coumestrol can induce toxicity in skin cancer cells by triggering caspase-mediated apoptosis. Moreover, it also inhibits the cancer cell's migration by impairing the MMP-2 and -9 expressions [79].

5.4.2 Generation of ROS and Oxidative Stress (DNA Damage)

Byproducts of mitochondrial metabolism produce ROS in eukaryotic cells and other organs that begin to use more oxygen. Certain ROS possess unpaired electrons, such as O_2^- and ^-OH. In addition to high concentrations, ROS can also exist at moderate concentrations, aiding the endoplasmic reticulum's protein folding and regulating apoptosis by regulating caspase activity [80]. Naphthoquinones can absorb electrons from extremely reactive species like semiquinones and hydroxyquinones. Furthermore, Stangel et al. found that this compound can cause apoptosis [81]. Curcumin is an example of a plant-based phenolic compound that acts as an antioxidant and generates ROS to persuade apoptosis in malignant cells [82].

5.4.3 Endoplasmic Reticulum Stress Generation and Induction of Apoptosis

Shikonin (a plant-based naphthoquinone found in the root of the *Lithospermum erythrorhizon*) has also been shown to have tumor growth suppressor properties since it inhibits the tumor proteasome in various melanoma cells, such as liver and prostate cancers [83]. Misfolded proteins cause endoplasmic reticulum stress due to tumor proteasome suppression [84]. According to a recent study, Shikonin has also been linked to the formation of ERS in brain cancer stem cells [85]. In addition, it compromises the cytotoxic effects by activating the JNK/c-Jun pathway concurrently.

5.4.4 Regulation of the Tumor Suppressor Factors p53 and Apoptosis

Activation of the p53 gene is an essential factor in anti-cancer therapy that aids in the subsequent manipulation of apoptosis [86]. Another plant-based compound that has been revealed to trigger p53 in brain tumor cells is called plubagin. Accordingly, this compound alternates the expression of the Bax/Bcl-2 ratio and stops the cell cycle at the G2/M stage [87]. However, in HepG2, it also upregulates the pro-apoptotic gene expression, for example, caspase-3 and caspase-7 [88].

5.4.5 Inhibition of the DNA Topoisomerases

Cancer treatments also target topoisomerase I and II to treat cancer cells. These proteins play a crucial role in breaking phosphodiester linkages to restore proper cell activity; in the absence of such breaks, apoptosis occurs differently [89]. In an investigation into the suppression of topoisomerase I activity, it was noted that β-lapachone was the first topoisomerase inhibitor to destroy cancer cells [90]. Certain derivatives of naphthoquinones can prevent the expression of topoisomerase enzymes [91]. A plant-derived compound, thujone, can inhibit the proliferation of lymphoma cells. It was examined in the in-silico docking study that thujone also has an inhibitory role against the expression of topoisomerase II [92] (Figure 5.3).

By suppressing matrix metalloprotease (MMP) expression and activating caspase proteins, coumestrol prevents the migration of malignant cells and causes them to endure apoptosis. Shikonin can also cause the generation of ROS, further induces apoptosis to eradicate cancer cells; Thujone has an anti-cancer property to trigger apoptosis in tumor cells by inhibiting the topoisomerase; plubagin can arrest the cell cycle in tumor cells to stop its uncontrolled growth; in a similar vein, curcumin can also kill cancer cells by releasing ROS.

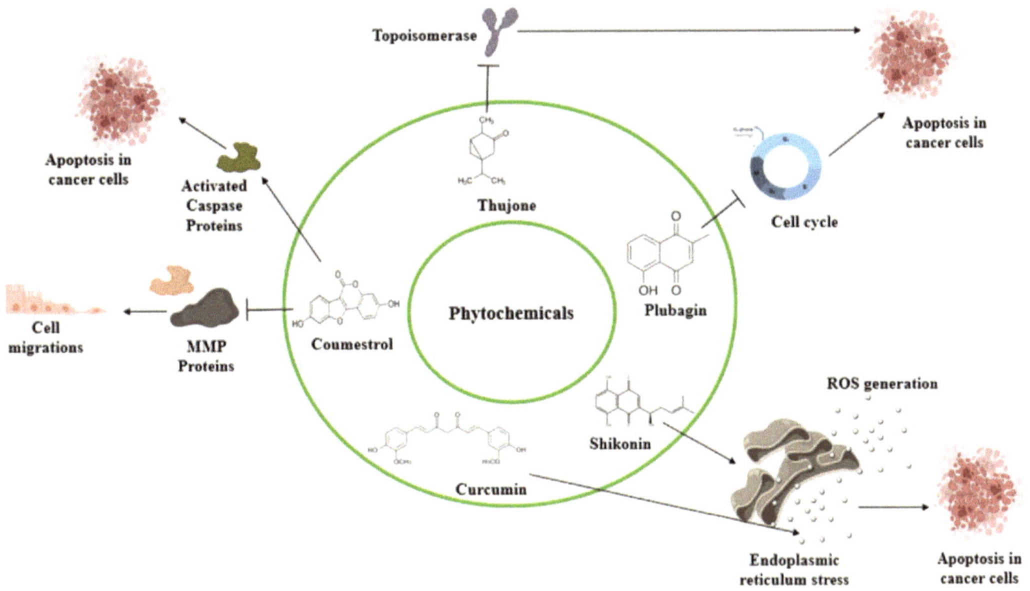

FIGURE 5.3 Phytochemical-mediated distinct anti-cancer mechanism.

5.5 LIST OF NATURALLY ISOLATED PLANT-DERIVED BIOACTIVE COMPOUNDS IN ANTI-CANCER THERAPY

The potential therapeutic properties of plant-based compounds against malignant cells have been the focus of recent research on cancer. Pre-clinical research based on phytochemicals is ongoing, and some are even in clinical trials. Some of the important phytochemicals effective in treating different kinds of cancer in humans are described below and listed in Table 5.5.

5.5.1 Andrographoide

Andrographis paniculata is an essential medicinal plant utilized in traditional medicine to treat numerous diseases in Southeast Asian countries. Andrographolide, a lactone-based compound that is the major component of *Andrographis paniculata* has exhibited several therapeutic potentials against multiple diseases, including inflammation, liver dysfunctions, common cough, and cold [93]. According to recent research, andrographolide can stop cancer cells growth, encourage apoptosis, and cause cell cycle arrest in various cancer cell lines, such as colorectal, colon, ovarian, lung, and prostate cancers [94, 95]. SAR studies revealed the andrographolide's cytotoxic properties against numerous cancer cell lines are attributed to its α-alkylidene γ-butyrolactone motifs and its three -OH at C-3, C-14, and C-19 position.

Recently, Khan and co-workers [96] observed that andrographolide showed anti-cancer properties against colon cancer HT-29 cells. Molecular mechanism studies indicated that andrographolide disrupted the mitochondrial membrane potential and arrested the cell cycle in the G2/M stage. In another study, Peng and co-workers examined the potentiality of andrographolide for inhibiting breast cancer cells. Andrographolide suppressed the COX-2 expression by attenuating p300 HAT activities and angiogenesis by deactivation of p300 signaling in cancer cell lines [97].

Andrographolide displayed anti-cancer activity in prostate cancer and inhibited cell viability and cell migration by suppressing the expression of CXCL11, CXCR3, CXCR7, and IL-16 [98]. A literature survey revealed that andrographolide also exhibited apoptosis against leukemic HL-60 cells and prostatic adenocarcinoma PC-3 cells [99]. Recently, Forestier-Román et al. reported that

TABLE 5.5
List of Some Important Phytochemicals Explored against Cancer Treatment

Sl. No	Name of the Plant	Parts of the Plant Used	Chemical Nature of the Compound, Molecular Formula of the Compound, and Phytochemicals	Specific Cancer Inhibited	Mechanism of Action
1.	*Andrographis paniculata*	Whole plant	Diterpenoid $C_{10}H_{30}O_5$ Andrographoide	Colon, breast, prostate cancer	Increasing intracellular ROS level [96] Inhibition of COX-2 and VEGF pathway [97] Facilitate DNA damage [100]
2.	*Azadirachta indica*	Leaves, flowers, seeds, fruits, roots and bark	Limonoid Triterpene $C_{27}H_{30}0_7$ Nimbolide	Pancreatic, colon cancer	Inhibition in PI3K/AKT/mTOR and ERK signaling [101] Suppression of Bcl-x, CXCR4, VEGF, and NF-B [102]
3.	*Artemisia annua*	Leaf	Alkaloid $C_{15}H_{22}O_5$ Artemisinin	Colon, kidney cancer	Increase in ROS production [105] Inhibition of AKT signaling [106]
4.	*Camellia sinensis*	Leaf	Catechin $C_{22}H_{18}O_{11}$ Epigallocatechin Gallate	Bile duct, lung, ovarian cancer	Inhibition of Notch1, MMP-2, and -9 signaling [109] Activation of AMPK signaling pathway [110]
5.	*Capsicum annuum*	Fruit	Capsaicinoid $C_{18}H_{27}NO_3$ Capsaicin	Colon, lung, breast, esophageal cancer	MMP-2 and -9 levels down-regulation [111] FBI-1-facilitated NF-B pathway down-regulation [112] Stabilization and activation of p53 [114]
6.	*Citrus limon*	Fruits	Flavonoid $C_{28}H_{34}O_{15}$ Hesperidin	Lung, liver, breast cancer	Apoptosis induction and cell cycle inhibition *via* down-regulating Bcl-xL and up-regulating Bax, Bak, and tBid proteins [116] Inhibition of PD-L1 expression by down-regulating Akt and NF-B signaling [73]
7.	*Curcuma longa*	Rhizome	Curcuminoids $C_{12}H_{20}O_6$ Curcumin	Breast, gastric, oral cancer	Up-regulation of PTEN/Akt signaling pathway [119]
8.	*Cucumis sativus*	Fruit, seed and stem	Triterpene $C_{32}H_{46}O_8$ Cucurbitacin	Colon, gastric, bile, breast cancer	Reduction of Hippo-YAP signaling pathway [124] Suppression of Akt expression [125]
9.	*Crocus sativus*		Alkaloid $C_{10}H_{14}O$ Safranal	Liver, prostate cancer	Down-regulating Bcl-xL and up-regulating Bax, Bak, and tBid proteins [128] Down-regulating FGF and NF-B signal transduction pathway [129]

TABLE 5.5 *(Continued)*
List of Some Important Phytochemicals Explored against Cancer Treatment

Sl. No	Name of the Plant	Parts of the Plant Used	Chemical Nature of the Compound, Molecular Formula of the Compound, and Phytochemicals	Specific Cancer Inhibited	Mechanism of Action
10.	*Dioscorea villosa*	Roots and rhizomes	Saponin $C_{27}H_{42}O_3$ Diosgenin	Brest, Liver cancer	Down-regulation of Skp2 [133] Reduction of Akt and up-regulating p21 and p27 expression [134]
11.	*Glycyrrhiza glabra*	Root and rhizome	Chalcone $C_{21}H_{22}O_4$ Licochalcone	Breast, lung cancer	Inhibition of PI3K/Akt/mTOR pathway [137] ERK induction and p38 initiation while suppresses JNK signaling [138]
12.	*Magnolia officinalis*	Bark	Lignan $C_{18}H_{18}0_2$ Magnolol	Lung, gallbladder, liver cancer	Inactivate AKT and mTOR signaling pathways [141]
13.	*Matricaria chamomilla*	Dried Leaf	Flavonoid $C_{15}H_{10}0_5$ Apigenin	Colon, liver, breast, bone cancer	Reduction of the Mcl-1, AKT, and ERK pro-survival regulators [144]
14.	*Papaver somniferum*	Unripe seed capsule	Alkaloid $C_{22}H_{23}NO_7$ Noscapine	Breast, prostate cancer	Suppression of microtubule dynamics [146] Decreases NF-B and increases IB expression [147]
15.	*Peganum harmala*	Whole plant	Alkaloid $C_{13}H_{12}N_2O$ Harmine	Breast, thyroid, gastric cancer	Down-regulation of TAZ [151]
16.	*Piper nigrum*	Root	Alkaloid $C_{17}H_{19}NO_3$ Piperine	Colon, breast cancer Lung cancer	Reduction of ERK-modulated metastatic process [152]
17.	*Polygonum cuspidatum*	Dry root	Stilbenoid $C_{16}H_{16}O_3$ Pterostilbene	Ovarian, lung, breast, colon cancer	Diminishes release of NF-B p50, and NF-B p65 [160] Down-regulating AKT and NF-B signaling [161] Increases generation of ROS, caspase-3 activity and ER stress [162]
18.	*Solanum lycopersicum*	Fruits	Carotenoid $C_{40}H_{56}$ Lycopene	Breast, colon, brain cancer	Initiation of caspase-3 pathway [164] Reduction of NF-B and JNK signaling [165]

treatment of prostate cancer cells with andrographolide affects cell motility, proliferation of cells, and apoptosis in the cancer cell by decreasing the PCa cell migration, increasing apoptosis, and lowering invasion [100].

5.5.2 Nimbolide

Nimbolide is a plant-derived chemical found in the flowers and leaves of the neem tree (*Azadirachta indica*). It has been widely utilized as a conventional folk remedy to treat several ailments. The significant cell cytotoxicity and anti-cancer effects of nimbolide were probably attributed to the α, β-unsaturated ketone unit. Additionally, it was proposed that a γ-lactone moiety was the source of its cytotoxicity. Subramani and colleagues have recently examined [101] nimbolide's anti-cancer properties in pancreatic malignant cell lines. Nimbolide caused the cancer cells to produce excessive amounts of ROS, which prevented both autophagy and death in pancreatic cells. According to *in vivo* research, nimbolide actively inhibits the growth and spread of pancreatic malignancy by preventing the migration of cells into the colony, inhibiting the epithelial-to-mesenchymal transition (EMT), and promoting the formation of colony.

Nimbolide can decrease colorectal cancer cells by inducing a proinflammatory milieu within the cancer cells, as noted by Gupta et al. [102]. The author noted that nimbolide inhibits NF-κB activation, which prevents proliferation and causes apoptosis.

5.5.3 Artemisinin

Chinese traditional medicine has employed artemisinin (ART), a noteworthy substance, to treat fever [103]. Currently, one of the main front-line medications used to treat malaria is ART and its variants. ART and its analogs have been shown to have cytotoxic effects on many human melanoma cells [104]. According to SAR research, a peroxide bridge inside the ART molecule is intimately associated with the cytotoxicity of the drug's anti-cancer properties.

Wang et al.'s study [105] looked at artemisinin's ability to inhibit colorectal cancer cells. Cancer cells respond faster to ART than healthy cells in terms of increased heme production. It has been discovered that adding aminolevulinic acid (ALA), a heme synthesis precursor that is clinically used, increases heme production levels in cells and amplifies the cytotoxicity of ART on colorectal cancer cells. The authors proposed a novel ART combination therapy that incorporated ALA, known to dramatically boost ART's anti-cancer properties.

One of the most common urologic cancers and a major source of cancer-associated mortality is clear cell renal cell carcinoma (ccRCC). Yu et al.'s recent observations [106] indicate that ART inhibits cell migration, metastasis, and proliferation, contributing to the anti-cancer actions of ccRCC. It has been documented that combining ART with an AKT inhibitor improved the anti-cancer activities of ART.

5.5.4 Epigallocatechin Gallate

Two important catechins in green tea are epigallocatechin and its ester epigallocatechin-3-gallate (EGCG) having different biological functions. The gallic acid group significantly enhanced the anti-cancer activities of catechins. Owing to its antioxidant qualities, EGCG's effects on several cancer cell types, for example, those from the brain, bladder, cervical, or breast, have been extensively studied [107]. According to Senggunprai et al., EGCG can suppress CCA proliferation of cells and cytokine-induced migration by regulating nitric oxide synthase [108]. Kwak et al. reported that EGCG can also reduce the growth and invasion of cholangiocarcinoma (CCA) cells [109].

It is also known that EGCG regulates a number of carcinogenic signal expressions, for example, EGFR, vascular endothelial and insulin-like growth factors, nuclear factor-κB reduction, cell cycle regulatory proteins, down-regulated PI3-K/Akt signaling, and activation of protein-1 signaling pathways. Chen et al. found in 2020 that EGCG suppressed lung cancer cells by the signaling pathway of AMP-initiated protein kinase [110].

5.5.5 Capsaicin

An essential phytochemical found in red pepper, capsaicin has various biological properties, for instance, anti-mutagenic, anti-cancer, anti-metastatic, anti-angiogenic, and chemopreventive properties. Prostatic, pancreatic, liver, leukemia, skin, lung, colon, bladder, and endothelial cell cancers are only a few of the cancer types that this chemical are being studied for its effectiveness in treating [111]. The substituents at 3 and 4 positions of the capsaicin A-ring are vital for strong agonist activity. The 4-OH moiety is very important in capsaicin derivatives, according to SAR studies. A hydrophobic component, such as an octyl chain or substituted benzyl, is compulsory for high effectiveness. Researchers discovered that capsaicin reduced the production of apoptosis and proliferation in breast cancer by down-regulating the NF-κB pathway facilitated by FBI-1 and controlling a variety of targets, including caspase-3, Racl, ROS, and HER-2 [112].

By persuading apoptosis in cancer cells having the p53 gene, sometimes referred to as the "Grandfather of the Genome," capsaicin is also beneficial [113]. It has recently been noted that capsaicin exhibits anti-cancer capabilities toward colon cancer cells by causing apoptosis and cell cycle G0/G1 stage arrest. Capsaicin's anti-tumor efficaciousness has been primarily attributed to its molecular mechanism of p53 activation and stabilization [114].

5.5.6 Hesperidin

Hesperidin is a flavonoid widely utilized in Chinese medicine and mostly derived from citrus species. While it is not harmful to healthy cells, it is highly effective against many cancer cells, for example, hepatocellular carcinoma, bladder, breast, and prostate cancers [115, 116]. Hisperidin's anti-cancer properties are attributed to the C4=C8 double bonds of the A and B rings, the 4-carbonyl of the C ring, and the 3'-*o*-hydroxy and 4'-*o*-methoxy moieties of the B ring, according to SAR studies. Hesperidin has been shown to have anti-cancer activities by promoting apoptosis and preventing cell invasion by controlling the expression of the MMP family and the EMT [117].

Hesperidin was shown to be more expressed in TNBC, MDA-MB231 cell lines, than in other breast cancer cells, MCF-7, according to research by Kongtawelert and colleagues [73]. By down-regulation of Akt and NF-κB signaling in TNBC, this chemical efficiently repressed proliferation of cells in MDA-MB231 cells and decreased the growth of breast cancer cells by PD-L1 expression suppression.

5.5.7 Curcumin

Curcumin is a highly bioactive compound that is present in the roots of the *Zingiberaceae* family plant Curcuma. It has a number of pharmacological characteristics, including anti-bacterial, anti-viral, anti-fungal, antioxidant, and anti-inflammatory properties. Having both phenolic -OH and -diketone moieties in a single compound gives curcumin an advantage over other antioxidants, which typically only have one functional group. Researchers surmised that curcumin's strong antioxidant content may be the reason for some of its biological effects, including its anti-cancer effects.

According to recent research, curcumin reduced the development of human glioblastoma cells *via* regulating a number of nuclear and cellular components. Curcumin has been reported to up-regulate the expression of multiple genes and proteins, such as p16, p21, and p53; Bcl-2; EIK-1; Erk enzyme; c-Jun N-terminal kinase; and caspase enzymes, such as caspase-3, -8, and -9. Additionally, it suppressed the expression of pRB, cyclin Dl, mammalian target of rapamycin (mTOR), NF-κB, Bcl-2, and p65 in malignant cells [118]. Wang et al.'s recent analysis [119] examined curcumin's anti-neoplastic properties against MCF-7 cells. According to the authors, curcumin demonstrated cytotoxic activity against MCF-7 cells by decreasing cell viability, inducing apoptosis, and enhancing caspase-3/9 actions. Furthermore, curcumin increased PTEN/Akt signaling and decreased expression of miR-21 in MCF-7 cells.

SAR analysis of several curcumin compounds has been studied in relation to cell lines that represent prostate cancer. The researchers discovered that in prostate cancer cells, the transcription of the androgen receptor (AR) was lowered by a few curcumin analogs. The conjugated β-diketone is necessary for curcumin's anti-AR property, according to the scientists' analysis of the SAR of curcumin analogs used as anti-AR reagents. The saturation or elimination of C=C bonds from the curcumin structure has reduced or eliminated the anti-AR effect of curcumin. Curcumin analogs' inhibitory activity was significantly increased when methoxycarbonyl ethyl was added to the central methylene motif and when the hydroxyl group in the phenyl ring was replaced with methoxy or methoxycarbonyl methoxy groups. However, the curcumin analogs' anti-AR activity remained unchanged when electron-withdrawing substitutions were made in the phenyl ring [120].

5.5.8 Cucurbitacin

Natural tetracyclic triterpenes, such as cucurbitacins, are found in several plant groups, including the Cruciferae and Cucurbitaceae. They demonstrated numerous pharmacological activities, for example, hepatoprotective, anti-inflammatory, antioxidant, anti-diabetic, and anti-cancer properties [121]. One of the most potent cucurbitacins, cucurbitacin B (CuB), has been utilized extensively for a variety of cancer treatments [122]. Cucurbitacin can successfully prevent the growth and invasion of several human malignancies, including glioma, breast and stomach cancer cell lines [123].

Cucurbitacin B (CuB) was investigated for its anti-cancer properties in CRC cell lines SW620 and HT29 by Chai et al. [124]. According to reports, CuB inhibited YAP expression, decreased invasion and proliferation, and down-regulated its downstream target genes, Cyr 61 and c-Myc, to induce apoptosis in CRC cells. Furthermore, CuB activated the Hippo-YAP signaling pathway for apoptosis induction and raised phospho-YAP in CRC cells by up-regulating LATS1 expression. Eight distinct cucurbitacins were tested for their anti-tumor properties against gastric cancer cells by Si and colleagues [125]. Cucurbitacin E, according to the authors, had the greatest impact on cellular proliferation and inhibited Akt activation in stomach cancer lines.

5.5.9 Safranal

The plant known as saffron, *Crocus sativus* L., has drawn much more interest from scientists due to its abundance of bioactive compounds, including safranal, crocin, carotenoids, and crocetin [126]. For human liver, lung, skin, colorectal, pancreatic, prostate, and breast cancers, safranal is indicated as a potent anti-cancer agent [127]. More recently, Jiang et al. [128] found that Safranal suppresses S-phase kinase-related protein 2 to prevent prostate cancer. Mechanistically, safranal increased the levels of cyclin-dependent kinase inhibitors, p21 and p27, through down-regulation of the G0/G1 cell cycle controlling proteins CDK2, CDK4, CDK6, and phospho-Rb. This prevented quiescent prostate cancer cells from being reactivated.

Another study used the HepG2 cell line to inspect the molecular mechanism through which safranal exhibits its anti-cancer activities on liver cancer. According to the study, safranal prevented DNA from mending itself and increased the risk of double-strand breaks in cancer cells' DNA. Furthermore, ER stress promoted safranal-induced apoptosis, and apoptosis happened in hepatocellular carcinoma cells [129].

5.5.10 Diosgenin

Numerous human cancer cells have shown anti-cancer capabilities for diosgenin, a steroidal sapogenin obtained from fenugreek seeds [130]. An electron-withdrawing ortho-substituted moiety or a simple phenyl moiety of the analogs linked to the original diosgenin was principally responsible for the substantial anti-proliferative effect, as per SAR investigation. Diosgenin was found to inhibit proliferation of cell in CCA by persuading cell cycle arrest and mitochondria-mediated apoptosis [131]. Diosgenin effectively stopped the development of breast cancer cells, raised JNK phosphorylation in breast cancer cells, and lowered fatty acid synthase expression by phosphorylating Akt and mTOR [132].

Liu et al. [133] reported that diosgenin inhibited the growth of breast cancer cells *via* down-regulating Skp2. By suppressing Skp2, this chemical has the capacity to reduce the viability and motility of cells and induction of apoptosis in breast cancer cells. Li and colleagues conducted a study in which they discovered [134] that diosgenin produced G2/M cell cycle arrest and death in hepatocellular carcinoma cells through the up-regulation of p21 and p27 levels and inhibiting Akt.

5.5.11 Licochalcone

The phenolic natural chemicals called chalconoids are obtained from the licorice plants Glycyrrhiza glabra and Glycyrrhiza inflata [135]. One chalconoid that is isolated in significant amounts from licorice is licochalcone A. Licochalcone A has been shown in prior studies to possess various biological characteristics, including anti-cancer, anti-bacterial, anti-leishmaniasis, anti-inflammatory, antioxidant, and anti-malarial properties [136]. It was found that certain synthetic analogs of licochalcones are more lethal to malignant cells compared to the naturally occurring compound. In comparison to Licochalcone A, trifluoromethyl-retrochalcone and bromo-retrochalcone exhibit strong anti-cancer action and can be employed as anti-neoplastic medications.

The anti-cancer activities of Licochalcone A against MCF-7 breast cancer cells were studied in 2018 by Xue et al. [137]. The authors noted that by inhibiting the phosphoinositide 3-kinase (PI3K)/ RAC-α serine-threonine-protein kinase (Akt)/ mTOR signaling pathway and activating the LC3-II signaling pathway, Licochalcone A markedly repressed the cell viability and increased apoptosis and autophagy. Furthermore, licochalcone A decreases the expression of B-cell lymphoma-2 and enhances caspase-3 activity. Therefore, Licochalcone A's possible ability to prevent breast cancer is due to its reduction of PI3K/Akt/mTOR activation, which in turn encourages MCF-7 cell death and autophagy.

Licochalcone A dramatically lowers the expression of c-IAP1, c-IAP2, XIAP, Survivin, c-FLIPL, and RIP1 in lung cancer cells without changing the mRNA level, according to a different study by the Liu group [138]. According to mechanistic investigations, licochalcone A reduces the activity of c-Jun N-terminal kinase (JNK) in lung cancer cells, increases ERK and p38 activation, and triggers cytoprotective autophagy.

5.5.12 Magnolol

One plant in the Magnolia family is *Magnolia officinalis*. The root, stem, and branch bark of this plant are all employed in Chinese herbal medicine. *Magnolia officinalis* bark contains two primary active chemicals that are extracted: magnolol and its structural counterpart, honokiol. These two phenolic compounds, magnolol, and honokiol, have demonstrated several pharmacological characteristics, viz. anti-inflammatory, antioxidant, and anti-cancer activities.

The position of one -OH moiety is the only difference between the two isomers, magnolol and honokiol. With only slight variations in activity, their biological characteristics were nearly identical. The intramolecular H-bond that forms between di-ortho-hydroxyl groups, catalyzes for radicals to extract the H atom from the -OH group, causes the activity difference [139]. Although both substances have confirmed remarkable anti-cancer activities, their actions in various cancer cell types vary [140].

Magnolol therapy substantially decreased G2/M progression and repressed proliferation of cell and colony formation in human NSCLC cells by the mitotic phase arrest, according to recent findings by Shen et al. [141]. They also noticed that magnolol therapy in NSCLC cells activated apoptosis in the malignant cells through a mechanism that was not dependent on p53, and autophagy induction happened by obstructing the Akt/mTOR pathway, which in turn inhibited the NSCLC.

5.5.13 Apigenin

Dietary flavonoid apigenin is extensively found in fruits and vegetables, including tea, oranges, chamomile, onions, and wheat sprouts. It has many biological properties, including the capability to hinder the development of several cancer cells. Apigenin treatment causes apoptosis and growth arrest in various cancer cell types by regulating several signaling pathways [142]. Because apigenin has the ability to block cancer-resistant proteins, it is thought that the -OH group in C-5 and the -OCH3 group in C-3 play a role. Additionally, a theoretical model investigating the anti-tumor properties of flavonoids suggests that 5,7-dimethoxy flavonoids have the most effectiveness.

Apigenin has been demonstrated in recent research to be effective against colon cancer. The administration of apigenin to colon cancer patients modulated many variables, including PI3K, AKT, ERK, STAT3, and JNK, leading to cell cycle arrest and death [143]. By lowering the levels of the pro-survival regulators Mcl-1, AKT, and ERK, apigenin may enhance the anti-cancer action of ABT-263 toward human colon cancer cells, according to research by Shao et al. [144].

5.5.14 Noscapine

A non-narcotic alkaloid called noscapine, a phthalide isoquinoline, was extracted from opium poppies and showed some antitussive, mild analgesic, and maybe anti-cancer properties. Noscapine has demonstrated strong anti-tumor properties against tumors of the bladder, breast, ovarian, malignant melanoma, and glioblastoma, according to recent investigations [145]. The halogenated, amino, nitro, and azido derivatives of noscapine derivatives were synthesized by changing the C-9 position of the isoquinoline ring of the noscapine moiety, and it was demonstrated that they exhibited more anti-cancer action than noscapine. O-alkylated and acylated noscapinoids, the derivatives obtained by altering the benzofuranone ring structure of noscapine, had better anti-cancer efficacy than noscapine. Noscapine attaches to tubulin efficiently, stopping cancer cells' mitosis.

Researchers discovered that giving noscapine orally significantly slows the development of tumors in cases of prostate cancer [146]. Three cell lines, including MCF-10F (a control breast cell line), MCF-7 (a luminal-like adenocarcinoma triple-positive breast cell line), and MDA-MB-231 (a breast cancer triple-negative cell line), were used in a different report [147] to examine the apoptotic effect of noscapine in breast cancer cell lines. Noscapine reduced the expression of the Bcl-xL gene and the Bcl-2 protein, leading breast cancer cells to apoptosis.

5.5.15 Harmine

Harmine, a β-carboline alkaloid, was extracted in 1847 using seeds from *Peganum harmala* and *Banisteriopsis caapi* [148]. It is also present in many therapeutic plants, and traditional medicine has long been used it in the Middle East and Asia. The compound's notable anti-cancer capabilities, for example, its capability to suppress proliferation, apoptosis induction, prevent tumorigenesis, and facilitate migration and invasion, have been corroborated by *in vitro* and *in vivo* research using harmine [149, 150]. According to a review of the literature, this substance may also prevent the spread of several malignancies, such as hepatic, lung, breast, and gastric cancers [149, 150]. The results of the SAR studies indicated that: (1) harmine's cytotoxic activities were increased when an alkoxy substituent was added at position-7; (2) harmine's cytotoxicity and cell line specificity were affected by alkoxy chain length; (3) N^9-alkylated harmine analogs demonstrated significant cytotoxicity; and (4) N^2-alkylated β-carboline analogs demonstrated specific cytotoxicity.

Ding et al. [151] investigated harmine's anti-tumor effectiveness and underlying mechanisms in breast cancer cell lines. Harmine slowed down cellular migration and proliferation, prevented cancer growth, and encouraged apoptosis, according to the research. Moreover, the molecule reduced the expression of Bcl-2, p-Erk, TAZ, and p-Akt while raising the expression of Bax. TAZ overexpression in breast cancer cells reduced the anti-tumor properties of harmine mechanistically.

5.5.16 Piperine

Piperine is an alkaloid that is derived from lysine. Studies conducted on piperine have revealed various biological and pharmacological characteristics, for example, anti-inflammatory, antioxidant, anti-convulsive, liver-protective, and anti-cancer properties [152]. Three components comprise the structure of piperine: a conjugated double bond, a piperidine moiety, and a methylenedioxyphenyl ring. Through the induction of several biological processes that have anti-cancer properties, these active moieties induce apoptosis in cancer cells.

Investigation with piperine has shown that it induces apoptosis through increasing p53 signaling in lung cancer cells [153], inhibits proliferation and metastasis in breast cancer cell lines [154], causes cell cycle arrest and encourages autophagy in prostate cancer cells. Piperine also inhibited NF-κβ, c-Fos, ATF-2, and CREB34 enzymes in tumor cells [155]. Researchers also found that piperine inhibits rectal cancer by increasing ROS levels and, increases apoptosis [156] and preventing proliferation of cells in colorectal cancer cells [157].

5.5.17 Pterostilbene

Pterostilbene is a naturally occurring polyphenolic that may be found in blueberries and grapes. It shares structural and biological characteristics with resveratrol [158]. Compared to resveratrol, pterostilbene has a longer half-life and improved oral bioavailability. Pterostilbene, like resveratrol, demonstrated several pharmacological characteristics, such as, anti-inflammatory, antioxidant, and anti-tumor effects, including gastric, breast, colon, melanoma, and prostate cancers [159]. The SAR study found that adding smaller methoxy or dimethylamino methyl side chains to the 3-position increases the biological activity of pterostilbene, whereas adding larger dialkylamino methyl side chains decrease activity.

SK-BR-3, ER-positive MCF-7, and triple-negative MDA-MB-468 cells were the breast cancer cell lines used by Wakimoto et al. [160] to examine the anti-cancer effect of pterostilbene. Pterostilbene demonstrated anti-proliferative properties, repressed the development of MDA-MB-468 cells, and promoted apoptosis by arresting cell cycle progression at the G0/G1 stage. Pterostilbene inhibited the phosphorylation of AKT, which was followed by the up-regulation of BAX. It also induced the start of extracellular signal-regulated kinase and the up-regulation of p21. Based on this finding, the author proposed that pterostilbene would be a viable option for treating triple-negative breast cancer in humans.

Pterostilbene-activated ERS in different research to show anti-cancer efficacy against NSCLC [161]. In NSCLC cell lines, pterostilbene increased ERS signaling, reducing cell viability and triggering apoptosis. Zhang's group found that the DNA repair pathway facilitated *via* topoisomerase 1 (Topl)/tyrosyl-DNA phosphodiesterase 1 (Tdpl) is highly successful in treating colorectal cancer [162]. Pterostilbene suppressed the DNA damage repair pathway as well as the activity of the Topl enzyme.

5.5.18 Lycopene

Lycopene has 40 C atoms connected by conjugated double bonds, and is a member of the carotenes class of chemical compounds. It is a primary constituent of fruits and vegetables that are red in color and a significant carotenoid found in human plasma [163]. In relation to stomach, lung, prostate, and breast malignancies, lycopene demonstrated anti-inflammatory, antioxidant, and anti-cancer activities [164]. The anti-cancer properties of lycopene are facilitated *via* its antioxidant capacity. Lycopene's special conjugated polyene structure gives it its red color and antioxidant characteristics.

The anti-inflammatory activities of lycopene were examined in relation to SW480 colorectal malignant cells by Cha et al. [165]. In SW480 colorectal cancer cells, lycopene inhibited the production of TNF-α, IL-1β, IL-6, COX-2, and iNOS while blocking NF-κB and activating the inflammatory-associated JNK.

Structures of some important phytochemicals are given in Figure 5.4.

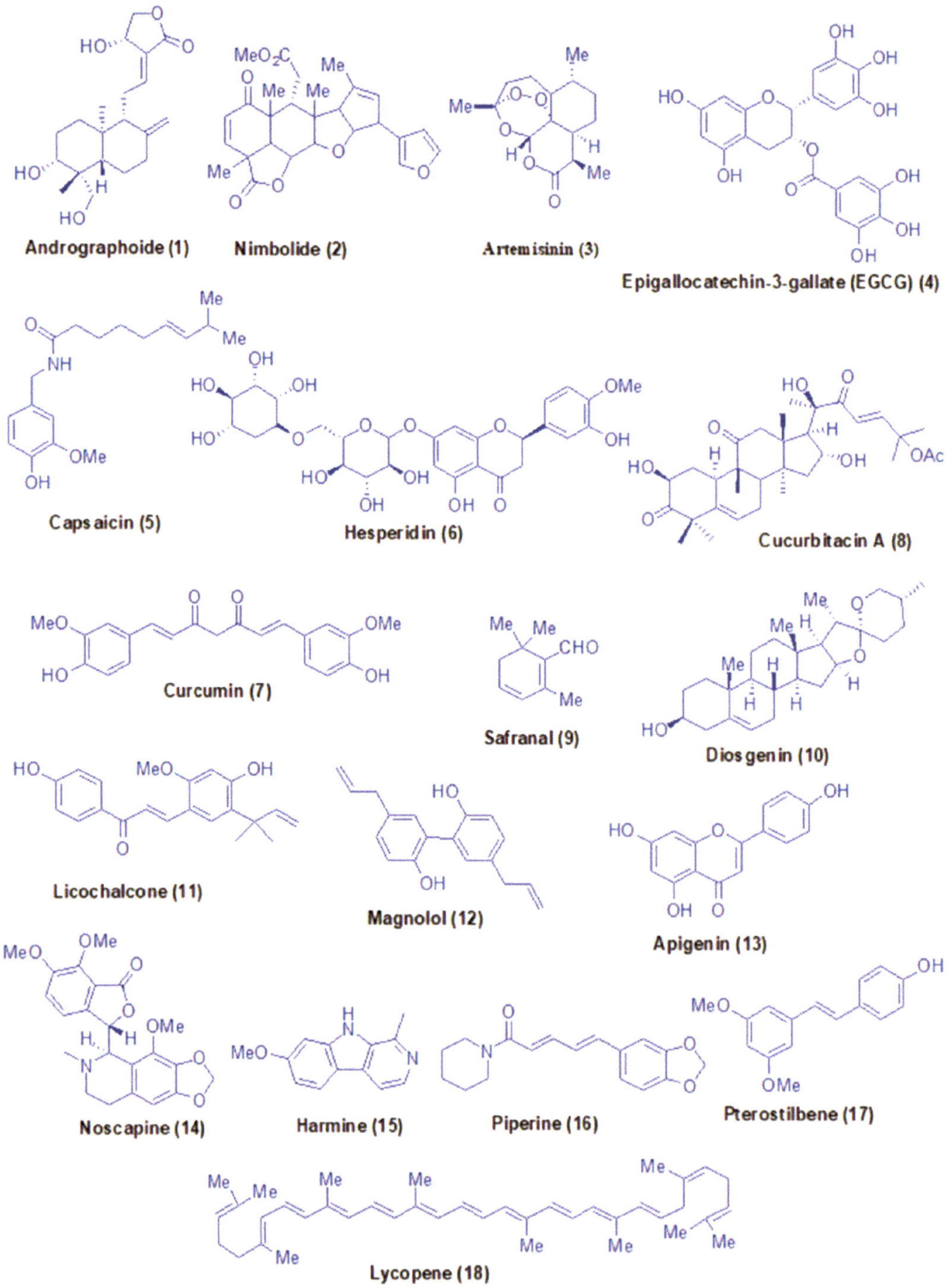

FIGURE 5.4 Structures of some important phytochemicals for cancer therapy.

TABLE 5.6
Phytochemical-Mediated Toxic Effects

Name of Plant Species	Compounds	Toxic Effects	References
Datura stramonium	Scopolamine	Respiratory failure	[168]
Gutierrezia sarothrae	Saponin	Causes abortion	[169]
Nicotiana tabacum	Nicotine	Vomiting, collapse slow pulse	[170]
***Dieffenbachia* sp.**	Asparagine	Damage to the cornea of the eye	[171]

5.6 PHYTOCHEMICALS OR PLANT-DERIVED COMPOUNDS AND THEIR SIDE EFFECTS

Nevertheless, research has shown that humans absorb these phytochemicals poorly. Certain phytochemicals are rapidly eliminated from the body due to limited absorption [166]. Furthermore, some phytochemicals are absorbed by healthy tissue or dispersed throughout the body's other main organs, where they further accumulate and unintentionally cause toxicity. Serious harm may result from the phytochemicals' unwanted biological activity. Numerous phytochemicals have harmful effects on humans as well as animals, making them function as phytotoxins. Certain phytochemicals affect the heart, the CNS, or the gastrointestinal tract. Furthermore, a few phytochemicals have been found to cause skin irritation when exposed to UV light [167]. Moreover, some phytochemicals released from the pollen of ryegrasses, birch, and hazel trees induce hay fever. Finally, it has been noted that some toxins from partially cooked beans can cause food poisoning (Table 5.6).

5.7 CONCLUSIONS AND FUTURE PROSPECTS

Plant-derived bioactive compounds or phytochemicals are now a valuable resource for developing novel therapeutics and treatment strategies. Consuming a wide range of plant-based foods can help lower the chance of developing chronic illnesses like cancer. Numerous synthetic pharmaceuticals have been discovered by researchers to cure cancer; however, synthetic anti-cancer medications are expensive and come with serious side effects including anemia, damage to important organs, and loss of hair and nails. Investigating plant-derived bioactive compounds for anti-cancer treatment has revealed diverse bioactive substances with various modes of action, therapeutic potential, and clinical uses. Plant-derived bioactive compounds, ranging from alkaloids and polyphenols to terpenoids and flavonoids, demonstrate diverse impacts on cellular processes, molecular targets, and cancer hallmarks. In some cases, to maximize the efficacy of chemotherapy medications while diminishing their side effects, plant-derived bioactive compounds work in concert with them. This presents a great opportunity to create innovative treatment approaches in the field of cancer biology research. However, there are inherent difficulties in translating plant-derived bioactive compounds into clinical practice, such as restrictions on bioavailability, safety profiles, drug interactions, and regulatory considerations. To maximize therapeutic efficacy, reduce hazards, and improve patient outcomes, thorough assessment, monitoring, and management of plant-derived bioactive compounds require interdisciplinary cooperation, rigorous clinical studies, standardized methodologies, and regulatory frameworks.

The majority of the supporting data demonstrated many advantages of developing anti-cancer drugs from medicinal plants, such as their exceptional anti-cancer activity and non-toxic side effects. For instance, curcumin and paclitaxel are very good phytochemicals used in clinical cancer therapy as natural anti-cancer medications. Moreover, earlier research has shown that phytochemicals can dramatically slow down the development of several cancers, including liver, breast, colorectal, lung, and others. This could be partially explained by their capacity to change the expression level of nucleic acids, such as miRNAs, which in turn increases the susceptibility of tumor cells to chemotherapeutic agents, and suppress the expression of specific signaling pathway-associated proteins that induce cancer or participate in immune response escape. Analogs of bioactive compounds originating from plants have a range of effects on several molecular targets of malignant cells, according to SAR research. These analogs may be developed into secure, efficient medications to treat various cancer forms. The world needs to rely more on natural resources and increase research on plant-derived bioactive compounds to find cost-effective anti-cancer drugs with minimal side effects. Since many medicinal plants have not yet been thoroughly investigated and can provide a variety of chemicals with varied structural features and biological efficacy for developing novel medications, further study is required in the quest for active anti-cancer drugs. The future of targeted cancer treatment looks quite promising in this research area. The improved biodegradability and biocompatibility of these phytochemicals have increased their effectiveness in the treatment of cancer. By looking at SAR, we can unleash the immense potential of plant-derived compounds to develop innovative and potent cancer treatments.

ACKNOWLEDGMENTS

S. Adhikari acknowledges support from Govt. Degree College (Dharmanagar, Tripura(N)-799253, India).

CONFLICT OF INTEREST

The authors declare no conflicts of interest.

REFERENCES

1. Siegel, R.L., Miller, K.D., Wagle, N.S., and Jemal, A. 2023. Cancer statistics, 2023. *CA: A Cancer Journal for Clinicians*, 73:17–48.
2. Cao, W., Chen, H.-D., Yu, Y.-W., Li, N., and Chen, W.-Q. 2021. Changing profiles of cancer burden worldwide and in China: A secondary analysis of the global cancer statistics 2020. *Chinese Medical Journal*, 134:783.
3. Compton, C., and Compton, C. 2020. *Cancer initiation, promotion, and progression and the acquisition of key behavioral traits* (pp. 25–48). Springer International Publishing.
4. Nath, P., Datta, A., Sen, T., and Adhikari, S. 2024. Emergence of metal-based anticancer therapeutics: A promising perspective. In *Biomarkers in Cancer Detection and Monitoring of Therapeutics* (pp. 411–450), Academic Press.
5. Adhikari, S., Nath, P., Das, A., Datta, A., Baildya, N., Duttaroy, A.K., and Pathak, S. 2024. A review on metal complexes and its anti-cancer activities: Recent updates from in vivo studies. *Biomedicine & Pharmacotherapy*, 171:116211.
6. Nath, P., Datta, A., and Adhikari, S. 2022. Recent advances of metal-based anticancer agents and their in vivo potential against various types of malignancies. *Handbook of Animal Models and Its Uses in Cancer Research*, 28:917–943.
7. Das, A., Deka, D., Baildya, N., Banerjee, A., Bisgin, A., Adhikari, S., Duttaroy, A.K., and Pathak, S. 2023. BMAP-27 peptide reduces proliferation and increases apoptosis in primary and metastatic colon cancer cell lines. *International Journal of Peptide Research and Therapeutics*, 29:100.
8. Das, A., Adhikari, S., Deka, D., Bisgin, A., Paul, S., Balidya, N., Boga, I., Banerjee, A., and Pathak, S. 2023. An updated review on recent advances in the usage of novel therapeutic peptides for breast cancer treatment. *International Journal of Peptide Research and Therapeutics*, 29:32.
9. Bhattacharjee, T., Adhikari, S., Sheikh, A.H., Mahmoudi, G., Mlowe, S., Akerman, M.P., Choudhury, N.A., Chakraborty, S., Butcher, R.J., Kennedy, A.R., and Demir, B.S. 2022. Syntheses, crystal structures, theoretical studies, and anticancer properties of an unsymmetrical Schiff base ligand N-2-(6-methylpyridyl)-2-hydroxy-1-naphthaldimine and its Ni (II) complex. *Journal of Molecular Structure*, 1269:133717.
10. Bhattacharjee, T., Adhikari, S., Bhattacharjee, S., Debnath, S., Das, A., Daniliuc, C.G., and Thirumoorthy, K. 2022. Exploring dithiolate-amine binary ligand systems for the supramolecular assemblies of Ni (II) coordination compounds: Crystal structures, theoretical studies, cytotoxicity studies, and molecular docking studies. *Inorganica Chimica Acta*, 543:121157.

11. Adhikari, S., Sheikh, A.H., Baildya, N., Mahmoudi, G., Choudhury, N.A., Okpareke, O., Sen, T., Verma, A.K., Singh, R.K., Pathak, S., and Kaminsky, W. 2023. Antiproliferative evaluation and supramolecular properties of a Pd (II) complex harvested from benzil bis (pyridyl hydrazone) ligand: Combined experimental and theoretical studies. *Inorganic Chemistry Communications*, 152:110646.
12. Soerjomataram, I., and Bray, F. 2021. Planning for tomorrow: Global cancer incidence and the role of prevention 2020–2070. *Nature Reviews Clinical Oncology*, 18:663–672.
13. Singh, M.K., Sutradhar, S., Paul, B., Adhikari, S., Laskar, F., Acharya, S., Chakraborty, D., Biswas, S., Das, A., Roy, S., and Frontera, A. 2018. Mixed-ligand complexes of zinc (II) with 1, 1-dicyanoethylene-2, 2-dithiolate and N-donor ligands: A combined experimental and theoretical study. *Journal of Molecular Structure*, 1164:334–343.
14. Das, A., Adhikari, S., Deka, D., Baildya, N., Sahare, P., Banerjee, A., Paul, S., Bisgin, A., and Pathak, S. 2023. An updated review on the role of nanoformulated phytochemicals in colorectal cancer. *Medicina*, 59:685.
15. Storz, M.A. 2020. Will the plant-based movement redefine physicians' understanding of chronic disease? *The New Bioethics*, 26:141–157.
16. Najmi, A., Javed, S.A., Al Bratty, M., and Alhazmi, H.A. 2022. Modern approaches in the discovery and development of plant-based natural products and their analogues as potential therapeutic agents. *Molecules*, 27:349.
17. Ijaz, S., Akhtar, N., Khan, M.S., Hameed, A., Irfan, M., Arshad, M.A., Ali, S., and Asrar, M. 2018. Plant derived anticancer agents: A green approach towards skin cancers. *Biomedicine & Pharmacotherapy*, 103:1643–1651.
18. Egbuna, C., Kumar, S., Ifemeje, J.C., Ezzat, S.M., and Kaliyaperumal, S. eds. 2019. *Phytochemicals as lead compounds for new drug discovery*, Elsevier.
19. Alamgir, A.N.M., and Alamgir, A.N.M. 2018. Phytoconstituents—Active and inert constituents, metabolic pathways, chemistry and application of phytoconstituents, primary metabolic products, and bioactive compounds of primary metabolic origin. *Therapeutic Use of Medicinal Plants and Their Extracts: Phytochemistry and Bioactive Compounds*, 2:25–164.
20. Lin, X., Peng, Z., and Su, C. 2015. Potential anti-cancer activities and mechanisms of costunolide and dehydrocostuslactone. *International Journal of Molecular Sciences*, 16:10888–10906.
21. Li, G., and Lou, H.X. 2018. Strategies to diversify natural products for drug discovery. *Medicinal Research Reviews*, 38:1255–1294.
22. Catalano, E. 2016. Role of phytochemicals in the chemoprevention of tumors. *arXiv preprint arXiv, 1605.04519.*
23. Garcia-Oliveira, P., Otero, P., Pereira, A.G., Chamorro, F., Carpena, M., Echave, J., Fraga-Corral, M., Simal-Gandara, J., and Prieto, M.A. 2021. Status and challenges of plant-anticancer compounds in cancer treatment. *Pharmaceuticals*, 14:157.
24. Wang, L.M., and Ren, D.M. 2010. Flavopiridol, the first cyclin-dependent kinase inhibitor: Recent advances in combination chemotherapy. *Mini-Reviews in Medicinal Chemistry*, 10:1058–1070.
25. Chunarkar-Patil, P., Kaleem, M., Mishra, R., Ray, S., Ahmad, A., Verma, D., Bhayye, S., Dubey, R., Singh, H.N., and Kumar, S. 2024. Anticancer drug discovery based on natural products: From computational approaches to clinical studies. *Biomedicines*, 12:201.
26. Newman, D.J., and Cragg, G.M. 2016. Natural products as sources of new drugs from 1981 to 2014. *Journal of Natural Products*, 79:629–661.
27. Sharma, A., Sharma, S., Kumar, A., Kumar, V., and Sharma, A.K. 2022. Plant secondary metabolites: An introduction of their chemistry and biological significance with physicochemical aspect. In *Plant Secondary Metabolites: Physico-Chemical Properties and Therapeutic Applications* (pp. 1–45), Springer Nature Singapore.
28. Park, I.W., Yoo, J., Adhikari, S., Park, J.S., Sessler, J.L., and Lee, C.H. 2012. Calix [4] pyrrole-based heteroditopic ion-pair receptor that displays anion-modulated, cation-binding behavior. *Chemistry–A European Journal*, 18:15073–15078.
29. Park, I.W., Yoo, J., Kim, B., Adhikari, S., Kim, S.K., Yeon, Y., Haynes, C.J., Sutton, J.L., Tong, C.C., Lynch, V.M., and Sessler, J.L. 2012. Oligoether-strapped calix [4] pyrrole: An ion-pair receptor displaying cation-dependent chloride anion transport. *Chemistry–A European Journal*, 18:2514–2523.
30. Adhikari, S., Bhattacharjee, T., Nath, P., Das, A., Jasinski, J.P., Butcher, R.J., and Maiti, D. 2020. Bimetallic and trimetallic Cd (II) and Hg (II) mixed-ligand complexes with 1, 1-dicyanoethylene-2, 2-dithiolate and polyamines: Synthesis, crystal structure, Hirshfeld surface analysis, and antimicrobial study. *Inorganica Chimica Acta*, 512:119877.

31. Bhattacharjee, T., Adhikari, S., Datta, A., Daniliuc, C.G., Montazerozohori, M., Naghiha, R., and Hayati, P. 2022. Cadmium (II) coordination polymer based on flexible dithiolate-polyamine binary ligands system: Crystal structure, Hirshfeld surface analysis, antimicrobial, and DNA cleavage potential. *Polyhedron*, 211:115544.
32. Adhikari, S., Bhattacharjee, T., Butcher, R.J., Porchia, M., De Franco, M., Marzano, C., Gandin, V., and Tisato, F. 2019. Synthesis and characterization of mixed-ligand Zn (II) and Cu (II) complexes including polyamines and dicyano-dithiolate (2-): In vitro cytotoxic activity of Cu (II) compounds. *Inorganica Chimica Acta*, 498:119098.
33. Wang, S., Dong, G., and Sheng, C. 2019. Structural simplification of natural products. *Chemical Reviews*, 119:4180–4220.
34. Mazumder, K., Aktar, A., Roy, P., Biswas, B., Hossain, M.E., Sarkar, K.K., Bachar, S.C., Ahmed, F., Monjur-Al-Hossain, A.S.M., and Fukase, K. 2022. A review on mechanistic insight of plant derived anticancer bioactive phytocompounds and their structure activity relationship. *Molecules*, 27:3036.
35. Goswami, N., Sethi, P., Jyoti, A., Nagar, G., Gupta, S.R., Singh, A., and Singh, I.K. 2022. Plant-derived bioactive compounds in neuroblastoma therapeutics: Current outlook and future perspective. *Chemical Biology Letters*, 9:400.
36. Vázquez, J., López, M., Gibert, E., Herrero, E., and Luque, F.J. 2020. Merging ligand-based and structure-based methods in drug discovery: An overview of combined virtual screening approaches. *Molecules*, 25:4723.
37. Bennouna, J., Delord, J.-P., Campone, M., and Nguyen, L. 2008. Vinflunine: A new microtubule inhibitor agent. *Clinical Cancer Research*, 14:1625–1632.
38. Thomford, N.E., Senthebane, D.A., Rowe, A., Munro, D., Seele, P., Maroyi, A., and Dzobo, K. 2018. Natural products for drug discovery in the 21st century: Innovations for novel drug discovery. *International Journal of Molecular Sciences*, 19:1578.
39. Ramos, P., and Bentires-Alj, M. 2015. Mechanism-based cancer therapy: Resistance to therapy, therapy for resistance. *Oncogene*, 34:3617–3626. https://doi.org/10.1038/onc.2014.314
40. Shigematsu, H., Lin, L., Takahashi, T., Nomura, M., Suzuki, M., Wistuba, I.I., Fong, K.M., Lee, H., Toyooka, S., and Shimizu, N. 2005. Clinical and biological features associated with epidermal growth factor receptor gene mutations in lung cancers. *JNCI*, 97:339–346. https://doi.org/10.1093/jnci/dji055
41. Reguart, N., and Remon, J. 2015. Common EGFR-mutated subgroups (Del19/L858R) *in advanced non-small-cell lung cancer: Chasing better outcomes with tyrosine kinase inhibitors. Future Oncology*, 11:8. https://doi.org/10.2217/fon.15.15
42. Lee, C.K., Wu, Y.L., Ding, P.N., Lord, S.J., Inoue, A., Zhou, C., Mitsudomi, T., Rosell, R., Pavlakis, N., and Links, M. 2015. Impact of specific epidermal growth factor receptor (EGFR) mutations and clinical characteristics on outcomes after treatment with EGFR tyrosine kinase inhibitors versus chemotherapy in EGFR-mutant lung cancer: A meta-analysis. *Journal of Clinical Oncology*, 33(17):1958–1965. http://doi.org/10.1200/JCO.2014.58.1736
43. Bean, J., Brennan, C., Shih, J.Y., Riely, G., Viale, A., Wang, L., Chitale, D., Motoi, N., Szoke, J., and Broderick, S. 2007. MET amplification occurs with or without T790M mutations in EGFR mutant lung tumors with acquired resistance to gefitinib or erlotinib. *Proceedings of the National Academy of Sciences of the United States of America*,104(52):20932–20937. https://doi.org/10.1073/pnas.0710370104
44. Paik, P.K., Arcila, M.E., Fara, M., Sima, C.S., Miller, V.A., Kris, M.G., Ladanyi, M., and Riely, G.J. 2011. Clinical characteristics of patients with lung adenocarcinomas harboring BRAF mutations. *Journal of Clinical Oncology*, 29(15):2046–2051. https://doi.org/10.1200/JCO.2010.33.1280
45. Giantonio, B.J., Catalano, P.J., Meropol, N.J., O'Dwyer, P.J., Mitchell, E.P., Alberts, S.R., Schwartz, M.A., and Benson, A.B. 2007. Bevacizumab in combination with oxaliplatin, fluorouracil, and leucovorin (FOLFOX4) for previously treated metastatic colorectal cancer: Results from the Eastern cooperative oncology group study E3200. *Journal of Clinical Oncology*, 25(12):1539–1544. https://doi.org/10.1200/JCO.2006.09.6305
46. Cunningham, D., Humblet, Y., Siena, S., Khayat, D., Bleiberg, H., Santoro, A., Bets, D., Mueser, M., Harstrick, A., and Verslype, C. 2004. Cetuximab monotherapy and cetuximab plus irinotecan in irinotecan-refractory metastatic colorectal cancer. *New England Journal of Medicine*, 351:337–345. https://doi.org/10.1056/NEJMoa033025
47. Van Cutsem, E., Köhne, C.-H., Hitre, E., Zaluski, J., Chang Chien, C.-R., Makhson, A., D'Haens, G., Pintér, T., Lim, R., and Bodoky, G. 2009. Cetuximab and chemotherapy as initial treatment for metastatic colorectal cancer. *New England Journal of Medicine*, 360:1408–1417. https://doi.org/10.1056/NEJMoa0805019

48. Inman, G.J. 2011. Switching TGF from a tumor suppressor to a tumor promoter. *Current Opinion in Genetics and Development*, 21:93–99. https://doi.org/10.1016/j.gde.2010.12.004
49. Neuzillet, C., Tijeras-Raballand, A., Cohen, R., Cros, J., Faivre, S., Raymond, E., and de Gramont, A. 2015. Targeting the TGF pathway for cancer therapy. *Pharmacology*, 147:22–31. https://doi.org/10.1016/j.pharmthera.2014.11.001
50. Herbertz, S., Sawyer, J.S., Stauber, A.J., Gueorguieva, I., Driscoll, K.E., Estrem, S.T., Cleverly, A.L., Desaiah, D., Guba, S.C., and Benhadji, K.A. 2015. Clinical development of galunisertib (Ly2157299 monohydrate), a small molecule inhibitor of transforming growth factor-beta signaling pathway. *Drug Design Development Therapy*, 9:4479–4499. https://doi.org/10.2147/DDDT.S86621
51. Crawford, E.D., Shore, N.D., Petrylak, D.P., Higano, C.S., and Ryan, C.J. 2017. Abiraterone acetate and prednisone in chemotherapy-naïve prostate cancer patients: Rationale, evidence and clinical utility. *Therapeutic Advances in Medical Oncology*, 9(5):319–333. https://doi.org/10.1177/1758834017698644
52. Prasad, V., De Jesús, K., and Mailankody, S. 2017. The high price of anticancer drugs: Origins, implications, barriers, solutions. *Nature Reviews Clinical Oncology*, 14:381–390. https://doi.org/10.1038/nrclinonc.2017.31
53. Gao, Q., Feng, J., Liu, W., Wen, C., Wu, Y., Liao, Q., Zou, L., Sui, X., Xie, T., Zhang, J., and Hu, Y. 2022. Opportunities and challenges for co-delivery nanomedicines based on combination of phytochemicals with chemotherapeutic drugs in cancer treatment. *Advanced Drug Delivery Reviews*, 188:114445. https://doi.org/10.1016/j.addr.2022.114445
54. Zhang, J., Chen, Z., Huang, X., Shi, W., Zhang, R., Chen, M., Huang, H., and Wu, L. 2019. Insights on the multifunctional activities of magnolol. *BioMed Research International*, 2019:1847130.
55. Rasheduzzaman, M., Jeong, J.K., and Park, S.Y. 2018. Resveratrol sensitizes lung cancer cell to TRAIL by p53 independent and suppression of Akt/NF-kappaB signaling. *Life Sciences*, 208:208–220.
56. Mani, R., and Natesan, V. 2018. Chrysin: Sources, beneficial pharmacological activities, and molecular mechanism of action. *Phytochemistry*, 145:187–196.
57. Yan, Y., Li, J., Han, J., Hou, N., Song, Y., and Dong, L. 2015. Chlorogenic acid enhances the effects of 5-fluorouracil in human hepatocellular carcinoma cells through the inhibition of extracellular signal-regulated kinases. *Anticancer Drugs*, 26(5):540–546.
58. Zhang, Z., Xu, S., Wang, Y., Yu, Y., Li, F., Zhu, H., Shen, Y., Huang, S., and Guo, S. 2018. Near infrared triggered co-delivery of doxorubicin and quercetin by using gold nanocages with tetradecanol to maximize anti-tumor effects on MCF-7/ADR cells. *Journal of Colloid and Interface Science*, 509:47–57.
59. Amoodizaj, F.F., Baghaeifar, S., Taheri, E., Jadid, M.F.S., Safi, M., Sani, N.S., Hajazimian, S., Isazadeh, A., and Shanehbandi, D. 2020. Enhanced anticancer potency of doxorubicin in combination with curcumin in gastric adenocarcinoma. *Journal of Biochemical Molecular Toxicology*, 34:e22486.
60. Ong, C.P., Lee, W.L., Tang, Y.Q., and Yap, W.H. 2019. Honokiol: A review of its anticancer potential and mechanisms. *Cancers (Basel)*, 12:48.
61. Zhao, Q., Wang, J., Zou, M.J., Hu, R., Zhao, L., Qiang, L., Rong, J.J., You, Q.D., and Guo, Q.L. 2010. Wogonin potentiates the antitumor effects of low dose 5-fluorouracil against gastric cancer through induction of apoptosis by down-regulation of NF-kappaB and regulation of its metabolism. *Toxicology Letters*, 197:201–210.
62. Halder, A., Jethwa, M., Mukherjee, P., Ghosh, S., Das, S., Helal Uddin, A.B.M., Mukherjee, A., Chatterji, U., and Roy, P. 2020. Lactoferrin tethered betulinic acid nanoparticles promote rapid delivery and cell death in triple negative breast and laryngeal cancer cells. *Artificial Cells Nanomedicine, and Biotechnology*, 48:1362–1371.
63. Amirsaadat, S., Pilehvar-Soltanahmadi, Y., Zarghami, F., Alipour, S., Ebrahimnezhad, Z., and Zarghami, N. 2017. Silibinin-loaded magnetic nanoparticles inhibit hTERT gene expression and proliferation of lung cancer cells. *Artificial Cells Nanomedicine, and Biotechnology*, 45:1649–1656.
64. Jagwani, S., Jalalpure, S., Dhamecha, D., Jadhav, K., and Bohara, R. 2022. Pharmacokinetic and phar macodynamic evaluation of resveratrol loaded cationic liposomes for targeting hepato cellular carcinoma. *International Journal of Pharmacy*, 618:121605.
65. Kumar, C.S., Thangam, R., Mary, S.A., Kannan, P.R., Arun, G., and Madhan, B. 2020. Targeted delivery and apoptosis induction of trans-resveratrol-ferulic acid loaded chitosan coated folic acid conjugate solid lipid nanoparticles in colon cancer cells. *Carbohydrate Polymers*, 231:115682.
66. Ge, P., Niu, B., Wu, Y., Xu, W., Li, M., and Sun, H. 2020. Enhanced cancer therapy of celastrol in vitro and in vivo by smart dendrimers delivery with specificity and biosafety. *Journal of Chemical Engineering*, 383:123228.
67. Dandawate, P.R., Subramaniam, D., Jensen, R.A., and Ananta, S. 2016. Targeting cancer stem cells and signaling pathways by phytochemicals: Novel approach for breast cancer therapy. *Seminars in Cancer Biology*, 40–41:192:208. https://doi.org/10.1016/j.semcancer.2016.09.001

68. Pandey, P., Khan, F., Oves, M., Alzahrani, F.A., and Qari, H.A. 2021. A novel approach to unraveling the apoptotic potential of rutin (bioflavonoid) via targeting *Jab1* in cervical cancer cells. *Molecules*, 26:5529. https://doi.org/10.3390/molecules26185529
69. Dalpatraj, N., Naik, A., and Thakur, N. 2023. Combination treatment of a phytochemical and a histone demethylase inhibitor—A novel approach towards targeting TGF-induced EMT, invasion, and migration in prostate cancer. *International Journal of Molecular Sciences*, 24:1860. https://doi.org/10.3390/ijms24031860
70. Lee, J., Han, Y., Wang, W., Kim, H., Kim, S., Yang, K.M., Kim, S.J., Dhanasekaran, D.N., and Song, Y.S. 2021. Phytochemicals in cancer immune checkpoint inhibitor therapy. *Biomolecules*, 11:1107. https://doi.org/10.3390/biom11081107
71. Mazewski, C., Kim, M.S., and de Mejia, E.G. 2019. Anthocyanins, delphinidin-3-O-glucoside and cyanidin-3-O-glucoside, inhibit immune checkpoints in human colorectal cancer cells in vitro and in silico. *Scientific Reports*, 9:11560.
72. Tajuddin, W.N.B.W.M., Lajis, N.H., Abas, F., Othman, I., and Naidu, R. 2019. Mechanistic understanding of Curcumin's therapeutic effects in lung cancer. *Nutrients*, 11:2989.
73. Kongtawelert, P., Wudtiwai, B., Shwe, T.H., Pothacharoen, P., and Phitak, T. 2020. Inhibitory effect of hesperidin on the expression of programmed death ligand (PD-L1) in breast cancer. *Molecules*, 25:252.
74. Verdura, S., Cuyàs, E., Cortada, E., Brunet, J., Lopez-Bonet, E., Martin-Castillo, B., Bosch-Barrera, J., Encinar, J.A., and Menendez, J.A. 2020. Resveratrol targets PD-L1 glycosylation and dimerization to enhance antitumor T-cell immunity. *Aging (Albany NY)*, 12:8–34.
75. Kang, D.Y., Sp, N., Jo, E.S., Rugamba, A., Hong, D.Y., Lee, H.G., Yoo, J.-S., Liu, Q., Jang, K.-J., and Yang, Y.M. 2020. The inhibitory mechanisms of tumor PD-L1 expression by natural bioactive gallic acid in non-small-cell lung cancer (NSCLC) cells. *Cancers (Basel)*, 12:727.
76. Nigam, M., Mishra, A.P., Deb, V.K., Dimri, D.B., Tiwari, V., Bungau, S.G., Bungau, A.F., and Radu, A.F. 2023. Evaluation of the association of chronic inflammation and cancer: Insights and implications. *Biomedical and Pharmacotherapy*, 164:115015. https://doi.org/10.1016/j.biopha.2023.115015
77. Sood, E., Deb, V.K., Verma, U., Sharma, N., and kumar, A. 2018. Hepatitis b virus (HBV) DNA quantification by real time PCR. *International Journal of Pharma Research & Health Sciences*, 6:2468–2471.
78. Hung, D.T., Jamison, T.F., and Schreiber, S.L. 1996. Understanding and controlling the cell cycle with natural products. *Chemical Biology*, 3:623–639. https://doi.org/10.1016/S1074-5521(96)90129-5
79. Kuang, W., Hu, W., Ren, H., Shao, Y., and Liu, B. 2021. Plant derived coumestrol phytochemical targets human skin carcinoma cells by inducing mitochondrial-mediated apoptosis, cell cycle arrest, inhibition of cell migration and invasion and modulation of m-TOR/PI3K/AKT signalling pathway. *Saudi Journal of Biological Sciences*, 28:2739–2746. https://doi.org/10.1016/j.sjbs.2021.03.043
80. Wellington, K.W. 2015. Understanding cancer and the anticancer activities of naphthoquinones—a review. *RSC Advances*, 5:20309–20338. https://doi.org/10.1039/C4RA13547D
81. Stangel, M., Zettl, U.K., Mix, E., Zielasek, J., Toyka, K.V., Hartung, H.P., and Gold, R. 1996. H_2O_2 and nitric oxide-mediated oxidative stress induce apoptosis in rat skeletal muscle myoblasts. *Journal of Neuropathology & Experimental Neurology*, 55:36–43. https://doi.org/10.1097/00005072-199601000-00004
82. Giordano, A., and Tommonaro, G. 2019. Curcumin and cancer. *Nutrients*, 11:2376. https://doi.org/10.3390/nu11102376.
83. Yang, H., Zhou, P., Huang, H., Chen, D., Ma, N., Cui, Q.C., Shen, S., Dong, W., Zhang, X., Lian, W., Wang, X., Dou, Q.P., and Liu, J. 2009. Shikonin exerts antitumor activity via proteasome inhibition and cell death induction in vitro and in vivo. *International Journal of Cancer*, 124:2450–2459. http://doi.org/10.1002/ijc.24195
84. Gara, R.K., Srivastava, V.K., Duggal, S., Bagga, J.K., Bhatt, M., Sanyal, S., and Mishra, D.P. 2015. Shikonin selectively induces apoptosis in human prostate cancer cells through the endoplasmic reticulum stress and mitochondrial apoptotic pathway. *Journal of Biomedical Science*, 22:26. https://doi.org/10.1186/s12929-015-0127-1
85. Liu, J., Wang, P., Xue, Y., Li, Z., Qu, C., and Liu, Y. 2015. Enhanced antitumor effect of shikonin by inhibiting endoplasmic reticulum stress via JNK/c-Jun pathway in human glioblastoma stem cells. *Biochemical and Biophysical Research Communications*, 466:103–110. https://doi.org/10.1016/j.bbrc.2015.08.115
86. Jang, S.Y., Hong, D., Jeong, S.Y., and Kim, J.-H. 2015. Shikonin causes apoptosis by up-regulating p73 and down-regulating ICBP90 in human cancer cells. *Biochemical and Biophysical Research Communications*, 465:71–76. https://doi.org/10.1016/j.bbrc.2015.07.131

87. Khaw, A.K., Sameni, S., Venkatesan, S., Kalthur, G., and Hande, M.P. 2015. Plumbagin alters telomere dynamics, induces DNA damage and cell death in human brain tumour cells. *Mutation Research: Genetic Toxicology and Environmental Mutagenesis*, 793:86–95. https://doi.org/10.1016/j.mrgentox.2015.06.004
88. Hwang, G.H., Ryu, J.M., Jeon, Y.J., Choi, J., Han, H.J., Lee, Y.-M., Lee, S., Bae, J.-S., Jung, J.-W., Chang, W., Kim, L.K., Jee, J.-G., and Lee, M.Y. 2015. The role of thioredoxin reductase and glutathione reductase in plumbagin-induced, reactive oxygen species-mediated apoptosis in cancer cell lines. *European Journal of Pharmacology*, 765:384–393. https://doi.org/10.1016/j.ejphar.2015.08.058
89. Da Silva, M.N., Ferreira, V.F., and De Souza, M.C.B.V. 2003. An overview of the chemistry and pharmacology of naphthoquinones with emphasis on β-lapachone and derivatives. *Quim Nova*, 26:407–416.
90. Krishnan, P., and Bastow, K.F. 2000. Novel mechanisms of DNA topoisomerase II inhibition by pyranonaphthoquinone derivatives—eleutherin, α lapachone, and β lapachone. *Biochemical Pharmacology*, 60:1367–1379. https://doi.org/10.1016/s0006-2952(00)00437-8
91. Chae, G.-H., Song, G.-Y., Kim, Y., Cho, H., Sok, D.-E., and Ahn, B.-Z. 1999. 2-or 6-(1-azidoalkyl)-5,8-dimethoxy-1,4-naphthoquinone: Synthesis, evaluation of cytotoxic activity; antitumor activity and inhibitory effect on DNA topoisomerase-I. *Archives of Pharmacal Research*, 22:507–514. https://doi.org/10.1007/bf02979161
92. Srinivasan, R., Aruna, A., Lee, J.S., Kim, M., Shivakumar, M.S., and Natarajan, D. 2020. Antioxidant and antiproliferative potential of bioactive molecules ursolic acid and thujone isolated from *Memecylon edule* and *Elaeagnus indica* and their inhibitory effect on topoisomerase II by molecular docking approach. *BioMed Research International*, *2020*. https://doi.org/10.1155/2020/8716927.
93. Khamphaya, T., Chansela, P., Piyachaturawat, P., Suksamrarn, A., Nathanson, M.H., and Weerachayaphorn, J. 2016. Effects of andrographolide on intrahepatic cholestasis induced by alpha-naphthylisothiocyanate in rats. *European Journal of Pharmacology*, 789:254–264.
94. Yang, S., Evens, A.M., Prachand, S., Singh, A.T., Bhalla, S., David, K., and Gordon, L.I. 2010. Mitochondrial-mediated apoptosis in lymphoma cells by the diterpenoid lactone andrographolide, the active component of Andrographis paniculata. *Clinical Cancer Research*, 16:4755–4768.
95. Chen, C.C., Wu, M.L., Doerksen, R.J., Ho, C.T., and Huang, T.C. 2015. Andrographolide induces apoptosis via down-regulation of glyoxalase 1 and HMG-CoA reductase in HL-60 cells. *Journal of Functional Foods*, 14:226–235.
96. Khan, I., Khan, F., Farooqui, A., and Ansari, I.A. 2018. Andrographolide exhibits anticancer potential against human colon cancer cells by inducing cell cycle arrest and programmed cell death via augmentation of intracellular reactive oxygen species level. *Nutrition and Cancer*, 70:787–803.
97. Peng, Y., Wang, Y., Tang, N., Sun, D., Lan, Y., Yu, Z., and Lv, C. 2018. Andrographolide inhibits breast cancer through suppressing COX-2 expression and angiogenesis via inactivation of p300 signaling and VEGF pathway. *Journal of Experimental & Clinical Cancer Research*, 37:1–14.
98. Mir, H., Kapur, N., Singh, R., Sonpavde, G., Lillard, J.W. Jr, and Singh, S. 2016. Andrographolide inhibits prostate cancer by targeting cell cycle regulators, CXCR3 and CXCR7 chemokine receptors. *Cell Cycle*, 15:819–826.
99. Chun, J.Y., Tummala, R., Nadiminty, N., Lou, W., Liu, C., Yang, J., and Gao, A.C. 2010. Andrographolide, an herbal medicine, inhibits interleukin-6 expression and suppresses prostate cancer cell growth. *Genes & Cancer*, 1:868–876.
100. Forestier-Román, I.S., López-Rivas, A., Sánchez-Vázquez, M.M., Rohena-Rivera, K., Nieves-Burgos, G., Ortiz-Zuazaga, H., and Martínez-Ferrer, M. 2019. Andrographolide induces DNA damage in prostate cancer cells. *Oncotarget*, 10:1085.
101. Subramani, R., Gonzalez, E., Arumugam, A., Nandy, S., Gonzalez, V., Medel, J., and Lakshmanaswamy, R. 2016. Nimbolide inhibits pancreatic cancer growth and metastasis through ROS-mediated apoptosis and inhibition of epithelial-to-mesenchymal transition. *Scientific Reports*, 6:19819.
102. Gupta, S.C., Prasad, S., Sethumadhavan, D.R., Nair, M.S., Mo, Y.Y., and Aggarwal, B.B. 2013. Nimbolide, a limonoid triterpene, inhibits growth of human colorectal cancer xenografts by suppressing the proinflammatory microenvironment. *Clinical Cancer Research*, 19:4465–4476.
103. Tu, Y. 2016. Artemisinin—a gift from traditional Chinese medicine to the world (Nobel lecture). *Angewandte Chemie International Edition*, 55:10210–10226.
104. Fröhlich, T., Ndreshkjana, B., Muenzner, J.K., Reiter, C., Hofmeister, E., Mederer, S., and Tsogoeva, S.B. 2017. Synthesis of novel hybrids of thymoquinone and artemisinin with high activity and selectivity against colon cancer. *ChemMedChem*, 12:226–234.
105. Wang, J., Zhang, J., Shi, Y., Xu, C., Zhang, C., Wong, Y.K., and Lin, Q. 2017. Mechanistic investigation of the specific anticancer property of artemisinin and its combination with aminolevulinic acid for enhanced anticolorectal cancer activity. *ACS central Science*, 3:743–750.

106. Yu, C., Sun, P., Zhou, Y., Shen, B., Zhou, M., Wu, L., and Kong, M. 2019. Inhibition of AKT enhances the anti-cancer effects of artemisinin in clear cell renal cell carcinoma. *Biomedicine & Pharmacotherapy*, 118:109383.
107. Min, K.J., and Kwon, T.K. 2014. Anticancer effects and molecular mechanisms of epigallocatechin-3-gallate. *Integrative Medicine Research*, 3:16–24.
108. Senggunprai, L., Kukongviriyapan, V., Prawan, A., and Kukongviriyapan, U. 2014. Quercetin and EGCG exhibit chemopreventive effects in cholangiocarcinoma cells via suppression of JAK/STAT signaling pathway. *Phytotherapy Research*, 28:841–848.
109. Kwak, T.W., Park, S.B., Kim, H.J., Jeong, Y.I., and Kang, D.H. 2016. Anticancer activities of epigallocatechin-3-gallate against cholangiocarcinoma cells. *OncoTargets and Therapy*, 10, 137–144.
110. Chen, B.H., Hsieh, C.H., Tsai, S.Y., Wang, C.Y., and Wang, C.C. 2020. Anticancer effects of epigallocatechin-3-gallate nanoemulsion on lung cancer cells through the activation of AMP-activated protein kinase signaling pathway. *Scientific Reports*, 10:5163.
111. Clark, R., and Lee, S.H. 2016. Anticancer properties of capsaicin against human cancer. *Anticancer Research*, 36:837–843.
112. Chen, M., Xiao, C., Jiang, W., Yang, W., Qin, Q., Tan, Q., and Wei, C. 2021. Capsaicin inhibits proliferation and induces apoptosis in breast cancer by down-regulating FBI-1-mediated NF-κB pathway. *Drug Design, Development and Therapy*, 125–140.
113. Sarkar, A., Bhattacharjee, S., and Mandal, D.P. 2015. Induction of apoptosis by eugenol and capsaicin in human gastric cancer AGS cells-elucidating the role of p53. *Asian Pacific Journal of Cancer Prevention*, 16:6753–6759.
114. Jin, J., Lin, G., Huang, H., Xu, D., Yu, H., Ma, X., and Jiang, H. 2014. Capsaicin mediates cell cycle arrest and apoptosis in human colon cancer cells via stabilizing and activating p53. *International Journal of Biological Sciences*, 10:285.
115. Banjerdpongchai, R., Wudtiwai, B., Khaw-On, P., Rachakhom, W., Duangnil, N., and Kongtawelert, P. 2016. Hesperidin from Citrus seed induces human hepatocellular carcinoma HepG2 cell apoptosis via both mitochondrial and death receptor pathways. *Tumor Biology*, 37:227–237.
116. Febriansah, R., Dyaningtyas, D.P., Nurulita, N.A., Meiyanto, E., and Nugroho, A.E. 2014. Hesperidin as a preventive resistance agent in MCF–7 breast cancer cells line resistance to doxorubicin. *Asian Pacific Journal of Tropical Biomedicine*, 4:228–233.
117. Lee, H.J., Im, A.R., Kim, S.M., Kang, H.S., Lee, J.D., and Chae, S. 2018. The flavonoid hesperidin exerts anti-photoaging effect by downregulating matrix metalloproteinase (MMP)-9 expression via mitogen activated protein kinase (MAPK)-dependent signaling pathways. *BMC Complementary and Alternative Medicine*, 18:1–9.
118. Vallianou, N.G., Evangelopoulos, A., Schizas, N., and Kazazis, C. 2015. Potential anticancer properties and mechanisms of action of curcumin. *Anticancer Research*, 35:645–651.
119. Wang, X., Hang, Y., Liu, J., Hou, Y., Wang, N., and Wang, M. 2017. Anticancer effect of curcumin inhibits cell growth through miR-21/PTEN/Akt pathway in breast cancer cell. *Oncology Letters*, 13:4825–4831.
120. Lin, L., and Lee, K.H. 2006. Structure-activity relationships of curcumin and its analogs with different biological activities. *Studies in Natural Products Chemistry*, 33:785–812.
121. Chen, J.C., Chiu, M.H., Nie, R.L., Cordell, G.A., and Qiu, S.X. 2005. Cucurbitacins and cucurbitane glycosides: Structures and biological activities. *Natural Product Reports*, 22:386–399.
122. Cai, F., Zhang, L., Xiao, X., Duan, C., Huang, Q., Fan, C., and Liu, Y. 2016. Cucurbitacin b reverses multidrug resistance by targeting CIP2A to reactivate protein phosphatase 2A in MCF-7/adriamycin cells. *Oncology Reports*, 36:1180–1186.
123. El-Senduny, F.F., Badria, F.A., El-Waseef, A.M., Chauhan, S.C., and Halaweish, F. 2016. Approach for chemosensitization of cisplatin-resistant ovarian cancer by cucurbitacin B. *Tumor Biology*, 37:685–698.
124. Chai, Y., Xiang, K., Wu, Y., Zhang, T., Liu, Y., Liu, X., and Si, Y. 2018. Cucurbitacin b inhibits the hippo-YAP signaling pathway and exerts anticancer activity in colorectal cancer cells. *Medical Science Monitor: International Medical Journal of Experimental and Clinical Research*, 24:9251.
125. Si, W., Lyu, J., Liu, Z., Wang, C., Huang, J., Jiang, L., and Ma, T. 2019. Cucurbitacin e inhibits cellular proliferation and enhances the chemo-response in gastric cancer by suppressing AKt activation. *Journal of Cancer*, 10:5843.
126. Hoshyar, R., and Mollaei, H. 2017. A comprehensive review on anticancer mechanisms of the main carotenoid of saffron, crocin. *Journal of Pharmacy and Pharmacology*, 69:1419–1427.
127. Lambrianidou, A., Koutsougianni, F., Papapostolou, I., and Dimas, K. 2020. Recent advances on the anticancer properties of saffron (Crocus sativus l.) and its major constituents. *Molecules*, 26:86.

128. Jiang, X., Li, Y., Feng, J.L., Nik Nabil, W.N., Wu, R., Lu, Y., and Xu, H.X. 2020. Safrana l prevents prostate cancer recurrence by blocking the re-activation of quiescent cancer cells via downregulation of S-phase kinase-associated protein 2. *Frontiers in Cell and Developmental Biology*, 8:598620.
129. Al-Hrout, A.A., Chaiboonchoe, A., Khraiwesh, B., Murali, C., Baig, B., El-Awady, R., and Amin, A. 2018. Safranal induces DNA double-strand breakage and ER-stress-mediated cell death in hepatocellular carcinoma cells. *Scientific Reports*, 8:16951.
130. Sethi, G., Shanmugam, M.K., Warrier, S., Merarchi, M., Arfuso, F., Kumar, A.P., and Bishayee, A. 2018. Pro-apoptotic and anti-cancer properties of diosgenin: A comprehensive and critical review. *Nutrients*, 10:645.
131. Mao, X.M., Zhou, P., Li, S.Y., Zhang, X.Y., Shen, J.X., Chen, Q.X., and Shen, D.Y. 2019. Diosgenin suppresses cholangiocarcinoma cells via inducing cell cycle arrest and mitochondria-mediated apoptosis. *OncoTargets and Therapy*, 12:9093.
132. Chiang, C.T., Way, T.D., Tsai, S.J., and Lin, J.K. 2007. Diosgenin, a naturally occurring steroid, suppresses fatty acid synthase expression in HER2-overexpressing breast cancer cells through modulating Akt, mTOR and JNK phosphorylation. *FEBS Letters*, 581:5735–5742.
133. Liu, Y., Zhou, Z., Yan, J., Wu, X., and Xu, G. 2020. Diosgenin exerts antitumor activity via downregulation of Skp2 in breast cancer cells. *BioMed Research International*, 2020: 8072639.
134. Li, Y., Wang, X., Cheng, S., Du, J., Deng, Z., Zhang, Y., and Ling, C. 2015. Diosgenin induces G2/M cell cycle arrest and apoptosis in human hepatocellular carcinoma cells. *Oncology Reports*, 33:693–698.
135. Kim, K.H., Yoon, G., Cho, J.J., Cho, J.H., Cho, Y.S., Chae, J.I., and Shim, J.H. 2015. Licochalcone a induces apoptosis in malignant pleural mesothelioma through downregulation of Sp1 and subsequent activation of mitochondria-related apoptotic pathway. *International Journal of Oncology*, 46:1385–1392.
136. Chen, M., Theander, T.G., Christensen, S.B., Hviid, L., Zhai, L.I.N., and Kharazmi, A. 1994. Licochalcone a, a new antimalarial agent, inhibits in vitro growth of the human malaria parasite plasmodium falciparum and protects mice from *P. yoelii* infection. *Antimicrobial Agents and Chemotherapy*, 38(7), 1470–1475.
137. Xue, L., Zhang, W.J., Fan, Q.X., and Wang, L.X. 2018. Licochalcone A inhibits PI3K/Akt/mTOR signaling pathway activation and promotes autophagy in breast cancer cells. *Oncology Letters*, 15:1869–1873.
138. Luo, W., Sun, R., Chen, X., Li, J., Jiang, J., He, Y., and Wen, H. 2021. ERK activation-mediated autophagy induction resists licochalcone A-induced anticancer activities in lung cancer cells in vitro. *OncoTargets and Therapy*, 13, 13437–13450.
139. Zhao, C., and Liu, Z.Q. 2011. Comparison of antioxidant abilities of magnolol and honokiol to scavenge radicals and to protect DNA. *Biochimie*, 93:1755–1760.
140. Ong, C.P., Lee, W.L., Tang, Y.Q., and Yap, W.H. 2019. Honokiol: A review of its anticancer potential and mechanisms. *Cancers*, 12:48.
141. Shen, J., Ma, H., Zhang, T., Liu, H., Yu, L., Li, G., and Hu, M. 2017. Magnolol inhibits the growth of non-small cell lung cancer via inhibiting microtubule polymerization. *Cellular Physiology and Biochemistry*, 42:1789–1801.
142. Patel, D., Shukla, S., and Gupta, S. 2007. Apigenin and cancer chemoprevention: Progress, potential and promise. *International Journal of Oncology*, 30:233–245.
143. Bruno, A., Siena, L., Gerbino, S., Ferraro, M., Chanez, P., Giammanco, M., and Pace, E. 2011. Apigenin affects leptin/leptin receptor pathway and induces cell apoptosis in lung adenocarcinoma cell line. *European Journal of Cancer*, 47:2042–2051.
144. Shao, H., Jing, K., Mahmoud, E., Huang, H., Fang, X., and Yu, C. 2013. Apigenin sensitizes colon cancer cells to antitumor activity of ABT-263. *Molecular Cancer Therapeutics*, 12:2640–2650.
145. DeBono, A., Capuano, B., and Scammells, P.J. 2015. Progress toward the development of noscapine and derivatives as anticancer agents. *Journal of Medicinal Chemistry*, 58:5699–5727.
146. Barken, I., Geller, J., and Rogosnitzky, M. 2010. Prophylactic noscapine therapy inhibits human prostate cancer progression and metastasis in a mouse model. *Anticancer Research*, 30:399–401.
147. Quisbert-Valenzuela, E.O., and Calaf, G.M. 2016. Apoptotic effect of noscapine in breast cancer cell lines. *International Journal of Oncology*, 48:2666–2674.
148. Patel, K., Gadewar, M., Tripathi, R., Prasad, S.K., and Patel, D.K. 2012. A review on medicinal importance, pharmacological activity and bioanalytical aspects of beta-carboline alkaloid "Harmine". *Asian Pacific Journal of Tropical Biomedicine*, 2:660–664.
149. Zhang, H., Sun, K., Ding, J., Xu, H., Zhu, L., Zhang, K., and Sun, W. 2014. Harmine induces apoptosis and inhibits tumor cell proliferation, migration and invasion through down-regulation of cyclooxygenase-2 expression in gastric cancer. *Phytomedicine*, 21:348–355.

150. Dai, F., Chen, Y., Song, Y., Huang, L., Zhai, D., Dong, Y., and Yi, Z. 2012. A natural small molecule harmine inhibits angiogenesis and suppresses tumour growth through activation of p53 in endothelial cells. *PLoS One*, 7:e52162.
151. Ding, Y., He, J., Huang, J., Yu, T., Shi, X., Zhang, T., and Peng, C. 2019. Harmine induces anticancer activity in breast cancer cells via targeting TAZ. *International Journal of Oncology*, 54:1995–2004.
152. Kim, H.G., Han, E.H., Jang, W.S., Choi, J.H., Khanal, T., Park, B.H., and Jeong, H.G. 2012. Piperine inhibits PMA-induced cyclooxygenase-2 expression through downregulating NF-κB, C/EBP and AP-1 signaling pathways in murine macrophages. *Food and Chemical Toxicology*, 50:2342–2348.
153. Lin, Y., Xu, J., Liao, H., Li, L., and Pan, L. 2014. Piperine induces apoptosis of lung cancer A549 cells via p53-dependent mitochondrial signaling pathway. *Tumor Biology*, 35:3305–3310.
154. Lai, L.H., Fu, Q.H., Liu, Y., Jiang, K., Guo, Q.M., Chen, Q.Y., and Shen, J.G. 2012. Piperine suppresses tumor growth and metastasis in vitro and in vivo in a 4T1 murine breast cancer model. *Acta Pharmacologica Sinica*, 33:523–530.
155. Pradeep, C.R., and Kuttan, G. 2004. Piperine is a potent inhibitor of nuclear factor-κB (NF-κB), c-Fos, CREB, ATF-2 and proinflammatory cytokine gene expression in B16F-10 melanoma cells. *International Immunopharmacology*, 4:1795–1803.
156. Yaffe, P.B., Doucette, C.D., Walsh, M., and Hoskin, D.W. 2013. Piperine impairs cell cycle progression and causes reactive oxygen species-dependent apoptosis in rectal cancer cells. *Experimental and Molecular Pathology*, 94:109–114.
157. Yaffe, P.B., Power Coombs, M.R., Doucette, C.D., Walsh, M., and Hoskin, D.W. 2015. Piperine, an alkaloid from black pepper, inhibits growth of human colon cancer cells via G1 arrest and apoptosis triggered by endoplasmic reticulum stress. *Molecular Carcinogenesis*, 54:1070–1085.
158. Estrela, J.M., Ortega, A., Mena, S., Rodriguez, M.L., and Asensi, M. 2013. Pterostilbene: Biomedical applications. *Critical Reviews in Clinical Laboratory Sciences*, 50:65–78.
159. Lin, W.S., Leland, J.V., Ho, C.T., and Pan, M.H. 2020. Occurrence, bioavailability, anti-inflammatory, and anticancer effects of pterostilbene. *Journal of Agricultural and Food Chemistry*, 68:12788–12799.
160. Wakimoto, R., Ono, M., Takeshima, M., Higuchi, T., and Nakano, S. 2017. Differential anticancer activity of pterostilbene against three subtypes of human breast cancer cells. *Anticancer Research*, 37:6153–6159.
161. Ma, Z., Yang, Y., Di, S., Feng, X., Liu, D., Jiang, S., and Li, X. 2017. Pterostilbene exerts anticancer activity on non-small-cell lung cancer via activating endoplasmic reticulum stress. *Scientific Reports*, 7:8091.
162. Zhang, Y., Li, Y., Sun, C., Chen, X., Han, L., Wang, T., and Zhao, D. 2021. Effect of pterostilbene, a natural derivative of resveratrol, in the treatment of colorectal cancer through Top1/Tdp1-mediated DNA repair pathway. *Cancers*, 13:4002.
163. Gerster, H. 1997. The potential role of lycopene for human health. *Journal of the American College of Nutrition*, 16:109–126.
164. Koh, M.S., Hwang, J.S., and Moon, A.R. 2010. Lycopene inhibits proliferation, invasion and migration of human breast cancer cells. *Biomolecules & Therapeutics*, 18:92–98.
165. Cha, J.H., Kim, W.K., Ha, A.W., Kim, M.H., and Chang, M.J. 2017. Anti-inflammatory effect of lycopene in SW480 human colorectal cancer cells. *Nutrition Research and Practice*, 11:90–96.
166. Johnson, I.T. 2007. Phytochemicals and cancer. *Proceedings of the Nutrition Society*, 66:207–215.
167. Friday, O.A. 2019. Plant toxins. *American Journal of Biomedical Science & Research*, 4:173–175.
168. Steenkamp, P.A., Harding, N.M., Van Heerden, F.R., and Van Wyk, B.E. 2004. Fatal Datura poisoning: Identification of atropine and scopolamine by high performance liquid chromatography/photodiode array/mass spectrometry. *Forensic Science International*, 145:31–39.
169. Ralphs, M.H., and McDaniel, K.C. 2011. Broom snakeweed (*Gutierrezia sarothrae*): Toxicology, ecology, control, and management. *Invasive Plant Science and Management*, 4:125–132.
170. Benowitz, N.L.1998. *Nicotine safety and toxicity*. Oxford University Press.
171. Strong F. M. (1974). Toxicants occurring naturally in foods. *Nutrition reviews*, *32*(8), 225–231. https://doi.org/10.1111/j.1753-4887.1974.tb06323.x

6 New Insights on Anti-Colorectal Cancer Effects of Plant-Based Bioactive Compounds

Akshatha SJ[1], *Anusha SN*[2], *Anjani*[3], *and Mahadesh Prasad AJ*[2]
[1]Department of Studies in Microbiology, School of Life Science, SBRR Mahajana First Grade College (Autonomous), PG wing, Pooja Bhagavat Memorial Mahajana Education Centre K.R.S Road, Metagalli, Mysore, India
[2]Department of Studies in Biochemistry, School of Life Science, SBRR Mahajana First Grade College (Autonomous), PG wing, Pooja Bhagavat Memorial Mahajana Education Centre K.R.S Road, Metagalli, Mysore, India
[3]Department of Biotechnology, SBRR Mahajana First Grade College (Autonomous) Jayalakshmipuram, Mysore, India

6.1 INTRODUCTION

Cancer cells are distinct from normal cells in several ways. They exhibit high proliferation rates, reduced differentiation, a high nucleus-to-cytoplasm ratio, increased mitotic activity, and an unusual energy-generating mechanism known as the Warburg effect. The field of cancer emerged in the 1920s when Otto Warburg introduced the Warburg effect, further studies confirmed that the cancer cells use glycolysis even in the presence of oxygen by aerobic glycolysis to shuttle the glycolytic intermediates and lactate to increase the biomass (Sedlak et al., 2023).

Colorectal cancer (CRC), the third most prevalent malignancies and ranks second in cancer-related diagnosis worldwide. Commonly it originates in the colon or rectum as polyps, which are initially benign but can gradually transforms into malignant tumor. CRC often remains asymptomatic instead "without any noticeable symptoms" in its early stages. Individuals over 50 years old are at a higher risk, and various factors contribute to its development, including unhealthy diets, alcohol consumption, smoking, lack of physical activity, obesity, and modern lifestyles (Sedlak et al., 2023).

Natural products are becoming a common source of novel anticancer medications, almost half of the anticancer medications in use today have some sort of natural product origin.

Products with different structures include unsaturated fatty acids, diterpenoid, polysaccharides, polyphenols, and alkaloids (Roviello et al., 2020). The ongoing identification of these items has led to the development of novel avenues for cancer prevention and therapy; the natural product-based medications that are now being utilized clinically to treat CRC.

Plant-based phytoconstituents have been demonstrated to have a substantial effect on carcinogenesis. These include phenolics, flavonoids, terpenoids, alkaloids, glycosides, steroids, saponins, quinones, and secondary metabolites. With improved cytotoxic potential, the phytochemicals

DOI: 10.1201/9781003486237-6

efficiently controlled the cellular processes. Higher concentrations of phenolic groups in plants or isolated compounds offer more potent antibacterial, antioxidant, anticancer, and anti-inflammatory effects. The most significant chemopreventive substances were discovered to be plant extracts and isolated chemicals, which also have a variety of biological functions. To put it more clearly, polyisoprenylated benzophenones are a naturally occurring compound with anti-inflammatory, antiviral, anticancer, and antioxidant properties.

The phytochemicals found in plant extracts promote metastasis, major proapoptotic, necroptotic, or autophagic processes, cell cycle arrest, DNA repair, and reactive oxygen species (ROS) and reactive nitrogen species (RNS) buildup as well as programmed cell death. Particularly, polyphenolic chemicals cause apoptosis by a variety of methods, including the activation of signaling cascades, the deactivation of inflammation, the alteration of cell cycle pathways, and the modulation of apoptotic proteins by destroying DNA and controlling gene transcription, the activity, and apoptosis were accomplished.

However, 5–6% CRC cases have epigenetic association via promoter methylation, histone modification (Macharia et al., 2022), and mutations in genes like KRAS (the homolog of Kirsten rat sarcoma proto-oncogene), BRAF (the homolog of rapidly accelerated fibrosarcoma proto-oncogene) also taking part in CRC etiology. Polymorphism of proteins NABP1 (nucleic acid-binding protein 1), cyclin D2, laminin 1, and transcription factors which target the (Wnt)/β-catenin pathway are connected with CRC incident (Costea et al., 2018). Epigenome of cancerous cells can be reversed by targeting co-activators, transcription factors, co-repressor and histone modification methods, which give insightful therapeutic effects (Macharia et al., 2022).

Although CRC is asymptomatic at early stages and diagnosed at advanced stages often with anemia in many cases, it is not incurable. Diagnosis by screening tests like sigmoidoscopy, colonoscopy, and treatments like use of medication, surgery, radiation therapy, and chemotherapies are ongoing to manage the risk of CRC (Grace et al., 2023). These chemotherapy and radiation therapies result in a number of undesirable side effects like altered balance of intestinal flora, lead to mucositis, gastrointestinal disorders such as vomiting, diarrhea, nausea reduce the life quality of patients (Memmola et al., 2022).

A survey proclaimed that more or less 31% of people worldwide preferred adopting traditional medicines prior to choosing conventional medicine (Kumar et al., 2016).

For centuries, people have explored and utilized medicinal plants, integrating them into traditional medicinal practices dating back to prehistoric eras. The field of drug-related research harnesses ethnobotany to seek pharmacologically active natural substances, leading to the discovery of hundreds of beneficial compounds through this approach. Among these are well-known medications such as aspirin, opium, digoxin, and quinine. Compounds extracted from plants belong to various biochemical classes, with the majority falling into four main categories: terpenes, alkaloids, glycosides, and polyphenols (Forbes-Hernández et al., 2014). Other creatures, including sheep, non-human primates, and monarch butterflies, also rely on herbs as medicine, indicating that humans were not the sole users of medicinal plants (Awuchii and Godswil, 2019).

In this chapter, we reviewed the contributions and benefits of plant byproducts against CRC. We discussed current global epidemiology, drug resistance, its challenges and risk factors, additionally effective multiple signaling pathways were demonstrated to overcome the drawbacks associated with conventional therapies. Furthermore, we emphasized the advanced applications, recommendations of important plant products for CRC treatment.

6.1.1 Epidemiology

CRC is among the top three most widespread cancers worldwide, accounting for 10% of global cancer cases and 9.4% of cancer-related deaths in 2020. The incidence and mortality of CRC have been decreasing globally over the past decades, primarily attributed to the implementation of screening programs. These programs involve the identification and removal of adenomatous polyps, thus breaking the adenoma-carcinoma sequence. The decline in CRC incidence started in the 1990s, coinciding with the widespread adoption of colonoscopy for screening average-risk CRC patients in various countries (Pignone et al., 2002).

The risk of developing CRC substantially rises with age. From 2015 to 2019, there's an 80%–100% rise in each 5-year age group until age 50, the increase diminishes to 20%–30% in the 55–59 age group and older. Yet, the rise is only 9% from ages 50–54 to 55–59 due to the disruption of the usual age-related pattern of identifying symptomatic CRC. This interruption is a result of screening, recommended to start at age 50 until 2018, detecting precancerous adenomas and prevalent cancers in asymptomatic individuals (Wolf et al., 2018).

Between 2015 and 2019, the average annual incidence rate for CRC was 33% higher in men (41.5 per 100,000) than in women (31.2 per 100,000). However, the gender difference in cancer rates varies by age and tumor location. For example, the ratio of male-to-female incidence ranges from 1.10 for proximal colon cancer to 1.61 for rectal cancer, indicating significant differences in tumor distribution. Similar gender variations are observed in other high-income countries, suggesting a link to environmental, carcinogenic, hormonal exposures, and distinct mechanisms across anatomical subsites (White et al., 2018).

Globally, in 2020, there were approximately 1.15 million new cases of colon cancer, 0.7 million new cases of rectal cancer, and 50,000 new cases of anal cancer, according to GLOBOCAN 2020 estimates. Predictions indicate that by 2040, these numbers are expected to rise to 1.92 million new cases of colon cancer, 1.16 million new cases of rectal cancer, and 78,000 new cases of anal cancer, reflecting a concerning increase with time (Xi and Xu, 2021).

The occurrence of CRC displays a diverse geographical pattern. Research suggests a positive correlation between the incidence of CRC and the Human Development Index (HDI), with countries boasting high HDI levels experiencing approximately four times higher CRC rates compared to low HDI nations. Developed countries currently exhibit stabilized or decreasing CRC incidence, while developing countries show an upward trend, likely associated with an increased exposure to CRC risk factors due to the adoption of a western lifestyle during economic transitions (Siegel et al., 2019).

Analyzing cancer trends in young adults provides a more accurate measure of progress in disease control, as it reflects current exposures rather than cumulative effects seen in older adults. Studying cancer rates in younger populations allows for a more immediate understanding of the effectiveness of interventions and changes in lifestyle or environmental factors influencing cancer development. From 2011 to 2019, cancer rates rose by 1.9% annually in people under 50 and those aged 50–54. There's a significant increase in young and middle-aged adults, while older adults show decreasing trends. This shift has made the patient population younger, with the proportion of newly diagnosed individuals under 55 nearly doubling from 11% in 1995 to 20% in 2019. For rectal cancer, a major driver of early-onset cases, the proportion increased from 15% to 28%. In the last decade, rectal cancer incidence in those under 50 rose by about 2% annually, compared to 0.5% and 1.6% for tumors in the proximal and distal colon, respectively (Siegel et al., 2023).

6.2 RISK FACTORS OF COLORECTAL CANCER

The lifetime risk of CRC is comparable between men and women in the United States, the age-adjusted incidence is 31% higher in men due to their shorter life expectancy. Globally, the incidence rate in men is 44% higher (23.4 cases per 100,000 persons) than in women (16.2 cases per 100,000 persons), with a more significant disparity in rectal cancer compared to colon cancer. Specific case rates for rectal and colon cancer in men are notably higher than in women. Men have a 75% higher rate of diagnosed rectal cancer cases and a 31% higher rate of colon cancer cases per 100,000 individuals compared to women. Additionally, a multinational study across European countries indicates a higher prevalence of proximal colon cancer in women (Siegel et al., 2020).

Sporadic and genetic factors: Approximately 60–65% of CRC cases are categorized as sporadic, lacking any family history or inherited genomic alterations (Tian et al., 2019). The onset of sporadic CRC is predominantly linked to acquired somatic genetic mutations or epigenetic changes induced by modifiable risk factors. The remaining 35–4.0% of CRC patients exhibit susceptibility to heritable components, including family history without obvious genetic predisposition, hereditary cancer syndromes such as Lynch syndrome, low-penetrance genetic variations, and other unknown inherited genomic abnormalities (Kastrinos et al., 2020).

Approximately 2–8% of CRC cases are attributed to inherited syndromes. The primary hereditary syndromes associated with CRC development are hereditary nonpolyposis CRC (HNPCC or Lynch syndrome) and familial adenomatous polyposis coli (FAP). Individuals with HNPCC face a 20% risk of developing CRC by the age of 50 and an 80% risk by the age of 85. It is presumed that almost all individuals with undetected and untreated FAP syndrome are expected to be diagnosed with CRC before reaching the age of 35–40. The elevated likelihood of developing CRC is also associated with the presence of MUTYH-associated polyposis (MAP), Peutz-Jeghers syndrome, PTEN hamartoma tumors syndrome, and Juvenile polyposis syndrome (Valle et al., 2019).

Lifestyle: Physical inactivity and a sedentary lifestyle correlate positively with colon cancer risk, with sedentary individuals facing a 25–50% higher risk compared to those who are physically active (Schmid and Leitzmann, 2014).

Overweight and obesity are established CRC risk factors, with obese men having a 50% higher risk of colon cancer and a 25% higher risk of rectal cancer than normal-weight men (Dong et al., 2017). In obese women, the risk of colon cancer increases by 10%, with no increase in rectal cancer risk (Liu et al., 2019). While an early body mass index (BMI) is considered a more significant risk factor for colon cancer in women, in contrast, late body weight gain has a greater impact on the risk of CRC in men (Xue et al., 2017).

Consuming 100 g of red meat daily is estimated to increase the risk of CRC by about 17%, and a daily intake of 50 g of processed meat is associated with an approximately 18% higher risk (Aykan, 2015). High dietary fiber consumption has been linked to a potential 50% reduction in the risk of developing CRC. Fruits and vegetables, rich in bioactive compounds like vitamins, minerals, folate, plant sterols, and protease inhibitors, demonstrate strong antioxidant and anti-inflammatory properties that may inhibit DNA and cellular damage. Numerous studies suggest that a diet high in fruits and vegetables is associated with a lower risk of CRC development (Murphy et al., 2019).

Research indicates that individuals who smoke cigarettes are at a 2–3 times higher risk of developing CRC compared to non-smokers, and this risk increases with the amount and duration of smoking (Giovannucci, 2001). Cigarette smoking is estimated to be responsible for around 12% of CRC-related deaths. The complex mixture of over 60 established carcinogens in tobacco smoke, such as N-nitrosamines, polycyclic aromatic hydrocarbons, aromatic amines, aldehydes, and metals, is known to cause DNA damage (Ahmed et al., 2014).

Consuming 2–3 alcoholic drinks daily is estimated to raise the risk of CRC by around 20%, and exceeding three drinks increases this risk to approximately 40%. Those accustomed to drinking four or more daily drinks further elevate their chances, with up to a 52% increased risk of developing CRC. Potential mechanisms by which alcohol may induce carcinogenesis include the generation of ROS and RNS, the production of mutagenic acetaldehyde, depletion of S-adenosylmethionine, inactivation of tumor suppressor genes, hormonal imbalance, reduction in folate concentration, and impairment of retinoic acid metabolism (Rossi et al., 2018).

Health history: Diabetes mellitus is a consistently linked risk factor for CRC, presenting a heightened relative risk of 30%. A Swedish study suggests that individuals with diabetes reach the ten-year cumulative risk for CRC, warranting general screening programs, five years earlier than those without diabetes (Ali Khan et al., 2020). Chronic inflammation associated with diabetes contributes to various stages of cancer development, including carcinogenesis, malignant transformation, tumor growth, invasion, and metastasis. This influence is exerted through the action of pro-inflammatory cytokines, specifically tumor necrosis factor-alpha (TNF-α) and interleukin-6 (IL-6) (Brzacki et al., 2019).

Being the third highest-risk factor for CRC, inflammatory bowel disease (IBD) increases the risk of developing CRC by 2–6 times compared to individuals without IBD. This elevated risk is influenced by the duration of IBD and the extent and severity of the disease, as chronic inflammation in IBD contributes to tumor growth (Keller et al., 2019).

Colon polyps, characterized as abnormal tissue growths in the colon's mucous layer, are believed to be the source of about 95% of CRCs. Even though the majority of cancers stem from adenomas, only around 5% of polyps evolve into CRC. Recognizing the significance of this, especially considering that about 40% of individuals aged 50 or above have one or more adenomatous polyps, it is crucial to identify and remove these polyps before they progress to cancer (Shussman and Wexner, 2014).

6.3 COLORECTAL CANCER DEVELOPMENT

CRC is genetically diverse which can develop through various mechanisms. When epithelial cells undergo a series of genetic or epigenetic modifications that allow for their hyperproliferative nature, CRC arises (Hossain et al., 2022). These quickly growing cells evolve into a benign adenoma, which can progress to cancer and spread by a number of different pathways, such as chromosomal instability (CIN), serrated neoplasia, and microsatellite instability (MSI). The term "adenoma-carcinoma sequence" refers to the progression of cancer. The great majority of sporadic CRC cases are caused by the conventional pathway (Alzahrani et al., 2021). A small adenoma first develops into a large adenoma and then cancer. When an adenocarcinoma becomes invasive it rapidly spread to multiple organs via blood stream and lymphatic tissues. CRC often grows slowly and usually does not cause symptoms until it reaches a significant size of several centimeters. At that point, it may obstruct the passage of feces and cause cramps, pain, or bleeding that may show up as dark, "tarry," or visible bleeding during bowel movements (Li et al., 2023).

These growths, or aggregates, are localized abnormal cell growths within the intestinal mucosa that extend into the intestinal lumen and are known as polyps. Polyps may undergo genetic alterations to enable them to invade the intestinal wall, a characteristic that is specific to CRC (Marmol et al., 2017). Further, they may undergo alteration and spread to nearby lymph nodes, and ultimately to distant metastatic sites. Adenomas and sessile serrated polyps (SSPs) are the two primary forms of polyps with malignant potential; these two types of polyps have differing risks of progressing to CRC (Kanthan et al., 2012). Most adenomas have a tubular histology in general, with small, rounded, atypical glands; however, as they grow, they frequently develop areas of long filamentous architecture, which pathology reports refer to as villous or tubulovillous (Sawicki et al., 2021). The larger the polyp, the higher the chance that an adenoma will turn into a CRC. While 60%–70% of CRCs originate from adenomas, only about 10% of even the most advanced adenomas those measuring ≥1 cm or exhibiting ≥25% villous component or high-grade dysplasia of any size is malignant (Koornstra et al., 2003).

A sequence of changes in genes and/or epigenetics leads to the histological progression from polyp to cancer. Mutations in DNA can be inherited or acquired. APC, MSH2, MLH1, and PMS2 gene mutations are examples of true inherited mutations linked to CRC, which are rare and cause about 5% of CRCs (Tariq and Ghias, 2016). However, research on these inherited mutations has shed light on the sequential genetic development from premalignant polyps to cancer, in addition to occasionally occurring APC and DNA mismatch repair mutation (Mariani et al., 2014).

Cytologic and histologic dysplasia reflects the accumulation of genetic mutations and epigenetic modifications that may occur as the polyp's cells multiply and grow larger (Lu et al., 2023). The characteristics of high-grade dysplasia may appear with increasing damage to cellular DNA over time, indicating a very high risk of invasive carcinoma development, these polyps have the potential to spread into surrounding tissue of the rectum walls (Hamouda and Benkhadir, 2023). This localized, malignant growth has the potential to become neovascularized, which would facilitate easy access to the circulatory and lymphatic systems and accelerate the spread of cancerous cells to other organs (Wang et al., 2024).

6.4 CHEMO-RADIOTHERAPEUTIC APPROACHES AGAINST CRC

6.4.1 Chemotherapy

Since the onset of the industrial revolution, there has been continuous progress in the scientific world. As we entered the 21st century, the treatment of certain diseases, including CRC, demands new and innovative therapeutic solutions. Depending on the confinement and progression of CRC, a combination of these treatments may be employed. Nevertheless, even with the use of combined therapies, more than half of patients experience a recurrence characterized by acquired multidrug-resistant CRC (Schuell al., 2005).

Chemotherapy plays a significant role in the treatment of CRC. It involves the use of drugs to either kill cancer cells or inhibit their ability to grow and divide as shown in Table 6.1. Approved cytotoxic drugs for CRC, such as fluoropyrimidines – irinotecan, oxaliplatin, tri-fluridine-tipiracil, capecitabine, and 5-fluorouracil (5-FU), play a crucial role in slowing down disease progression and extending an individual's lifespan (Kumar et al., 2023).

TABLE 6.1
Pharmacological Profiles of Synthetic Drugs: Insights Into Administration, Mechanism of Action, and Side Effects

Synthetic Drug	Administration	Mechanism of Action	Side Effects
		Cytotoxic Drugs	
5-Fluorouracil (5-FU)	Intravenous infusion	Analog of uracil Inhibitor of thymidylate synthase (TS) Decreases DNA and RNA syntheses by inhibiting the production of dTMPs Induce cell cycle arrest and leads to apoptosis	Fever, stomatitis, mucositis, leukopenia, thrombocytopenia (Blondy et al., 2020)
Capecitabine	Oral	Prodrug of 5-FU	Hand foot syndrome, nausea, vomiting, diarrhea, fatigue, leukopenia, thrombocytopenia, indigestion, mucositis, change in taste (Comella, 2007)
Oxaliplatin	Intravenous infusion	Generates DNA adducts, inhibits DNA damage repair	Hypersensitive reaction, anemia, neutropenia, nausea, vomiting, fatigue, leukopenia (Comella et al., 2009)
Irinotecan	Intravenous infusion	Inhibits topoisomerase I, affects DNA replication and transcription, leads to cell death	Nausea, vomiting, diarrhea, fatigue, change in blood count (Fujita et al., 2015)
Trifluridine-tipiracil	Oral	Trifluridine is a nucleoside analog incorporate in DNA leads to chain termination and cell cycle arrest and apoptosis Tipiracil inhibits thymidine phosphorylase, thereby enhances trifluridine	Nausea, fatigue, anemia, neutropenia (Mulet et al., 2018)
tegafur/uracil (UFT)	Oral	Prodrug of 5-FU	Cardiotoxicity, allergic reactions, peripheral neuropathy, decreased blood cell count, fatigue (Yu et al., 2023)

(Continued)

TABLE 6.1 *(Continued)*
Pharmacological Profiles of Synthetic Drugs: Insights Into Administration, Mechanism of Action, and Side Effects

Synthetic Drug	Administration	Mechanism of Action	Side Effects
		Molecular-Targeted Drugs	
Bevacizumab (BEV)	Intravenous infusion	Binds to VEGF thereby reduces the blood flow to tumor cells, inhibits angiogenesis	Hypertension, proteinuria, gastrointestinal perforations, impaired wound healing, thromboembolism, nausea, diarrhea, and fatigue, decreased white blood cell count (Tappenden et al., 2007)
Ramucirumab (RAM)	Intravenous infusion	This monoclonal antibody inhibits VEGFR-2, thereby inhibits angiogenesis and blood flow to cancer cells	Hypertension, fatigue, proteinuria, leukopenia, thrombocytopenia, impaired wound healing, thromboembolism, gastrointestinal perforations (Ju et al., 2020)
Aflibercept beta (AFL)	Intravenous infusion	Binds to VEGF and PlGF, preventing them from interacting with their receptors, thus inhibits angiogenesis and reduces blood supply to tumor	Hypertension, gastrointestinal issues, proteinuria, neutropenia, fatigue, hand foot syndrome, thromboembolism (Muro et al., 2020)
Cetuximab (CET)	Intravenous infusion	Binds to the EGFR on the surface of cancer cells, thus inhibits the proliferation of cancer cells, stimulates immune responses to target and destroys cancer cells	Skin rash, diarrhea, nausea, vomiting, hypomagnesemia, electrolyte imbalances, fatigue, hypersensitivity reactions (Wei et al., 2022)
Panitumumab (PANI)	Intravenous infusion	Binds and inhibits EGFR, thus inhibits cancer cell proliferation	Skin toxicity (rash, dry skin), hypomagnesemia, fatigue, gastrointestinal symptoms (diarrhea, nausea), infusion reactions (Mano and Humblet, 2008)
Regorafenib (REG)	Oral	Inhibits VEGFR, PDGFR and inhibits angiogenesis and reduces blood supply to tumor cells. Inhibits RAF and RET, hence inhibits cell proliferation. Enhances immunity against cancer.	Hand foot skin reaction, fatigue, hypertension, gastrointestinal issues, decreased appetite, elevated liver enzymes (Matsumoto et al., 2023)
Encorafenib	Oral	Inhibits mutated BRAF kinase, hence inhibits cell growth. Suppresses MAPK pathways.	Fatigue, nausea, diarrhea, skin rashes, itching, abdominal pain, joint pain (Tabernero et al., 2021)
Binimetinib	Oral	Inhibits MEK, thus dysregulates MAPK signaling pathway thereby inhibits uncontrolled growth of tumor	Fatigue, nausea, diarrhea, abdominal pain, peripheral edema, skin rashes and itching (Roviello et al., 2020)
		Immune Checkpoint Inhibitor	
Pembrolizumab (Pembro)	Intravenous infusion	It is an immune checkpoint inhibitor, binds to the PD-1 receptor on T cells. Enhances the immune system against cancer cells.	Immune-related adverse events, joint pain, fatigue, nausea, diarrhea, rashes (Coupez et al., 2020)

6.4.2 Radiotherapy

The main goal of radiotherapy is to reduce the risk of local recurrence and enhance overall survival in cancer treatment. However, research indicates that preoperative or neoadjuvant (before surgery) radiotherapy is more effective than postoperative or adjuvant (after surgery) therapy in decreasing local recurrence, it does not necessarily lead to an improvement in overall survival (Gijn et al., 2011).

The study compared two adjuvant radiotherapies for CRC: short-course (SC) and long-course (LC) radiotherapy. Results showed that the three-year local recurrence rates were not statistically significantly different between SC and LC. The cumulative incidence suggested either no clinically important difference or potential advantages for LC. LC was deemed more effective in reducing local recurrence for distal tumors. However, there were no differences in rates of distant recurrence, relapse-free survival, overall survival, or late toxicity between the two treatments (Ngan et al., 2012).

Contact therapy for rectal cancer, initiated in the 1960s by Papillon in Lyon, utilized a portable x-ray machine operating at 50 kV with a 4-cm source-to-skin distance. The treatment involved delivering a significant endocavitary dose of 10–40 Gy per fraction weekly or biweekly, totaling 45–100 Gy. Studies from the 1980s to the 1990s suggested that contact therapy was effective in achieving disease control and rectal preservation for early-stage rectal tumors (Kovalic, 1988).

High-dose-rate (HDR) endorectal brachytherapy is a form of endocavitary radiotherapy that employs real-time fluoroscopy guidance. It offers precise dose conformality with a steep fall-off, enabling dose escalation without compromising Organ at Risk (OAR) doses. Using iridium-192 in a remote after-loading system, this technique emits gamma radiation and employs special rectal implants based on the Paris system for dose specification. Initially applied for adjuvant or palliative treatment in rectal cancer, this technique enhances the precision and effectiveness of radiation therapy (Kaufman et al., 1989).

Intraoperative radiotherapy (IORT) is a specialized technique aiming to enhance the treatment outcomes for locally advanced rectal tumors by delivering precise radiation during surgery. It enables targeted treatment, allows for dose escalation beyond external beam radiotherapy (EBRT) limits, and supports reirradiation in cases of recurrent tumors when additional EBRT is not feasible. Administered as a supplementary dose alongside EBRT, IORT utilizes methods such as electron beam (IOERT), high-dose-rate brachytherapy (HDR-IORT), or orthovoltage x-ray (Pilar et al., 2017).

Stereotactic ablative body radiotherapy (SABR) is a potent local therapy for colorectal liver metastases (CRLM), triggering significant immunological changes in the tumor microenvironment. SABR activates cytoprotective and cytotoxic pathways in both malignant and non-malignant cells, causing direct tumor cell death. Combining SABR with immunotherapy may enhance the abscopal response in distant lesions, addressing different stages of the immune-mediated response alongside inducing intratumoral cell death (Wang et al., 2021).

Radioembolization is a treatment option for metastatic liver disease, particularly when the metastasis is confined to the liver. This approach utilizes yttrium-90 (Y-90)-impregnated microspheres, such as TheraSpheres and SIR-Spheres, to deliver localized radiation. Metastatic liver tumors receive their primary blood supply from the hepatic artery, allowing the microspheres to be trapped within the tumors when infused through this artery. Y-90 emits beta radiation with a range of 2.5 mm in the tissue. This localized radiation delivery technique is also referred to as intra-arterial brachytherapy and offers an option for treating liver-confined metastatic tumors (Tam and Wu, 2019).

The study suggests the use of carbon-ion radiotherapy (CIRT) as a potentially effective treatment for oligometastatic CRC that has spread to the liver and lungs. C-ion RT produces clinical outcomes comparable to stereotactic body radiotherapy (SBRT). Notably, C-ion RT is deemed a safe option based on the absence of radiation-induced liver disease (RILD) and radiation-induced pneumonitis. As a result, the study recommends considering C-ion RT for patients with oligometastatic CRC in the liver and lungs who may not be suitable candidates for surgery (Shiba et al., 2021).

CIRT is highlighted as a superior choice compared to x-ray radiation due to its physio-biological characteristics, offering increased effectiveness and reduced toxicity. The study suggests that using a single-fraction of CIRT for treating CRC liver metastasis is safe up to a dose of 58 Gray (RBE), as long as precautions are taken to avoid the central hepatic portal region during the treatment (Makishima et al., 2019).

6.5 ANTICANCER EFFECTS OF PLANT-DERIVED BIOACTIVE COMPOUND

Phytochemicals from plants can stop the growth of tumors. Numerous phytochemicals and bioactive substances, including luteolin, rosmarinic acid, carnosic acid, emodin, eugenol, kaempferol, genistein, resveratrol, caffeic acid, epigallocatechin, saponins, polysaccharides, triterpenoids, alkaloids, glycosides, phenols, and quercetin, are found in medicinal plants. The seeds, fruits, roots, and leaves of the plants are the sources of these phytochemicals. They can trigger caspase activation or the release of cytochrome C to cause apoptosis via the mitochondrial pathway. Certain substances have the ability to inhibit anti-apoptotic factors while promoting the synthesis of pro-apoptotic proteins including caspase-3 and BAX. They can block the activation of nuclear factor κ-B (NF-kB) which is a pro-inflammatory molecule. Some of the plant active compound inhibits cyclooxygenase 2 activity, insulin growth factor receptor (IGFR). They are also involved in the regulation of migration and invasion of epithelial mesenchymal transition by stimulating polyadenosine diphosphate ribose polymerase enzyme. Few of phytoconstituents such as flavonoids, caffeic acid, epigallocatechin, saponins acts as epigenetic regulators causing DNA degradation of DNA methyl transferase and histone deacetylase to suppress CRC.

Purification of herbal components and clinical studies and relevant in vitro models confirming their bioactivity could result in the development of efficacious therapeutics for the treatment of colon cancer. Plants of grape, soybean, green tea, garlic, olive, and pomegranate are the most efficient for treating colon cancer owing to their anticancer properties. Studies on fruits, seeds leaves, rhizospheres, root and stem of various plants have already validated using different in vitro and in vivo techniques. Plant-derived bioactive compounds have the ability to affect carcinogenesis at different phases by preventing, postponing, or reversing the growth of the tumor before it becomes invasively malignant. Since traditional medicines have anti-angiogenic, anti-invasive, anti-proliferative, antioxidant, anti-inflammatory, and anti-mutagenic qualities, they have been used to treat cancer all over the world.

6.6 MEDICINAL PLANTS IN THE PREVENTION, TREATMENT OF CRC, AND PHYTOCHEMICALS ON CANCER CELL METABOLISM

CRC treatment encompasses surgery for early-stage cases, along with chemotherapy, radiotherapy, and more recently, targeted therapies. Although targeted therapies have shown improvements in outcomes for metastatic CRC, the use of these treatments is accompanied by observed limitations and side effects.

The utilization of compounds from medicinal plants shows promise in both preventing and treating CRC, demonstrated through experiments conducted in the laboratory (in vitro) and living organisms (in vivo). Herbal extracts impede CRC growth by triggering processes like apoptosis, autophagy, cell cycle arrest, and modulation of signaling pathways as shown in Table 6.2. Phenolic chemicals, identified as key active components, play a pivotal role in these anti-colorectal actions. The combined effects of herbal formulations suggest their potential as treatments for CRC. However, further investigation is crucial to isolate, identify, and utilize various phytochemicals in the prevention or therapy of CRC, with the aim of enhancing patient outcomes (Benarba and Pandiella, 2018).

A wide range of nature phytochemicals can be employed in cancer prevention and treatment to overcome obstacles associated with traditional techniques (chemotherapy and surgery). A number of advantages of phytochemicals have been discovered, such as immune-modulatory activity, synergetic effects with combination of other drugs and selective toxicity. With the vast array of advantages, phytochemicals are gaining attention and progressively being recognized as medications beneficial in treatment of cancer (Shannar et al., 2022). Nutritional intervention can improve cancer chemotherapy by influencing cancer metabolism in positive aspects. Some of the bioactive phytochemicals that can play significant role in metabolism of cancer cell are discussed in this section.

TABLE 6.2
Plant Pharmacology – Unveiling Names, Utilized Parts, Bioactive Compounds, and Modes of Action

Sl. No.	Name of the Plant	Parts Used	Bioactive Compound	Mode of Action	References
1.	*Curcuma longa* (Turmeric)	Root	Curcumin, turmerones	Neutralize free radicals, induce apoptosis, arrest cell cycle, suppression of NF-κB pathway.	Dimas et al. (2015)
2.	*Camellia sinensis* (Green Tea)	Leaf	Catechin, epigallocaechin gallate, theaflavins, flavon-3-ol, teadenol B	Suppress ERK1, ERK2, and VEGF. Express caspases 3/7, 8, 9. Induce apoptosis.	Esghaei et al. (2018)
3.	*Zingiber officinale* (Ginger)	Peel	6-Gingerol, 6-shogaol, 6-paradol, zingerone, dehydrozingerone	Inhibit proliferation, metastasis. Enhance drug effect, apoptosis, and cell cycle arrest. Anti-inflammatory, antioxidant.	Xiang et al. (2024)
4.	*Allium sativum* (Garlic)	Root	Allicin, Z-ajoene, allyl methyl sufide (AMS), E-ajoene, 2-vinyl-4H-1,3-dithiin, diallyl trisulfide, alliin, diallyl sulfide (DAS), diallyl disulfide (DADS)	Upregulation of PTEN, downregulate Akt, P Akt expression, inhibit PI3k/ Akt pathway. Free radical scavenging, DNA repair, anti-inflammatory, cell cycle arrest, anti-angiogenic.	Tesfaye (2021) and Dong et al. (2014)
5.	*Panax ginseng* (Ginseng)	Root	Ginsenosides, ginseng polysaccharides, flavonoids, polysaccharides, ginseng polypeptides	Suppress C/EBPβ/NF-κB pathways, downregulate lncRNA CCAT1, inhibit migration, proliferation and invasion of CRC cells, and promote apoptosis.	Zhao et al. (2022)
6.	*Boswellia serrata* (Boswellia)	Resin	Boswellic acid	Increased caspase 3 activity, inhibit VEGF, HIF1, mPGES1, CXCR4, MMP$, MMP9, PGE2 expression. Inhibit cell migration and angiogenesis.	Ranjbarnejad et al. (2017)
7.	*Thymus vulgaris* (Thyme)	Leaf	Thymol, carvacrol	Activate BAX/Bcl2 signaling pathway and induce apoptosis. Suppress EMT, invasion, metastasis by inhibiting Wnt/β-catenin pathway.	Zeng et al. (2020)
8.	*Rosmarinus officinalis* (Rosemary)	Leaf	Carnosic acid, carnosol, rosmarinic acid, rosmanol, betulinic acid	Activate Nrf2, XBP1. Increase intracellular ROS. Downregulate TYMS, TK1-enzymes regulate 5-FU. Cell cycle arrest.	Pérez Sánchez et al. (2019)
9.	*Vitis vinifera* (Grape)	Fruit, Seed	Hydroxy cinnamic acids, stilbenoids, proanthocyanidins, catechin, epicatechin, quercetin, gallic acid, procyanidins	Inhibit Wnt/β- catenin pathway. Increase p53, p21, Bax/Bcl2. Activate Nrf2. Cell cycle arrest, inhibit cell proliferation, activate apoptosis.	Adebesin et al. (2022)
10.	*Glycine max* (Soybean)	Seed	Anthoxanthin, saponin, cyanidin-3-O-glycoside	Enhance Rab6 protein, lead to cell death. Suppress PKC. Enhance alkaline phosphatase.	Yoshida et al. (2020) and Oh and Sung (2001)

TABLE 6.2 *(Continued)*
Plant Pharmacology – Unveiling Names, Utilized Parts, Bioactive Compounds, and Modes of Action

Sl. No.	Name of the Plant	Parts Used	Bioactive Compound	Mode of Action	References
11.	*Punica granatum* (Pomegranate)	Peel, seed, leaves, juice	Ellagic acid, alpha linolenic acid, linoleic acid, oleic acid, punicalagin, gallic acid, caffeic acid, chlorogenic acid, p-coumaric acid, aglycone, ferulic acid, anthocynosides, alkaloids, organic acids, lignans, minerals, ursane, triterpenic acids, steroids	Increased level of caspase 3, lead to apoptosis of cancer cells. Modulate gene expression of TYMs, CD44, CDKN, CTNNB. Reduce LBP levels in CRC patients.	Kusmardi et al. (2021) and Eghbali et al. (2021)
12.	*Glycyrrhiza glabra* (Licorice)	Root	Glycyrrhizin (G), glycyrrhetinic acid (GA),	Inhibit ROS generation, thus reduce inflammation. Enhance IL-10. Inhibit p38-MAPK, p-JNK, NF-κB pathways.	Leite et al. (2022)
13.	*Opuntia ficus-indica* (Nopal cactus)	Fruit	Indicaxanthin pigment	Demethylation of p16INK4a gene promoter, leads to accumulation of p16, and arrests cell cycle. Reactivation of silenced mRNA.	Naselli et al. (2014)
14.	*Sasa quelpaertensis* (Dvergbambus)	Leaf	Tricin, p-coumaric acid	Downregulate β-catenin and phosphorylated GSK3β. Upregulate CK20, thereby enhance cell differentiation. Suppress HIF-1α and Wnt/β-catenin pathway	Min et al. (2015)
15.	*Salvia chinensis Benth* (Shijianchuan)	Stem	Quersetin, resveratrol, ursolic acid, beta-sitosterol, emodin, deoxyschizandrin-A, daucosterol, chrysophanol, vanillin	Reduce oxidative stress markers like SOD, LPO, NO, G6PD, GSH. Regulate β1 integrin and promote apoptosis. Caspase-dependent apoptosis. Inhibit PI3k/ Akt pathway	Zheng et al. (2023)
16.	*Aloe barbadensis miller* (*Aloe vera*)	Leaf	Aloin, aloe-emodin	Inhibit proliferation, induce apoptosis. Block STAT3 and inhibit angiogenesis. ROS generation.	Manirakiza et al. (2021)
17.	*Trigonella foenum* (Fenugreek)	Seeds, Green leaves	Diosgenin	Induce apoptosis by suppressing Bcl 2 and activating caspase 3 expression	Hossain et al. (2022)
18.	*Sesamum indicum* (Sesame)	Seeds	Sesamin, sesamol, sesamolin, sesaminol, pedaliin	Target p53, JNK, MAPK, PI3K/Akt, NF-B, TNF, PPAR, caspase 3, Nrf-2, eNOS, and LOX signaling. Inhibit NF-κB pathway. Induce apoptosis, arrest cell cycle	Kim et al. (2021)
19.	*Linum usitatissimum* L. (Flaxseed)	Seed	Flavanoids, phytoestrogens, lenolenic acid, eicosapentaenoic acid, docosahexaenoic acid	Inhibit cell proliferation and induce apoptosis. Upregulate caspase 3, downregulate Bcl2 and PCNA.	Hossain et al. (2022)
20.	*Cinnamomum verum* (Cinnamon)	Bark	2-Methoxycinnamaldehyde (2-MCA)	Upregulation of proapoptotic molecule, downregulate cell growth marker topoisomerase I and II.	Tsai et al. (2016)

(Continued)

TABLE 6.2 *(Continued)*
Plant Pharmacology – Unveiling Names, Utilized Parts, Bioactive Compounds, and Modes of Action

Sl. No.	Name of the Plant	Parts Used	Bioactive Compound	Mode of Action	References
21.	*Humulus lupulus* (Hops)	Hop Cones	Xanthohumol (XN), isoxanthohumol (IXN), tetrahydroxanthohumol (TXN)	Increase autophagy in CRC cells, induction of S phase arrest, ROS release, and ER stress. Decrease in cyclin A/CDK2. Increased PARP cleavage. Increased CHOP, ATF4, IRE1α.	Tamia et al. (2024)
22.	*Withaniasomnifera* (Ashwagandha)	Roots, Leaves	Withaferin, withanolides	Anti-inflammatory, antioxidant that neutralize free radicals, cell cycle regulation, induce apoptosis, Inhibit angiogenesis.	Singh et al. (2021)
23.	*Rubus occidentalis* (Black raspberry)	Berry	Phenolic acid, flavonoids, anthocyanin, tannic acid, styrene, terpene, ellagic acid	Demethylation of SFRP2 gene promoter, increased SFRP2 expression. Downregulate the expression of DNMT31, DNMT3B, p-STAT3	Chen et al. (2018)
24.	*Echinacea purpurea* (Echinacea)	Root	Alkamides, ketoalkene, caffeic acid derivatives, polysaccharides, glycoproteins, cichoric acid, n-hexane	Increased cAMP, p38-MAPK, cJNK, NF-κB, ATF-2/CREB-1. Decreased telomerase activity, lead to apoptosis.	Manayi et al. (2015)
25.	*Boswellia sacra* (Boswellia)	Gum resin	Acetyl-11-keto-beta-boswellic acid (AKBA)	Suppress NF-κB, inhibit the expression of COX2, bcl2, bcl-xL, IAP1, surviving, cyclin D1, ICAM1, MMP9. Inhibit distant organ metastasis.	Yadav et al. (2012)
26.	*Moringa oleifera* (Moringa)	Seeds, leaves, root, stem	Alkaloids, glucosinolates, isothiocyanates, thiocarbamates, vitamins, carotenoids, polyphenols, phenolic acids, flavanoids, glucosine, tannin, saponin, protocyanogen	Induce apoptosis by inducing mitochondrial dysfunction. Regulate the expression of Bcl2 and sirtuin 1.	Wu et al. (2021)
27.	*Ganoderma lucidum*	Broken spores	Polysaccharides, triterpenoids, alkaloids, fatty acids, lactones, steroids	Downregulate cyclin B1 and cyclin A2; Upregulate p21, hence disrupt cell cycle progression. Decrease Bcl2 and survivin, PARP, pro-caspase 3, pro-caspase 9, hence induce apoptosis. Reduction in Ki67, PCNA.	Na et al. (2017)
28.	*Ocimum sanctum* (Tulsi)	Leaves, Essential oil	Eugenol, β-caryophyllene, ursolic acid, rosmarinic acid, apigenin, carvacrol	Triggering cell death; cell cycle arrest; inhibit migration, metastasis, angiogenesis. Express Bax, decrease Bcl2, induce apoptosis. Inhibit MMPs, and reduce invasion of cancer cells. Inhibit Wnt/β-catenin pathway. Enhance thymidylate synthase expression, hence apoptosis by 5-FU. Production of ROS and activate apoptotic signal.	Hasan et al. (2023)
29.	*Crocus sativus* (Saffron)	Dried stigma	Crocin, crocetin, picrocrocin, safranal	Cell cycle arrest, upregulate p21, inhibit cell proliferation. Downregulate Beclin 1, Atg7.	Lambrianidou et al. (2020)
30.	*Carica papaya* (Papaya)	Seeds	Benzyl glucosinolate, isocyanate, lycopene	G2-M phase arrest, increased expression of caspase 3, p53. Downregulate Bcl2 gene expression. Trigger apoptosis of cancer cell.	Mahrous and Noseer (2023)

Phytochemicals can affect lactate production, pentose phosphate pathway, tricarboxylic acid cycle, and can promote apoptosis of cancer cells (Esmeeta et al., 2022). Some of the examples such as polyphenols of green tea EGCG activate AMP protein kinase pathway which can control cellular energy, protein synthesis, cell cycle regulation, growth of cell, and leads to apoptosis of cancer cells (Tamara et al., 2014). On the other hand, resveratrol, which is present in fruits and vegetables, has a direct impact on cell oxidation and ROS release via inhibiting the formation of tumor growth's microenvironment. Moreover, the metabolism in different types of cancer cells is modulated by phenolic compounds (Han et al., 2007).

Flavones from fruit and vegetables such as apigenin is found as significant anticancerous and antioxidant agent which can affect tumor growth by downregulating glucose uptake of GLUT-1 in pancreatic cells. As a result, there is a decrease in ATP and glycolytic flux. Apigenin can induce apoptosis without affecting other glycolytic enzymes and it increase sensitivity toward cancer chemotherapy (Xu et al., 2014). A common dietary spice, curcumin also works on similar mechanism of GLUT-1 downregulation and reduces availability of carbohydrate from tumor cells. This suggests mitochondrial-mediated apoptosis by dissociation of hexokinase-2 from mitochondria by the induction of curcumin (Wang et al., 2015).

6.7 COMBINED THERAPEUTIC APPROACHES IN CRC TREATMENT USING PLANT-DERIVED NATURAL PRODUCTS

A novel application of natural substances is the so-called combination therapy, which involves administering two or more substances – conventional chemotherapeutics and one or more natural chemicals at the same time. Some natural chemicals may sensitize to conventional cytotoxic therapy, increase the drug's effective concentration, enhance the combined effect of both delivered medicines, or exert cytotoxic effects only on tumor cells. Furthermore, combination therapy, which targets numerous signaling pathways, employs a variety of strategies to diminish resistance to anticancer medicines (Figure 6.1).

The desired impact could be to reduce the strain on the patient's body by replacing a portion of the dose of a standard chemotherapy with a natural material with a specific effect. Many natural chemicals are well tolerated by patients and do not have harmful effects even at high concentrations. The interaction of conventional chemotherapeutics with natural chemicals opens up new avenues

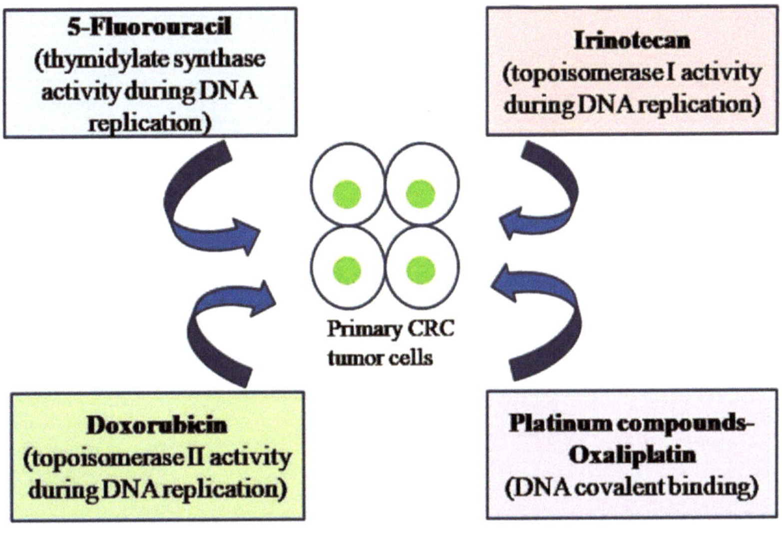

FIGURE 6.1 Chemotherapy of CRC and mechanism of action.

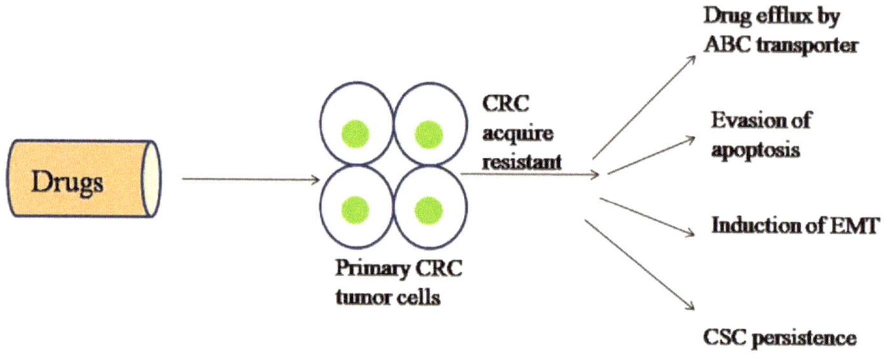

FIGURE 6.2 Mechanism of drug resistance by CRC cells.

for cancer research and treatment. It could be a viable method to potentially improving outcomes while minimizing the negative effects associated with traditional chemotherapy. Plant-derived bioactive components typically target numerous apoptotic processes, making them more effective than single medication treatments (Figure 6.2). This section discusses how CRC tumor cells develop drug resistance to systemic chemotherapy and how plant-derived compounds can be used in combination therapies to re-sensitize cancer cells shown in Figure 6.3 (Gavrilas et al., 2022).

6.8 CURRENT STATUS AND LIMITATIONS OF PLANT BIOACTIVE COMPOUNDS AGAINST CRC

The global incidence of CRC, often known as colon cancer, is concerning. It is the second most common cancer in women (5,71,000 cases, or 9.4% of all cancer cases), and the third most common in men (6,63,000 cases, or 10% of all cancer cases). According to WHO in 2020, incidence and mortality global cancer statistics is shown in Figures 6.4 and 6.5, respectively (Xi and Xu, 2021).

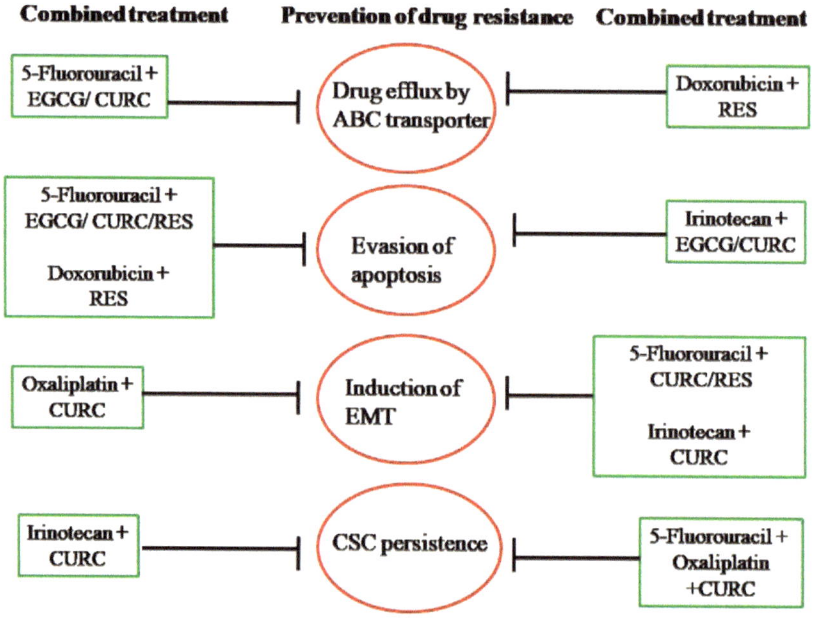

FIGURE 6.3 Combined treatment by combination of chemotherapy and natural plant products in prevention of drug resistance in CRC. CURC, curcumin; EGCG, epigallocatechin gallate; RES, resveratrol.

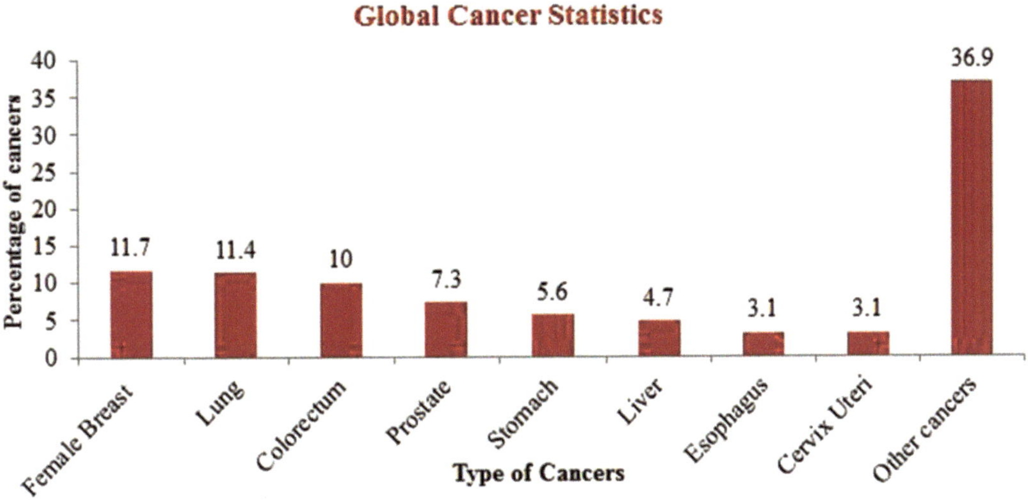

FIGURE 6.4 Graph showing incidence of global cancer.

The high incidence of CRC, which contributes to high mortality rates, and the growing global demand for more potent treatments are also major motivators. CRC medications have advanced significantly in research and development since they were initially made available to the general population. Since then, a number of novel treatments have been released, with promising results shown in Figures 6.4 and 6.5.

There is insufficient clinical research on phytochemical toxicity. Therefore, phytochemicals are not yet permitted for clinical use in cancer treatment. Some phytochemicals can be hazardous to humans and behave as antinutrients, preventing absorption of critical nutrients.

Phytochemicals include methodological issues such as small sample size, trial duration, and lack of control throughout clinical studies. Using traditional plants as a supplement with other drugs might lead to serious pharmacological responses, sometimes life-threatening.

Preclinical studies on phytochemicals for anticancer activities have shown poor molecular interactions with several signaling molecules. Nanoparticle-attached targeted medication delivery systems are being developed to make bioactive substances more accessible to tissues. The study of phytochemicals for cancer treatment is still in its early stages.

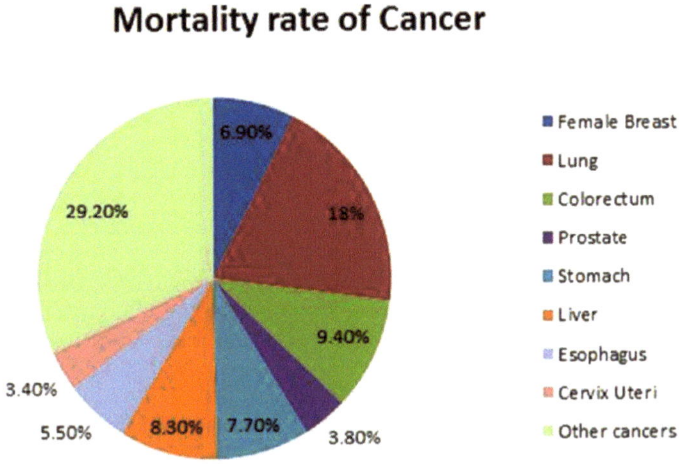

FIGURE 6.5 Chart showing mortality rate of different cancer globally.

Clinical investigations on phytochemicals must demonstrate their safety, effects, and efficacy. Standardization is crucial for ensuring consistency, composition, manufacturing process efficiency, and quality. To generate pharmaceutical compositions that meet international standards, phytochemicals require thorough clinical trials. Researching the anticancer benefits of plant metabolites and probiotics can be challenging due to differences in herbs, biochemical tools, and study methods. The study of a plant metabolite is limited by its inability to produce desired effects in all combinations (Thiengsusuk et al., 2020).

6.9 CONCLUSION AND FURTHER PROSPECTIVE

Plant-derived bioactive chemicals are promising options for overcoming the rising prevalence of acquired drug resistance in cancer therapy. When used in combination with standard chemotherapeutic drugs, they frequently produce enhanced antitumor effects and can block drug-resistant mechanisms. According to the research literature included in this chapter, bioactive dietary components can improve the chemotherapeutic impact of medications and diminish chemoresistance by lowering efflux proteins from the ABC family transporters, inducing apoptotic cell death, and reversing EMT. One of the major issues in clinical translation is the low bioavailability of most compounds, such as curcumin or resveratrol, which are quickly metabolized and eliminated by the body, making maintaining therapeutically appropriate concentrations in the circulation difficult. The advancement of nanotechnology brings up new avenues for increasing the medicinal effects of these plant-derived bioactive chemicals. A well-formulated nano strategy boosts bioavailability and bioactivity by reducing particle size, changing the surface, and entrapping the drug with various nanocarriers, as documented in detail in recent reviews in the instance of curcumin (Gera et al., 2017; Wong et al., 2019), resveratrol (Sharifi-Rad et al., 2021) or EGCG (Granja et al., 2016). As more information is obtained in the creation of innovative drug delivery vehicles, it is critical that these natural substances be clinically evaluated in a large number of studies to validate the efficacy found in vitro and in animal models. Furthermore, each compound's safety profile must be determined when provided to cancer patients. Despite the challenges that lay ahead, the potential of these plant-derived bioactive components is great, and it is well worth the scientific community's joint efforts to assist oncological patients.

ACKNOWLEDGMENT

We thank DST-FIST, KSTA, and Mahajana Education Society for generous support. Also, we would like to thank Dr. Ramesh SR for critical comment on the manuscript.

REFERENCES

Adebesin AO, Ayodele AO, Omotoso O, Akinnusi PA, Olubode SO (2022) Computational evaluation of bioactive compounds from *Vitis vinifera* as a novel β-catenin inhibitor for cancer treatment. Bull Natl Res Cent 46:183

Ahmed S, Johnson K, Ahmed O, Iqbal N (2014) Advances in the management of colorectal cancer: From biology to treatment. Int J Colorectal Dis 29:1031–1042

Ali Khan U, Fallah M, Sundquist K, Sundquist J, Brenner H, Kharazmi E (2020) Risk of colorectal cancer in patients with diabetes mellitus: A Swedish nationwide cohort study. PLoS Med 17:e1003431

Alzahrani SM, Doghaither HA, Ghafari AB (2021) General insight into cancer: An overview of colorectal cancer. Mol Clin Oncol 21:102–111

Aykan NF (2015) Red meat and colorectal cancer. Oncol Rev 9:288

Awuchi, Godswil (2019) Medicinal plants: The medical, food and nutritional biochemistry and uses. Int Journal of Advanced and Academic Research (IJAR) 5:1–22.

Benarba B, Pandiella A (2018) Colorectal cancer and medicinal plants: Principle findings from recent studies. Biomed Pharma 107:408–423

Blondy S, David V, Verdier M, Mathonnet M, Perraud A, Christou N (2020) 5-Fluorouracil resistance mechanisms in colorectal cancer: From classical pathways to promising processes. Cancer Sci 111(9): 3142–3154

Brzacki V, Nagorni A, Kallistratos M, Manolis A, Lovic D (2019) Diabetes mellitus: A clinical condition associated with metabolic syndrome and colorectal cancer risk. Curr Pharmacol Rep 5:205–209

Chen L, Jiang B, Zhong C, Guo J, Zhang L, Mu T, Zhang Q, Bi X (2018) Chemoprevention of colorectal cancer by black raspberry anthocyanins involved the modulation of gut microbiota and SFRP2 demethylation. Carcinogenesis 39(3):471–481

Comella P (2007) A review of the role of capecitabine in the treatment of colorectal cancer. Ther Clin Risk Manag 3(3):421–431

Comella P, Casaretti R, Sandomenico C, Avallone A, Franco L (2009) Role of oxaliplatin in the treatment of colorectal cancer. Ther Clin Risk Manag 5(1):229–238

Costea T, Hudiţă A, Ciolac O-A, Gălăţeanu B, Ginghină O, Costache M, Ganea C, Mocanu M-M (2018) Chemoprevention of colorectal cancer by dietary compounds. Int J Mol Sci 19(12):3787

Coupez D, Hulo P, Touchefeu Y, Bossard C, Bennouna J (2020) Pembrolizumab for the treatment of colorectal cancer. Expert Opin Biol Ther 20(3):219–226

Dimas K, Tsimplouli C, Houchen C, Pantazis P, Sakellaridis N, Tsangaris GT, Anastasiadou E, Ramanujam RP (2015) An ethanol extract of Hawaiian turmeric: Extensive in vitro anticancer activity against human colon cancer cells. Altern Ther Health Med 2:46–54

Dong M, Yang G, Liu H, Liu X, Lin S, Sun D, Wang Y (2014) Aged black garlic extract inhibits Ht29 colon cancer cell growth via the PI3K/Akt signaling pathway. Biomed Rep 2:250–254

Dong Y, Zhou J, Zhu Y, Luo L, He T, Hu H, Liu H, Zhang Y, Luo D, Xu S, Xu L, Liu J, Zhang J, Teng Z (2017) Abdominal obesity and colorectal cancer risk: Systematic review and meta-analysis of prospective studies. Bio Sci Rep 37(6):BSR20170945

Eghbali S, Askari SF, Avan R, Sahebkar A (2021) Therapeutic effects of *Punica granatum* (pomegranate): An updated review of clinical trials. J Nutr Metab 2021:5297162

Esghaei M, Ghaffari H, Rahimi Esboei B, Tapeh E, Salim B, Motevalian F (2018) Evaluation of anticancer activity of Camellia Sinensis in the Caco-2 colorectal cancer cell line. Asian Pac J Cancer Prev 19(6):1697–1701

Esmeeta A, Adhikary S, Dharshnaa V, Swarnamughi P, Ummul Maqsummiya Z, Banerjee A, Pathak S, Duttaroy AK (2022) Plant-derived bioactive compounds in colon cancer treatment: An updated review. Biomed Pharmacother 153:113384

Forbes-Hernández TY, Giampieri F, Gasparrini M, Mazzoni L, Quiles JL, Alvarez-Suarez JM, Battino M (2014) The effects of bioactive compounds from plant foods on mitochondrial function: A focus on apoptotic mechanisms. Food Chem Toxicol 68:154–182

Fujita K, Kubota Y, Ishida H, Sasaki Y (2015) Irinotecan, a key chemotherapeutic drug for metastatic colorectal cancer. World J Gastroenterol 21(43):12234–12248

Gavrilas LI, Cruceriu D, Mocan A, Loghin F, Miere D, Balacescu O (2022) Plant-derived bioactive compounds in colorectal cancer: Insights from combined regimens with conventional chemotherapy to overcome drug-resistance. Biomedicines 10(8):1948

Gera M, Sharma N, Ghosh M, Huynh DL, Lee SJ, Min T, Kwon T, Jeong DK (2017) Nanoformulations of curcumin: An emerging paradigm for improved remedial application. Oncotarget 8(39):66680–66698

Giovannucci E (2001) An updated review of the epidemiological evidence that cigarette smoking increases risk of colorectal cancer. Cancer Epidemiol Biomark Prev 10:725

Grace I AG, Stephen AE, Ngozi LU, Obidola SM, Oluwafemi AO, Ika MD, Akachukwu O, Grace IJ, Abayomi EA, Titilayo OJ (2023) Molecular modeling of the interactions of Curcuma longa compounds with VEGFR towards colorectal cancer drug development. Elsevier Ltd. 2352–9148

Granja A, Pinheiro M, Reis S (2016) Epigallocatechin gallate nanodelivery systems for cancer therapy. Nutrients 8(5):307

Hamouda SB, Benkhadir KE (2023) Interplay between signaling pathways and tumor microenvironment components: A paradoxical role in colorectal cancer. Int J Mol Sci 24:1–23

Han X, Shen T, Lou H (2007) Dietary polyphenols and their biological significance. Int J Mol Sci 8(9): 950–988

Hasan MR, Alotaibi BS, Althafar ZM, Mujamammi AH, Jameela J (2023) An update on the therapeutic anticancer potential of *Ocimum sanctum* L.: "Elixir of life". Molecules 28(3):1193

Hossain MS, Kader MA, Goh KW, Islam M, Khan MS, Harun-Ar Rashid M, Ooi J, Melo Coutinho HD, Al-Worafi YM, Moshawih S, Lim YC, Kibria KMK, Ming LC (2022) Herb and spices in colorectal cancer prevention and treatment: A narrative review. Front Pharmacol 13:865801

Hossain MS, Karuniawati H, Jairoun AA, Urbi Z, Ooi DJ (2022) Colorectal cancer: A review of carcinogenesis, global epidemiology, current challenges, risk factors, preventive and treatment strategies. Cancer 14:1732–1741
Ju M, Cheng H, Qu K, Lu X (2020) Efficacy and safety of ramucirumab treatment in patients with advanced colorectal cancer: A protocol for systematic review and meta analysis. Medicine (Baltimore) 99(24):e20618
Kanthan R, Senger JL, Kanthan SC (2012) Molecular events in primary and metastatic colorectal carcinoma: A review. Pathol Res Int 21:1–17
Kastrinos F, Jewel Samadder NJ, Burt RW (2020) Use of family history and genetic testing to determine risk of colorectal cancer. Gastroenterology 158:389–403
Kaufman N, Nori D, Shank B, Linares L, Harrison L, Fass D, Enker W (1989) Remote afterloading intraluminal brachytherapy in the treatment of rectal, rectosigmoid, and anal cancer: A feasibility study. Int J Radiat Oncol Biol Phys 17(3):663–668
Keller DS, Windsor A, Cohen R, Chand M (2019) Colorectal cancer in inflammatory bowel disease: Review of the evidence. Tech Coloproctol 23:3–13
Kim S, Yang HY, Lee HJ, Ju J (2021) In vitro antioxidant and anti-colon cancer activities of *Sesamum indicum* L. Leaf extract and its major component, pedaliin. Foods 10(6):1216
Koornstra JSD, Hollema H, Vries EG, Kleibeuker JH (2003) Changes in apoptosis during the development of colorectal cancer: A systematic review of the literature. Crit Rev Oncol Haematol 44:1–19
Kovalic JJ (1988) Endocavitary irradiation for rectal cancer and villous adenomas. Int J Radiat Oncol Biol Phys 14:261–264
Kumar A, Gautam V, Sandhu A, Rawat K, Sharma A, Saha L (2023) Current and emerging therapeutic approaches for colorectal cancer: A comprehensive review. World J Gastrointest Surg 15(4):495–519
Kumar D, Goel NK, Pandey AK, Sarpal SS (2016) Complementary and alternative medicine use among the cancer patients in northern India. South Asian J Cancer 05(01):008–011
Kusmardi K, Baihaqi LA, Estuningtyas A, Sahar N, Sunaryo H, Tedjo A (2021) Ethanol extract of pomegranate (*Punica granatum*) peel in increasing the expression of Caspase-3 in DSS-induced mice. Int J Inflammation 49:1–16, 4919410
Lambrianidou A, Koutsougianni F, Papapostolou I, Dimas K (2020) Recent advances on the anticancer properties of saffron (*Crocus sativus* L.) and its major constituents. Molecules 27;26(1):86
Leite CDS, Bonafé GA, Carvalho Santos J, Martinez CAR, Ortega MM, Ribeiro ML (2022) The anti-inflammatory properties of licorice (*Glycyrrhiza glabra*)-derived compounds in intestinal disorders. Int J Mol Sci 23(8):4121
Li J, Ma X, Chakravarthi D, Shalapour S, Depinho R (2023) Genetic and biological hallmarks of colorectal cancer. Genes Dev 31:11–22
Liu PH, Wu K, Ng K, Zauber AG, Nguyen LH, Song M, He X, Fuchs CS, Ogino S, Willett WC, Chan AT, Giovannucci EL, Cao Y (2019) Association of obesity with risk of early-onset colorectal cancer among women. JAMA Oncol 5:37–44
Lu J, Kornmann M, Traub B (2023) Role of epithelial to mesenchymal transition in colorectal cancer. International J Mol Sci 24:1–20
Macharia JM, Mwangi RW, Rozmann N, Zsolt K, Uchechukwu TPO, Wagara IN, Raposa BL (2022) Medicinal plants with anti-colorectal cancer bioactive compounds: Potential game-changers in colorectal cancer management. Elsevier Masson SAS. 0753–3322
Mahrous NS, Noseer EA. Anticancer potential of Carica papaya Linn black seed extract against human colon cancer cell line: In vitro study. BMC Complement Med Ther 23–271
Makishima H, Yasuda S, Isozaki Y, Kasuya G, Okada N, Miyazaki M, Mohamad O, Matsufuji N, Yamada S, Tsuji H, Kamada T (2019) Liver Cancer Working Group. Single fraction carbon ion radiotherapy for colorectal cancer liver metastasis: A dose escalation study. Cancer Sci 110(1):303–309
Manayi A, Vazirian M, Saeidnia S (2015) Echinacea purpurea: Pharmacology, phytochemistry and analysis methods. Pharmacogn Rev 9(17):63–72
Manirakiza A, Irakoze L, Manirakiza S (2021) Aloe and its effects on cancer: A narrative literature review. East Afr Health Res J 5(1):1–16
Mano M, Humblet Y (2008) Drug insight: Panitumumab, a human EGFR-targeted monoclonal antibody with promising clinical activity in colorectal cancer. Nat Clin Pract Oncol 5(7):415–425
Mariani F, Sena P, Roncucci L (2014) Inflammatory pathways in the early steps of colorectal cancer development. World J. Gastroenterol 7:9716–9731
Marmol I, Diego CS, Dieste AP, Cerrada E, Yoldi MJ (2017) colorectal carcinoma: A general overview and future perspectives in colorectal cancer. Int J Mol Sci 18:1–16

Matsumoto T, Ikoma T, Yamamura S, Miura K, Tsuduki T, Watanabe T, Nagai H, Takatani M, Yasui H (2023) Regorafenib is suitable for advanced colorectal cancer patients who have previously received trifluridine/tipiracil plus bevacizumab. Sci Rep 13:2433

Memmola R, Petrillo A, Di Lorenzo S, Altuna SC, Habeeb BS, Soggiu A, Bonizzi L, Garrone O, Ghidini M (2022) Correlation between olive oil intake and gut Microbiota in colorectal cancer prevention. Nutrients 14(18):3749

Min SJ, Lim JY, Kim HR, Kim SJ, Kim Y (2015) Sasa quelpaertensis leaf extract inhibits colon cancer by regulating cancer cell stemness in vitro and in vivo. Int J Mol Sci 16(5):9976–9997

Mulet N, Matos I, Noguerido A, Martini G, Élez ME, Argilés G, Tabernero J (2018) Evaluating trifluridine + tipiracil hydrochloride in a fixed combination (TAS-102) for the treatment of colorectal cancer. Expert OpinPharmacother 19(6):623–629

Muro K, Salinardi T, Singh AR, Macarulla T (2020) Safety of aflibercept in metastatic colorectal cancer: A literature review and expert perspective on clinical and real-world data. Cancers (Basel) 12(4):844

Murphy N, Moreno V, Hughes DJ, Vodicka L, Vodicka P, Aglago EK, Gunter MJ, Jenab M (2019) Lifestyle and dietary environmental factors in colorectal cancer susceptibility. Mol Asp Med 69:2–9

Na K, Li K, Sang T, Wu K, Wang Y, Wang X (2017) Anticarcinogenic effects of water extract of sporoderm-broken spores of Ganoderma lucidum on colorectal cancer in vitro and in vivo. Int J Oncol 50(5):1541–1554

Naselli F, Tesoriere L, Caradonna F, Bellavia D, Attanzio A, Gentile C, Livrea MA (2014) Anti-proliferative and pro-apoptotic activity of whole extract and isolated indicaxanthin from Opuntia ficus-indica associated with re-activation of the onco-suppressor p16(INK4a) gene in human colorectal carcinoma (Caco-2) cells. BiochemBiophys Res Commun 450(1):652–658

Ngan SY, Burmeister B, Fisher RJ, Solomon M, Goldstein D, Joseph D, Ackland SP, Schache D, McClure B, McLachlan SA, McKendrick J, Leong T, Hartopeanu C, Zalcberg J, Mackay J (2012) Randomized trial of short-course radiotherapy versus long-course chemoradiation comparing rates of local recurrence in patients with T3 rectal cancer: Trans-Tasman Radiation Oncology Group trial 01.04. J Clin Oncol 30(31):3827–3833

Oh YJ, Sung MK (2001) Soybean saponins inhibit cell proliferation by suppressing PKC activation and induce differentiation of HT-29 human colon adenocarcinoma cells. Nutr Cancer 39(1):132–138

Pérez Sánchez A, Barrajón Catalán E, Ruiz Torres V, AgullóChazarra L, Herranz López M, Valdés A, Cifuentes A, Micol V (2019) Rosemary (*Rosmarinus officinalis*) extract causes ROS-induced necrotic cell death and inhibits tumor growth in vivo. Sci Rep 9:808

Pignone M, Rich M, Teutsch SM, Berg AO, Lohr KN (2002) Screening for colorectal cancer in adults at average risk: A summary of the evidence for the U.S. Preventive Services Task Force. Ann Intern Med 137:132–141

Pilar A, Gupta M, Ghosh Laskar S, Laskar S (2017) Intraoperative radiotherapy: Review of techniques and results. E cancer medical science 11:750

Ranjbarnejad T, Saidijam M, Moradkhani S, Najafi R (2017) Methanolic extract of Boswellia serrata exhibits anti-cancer activities by targeting microsomal prostaglandin e synthase-1 in human colon cancer cells. Prostaglandins Other Lipid Mediat 131:1–8

Rossi M, Jahanzaib Anwar M, Usman A, Keshavarzian A, Bishehsari F (2018) Colorectal cancer and alcohol consumption-populations to molecules. Cancers 10:38

Roviello GD, Angelo A, Petrioli R, Roviello F, Cianchi F, Nobili S, Mini E, Lavacchi D (2020) Encorafenib, binimetinib, and cetuximab in BRAF V600E-mutated colorectal cancer. Transl Oncol 13(9):100795

Sawicki T, Ruszkowska M, Danielewicz A, Niedzwiedzka E, Arlukowicz T, Przybylowicz KE. (2021). A review of colorectal cancer in terms of epidemiology, risk factors, development, symptoms and diagnosis. Cancers 13: 1–14

Schmid D, Leitzmann MF (2014) Television viewing and time spent sedentary in relation to cancer risk: A meta-analysis. J Natl Cancer Inst 4:106–112.

Schuell B, Gruenberger T, Kornek GV, Dworan N, Depisch D, Lang F, Schneeweiss B, Scheithauer W (2005) Side Effects during chemotherapy predict tumour response in advanced colorectal cancer. Br J Cancer 93:744–748

Sedlak JC, Yilmaz OH, Roper J (2023) Metabolism and colorectal cancer. Annu Rev Pathol 18:467–492

Shannar A, Sarwar Md S, Kong A-, Ng T (2022) A new frontier in studying dietary phytochemicals in cancer and in health: Metabolic and epigenetic reprogramming. Prev Nutr Food Sci 27(4):335–346

Sharifi-Rad J, Quispe C, Mukazhanova Z, Knut E, Turgumbayeva A, Kipchakbayeva A, Seitimova G, Mahomoodally MF, Lobine D, Koay A, Wang J, Sheridan H, Leyva-Gómez G, Prado-Audelo MLD, Cortes H, Rescigno A, Zucca P, Sytar O, Imran M, Rodrigues CF, Cruz-Martins N, Ekiert H, Kumar M,

Abdull Razis AF, Sunusi U, Kamal RM, Szopa A (2021) Resveratrol-based nanoformulations as an emerging therapeutic strategy for cancer. Front Mol Biosci 8:649395

Shiba S, Shibuya K, Okamoto M, Okano N, Kubo N, Kaminuma T, Sato H, Okazaki S, Miyasaka Y, Kawamura H, Ohno T (2021) Carbon-ion radiotherapy for oligometastatic colorectal cancer in the liver or lung. Anticancer Res 41(4):1997–2005

Shussman N, Wexner SD (2014) Colorectal polyps and polyposis syndromes. Gastroenterol Rep 2:1–15

Siegel RL, Miller KD, Goding SA, Fedewa SA, Butterly LF, Anderson JC, Cercek A, Smith RA, Jemal A (2020) Colorectal cancer statistics, 2020. CA Cancer J Clin 70:145–164

Siegel RL, Torre LA, Soerjomataram I, Hayes RB, Bray F, Weber TK, Jemal A (2019) Global patterns and trends in colorectal cancer incidence in young adults. Gut 68:2179–2185

Siegel RL, Wagle NS, Cercek A, Smith RA, Jemal A (2023) Colorectal cancer statistics, 2023. CA Cancer J Clin 73:233–254

Singh N, Yadav SS, Rao AS, Nandal A, Kumar S, Ganaie SA, Narasihman B (2021) Review on anticancerous therapeutic potential of *Withania somnifera* (L.) Dunal. J Ethnopharmacol 270:113704

Tabernero J, Grothey A, Van Cutsem E, Yaeger R, Wasan H, Yoshino T, Desai J, Ciardiello F, Loupakis F, Hong YS, Steeghs N, Guren TK, Arkenau HT, Garcia-Alfonso P, Elez E, Gollerkeri A, Maharry K, Christy-Bittel J, Kopetz S (2021) Encorafenib plus cetuximab as a new standard of care for previously treated BRAF V600E-mutant metastatic colorectal cancer: Updated survival results and subgroup analyses from the BEACON study. J Clin Oncol 39(4):273–284

Tam SY, Wu VWC (2019) A review on the special radiotherapy techniques of colorectal cancer. Front Oncol 9:208

Tamara P, Jette N, Frolkis A, Steeves TD (2014). The prevalence of parkinson's disease: a systematic review and meta analysis. Mov Disord 29(13):1583–1590.

Tamia G, Amarakoon D, Wei CI, Lee SH (2024) Evaluation of *Humulus lupulus* (hops) compound anti-proliferative properties with mechanisms in human colorectal cancer cells. Food Bioscience 57:103493

Tappenden P, Jones R, Paisley S, Carroll C (2007) Systematic review and economic evaluation of bevacizumab and cetuximab for the treatment of metastatic colorectal cancer. Health Technol Assess 11(12):1–128

Tariq K, Ghias K (2016) Colorectal cancer carcinogenesis: A review of mechanisms. Cancer Biol Med 13:120–136

Tesfaye A (2021) Revealing the therapeutic uses of garlic (Allium sativum) and its potential for drug discovery. Sci World J 2021:8817288

Thiengsusuk A, Boonprasert K, Bangchang K (2020) A systematic review of drug metabolism studies of plants with anti-cancer properties: Approaches applied and limitations. Eur J Drug Metab Pharmacokinet 45(2):173–225.

Tian Y, Kharazmi E, Sundquist K, Sundquist J, Brenner H, Fallah M (2019) Familial colorectal cancer risk in half siblings and siblings: Nationwide cohort study. BMJ 364:l803

Tsai K, Cherng J, Liu YH, Chen TW, Wong HY, Yang S, Chou KS, Cherng JM (2016) Cinnamomum verum component 2-methoxycinnamaldehyde: A novel antiproliferative drug inducing cell death through targeting both topoisomerase I and II in human colorectal adenocarcinoma COLO 205 cells. Food Nutr Res 60:1–18.

Valle L, Vilar E, Tavtigian SV, Stoffel EM (2019) Genetic predisposition to colorectal cancer: Syndromes, genes, classification of genetic variants and implications for precision medicine. J Pathol 247:574–588

van Gijn W, Marijnen CA, Nagtegaal ID, Kranenbarg EM, Putter H, Wiggers T, Rutten HJ, Påhlman L, Glimelius B, van de Velde CJ (2011) Preoperative radiotherapy combined with total mesorectal excision for resectable rectal cancer: 12-year follow-up of the multicentre, randomised controlled TME trial. Lancet Oncol 12(6):575–582

Wang K, Fan H, Chen Q et al (2015) Curcumin inhibits aerobic glycolysis and induces mitochondrial-mediated apoptosis through hexokinase II in human colorectal cancer cells in vitro. Anti-Cancer Drugs 26 (1): 15–24.

Wang T, Li X, Ma R, Huang JS, Sun Z, Wang M (2024) Advancement in colorectal cancer research: Unveiling the cellular and molecular mechanisms of neddylation. Int J Oncol 66:1019–6439

Wei L, Lin Z, Xie S, Ruan D, Jiang W, Cui Y, Liu S, Wang T, Chen Z, Lin Q (2022) Complete response with cetuximab-based treatment of metastatic colorectal cancers: Two case reports and literature review. Front Oncol 12:798515

White A, Ironmonger L, Steele RJC, Ormiston-Smith N, Crawford C, Seims A (2018) A review of sex-related differences in colorectal cancer incidence, screening uptake, routes to diagnosis, cancer stage and survival in the UK. BMC Cancer 18(1):906

Wolf AMD, Fontham ETH, Church TR, Flowers CR, Guerra CE, LaMonte SJ, Etzioni R, McKenna MT, Oeffinger KC, Shih YT, Walter LC, Andrews KS, Brawley OW, Brooks D, Fedewa SA, Manassaram-Baptiste D, Siegel RL, Wender RC, Smith RA (2018) Colorectal cancer screening for average-risk adults: 2018 guideline update from the American Cancer Society. CA Cancer J Clin 68(4):250–281

Wong KE, Ngai SC, Chan KG, Lee LH, Goh BH, Chuah LH (2019) Curcumin nanoformulations for colorectal cancer: A review. Front Pharmaco 10:152

Wu YY, Xu YM, Lau ATY (2021) Anti-cancer and medicinal potentials of moringa isothiocyanate. Molecules 26(24):7512

Xi Y, Xu P (2021) Global colorectal cancer burden in 2020 and projections to 2040. Translational Oncol 14(10):101174

Xiang S, Jian Q, Chen W, Xu Q, Li J, Wang C, Wang R, Zhang D, Lin J, Zheng C (2024) Pharmacodynamic components and mechanisms of ginger (*Zingiber officinale*) in the prevention and treatment of colorectal cancer. J Ethnopharmacol 324:117733

Xu YY, Wu TT, Zhou SH, Bao YY, Wang QY, Fan J, Huang YP (2014) Apigenin suppresses GLUT-1 And p-AKT expression to enhance the chemosensitivity to cisplatin of laryngeal carcinoma hep-2 cells: An in vitro study. Int J Clin Exp Pathol 7(7):3938–3947

Xue K, Li FF, Chen YW, Zhou YH, He J (2017) Body mass index and the risk of cancer in women compared with men: A meta-analysis of prospective cohort studies. Eur J Cancer Prev 26:94–105

Yadav VR, Prasad S, Sung B, Gelovani JG, Guha S, Krishnan S, Aggarwal BB (2012) Boswellic acid inhibits growth and metastasis of human colorectal cancer in orthotopic mouse model by downregulating inflammatory, proliferative, invasive and angiogenic biomarkers. Int J Cancer 130(9):2176–2184

Yoshida K, Teppabut Y, Sawaguchi R, Nakane Y, Hayashi E, Oyama K, Nishizaki Y, Goda Y, Kondo T (2020) 5,7,3′,4′-tetrahydroxyflav-2-en-3-ol 3-O-glucoside, a new biosynthetic precursor of cyanidin 3-o-glucoside in the seed coat of black soybean, glycine max. Sci Rep 10:17184

Yu YL, Tseng WK, Liao CK, Yeh CY, Chen HH, Liu YH, Liaw YW, Fan CW (2023) Using oral tegafur/uracil (UFT) plus leucovorin as adjuvant chemotherapy in stage II colorectal cancer: A propensity score matching study from Taiwan. BMC Cancer 23:900

Zeng Q, Che Y, Zhang Y, Chen M, Guo Q, Zhang W (2020) Thymol isolated from Thymus vulgaris l. Inhibits colorectal cancer cell growth and metastasis by suppressing the Wnt/β-catenin pathway. Drug Des Devel Ther 14:2535–2547

Zhao L, Zhao H, Zhao Y, Sui M, Liu J, Li P, Liu N, Zhang K (2022) Role of ginseng, quercetin, and tea in enhancing chemotherapeutic efficacy of colorectal cancer. Front Med 9:939424

Zheng Q, Wang X, Gao T, Zhang B, Zhao N, Du R, Zhao Z (2023) Exploring the pharmacological and molecular mechanisms of *Salvia chinensis Benth* in colorectal cancer: A network pharmacology and molecular docking study. Medicine (Baltimore) 102(50):e36602

7 Evaluation of Nutraceuticals in Targeting Oral Cancer

Murugesan Amirtha Varshini, Lohapriya KB, Jayadevan Pooja, and Arikketh Devi
Cancer and Stem Cell Biology Laboratory, Department of Genetic Engineering, School of Bioengineering
SRM Institute of Science and Technology
Kattankulathur, Chengalpattu District, India

7.1 INTRODUCTION

Oral cancer – development of malignant tumors in the mouth and throat ranks as the 13th most common cancer worldwide according to WHO with 3.5 billion as of 2023. It is classified under the category of head and neck cancer which involves lips, tongue, gums, cheeks, base of mouth, with symptoms like throat pain, sticky saliva, xerostomia (dryness), and feeding and speaking difficulty. Approximately buccal mucosa accounts for 32%, the lower lip for 11%, the tongue for 22%, the palate for 11%, the floor of the mouth for 5%, the gingiva for 3%, the vestibule for 8%, and the alveolus for 5% of cases (Kumar & Jha, 2023, Noguti, J, *et al.*, 2012). Specifically, for every 59 men, 1 is likely to develop oral cancer, whereas for every 139 women, 1 is likely to develop oral cancer according to the American Cancer Society (2023). Oral Squamous Cell Carcinoma (OSCC) is the most prevalent kind of oral cancer and the risk factors involve both external (chemicals, radiation, infectious organisms (HPV) and tobacco consumption) and internal (hormones, mutation, immune condition, age, gender, genetic factors) causes. Treatments include chemotherapy, radiotherapy, immunotherapy, and surgical removal of tumors or a combination of the treatments (Danaraddi, Koneru, Hunasgi, Ramalu, & Vanishree, 2014). OSCC is invasive, with a high chance of recurrence and distant lymphatic metastasis. Therefore, it is important to comprehend about the mechanism of invasion for the development of potential treatment (Ellis, 2021).

Oral cancer metastasis is a multi-step process that begins with cell detachment from tumor tissue and continues through cell invasion control, proliferation, and evasion *via* lymphatics or blood arteries. The normal oral mucosal keratinocytes when exposed to toxic substances like tobacco and smoke, breaks the homeostatic nature of the cells and causes genetic instability, which genetically alters Tumor Suppressor Protein 53 (TP53), NOTCH 1, Epidermal Growth Factor Receptor (EGFR), Cyclin-Dependent Kinase Inhibitor 2a genes (CDKN2A), Signal Transducer and Activator of Transcription 3 (STAT3), Cyclin D1, Retinoblastoma (Rb) and Loss of Heterozygosity (LOH). The alteration in these genes causes E-cadherin to lose its function resulting in Epithelial-Mesenchymal transition (EMT). EMT is a highly dynamic process that can transform epithelial cells into mesenchymal phenotypes, which can lead to the spread of cancer cells to other areas of the body. It results in loss of adhesion between cells leading to the release of components like the Matrix Metalloproteinases 2/9/13 (MMP), Reactive Oxygen Species (ROS), Vascular Endothelial Growth Factor (VEGF), Chemokine Ligand 1/8 (CXCL), Platelet-Derived Growth Factor (PDGF), Interleukin 8,1 (IL), Fibroblast Growth Factor 2 (FGF-2), Transforming Growth Factor-β (TGF-β), Tumor Necrosis Factor-α (TNF-α), and Granulocyte-Macrophage Colony-Stimulating Factor (GM-CSF) (Rivera, 2015) which breaks the extracellular matrix (ECM) causing the cancer cells to move around freely and invade nearby cells (Irani, 2016). This stage is critical because it often makes treatment more challenging and can worsen the prognosis for patients.

 DOI: 10.1201/9781003486237-7

Treating patients with therapies like chemotherapy and radiotherapy ends up in destroying or damaging the normal cells and therefore, an alternative approach for treating patients is essential and this can be achieved by the use of nutraceuticals.

Nutraceuticals, a term introduced by Dr. Stephen DeFelice is the fusion of nutrients and pharmaceuticals. Nutraceuticals are the elements that carry specific health benefits including prevention or treatment for any kind of health issues (Ranzato, Martinotti, Calabrese, & Calabrese, 2014, Das, L, *et al.*, 2010). For example, fruits and vegetables contain various compounds like polyphenols (a strong antioxidant), phytoestrogens (in less quantity-acts as an anticancer agent), phytosterol (lowers cholesterol), lectins (improves immune system), oligosaccharides (enhances bone calcium), and these play an important role in managing and treating the diseases (Orzechowski, Ostaszewski, Jank, & Berwid, 2002). Nutraceuticals are said to potentially increase the function of Natural Killer (NK) cells and TNF-α in late-stage cancer patients (See Mason & Roshan, 2002). Moreover, it greatly reduces the toxicity caused by chemotherapy and radiotherapy (Grimble, 2003). Research states that approximately one-third of the oral cancers are caused due to lifestyle changes (Danaei, Vander Hoorn, Lopez, Murray, & Ezzati, 2005) and regardless of the advancement in the field of medicine, cancer is a worldwide threat to doctors and researchers. The use of nutraceuticals which are obtained from plants and animals along with the therapies could greatly aid in reducing the mortality rate caused by oral cancer.

7.2 NUTRACEUTICALS

Innumerable studies and research have been conducted on naturally occurring compounds. Several compounds have been discovered to be potential candidates for the treatment of cancer due to their bioactivities and structural and chemical uniqueness. These nutraceuticals are emerging as a beacon of hope in cancer treatment due to their antioxidant, anti-inflammatory, and anti-proliferative properties. Out of the numerous nutraceuticals, the ones with active anticancer prospects have been mentioned in Figure 7.1. These are grouped on the basis of their respective functional groups and are listed below along with their sources.

7.2.1 Phenols

The characteristic feature of phenols are their antioxidant properties and they have been exhibiting anticancer activity against various types of cancers using a variety of mechanisms like modulating the progression of the cell cycle and suppression of angiogenesis. However, the majorly used mechanism is scavenging. Due to their antioxidant capacity, these phenols act as scavengers of free radicals, leading to oxidative stress and DNA damage, which are the hallmarks of cancer. Some of the phenols showing promising properties are:

7.2.1.1 Curcumin

It is a compound extracted from turmeric and has been known to be effective against colon, prostate, and breast cancer. All this is attributed to the antiproliferative, anti-inflammatory, and anti-angiogenic properties of curcumin (Zlotogorski, Dayan, Dayan, Chaushu, Salo, & Vered, 2013).

7.2.1.2 Resveratrol

Resveratrol is found in grapes and red wine and is known for its antitumorigenic activity. Exhibiting appreciable activity against breast, colon, and gastric cancer, resveratrol has been regulating cell cycle progression. This induces apoptosis and inhibits angiogenesis (Almeida *et al.*, 2021).

7.2.1.3 Caffeic Acid

Extracted mainly from honey bee Propolis, caffeic acid has shown prominent antiproliferative activity against cervical and breast cancer cell lines (Alam *et al.*, 2022).

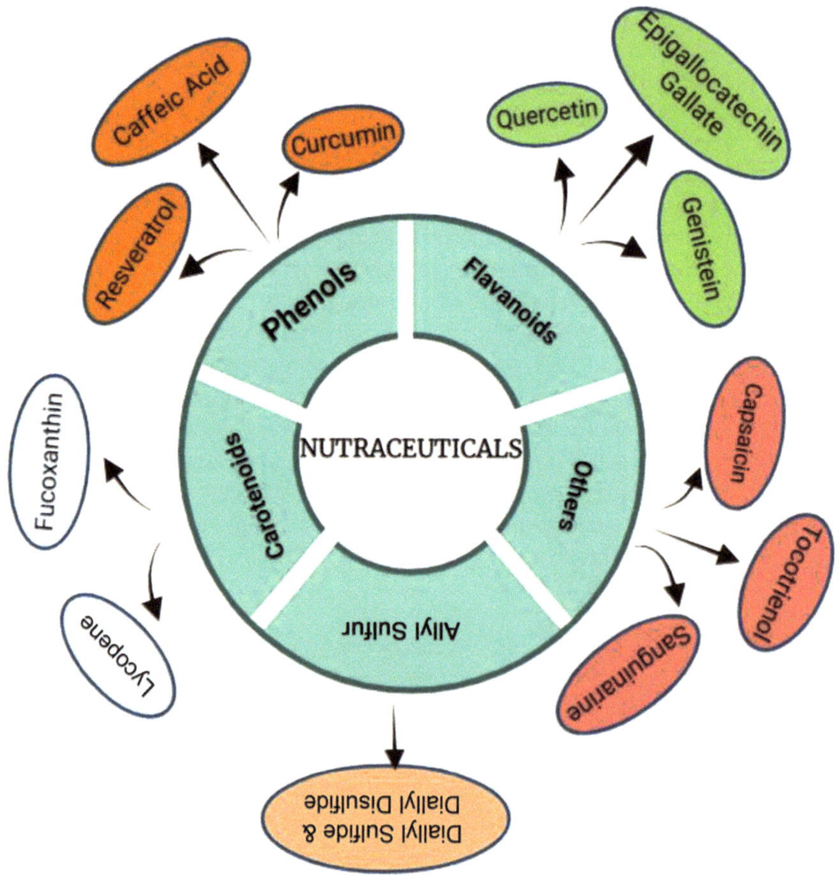

FIGURE 7.1 Nutraceuticals known for their active anti-cancer properties. (Created with BioRender.com.)

7.2.2 Carotenoids

Carotenoids, responsible for the vibrantly colored fruits and vegetables, have diverse health benefits. The major reason behind the anticancer effect expressed is due to their antioxidant activity which protects cells from oxidative damage by scavenging free radicals.

7.2.2.1 Fucoxanthin

This is a marine carotenoid, extracted mainly from brown seaweed. It has exhibited proapoptotic effects and antiproliferative effects on breast cancer cell lines (Martin, 2015).

7.2.3 Flavonoids

These are a diverse group of plant pigments possessing equally diverse health benefits including potential anticancer property. They can modulate cell signaling pathways involved in cell proliferation, differentiation, and survival and also possess the inherent ability to induce apoptosis.

7.2.3.1 Genistein

Observed majorly in soy products, studies have shown that it has a strong anticancer activity against breast, ovarian, and prostate cancer (Jeune, Kumi-Diaka, & Brown, 2005, Lee, J.Y, *et al.*, 2012).

7.2.3.2 Quercetin

It is a versatile compound known majorly for its antioxidant activities. It is found abundantly in onions, grapes, cherries, and berries. It has been observed to possess strong anticancer activities against breast, colon, prostate, ovarian cancer (Azeem, Hanif, Mahmood, Ameer, Chughtai, Abid, 2023).

7.2.3.3 Epigallocatechin Gallate (EGCG)

For numerous centuries, green tea has been a widely consumed beverage. Green tea is primarily composed of catechins, of which the most important are EGCG which accounts for 59% of all catechins (Narotzki, Reznick, Aizenbud, & Levy, 2012). Enriching green tea with its anti-inflammatory and potent antioxidant properties, studies prove EGCG to be beneficial against prostate, breast, pancreatic cancer and colon cancer (Kwon, O.S, *et al.*, 2020).

7.2.4 Allyl Sulfur

These compounds exhibit various biological activities like enhancing detoxification and suppression of tumor growth. It has been observed primarily in garlic, a commonly used culinary spice, containing unique sulfur-containing compounds.

7.2.4.1 Diallyl Sulfide and Diallyl Disulfide

It is the primary organosulfur compound in garlic, the most used of the kitchen remedies. It has been shown to have a significant effect on lung, prostate, breast, and gastric cancer (Mitra *et al.*, 2022).

7.2.5 Capsaicin

The chief contributor to the spice factor of chili peppers, capsaicin has been used traditionally for its analgesic properties. But lately, the anticancer activity of this bioactive compound has been hogging the limelight and it has been shown to have substantial action against colon, bladder, liver, and breast cancer (Friedman *et al.*, 2018, Friedman, J. R, *et al.*, 2019).

7.2.6 Sanguinarine

Used widely in toothpastes and mouthwash, it is an alkaloid derived from Bloodroot and the Mexican prickly poppy. Studies have shown a strong anticancer activity against lung, liver, bone, and brain cancer (Malla, Bhamidipati, & Adem, 2023).

7.2.7 Tocotrienol

Observed in the seed endosperm of monocots such as wheat, barley and rice, tocotrienols are the primary form of vitamin E. They possess proapoptotic and antioxidative functions along with chemosensitization and anticancer stem cell effects. They have shown strong effects against colon, prostate, mammary, and lung cancer (Ling, Luk, Al-Ejeh, & Khanna, 2012).

7.3 NUTRACEUTICALS TARGETING ORAL CANCER

Though cancer remains a mystery that continues to entangle as we try to solve it, oral cancer is one of the most significant cancers which exhibits low survival rate and high recurrence. This is mainly due to lack of awareness among patients and most of the cases are diagnosed at the last stages. Oral cancer has become a significant global health concern. Even if conventional therapies like surgery, radiation, and chemotherapy offer management options, they are often associated with adverse side effects and have limitations. This is where the nutraceuticals gain their prominence, as they have minimal side effects and their potential as an alternative to conventional methods makes them an emerging candidate to be explored in detail.

This chapter mainly focuses on the nutraceuticals which are utilized against oral cancer. The previous section gave an insight into the nutraceuticals currently employed in cancer. In this section, we delve deeper into the nutraceuticals targeting oral cancer specifically, with the mechanisms of each in the upcoming sections. A major percentage of nutraceuticals against oral cancer are phenols. This can be credited to the characteristic antioxidant property of phenols. However, there are other classes of nutraceuticals which have been explored, such as isoflavones, carotenoids, thiolic acids, and polyunsaturated fatty acids. Each of these has been discussed in detail in this chapter (Tables 7.1–7.2).

7.3.1 Curcumin

Curcumin exhibits anti-neoplastic activity in various types of cancers. It is the major constituent of turmeric. The anticancer characteristics can be attributed to the ability of curcumin to modulate signaling pathways involved in cell proliferation, survival, and migration. This polyphenol has been proven to be effective in the management of oral cancer with its antitumor effects against Oral Squamous Cell Carcinoma (Zlotogorski *et al.*, 2013).

7.3.2 Epigallocatechin Gallate (EGCG)

The primary component of green tea, epigallocatechin gallate is another polyphenol with a promising anticancer potential, especially on oral cancer. This is often attributed to its antioxidant, DNA methyltransferase (DNMT) and histone deacetylase (HDAC) inhibitor activity. These enzymes are associated with gene regulation and inhibiting these enzymes can influence cancer cell growth and differentiation (Min & Kwon, 2014).

7.3.3 Resveratrol

Exhibiting antioxidant, DNMT and HDAC inhibitor activity, this polyphenol has been detected in grapes, peanuts, cranberries, blueberries, and mulberries. Similar to EGCG, resveratrol has shown reassuring activity against oral cancer (Almeida *et al.*, 2021).

7.3.4 Quercetin

This flavonoid binds to various protein regulators, induces cell cycle arrest in various stages, and also reduces development of blood vessels. Additionally, the antioxidant and anti-inflammatory activity of quercetin contributes to its anticancer activity against oral cancer (Haghiac & Walle, 2005).

7.3.5 Berberine

This alkaloid, derived from barberry root bark, has been a part of Chinese and ayurvedic medicine for a long time. The anticancer activity of berberine has been attributed to its inhibiting action on cancer cell proliferation and in inducing apoptosis. Berberine has been proven to have potent activity against HSC-3 oral cancer cells (Lin *et al.*, 2007).

TABLE 7.1
Nutraceuticals Targeting Oral Cancer Pathways – *In Vitro* Studies

S. No.	Nutraceuticals	Sources	*In Vitro* Models	Pathways	Findings	References
1.	Curcumin	Curcumin (*Curcuma longa* L.)	Ca9–22 and HSC4	c-Met/extracellular signal-regulated kinase (Met/ERK) pathway	Curcumin treatment in OSCC cells inhibited cell motility and hepatocyte growth factor – induced EMT via c-Met blockade which inhibited vimentin levels	Ohnishi, Sakamoto, Zhengguang, Yasui, Hamada, Kubo, and Nakajima (2020)
			SCC-9 cells	Wnt/ β-catenin signaling	Curcumin inhibited OSCC cells proliferation through upregulating miR-9 expression, and suppressing Wnt/β-catenin signaling by increasing the expression levels of the GSK-3b, and decreasing the cyclin D1 level.	Xiao, Wang, Zhu, Zhang, and Zhou (2014)
2.	Malvidin	Blueberries	SCC-131	JAK/STAT-3 signaling pathway	Blueberry and malvidin suppressed STAT-3 phosphorylation in SCC-131 oral cancer cell line and induced mitochondrial-mediated apoptosis and G1/S phase cell cycle arrest.	Baba, A. B., Nivetha, R., and Nagini, S. (2017)
3.	Resveratrol	Red grapes	CAL 27 cell line	AMPK and PI3K/AKT/ mTOR pathways	Resveratrol appears to influence autophagy, and apoptosis, affecting oral cancer cell resistance to cisplatin through the AMPK and PI3K/AKT/ mTOR pathways	Chang *et al.* (2017)
			SCC-VI, SCC-25, and YD38	Apoptosis and G2/M phase cell cycle arrest	Resveratrol inhibits proliferation of OSCC cell lines by enhancing cell cycle arrest in the G2/M phase and by inducing apoptosis	Yu, Yang, Zhang, and Liu (2016)
4.	Pterostilbene	Blueberries	Cisplatin resistant CAL 27 cells	AKT pathway	Pterostilbene modulates the suppression of multidrug resistance protein 1 and triggers autophagic and apoptotic mechanisms in cisplatin-resistant human oral cancer CAR cells via AKT signaling	Chang *et al.* (2018)
5.	Epigallocatechin-3 gallate (EGCG)	Green tea	SCC-9	EMT, Src pathway	Inhibition on invasion by EGCG may be through downregulation of MMPs and u-PA (urokinase-type plasminogen activator) expression of oral cancer cells	Chen, Chu, Kuo, Chou, Lin, and Hsieh (2011)
6.	Bromelain (BR)	Ananas comosus	Ca9-22 and SCC-25	G1 cell cycle arrest	Bromelain treatment inhibited cell growth and proliferation and induced apoptosis in Ca9-22 and SCC-25 human oral squamous carcinoma cell lines through various pathways and G1 cell cycle arrest	Lee, Lee, Park, and Kim (2019)
7.	Vietnamese coriander	Vietnamese coriander	SAS cells	Akt/mTOR pathway	Vietnamese coriander inhibits cell proliferation, survival and migration *via* suppression of Akt/mTOR pathway in oral squamous cell carcinoma	Devi Khwairakpam *et al.* (2019)
8.	Sodium danshensu or salvianic acid A	*Salvia miltiorrhiza* (Danshen)	FaDu and Ca9-22 cell lines.	P38 signaling pathway	Sodium danshensu inhibits oral cancer cell migration and invasion by modulating p38 signaling pathway	Kumar *et al.* (2020)
9.	Genistein and biochanin A	*Indigofera suffruticosa*	SCC-15 and SCC-25	MAP kinase and Akt pathways	Inhibition of cell proliferation and MAP kinase and Akt pathways in oral squamous cell carcinoma by genistein and biochanin A	Johnson, Lai, Lai, and Bhushan (2010)
10.	Piperine (1-piperoyl piperidine)	Black pepper (*Piper nigrum* Linn.), long pepper (*Piper longum* Linn.)	KB cell line	Cell cycle arrest and mitochondrial oxidative stress	Piperine inhibits tumor growth by inducing cell apoptosis, depolarization of MMPs, cell cycle blockage, and caspase-3 activation and suppresses KB tumor growth by inducing nuclear condensation and cell shrinkage *in vitro*	Siddiqui, Ahamad, Jafri, Afzal, and Arshad (2017)

TABLE 7.2
Nutraceuticals Targeting Oral Cancer Pathways – *In Vivo* Studies

S. No.	Nutraceuticals	Sources	*In Vivo* Model	Pathways	Findings	References
1.	Curcumin	Curcumin (*Curcuma longa* L.)	Sprague Dawley rats	NF-κB pathway	Curcumin was effective in suppressing nuclear factor kappa B and cyclooxygenase 2 expression in experimentally induced oral squamous cell carcinoma	Maulina, Hadikrishna, Hardianto, Sjamsudin, Pontjo, and Yusuf (2019)
2.	Blueberry supplementation	Blueberries (*Vaccinium corymbosum* L.)	Hamsters	PI3K/Ak, NF-κB, transforming growth factor β and JAK/ STAT-3 signaling pathway	Blueberry inhibited the development and progression of hamster buccal pouch carcinomas by abrogating TGF-β and PI3K/Akt pathways and it suppressed NF-κB and by activation of anti-invasive and anti-angiogenic phenotype by downregulating MMPs and VEGF	Baba, Kowshik, Krishnaraj, Sophia, Dixit (2016), Baba, A.B *et al.* (2017).
3.	Black raspberries extract	Black raspberries	F344 rats	NF-κB /MAPK pathway	Blackberries have also been reported to inhibit oxidative stress in rat esophageal squamous cell carcinogenesis models and the hydrogen peroxide-activated NF-κB /MAPK pathway	Oghumu *et al.* (2017)
4.	Garcinol	*Garcinia indica* fruit	Athymic nu/nu mice	STAT3/NF-κB pathway	Garcinol reduced the growth of HNSCC mouse without exhibiting any significant toxicity, and downregulated the expression of p-signal transducer and activator of transcription (STAT)-3, p65, Ki-67, and CD31 in the treated groups as compared with the control group	Li *et al.* (2013)
5.	Epigallocatechin-3 gallate (EGCG)	Green tea	BALB/c nude (nu/nu) mice	Cell cycle arrest and apoptosis	EGCG induces cell cycle arrest and apoptosis in both *in vitro* and *in vivo*: in a mouse model, significant growth inhibition of the OSCC tumor was observed in EGCG-treated mice without a loss of body weight	Yoshimura *et al.* (2019)
6.	Apigenin	Fruits and vegetables such as parsley, onions, orange, tea, and wheat sprouts	Hamsters	Antioxidant and antiproliferative mechanism	Apigenin inhibited oral carcinogenesis by improving the status of antioxidant defense mechanism and prevented DMBA (7,12-dimethylbenz[a]anthracene) induced oral carcinogenesis in hamster buccal mucosa through its anti-cell proliferative and antioxidant potential as well as by modulating the status of Phases I and II detoxification agents	Silvan, Manoharan, Baskaran, Anusuya, Karthikeyan, Prabhakar (2011)
7.	Nimbolide	Neem tree (*Azadirachta indica*)	Syrian hamsters	PI3K/Akt/ GSK-3β signaling pathway	Nimbolide augments apoptosis by overcoming the shielding effects of cytoprotective autophagy through modulation of the PI3K/Akt signaling cascade by altering the phosphorylation status of Akt and GSK-3β	Sophia *et al.* (2018)
8.	Gedunin	Neem tree (*Azadirachta indica*)	Syrian hamsters	PI3K/Akt/ mTOR, and NF-κB signaling pathways	Gedunin prevents progression of HBP carcinomas *via* inhibition of kinases such as Akt, IKK, and AR, and the oncogenic transcription factors NF-κB and HIF-1α to block angiogenesis	Kishore *et al.* (2016)
9.	Bitter melon extract (BME)	Bitter melon (*Momordica charantia*)	C57BL/6 mice	Immune signaling	Bitter melon prevents the development of 4-NQO-induced oral squamous cell carcinoma in an immunocompetent mouse model by modulating immune signaling	Sur, Steele, Aurora, Varvares, Schwetye, and Ray (2018)

7.3.6 Genistein

A nutraceutical in soybeans and tofu primarily, genistein is an isoflavone with antiproliferative and proapoptotic potential. Genistein alters the angiogenesis and cell cycle and inhibits metastasis, thus making it a promising chemotherapeutic agent against Oral Squamous Cell Carcinoma (Park S.J *et al.*, 2010).

7.3.7 Silymarin

Extracted from the seeds of the milk thistle plant, silymarin is a flavanoligan with a steroid structure. The anticancer feature of silymarin regulates various hallmarks of cancers within the tumor microenvironment, involving tumor cell growth, apoptosis, metastasis, angiogenesis, and autophagy (Won, Kim, & Jang, 2018).

7.3.8 Lycopene

Tomatoes are known for their antioxidant activity, and this characteristic feature can be attributed primarily due to the presence of lycopene. This carotenoid has powerful singlet-oxygen characteristic, stimulates the production of the detoxifying/antioxidant enzymes, initiates apoptosis, and inhibits cell proliferation. The presence of these features assists the antitumor activity against oral cancer (Xinxing Lu, *et al.*, 2020, Wang, R, *et al.*, 2020).

7.3.9 Malvidin

Identified primarily in blueberries, red grapes, and red wine, malvidin is a highly biologically active polyphenol component containing anthocyanins. Anthocyanins are known as smart molecules for cancer prevention. They inhibit H_2O_2 and TNF-α-induced VEGF expression in oral cancer cells (Diaconeasa *et al.*, 2018).

7.3.10 Pterostilbene

The primary antioxidant compound in blueberries, pterostilbene has an increased bioavailability compared to other stilbene compounds. It is known for its antioxidative, anticancer, and antidiabetic properties. It suppresses cancer growth by acting on signaling pathways dependent as well as independent of apoptosis (Lin *et al.*, 2014).

7.3.11 Bromelain

Natives of America have been using pineapple for a variety of ailments. This is due to a group of enzymes found in the fruit and stem of the pineapple, bromelain. These proteolytic enzymes induce cell apoptosis in different cancers through different pathways. They also have been showing significant activity against Ca9-22 and SCC-25 oral cancer cell lines (Lee, Lee, Park, & Kim, 2019).

7.3.12 Garcinol

Used traditionally for its antioxidant and anti-inflammatory properties, garcinol is extracted from the kokum *(Garcinia indica)* fruit peel. This polyphenol has been utilized for a long time for its antioxidant, antiulcer, anti-inflammatory, and antibacterial properties. Recently the antitumor properties of the kokum peel have been explored. Garcinol present in the peel was found to suppress cell proliferation which ultimately led to cancer repression (Saadat & Gupta, 2012).

7.3.13 Apigenin

This flavonoid is abundantly present in common fruits and vegetables. It has a plethora of health benefits including antioxidant, anti-inflammatory, and chemoprevention activity. Additionally, it inhibits cancer cell migration and invasion, making it chemopreventive against various types of cancer including oral cancer (Maggioni *et al.*, 2013).

7.3.14 Nimbolide

The primary limonoid constituent of neem has a wide spectrum of pharmacological property, but predominantly it has been gaining prominence due to its anticancer activity. Scavenging free radicals, inducing programmed cellular death, inhibiting transformed cell proliferation are all some characteristics that imparts nimbolide its anticancer potential (Wang *et al.*, 2016).

7.3.15 Gedunin

Another compound extracted from Indian neem tree, gedunin is a tetranortriterpenoid which has been used traditionally for treating malaria. It is a cell proliferation inhibitor and apoptosis inductor, making it an anticancer agent (Priya & Abha, 2023). It is also an inhibitor of Hsp90 activity and activator of oncogenic signaling pathways, having a strong chemopreventive effect on various cancers including oral cancer (Kishore, Ganugula, Gade, Reddy, & Nagini, 2016).

7.4 MECHANISMS OF NUTRACEUTICALS AGAINST ORAL CANCER

The important hallmarks of cancer, such as invasion, angiogenesis, cell proliferation, evasion, and apoptosis have always been attributed to aberrant induction of oncogenic signaling pathways (Yedida, Nagini, & Mishra, 2013). Nutraceutical compounds possess the ability to prevent the development and progression of oral cancer by inhibiting tumor proliferation and inducing apoptosis *via* the modulation of multiple cellular pathways implicated in the process of carcinogenesis. Nutraceuticals affect various pathways in oral cancer, including the Phosphoinositide 3-Kinase (PI3K)/Akt, a serine/threonine protein kinase/Mammalian Target of Rapamycin (mTOR)/Nuclear factor-kappa B (NF-κB) signaling, Janus Kinase/Signal Transducer and Activator of Transcription (JAK/STAT), Mitogen-activated Protein Kinase (MAPK), mitochondrial pathway, and EMT. Inhibition of molecular mechanisms by phytochemicals in oral cancer has been elucidated in Figure 7.2. Understanding the molecular mechanisms that lead to oral cancer can enhance prevention and early detection.

7.4.1 Cellular Signaling Pathways Regulated by Phytochemicals in Oral Cancer

7.4.1.1 Mitochondrial Pathway

The mitochondrial apoptosis signaling pathway is essential for the purpose of managing cell death in response to various stimuli. Members of the Bcl-2 family cause a change in the permeability of the mitochondrial outer membrane. There are two main mechanisms that might cause programmed cell death, or apoptosis: the intrinsic pathway, which involves the mitochondria and the extrinsic pathway, which involves death receptors. It has been suggested that concentrating on apoptosis is a viable strategy for eliminating cancer cells. The proapoptotic protein Bax (Bcl-2 associated protein x), a member of the Bcl-2 (B-cell lymphoma-2) family, initiates the mitochondrial pathway and causes the cytosol to secrete cytochrome c. Cytochrome C joins forces with Apoptotic protease

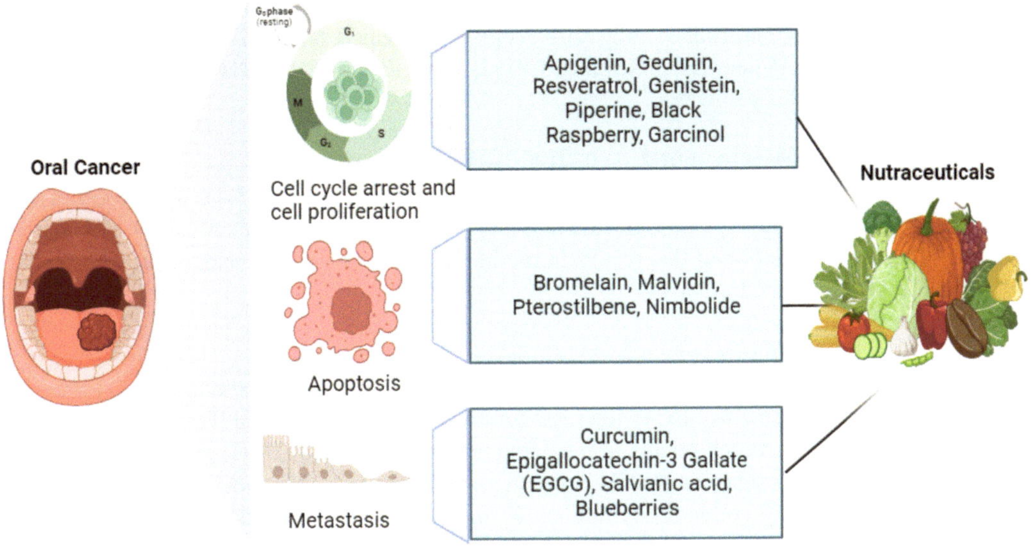

FIGURE 7.2 Inhibition of molecular mechanisms by phytochemicals in oral cancer. (Created with BioRender.com.)

activating factor-1 (Apaf-1) to produce a complex that, when combined with caspase-9, forms the apoptosome. Caspases are essential executors and the apoptosome activates caspase-3 and caspase-9, which cleave critical cellular proteins and trigger apoptosis (Kim *et al.*, 2018).

7.4.1.2 PI3K/Akt/mTOR Signaling Pathway

It controls metabolism, cell division, and in the majority of malignancies, the malignant phenotype. PI3K/AKT/mTOR signaling pathways inhibited Phosphatase and tensin homolog (PTEN) which in turn resulted in the uncontrolled cell proliferation of cancer tumors (Davis, Lehmann, & Li, 2015; Kumari, Chhikara, Singh, Rathi, & Singh, 2024; Simpson, Mell, & Cohen, 2015). Compounds and plants which potentially inhibit PI3K/AKT/mTOR signaling pathways are resveratrol, gedunin, and blueberry.

7.4.1.3 NF-κB Signaling Pathway

Playing a role in cellular growth, inflammation, transformation, apoptosis, metastasis, cancer progression, etc., NF-κB has five subunits which form dimers and regulates gene transcription by binding to specific sequences which increases IL-6 and overexpresses cyclooxygenase-2 (COX-2) resulting in cancer cell proliferation (Kumari *et al.*, 2024, Kaushik, M., & Tiku, A. B. 2022). Natural compounds such as curcumin, resveratrol, EGCG, lycopene, genistein, and luteolin are potent inhibitors of NF-κB.

7.4.1.4 JAK/STAT Pathway

It is reported to be an immune-suppressor and protects tumor cells from being destroyed by cytotoxic T lymphocytes (Usman, Jamal, Muy, Teh, & Waseem, 2021). Green tea, resveratrol, genistein, and curcumin have been identified to reduce the activity of STAT in cancer cells.

7.4.1.5 MAPK Signaling Pathway

It plays a role in extracellular signaling and controls expression of genes by transcription factors. The involvement of the proteins c-Jun (cellular Jun) and C-Fos (cellular FBT), which includes transcription factor Activator Protein 1 (AP-1) in the form of dimers. These factors attach to promoter regions of DNA and increase cyclin D1 expression, which results in the induction of cell proliferation (Peng, Deng, Pan, Gu, Liu, & Tang, 2017). Olive oil (Peluso, Sastry Yarla, Ambra, Pastore, & Perry, 2017), silibinin (Chen, Hsieh, Chiang, Chiou, Yang, & Chu, 2006) are shown to have antitumor activity against this pathway.

7.4.1.6 Epithelial–Mesenchymal Transition (EMT) Pathway

The hallmark of EMT is the lack of cell adhesion (Gotzmann *et al.*, 2004). It is reported to be associated with the loss of E-cadherin functionality which is important in cell adhesion and polarization. The downregulation of this results in the induced invasion and development of tumor cells (Chang, Wright, & Svoboda, 2007).

7.4.1.7 Wnt Pathway

Wnt is a secreted glycoprotein which is important in cell proliferation, migration, differentiation, and communication. Wnt proteins bind to the receptor and co-receptor Frizzled (Fzd) and Lipoprotein-Related Protein (LRP), which results in the activation of Disheveled (DSH) protein and stabilizes the β-catenin *via* dephosphorylation. This β-catenin accumulates and translocates to the nucleus, where it reacts with the T-cell factor/lymphocyte enhancer factor which induces the transcription of genes like cyclin D1, cellular myelocytomatosis (c-myc) and B-cell lymphoma-w (Bcl-w) which are important for the development and spread of cancer (Marimuthu, Andiappan, Wahab, Muthusekhar, Balakrishnan, & Shanmugam, 2018).

7.5 CLINICAL TRIALS

Nutraceuticals play a significant role in preventing the occurrence of oral cancer in healthy patients and apart from that, it also greatly reduces the side effects caused by chemotherapy and radiotherapy. There are a lot of compounds that are being utilized as dietary supplements for patients, for example: kaempferol, quercetin, caffeic acid (Dutta, Banerje, & Mitra, 2012), curcumin (Zlotogorski, Dayan, Dayan, Chaushu, Salo, & Vered, 2013), green tea, resveratrol (Zlotogorski *et al.*, 2013, Almatroodi, S. A, *et al.*, 2022), black raspberries (Knobloch *et al.*, 2015), garlic and onion (Lee & Tseng, 2020), beta carotene, aloe vera (Nagi, Rakesh, Reddy, Konidena, Makkad, & Vyas, 2021). In addition to these, researchers are focusing on unraveling and exploring new supplements which can efficiently reduce the spread of cancer and some of the clinical trials performed with certain dietary supplements are mentioned in Table 7.3.

7.6 CONCLUSION AND FUTURE PERSPECTIVES

There are a plethora of studies which delve into the potential of nutraceuticals for preventing oral cancer, each revealing potential molecular mechanisms while identifying a promising model in *in vitro* and preclinical trials. Despite all these insights with the innumerable studies, the transition from the preclinical trials to the clinical trials becomes more and more challenging, further reducing the avenue for development. It is therefore crucial to optimize the frequency, dosage, and duration of the phytochemical administration for the practical application as dietary regulators of oral cancer. There is great promise for the use of phytochemicals as a therapeutic tool to enhance human immune systems, since these compounds have demonstrated pharmacological and physiological activities through altering certain binding proteins. Though the non-toxic doses have been determined in preclinical studies, further investigation is the key to human translation. The convergence of contemporary scientific methodologies with traditional understanding of medicinal plants may yield novel and efficacious therapeutic interventions for this challenging disease. Additional research and clinical trials are imperative in order to authenticate the effectiveness and safety of these compounds derived from plants. This will facilitate the exploration of novel therapeutic alternatives for oral cancer that are less hazardous to health.

TABLE 7.3
Clinical Trials of Nutraceuticals Effects on Head and Neck Squamous Cell Carcinoma

S. No.	Dietary Supplements	Title	Intervention/Treatment	Status	ClinicalTrial ID
1.	Black raspberry confection	Black Raspberry Confection in Preventing Oral Cancer in Healthy Volunteers	Prevention of oral cancer in healthy patients	Phase 1 Active	NCT01961869
2.	Black raspberry confection	Freeze-Dried Black Raspberries in Preventing Oral Cancer Recurrence in High At-Risk Appalachian Patients Oral Cancer Survivors	Prevent oral cancer recurrence	Completed	NCT01504932
3.	Black raspberry	Prevention of Oral DNA Damage by Black Raspberries	Prevention of oral DNA damage caused to mouth cells due to smoking	Recruiting	NCT04372914
4.	Prodentis lozenge	The Oral Microbiome in OSCC	To determine the effect of probiotics in OSCC	Completed	NCT04925700
5.	β-Glucan	Myeloid-Derived Suppressor Cells (MDSCs) in OSCC Patients	Detection of myeloid-derived suppressor cells in OSCC patients	Unknown	NCT04387682
6.	Coffee	Assessment of Anticancer Effect of Green, Roasted and Decaffeinated Coffee on Oral Squamous Cell Carcinoma Cell Line	Anticancer effect on oral cancer cell line	Unknown	NCT03619304
7.	Licochalcone A	Effects of (Licochalcone A) and Paclitaxel on Human Oral Squamous Cell Carcinoma Cell Line	Antiproliferation effect on oral cancer	Phase 1 Unknown	NCT03292822
8.	Aloe vera juice	Effects of Oral Aloe Vera Juice on Chemotherapy and Radiation-Induced Oral Mucositis and Esophagitis	Effect on oral mucositis	Phase 3 Recruiting	NCT05369234
9.	Arginine/omega-3 fatty acids	Arginine/Omega-3 Fatty Acids/Nucleotides Nutritional Supplement in Treating Patients With Stage III or Stage IV Head and Neck Cancer Undergoing Chemotherapy and Radiation Therapy	Treating patients undergoing chemotherapy for head and neck cancer	Phase 2 Completed	NCT00559156
10.	Oral moisturizing jelly	Efficacy of Oral Moisturizing Jelly on Oral Health and Nutrition in Post-Radiotherapy Head and Neck Cancer Patients	Treating Xerostomia for head and neck cancer patients	Completed	NCT03035825
11.	Glutamine, placebo	Effects of Oral L-Glutamine in Head and Neck Cancer Patients During Radiotherapy	Effect on Radiation induced toxicity and maintain patient's nutritional status	Completed	NCT03015077
12.	Zinc, Vit C, arginine	Pre- and Postoperative Nutrition in Head and Neck Cancer Patients	Effect of wound healing on pre and post operated head and neck cancer patients	Completed	NCT03531190
13.	Omega-3, anthocyanins	Administration of ONS Based on Purple Sweet Potato in Head and Neck Cancer Patients With Radiotherapy	Effect on nutritional status and inflammatory response in head and neck cancer patients	Recruiting	NCT05740878
14.	L-lysine	L-Lysine in Treating Oral Mucositis in Patients Undergoing Radiation Therapy With or Without Chemotherapy for Head and Neck Cancer	Treating oral mucositis for head and neck cancer patients	Completed	NCT01155609

(Continued)

TABLE 7.3 (*Continued*)
Clinical Trials of Nutraceuticals Effects on Head and Neck Squamous Cell Carcinoma

S. No.	Dietary Supplements	Title	Intervention/Treatment	Status	ClinicalTrial ID
15.	Honey	Honey Products for Cancer Patients With Oral Mucositis: A Randomized Controlled Trial	Effect on treating patients with oral mucositis	Unknown	NCT04382079
16.	Hydrogen tablet	Hydrogen-Rich Water and Treatment of Mucositis Caused by Radiation Therapy in Head and Neck Cancer Patients	Treating oral mucositis	Recruiting	NCT05278260
17.	L-Glutamine	Effects of Oral L-Glutamine in Head and Neck Cancer Patients During Radiotherapy	Reduce the side effect of radiotherapy in head and neck cancer patients	Completed	NCT03015077
18.	Honey and thyme-based spray	The Effectiveness of a Thyme and Honey Spray for Oral Toxicities (HONEY)	Effect on oral mucositis in head and neck cancer patients	Recruiting	NCT04880148
19.	Zinc sulfate	Zinc Sulfate in Preventing Loss of Sense of Taste in Patients Undergoing Radiation Therapy for Head and Neck Cancer	Preventing the loss of taste in head and neck cancer patient undergoing radiotherapy	Phase 3 Completed	NCT00036881
20.	Honey mouthwash	The Effect of Honey on Xerostomia and Oral Mucositis	Effect on xerostomia and oral mucositis in head and neck cancer patients	Phase 2 Completed	NCT01465308
21.	Glutamine, placebo	Oral Glutamine and Mucositis of Head and Neck Cancer Patients Undergoing Radiation	Treating oral mucositis in head and neck cancer patients	Phase 3 Unknown	NCT02282839
22.	Manuka honey	A Randomized Placebo-Controlled Trial of Manuka Honey for Oral Mucositis Due to Radiation Therapy for Cancer	Treating oral mucositis in head and neck cancer patients	Phase 3 Completed	NCT00615420
23.	Capsaicin, placebo	Capsaicin Lozenges in Treating Patients With Mucositis Caused by Radiation Therapy	Treating mucositis in head and neck cancer patients	Phase 3 Completed	NCT00003610

Source: https://clinicaltrials.gov/

REFERENCES

Alam, M., Ahmed, S., Elasbali, A. M., Adnan, M., Alam, S., Hassan, M. I., & Pasupuleti, V. R. 2022. Therapeutic implications of caffeic acid in cancer and neurological diseases. *Frontiers in Oncology* 12:860508.

Almatroodi, S. A., Alsahli, M. A., Aljohani, A. S. M., Alhumaydhi, F. A., Babiker, A. Y., Khan, A. A., & Rahmani, A. H. 2022. Potential therapeutic targets of resveratrol, a plant polyphenol, and its role in the therapy of various types of cancer. *Molecules (Basel, Switzerland)* 27(9):2665.

Almeida, T.C., da Silva, G.N., de Souza, D.V., Andrea Cristina de Moraes Malinverni, Odair Aguiar Jr, Debora Estadella & Daniel Araki Ribeiro. 2021. Resveratrol effects in oral cancer cells: A comprehensive review. *Medical Oncology* 38: 97.

Azeem, M., Hanif, M., Mahmood, K., Ameer, N., Chughtai, F. R. S., & Abid, U. 2023. An insight into anticancer, antioxidant, antimicrobial, antidiabetic and anti-inflammatory effects of quercetin: A review. *Polymer Bulletin (Berlin, Germany)* 80(1):241–262.

Baba, A. B., Kowshik, J., Krishnaraj, J., Sophia, J., Dixit, M., & Nagini, S. 2016. Blueberry inhibits invasion and angiogenesis in 7, 12-dimethylbenz [a] anthracene (DMBA)-induced oral squamous cell carcinogenesis in hamsters via suppression of TGF-β and NF-κB signaling pathways. *The Journal of Nutritional Biochemistry* 35:37–47.

Baba, A. B., Nivetha, R., Chattopadhyay, I., & Nagini, S. 2017. Blueberry and malvidin inhibit cell cycle progression and induce mitochondrial-mediated apoptosis by abrogating the JAK/STAT-3 signalling pathway. *Food and Chemical Toxicology* 109: 534–543.

Chang, C. H., Lee, C. Y., Lu, C. C., Tsai, F. J., Hsu, Y. M., Tsao, J. W., … Wang, C. C. 2017. Resveratrol-induced autophagy and apoptosis in cisplatin-resistant human oral cancer CAR cells: A key role of AMPK and Akt/mTOR signaling. *International Journal of Oncology* 50(3):873–882.

Chang, H. P., Lu, C. C., Chiang, J. H., Tsai, F. J., Juan, Y. N., Tsao, J. W., … Yang, J. S. 2018. Pterostilbene modulates the suppression of multidrug resistance protein 1 and triggers autophagic and apoptotic mechanisms in cisplatin-resistant human oral cancer CAR cells via AKT signaling. *International Journal of Oncology* 52(5):1504–1514.

Chang, J. Y. F., Wright, J. M., & Svoboda, K. K. H. 2007. Signal transduction pathways involved in epithelial-mesenchymal transition in oral cancer compared with other cancers. *Cell Tissue Organ* 185(1–3): 40–47.

Chen, P. N., Chu, S. C., Kuo, W. H., Chou, M. Y., Lin, J. K., & Hsieh, Y. S. 2011. Epigallocatechin-3 gallate inhibits invasion, epithelial–mesenchymal transition, and tumor growth in oral cancer cells. *Journal of Agricultural and Food Chemistry* 59(8):3836–3844.

Chen, P.-N., Hsieh, Y.-S., Chiang, C.-L., Chiou, H.-L., Yang, S.-F., & Chu, S.-C. 2006. Silibinin inhibits invasion of oral cancer cells by suppressing the MAPK pathway. *Journal of Dental Research* 85(3):220–225.

Dagar, P. and Mishra, A., 2023. Anti-cancer and anti-venom analysis of a gedunin. *Current Bioactive Compounds* 19(5): 26–46.

Danaei, G., Vander Hoorn, S., Lopez, A. D., Murray, C. J., & Ezzati, M. 2005. Causes of cancer in the world: Comparative risk assessment of nine behavioral and environmental risk factors. *Lancet* 366:1784–1793.

Danaraddi, S., Koneru, A., Hunasgi, S., Ramalu, S., & Vanishree, M. 2014. Natural ways to prevent and treat oral cancer. *Journal of Oral Research and Review* 6(1):34.

Das, L., Bhaumik, E., Raychaudhuri, U., & Chakraborty, R. 2010. Role of nutraceuticals in human health. *Journal of Food Science and Technology* 49(2):173–183.

Davis, W. J., Lehmann, P. Z., & Li, W. 2015. Nuclear PI3K signaling in cell growth and tumorigenesis. *Frontiers in Cell and Developmental Biology* 3:24.

Devi Khwairakpam, A., Monisha, J., Roy, N. K., Bordoloi, D., Padmavathi, G., Banik, K., … Kunnumakkara, A. B. 2019. Vietnamese coriander inhibits cell proliferation, survival and migration via suppression of Akt/mTOR pathway in oral squamous cell carcinoma. *Journal of Basic and Clinical Physiology and Pharmacology* 31(3):20190162.

Diaconeasa, Z. M., Frond, A. D., Ştirbu, I., Rugina, D., & Socaciu, C. (2018). Anthocyanins-smart molecules for cancer prevention. Phytochemicals-Source of Antioxidants and Role in Disease Prevention.

Dutta, K. R., Banerjee, S., & Mitra, A. 2012. Medicinal plants of West Midnapore, India: Emphasis on phytochemical containment having role on oral cancer. *International Journal of Phytopharmacology* 3(2):198–208.

Ellis, I. R. 2021. The migration and invasion of oral squamous carcinoma cells: Matrix, growth factor and signalling involvement. *Cancer* 13:2633.

Friedman, J. R., Nolan, N. A., Brown, K. C., Miles, S. L., Akers, A. T., Colclough, K. W., Seidler, J. M., Rimoldi, J. M., Valentovic, M. A., & Dasgupta, P. 2018. Minireview: Capsaicin analogs in cancer therapy. *Journal of Pharmacology and Experimental Therapeutics* 364(3):462–473.

Friedman, J. R., Richbart, S. D., Merritt, J. C., Brown, K. C., Denning, K. L., Tirona, M. T., Valentovic, M. A., Miles, S. L., & Dasgupta, P. 2019. Capsaicinoids: Multiple effects on angiogenesis, invasion and metastasis in human cancers. *Biomedicine & Pharmacotherapy* 118:109317.

Gotzmann, J., Mikula, M., Eger, A., SchulteHermann, R., Foisner, R., Beug, H., & Mikulits, W. 2004. Molecular aspects of epithelial cell plasticity: Implications for local tumor invasion and metastasis. *Mutation Research* 566:9–20.

Grimble, R. F. 2003. Nutritional therapy for cancer cachexia. *Gut* 52:1391–1392.

Haghiac, M., & Walle, T. 2005. Quercetin induces necrosis and apoptosis in SCC-9 oral cancer cells. *Nutrition and Cancer* 53(2):220–231.

Irani, S. 2016. Distant metastasis from oral cancer: A review and molecular biologic aspects. *Journal of International Society of Preventive and Community Dentistry* 6(4):265.

Jeune, M. A. L., Kumi-Diaka, J., & Brown, J. 2005. Anticancer activities of pomegranate extracts and genistein in human breast cancer cells. *Journal of Medicinal Food* 8(4):469–475.

Johnson, T. L., Lai, M. B., Lai, J. C., & Bhushan, A. 2010. Inhibition of cell proliferation and MAP kinase and Akt pathways in oral squamous cell carcinoma by genistein and biochanin A. *Evidence-Based Complementary and Alternative Medicine* 7:351–358.

Kaushik, M., & Tiku, A. B. 2022. Molecular pathways modulated by phytochemicals in head and neck cancer. *Journal of Cell Communication and Signaling* 17(3):469–483.

Kim, S.-E., Shin, S.-H., Lee, J.-Y., Kim, C.-H., Chung, I.-K., Kang, H.-M., Park, H.-R., Park, B.-S., & Kim, I.-R. 2018. Resveratrol induces mitochondrial apoptosis and inhibits epithelial-mesenchymal transition in oral squamous cell carcinoma cells. *Nutrition and Cancer* 70(1):125–135.

Kishore, T. K. K., Ganugula, R., Gade, D. R., Reddy, G., & Nagini, B. 2016. Gedunin abrogates aldose reductase, PI3K/Akt/mToR, and NF-κB signaling pathways to inhibit angiogenesis in a hamster model of oral carcinogenesis. *Tumor Biology* 37:2083–2093.

Knobloch, T. J., Uhrig, L. K., Pearl, D. K., Casto, B. C., Warner, B. M., Clinton, S. K., & Sardo-Molmenti, C. L. 2015. Suppression of proinflammatory and prosurvival biomarkers in oral cancer patients consuming a black raspberry phytochemical-rich troche. *Cancer Prevention Research* 9(2):159–171.

Kumar, M., & Jha, A. K. 2023. Exploring the potential of dietary factors and plant extracts as chemopreventive agents in oral squamous cell carcinoma treatment. *Frontiers Oral Health* 4(4):1246873.

Kumar, V. B., Lin, S. H., Mahalakshmi, B., Lo, Y. S., Lin, C. C., Chuang, Y. C., … Chen, M. K. 2020. Sodium danshensu inhibits oral cancer cell migration and invasion by modulating p38 signaling pathway. *Frontiers in Endocrinology* 11:568436.

Kumari, M., Chhikara, B. S., Singh, P., Rathi, B., & Singh, G. 2024. Signaling and molecular pathways implicated in oral cancer: A concise review. *Chemical Biology Letters* 11(1):652.

Kwon, O. S., Jung, J. H., Shin, E. A., Park, J. E., Park, W. Y., & Kim, S. H. 2020. Epigallocatechin-3-gallate induces apoptosis as a TRAIL sensitizer via activation of caspase 8 and death receptor 5 in human colon cancer cells. *Biomedicines* 8(4):84.

Lee, J. Y., Kim, H. S., & Song, Y. S. 2012. Genistein as a potential anticancer agent against ovarian cancer. *Journal of Traditional and Complementary Medicine* 2(2):96–104.

Lee, J. H., Lee, J. T., Park, H. R., & Kim, J. B. 2019. The potential use of bromelain as a natural oral medicine having anticarcinogenic activities. *Food Science & Nutrition* 7(5):1656–1667.

Lee, T.-Y., & Tseng, Y.-H. 2020. The potential of phytochemicals in oral cancer prevention and therapy: A review of the evidence. *Biomolecule* 10(8):1150.

Li, F., Shanmugam, M. K., Chen, L., Chatterjee, S., Basha, J., Kumar, A. P., … Sethi, G. 2013. Garcinol, a polyisoprenylated benzophenone modulates multiple proinflammatory signaling cascades leading to the suppression of growth and survival of head and neck carcinoma. *Cancer Prevention Research* 6(8):843–854.

Lin, C.-C., Yang, J.-S., Chen, J.-T., Fan, S., Yu, F.-S., Yang, J.-L., Lu, C.-C., Kao, M.-C., Huang, A.-C., Lu, H.-F., & Chung, J.-G. 2007. Berberine induces apoptosis in human HSC-3 oral cancer cells via simultaneous activation of the death receptor-mediated and mitochondrial pathway. *Anticancer Research* 27(5A):3371–3378.

Lin, C.-W., Chou, Y.-E., Chiou, H.-L., Chen, M.-K., Yang, W.-E., Hsieh, M.-J., & Yang, S.-F. 2014. Pterostilbene suppresses oral cancer cell invasion by inhibiting MMP-2 expression. *Expert Opinion on Therapeutic Targets* 18(10):1109–1120.

Ling, M. T., Luk, S. U., Al-Ejeh, F., & Khanna, K. K. 2012. Tocotrienol as a potential anticancer agent. *Carcinogenesis* 33(2):233–239.

Maggioni, D., Garavello, W., Rigolio, R., Pignataro, L., Gaini, R., & Nicolini, G. 2013. Apigenin impairs oral squamous cell carcinoma growth in vitro inducing cell cycle arrest and apoptosis. *International Journal of Oncology* 43(5):1675–1682.

Malla, R. R., Bhamidipati, P., & Adem, M. 2023. Insights into the potential of sanguinarine as a promising therapeutic option for breast cancer. *Biochemical Pharmacology* 212:115565.

Marimuthu, M., Andiappan, M., Wahab, A., Muthusekhar, M. R., Balakrishnan, A., & Shanmugam, S. 2018. Canonical Wnt pathway Gene expression and their clinical correlation in oral squamous cell carcinoma. *Journal of Dental Research* 29:291–7.

Martin, L. J. 2015. Fucoxanthin and its metabolite fucoxanthinol in cancer prevention and treatment. *Marine Drugs* 13(8):4784–4798.

Maulina, T., Hadikrishna, I., Hardianto, A., Sjamsudin, E., Pontjo, B., & Yusuf, H. Y. 2019. The therapeutic activity of curcumin through its anti-cancer potential on oral squamous cell carcinoma: A study on Sprague Dawley rat. *SAGE Open Medicine* 7:2050312119875982.

Min, K. J., & Kwon, T. K. 2014. Anticancer effects and molecular mechanisms of epigallocatechin-3-gallate. *Integrative Medicine Research* 3(1):16–24.

Mitra, S., Das, R., Emran, T. B., Labib, R. K., Noor-, E.-, Tabassum, Islam, F., Sharma, R., Ahmad, I., Nainu, F., Chidambaram, K., Alhumaydhi, F. A., Chandran, D., Capasso, R., & Wilairatana, P. 2022. Diallyl disulfide: A bioactive garlic compound with anticancer potential. *Frontiers in Pharmacology* 13:943967.

Nagi, R., Rakesh, N., Reddy, S. S., Konidena, A., Makkad, R. S., & Vyas, T. 2021. Therapeutic role of phytochemicals in the prevention of oral potentially malignant disorders and oral cancer. *Journal of Evolution of Medical and Dental Sciences* 16:1156–1165.

Narotzki, B., Reznick, A. Z., Aizenbud, D., & Levy, Y. 2012. Green tea: A promising natural product in oral health. *Archives of oral Biology* 57(5):429–435.

Noguti, J., De Moura, C. F. G., Jesus, G. P. P. D., Silva, V. H. P. D., Hossaka, T. A., Oshima, C. T. F., & Ribeiro, D. A. 2012. Metastasis from oral cancer: An overview. *Cancer Genomics and Proteomics* 9(5): 329–335.

Oghumu, S., Casto, B. C., Ahn-Jarvis, J., Weghorst, L. C., Maloney, J., Geuy, P., … Knobloch, T. J. 2017. Inhibition of pro-inflammatory and anti-apoptotic biomarkers during experimental oral cancer chemoprevention by dietary black raspberries. *Frontiers in Immunology* 8:1325.

Ohnishi, Y., Sakamoto, T., Zhengguang, L., Yasui, H., Hamada, H., Kubo, H., & Nakajima, M. 2020. Curcumin inhibits epithelial-mesenchymal transition in oral cancer cells via c-Met blockade. *Oncology Letters* 19(6):4177–4182.

Orzechowski, A., Ostaszewski, P., Jank, M., & Berwid, S. J. 2002. Bioactive substances of plant origin in food–impact on genomics. *Reproduction Nutrition Development* 42:461–477.

Park, S. J., Kim, M. J., Kim, Y. K., Kim, S. M., Park, J. Y., & Myoung, H. 2010. Combined cetuximab and genistein treatment shows additive anti-cancer effect on oral squamous cell carcinoma. *Cancer Letters* 292(1): 54–63.

Peluso, I., Sastry Yarla, N., Ambra, R., Pastore, G., & Perry, G. 2017. MAPK signalling pathway in cancers: Olive products as cancer preventive and therapeutic agents. *Seminars in Cancer Biology* 56:185–195.

Peng, Q., Deng, Z., Pan, H., Gu, L., Liu, O., & Tang, Z. 2017. Mitogen-activated protein kinase signaling pathway in oral cancer. *Oncology Letter* 15(2):1379–1388.

Ranzato, E., Martinotti, S., Calabrese, C. M., & Calabrese, G. 2014. Role of nutraceuticals in cancer therapy. *Journal of Food Research* 8:18–25.

Rivera, C. 2015. Essentials of oral cancer. *International Journal of Clinical and Experimental Pathology* 8(9):11884.

Saadat, N., & Gupta, S. V. 2012. Potential role of garcinol as an anticancer agent. *Journal of Oncology* 2012: 647206.

See, D., Mason, S., & Roshan, R. 2002. Increased tumor necrosis factor alpha (TNF-alpha) and natural killer cell (NK) function using an integrative approach in late stage cancers. *Immunological Investigations* 31:137–153.

Siddiqui, S., Ahamad, M. S., Jafri, A., Afzal, M., & Arshad, M. 2017. Piperine triggers apoptosis of human oral squamous carcinoma through cell cycle arrest and mitochondrial oxidative stress. *Nutrition and Cancer* 69(5):791–799.

Silvan, S., Manoharan, S., Baskaran, N., Anusuya, C., Karthikeyan, S., & Prabhakar, M. M. 2011. Chemopreventive potential of apigenin in 7, 12-dimethylbenz (a) anthracene induced experimental oral carcinogenesis. *European Journal of Pharmacology* 670(2–3):571–577.

Simpson, D. R., Mell, L. K., & Cohen, E. E. 2015. Targeting the PI3K/AKT/mTOR pathway in squamous cell carcinoma of the head and neck. *Oral Oncology* 51(4):291–298.

Sophia, J., Kowshik, J., Dwivedi, A., Bhutia, S. K., Manavathi, B., Mishra, R., & Nagini, S. 2018. Nimbolide, a neem limonoid, inhibits cytoprotective autophagy to activate apoptosis via modulation of the PI3K/Akt/GSK-3β signalling pathway in oral cancer. *Cell Death & Disease* 9(11):1087.

Sur, S., Steele, R., Aurora, R., Varvares, M., Schwetye, K. E., & Ray, R. B. 2018. Bitter melon prevents the development of 4-NQO–induced oral squamous cell carcinoma in an immunocompetent mouse model by modulating immune signaling. *Cancer Prevention Research* 11(4):191–202.

Usman, S., Jamal, A., Muy-, T., Teh, A., & Waseem 2021. Major molecular signaling pathways in oral cancer associated with therapeutic resistance. *Frontiers in Oral Health* 25(1):603160.

Wang, L., Phan, D. D., Zhang, J., Ong, P. S., Thuya, W. L., Soo, R., Wong, A. L., Yong, W. P., Lee, S. C., Ho, P. C., Sethi, G., & Goh, B. C. 2016. Anticancer properties of nimbolide and pharmacokinetic considerations to accelerate its development. *Oncotarget* 7(28):44790–44802.

Wang, R., Lu, X., & Yu, R. 2020. Lycopene inhibits epithelial–mesenchymal transition and promotes apoptosis in oral cancer via PI3K/AKT/m-TOR signal pathway. *Drug Design, Development and Therapy* 14:2461–2471.

Won, D.-H., Kim, L.-H., & Jang, B., et al. 2018. In vitro and in vivo anti-cancer activity of silymarin on oral cancer. *Tumor Biology* 40(5):1010428318776170.

Xiao, C., Wang, L., Zhu, L., Zhang, C., & Zhou, J. 2014. Curcumin inhibits oral squamous cell carcinoma SCC-9 cells proliferation by regulating miR-9 expression. *Biochemical and Biophysical Research Communications* 454(4):576–580.

Yedida, G. R., Nagini, S., & Mishra, R. 2013. The importance of oncogenic transcription factors for oral cancer pathogenesis and treatment. *Oral Surgery, Oral Medicine, Oral Pathology and Oral Radiology* 116(2):179–188.

Yoshimura, H., Yoshida, H., Matsuda, S., Ryoke, T., Ohta, K., Ohmori, M., ... Sano, K. 2019. The therapeutic potential of epigallocatechin-3-gallate against human oral squamous cell carcinoma through inhibition of cell proliferation and induction of apoptosis: In vitro and in vivo murine xenograft study. *Molecular Medicine Reports* 20(2):1139–1148.

Yu, X. D., Yang, J. L., Zhang, W. L., & Liu, D. X. 2016. Resveratrol inhibits oral squamous cell carcinoma through induction of apoptosis and G2/M phase cell cycle arrest. *Tumor Biology* 37:2871–2877.

Zlotogorski, A., Dayan, A., Dayan, D., Chaushu, G., Salo, T., & Vered, M. 2013. Nutraceuticals as new treatment approaches for oral cancer–I: Curcumin. *Oral Oncology* 49(3):187–191.

8 Role of Resveratrol in Molecular Targeting of Melanoma

Girish Sharma[1,2], *Nishtha Chauhan*[1], *and Tanishqua Chopra*[1]
[1]Centre for Medical Biotechnology, Amity Institute of Biotechnology, Amity University, Noida, India
[2]Amity Centre for Cancer Epidemiology & Cancer Research, Amity University, Noida, India

8.1 INTRODUCTION

Sunlight exposure has been predicted as the major cause of skin cancer, especially the UV-B component (290–320 nm) of UVR, which is responsible for 90% of skin cancers [1]. Melanoma is the most aggressive of all skin tumors and is associated with a high risk of mortality from metastatic progression. According to the International Agency for Research on Cancer (IARC), there are an estimated 325,000 incidences of melanoma in 2020 in both males and females which is predicted to reach about 424,000 by the year 2040, of which there were 57,000 incidences of mortality in 2020, and this number is estimated to reach about 84,600 in the year 2040. Melanoma is a rare but fatal form of skin cancer treated with synthetic molecules and phytochemicals [2]. Alteration in gene functioning and signaling pathways in melanoma cells controls its progression and malignancy [3]. Most frequently altered genes identified in melanocytes are BRAF [4, 5] NRAS [6, 7], PTEN [phosphatase and tensin homolog, a tumor suppressor gene] [8-11], p53 [12,13], and p16 [14]. Diagnosis and staging of melanoma depend on circulating melanoma cells [15].

At present, there is no established method for detecting cutaneous melanoma. Methods proposed to date are costly for the identification of melanoma. Although various treatment options are available for melanoma, they have significant drawbacks, including various side effects and multiple drug resistance. Chemoprevention is the most cost-effective option that produces minimal or no side effects in treating melanoma. Most of the therapeutic agents target histone deacetylases (HDAC) [16]. In one such study, Vorinostat [a HDAC], in combination with proteasome inhibitor Morizomib, was assessed in a phase I study [17]. Although many chemopreventive agents, including phytochemicals like fisetin [4, 18], epigallocatechin-3-gallate [19], curcumin [20], capsaicin [21], genistein [22, 23] indole-3-carbinole [I3C] [24, 25] and luteolin [26] have been available for the treatment of various types of cancer, still resveratrol has diverse applications in clinical research [27]. It can be used alone and in combination with synthetic compounds such as Paclitaxel [28] or natural compounds like curcumin [29].

Resveratrol (trans-3,5,4′-trihydroxystilbene) was initially isolated as a constituent of the roots of white hellebore (*Veratrum grandiflorum O. Loes*) and after that, has been found in various plants including grapes, berries, peanuts, lily, pines (Table 8.1). Resveratrol has been known to be a reservoir of numerous polyphenolic compounds, including curcumin [30], rottlerin [31], genistein [32], and quercetin [33], which shows strong evidence of the inhibitory effect exhibited by resveratrol on the proliferation of carcinoma cells through the induction of apoptosis and cell cycle arrest in G2/M phase. Besides cardioprotective effects, resveratrol has exhibited anticancer properties, as suggested

DOI: 10.1201/9781003486237-8

TABLE 8.1
Sources of Resveratrol from Plants

Common Name	Botanical Name	Family
Grapes	*Vitis vinifera*	Vitaceae
Peanuts	*Arachis hypogeal*	Fabaceae
Mulberry	*Morus rubra*	Moraceae
Blueberry	*Vaccinium sp.*	Ericaceae
Cranberry	*Vaccinium sp.*	Ericaceae
White hellebore	*Veratrum grandiflorum O. Loes*	Ranunculaceae
Lily	*Veratrum spp.*	Liliaceae
Legumes	*Cassia spp.*	Fabaceae
Ko-jo-kon	*Polygonum cuspidatum*	Polygonaceae
Jackfruit	*Artocarpus heterophyllus*	Moraceae
Grasses	*Festuca, Hordeum, Poa, Stipa* and *Lolium spp.*	Poaceae
Eucalyptus	*Eucalyptus oblique*	Myrtaceae
Pine	*Pinus sylvestris*	Pinaceae
Rheum	*Rheum rhaponticum*	Polygonaceae
spruce	*Picea spp.*	Pinaceae

by its ability to suppress the proliferation of a wide variety of tumor cells, including melanoma. Resveratrol has also been effective in treating other types of cancer such as prostate, breast, lung, and colorectal [34]. Studies have shown that the route of administration of resveratrol impacts the outcome depending on the tumor type. Resveratrol has been shown to work synergistically, exhibiting a cumulative effect on the addition of other compounds, such as chloroquinine, thereby decreasing the cell viability of melanoma. Extensive research is being carried out to enhance the interaction of resveratrol with tumor cells and its efficient, targeted delivery using nanoparticle formulations [35]. Trans-resveratrol suppresses the growth of melanoma cells by imposing cytotoxic effects on them, which have been correlated with the reduced activity of telomerase and levels of *hTERT* gene transcript [36]. Resveratrol also suppressed AKT phosphorylation in melanoma cells [37] and thus has been effective in treating advanced melanoma [38].

8.2 MOLECULAR TARGETS OF RESVERATROL FOR MELANOMA

8.2.1 Activating Mitochondrial Apoptotic Pathway

Resveratrol has dual functionality, i.e., it can cause pro-apoptotic and antiapoptotic effects. In the case of melanoma, it acts as a pro-apoptotic agent by inducing apoptosis in these cells [39]. Studies have revealed that resveratrol inhibits uveal melanoma cells by activating the intrinsic mitochondrial pathway [40]. Loss of membrane potential due to mitochondrial dysfunction can signal apoptosis as a part of the intrinsic apoptotic pathway (Figure 8.1).

The oral administration of resveratrol has been shown to inhibit tumor growth in two xenograft models of human uveal melanoma, namely, primary tumor and metastatic lesions. Direct injection of resveratrol to the area adjacent to the tumor increases the concentration of the drug and thereby resulting in tumor cell death [41].

8.2.2 Inhibitory Effect of Resveratrol on NF-κB, COX-2, and AP-1

Nuclear factor–kappa B (NF-κB), a redox-sensitive transcription factor, has been found to mediate inflammatory response and the promotional stage of carcinogenesis. NF-κB driven transactivation of

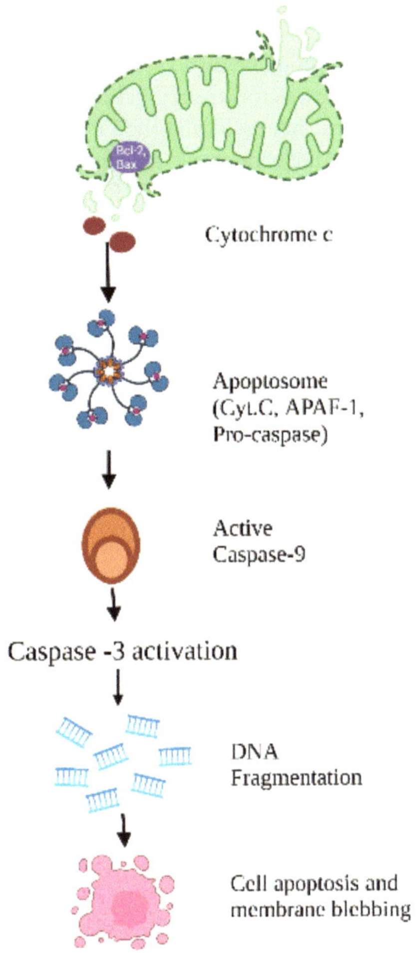

FIGURE 8.1 Intrinsic apoptotic pathway leading to cell death.

cyclooxygenase-2 (COX-2) leads to prostaglandin biosynthesis and inflammation. As inappropriate regulation of COX-2 results in both pre-malignant and malignant tissue, targeted inhibition of COX-2 can prevent cancer. Kundu *et al.* [42] reported that when O-tetradecanoyl phorbol 13-acetate (TPA), a prototype tumor promoter, is applied topically on mouse skin, it induces two transcription factors *viz* NF-κB and Activator Protein-1 (AP-1). These factors, in turn, activated upstream kinases such as Mitogen-Activated Protein (MAP) kinases. Further, they also reported that TPA-induced mouse skin increases I-κB kinase, and the topical application of resveratrol inhibited TPA-induced COX-2 expression by down-regulating NF-κB and blocking IKK signaling.

NF-κB plays an essential role in melanoma cell proliferation and cell survival. Studies [39] have reported resveratrol as an effective *in-vitro* inhibitor of NF-κB, resulting in apoptosis and reduced proliferation of melanoma cells. Thus, resveratrol overcomes chemoresistance through downregulation of NF-κB and other cell survival gene products (STAT3) in melanoma cells [39]. The role of resveratrol in the NF-κB signaling pathway has been reported through suppression of p65 and I-kappa B kinase activities [43]. Recent studies have suggested that downregulation of NF-κB mediated by resveratrol has been responsible for cell apoptosis. This has been associated with the downregulation of microphthalmia-associated transcription factors (MTF) in melanoma cells [44].

8.2.3 APE/Ref1

Apurinic/apyrimidinic endonuclease/redox factor 1 (APE/Ref-1) has been described as a multifunctional enzyme with two main activities, namely, enzymatic activity (catalyzes phosphodiester bond of 5′ or 3′ to an apurinic/apyrimidinic site) [45] and the DNA repair activity [46]. APE/Ref-1 acts as both redox-dependent and redox-independent. It activates several transcription factors such as Jun/Jun homodimers, Fos/Jun heterodimers, Myc, NF-κB, p53, members of cyclic AMP-responsive element-binding (CREB) protein, and the Pax family of proteins.

Yang *et al.* [47] found that Ref-1 content was higher in the nucleus and cytoplasm of melanoma cell lines than in normal melanocytes. Their Immunohistofluoroscence studies also found an increased Ref-1 expression in both nevi and malignant melanoma biopsy compared to the Ref-1 level of the normal melanocytes. Further, they transfected an adenovirus 'Adref-1' into a human melanocyte cell line that offered protection against UV-B-induced apoptosis and prevented increased foci formation in cell cultures. Adref-1 containing the sense construct of Ref-1 transfected into melanoma cell line offered protection from cisplatin or H2O2-induced apoptosis. Finally, when the antisense construct of Ref-1 was transfected into the melanoma cell line, it induced apoptosis. They also reported that when anti-APE/Ref-1 antibody bound to the Ref-1 protein, it blocked AP-1 and NF-κB. They also screened the natural analogs that bound and blocked Ref-1 protein. Further, 'resveratrol' bound to Ref-1 protein and blocked AP-1 and NF-κB. It also increased the sensitivity of melanoma cells to dacarbazine treatment.

Studies on human melanoma cells have revealed that when used in a dose-dependent manner, resveratrol acted as an inhibitor in the APE1/Ref-1-mediated DNA-binding of AP-1 and APE1/Ref-1 endonuclease activity and sensitized melanoma cells to treatment with the alkylating agent dacarbazine. These studies revealed the role of resveratrol in APE1/Ref-1 redox and repair activity [46].

8.2.4 Nitric Oxide and Downstream Targets

Nitric oxide is involved in tumor progression, including melanoma [48]. Yang *et al.* [49] analyzed the effect of diethylenetriamine/nitric oxide (DETA/NO) on the levels of APE/Ref-1 and downstream targets such as AP-1/Jun P, matrix metalloproteinase (MMP-1), Bcl-2, inducible nitric oxide synthase (iNOS) and had found that DETA/NO treatment induced an increase in levels of APE/Ref-1 and downstream targets. Further, siRNA designed to block the APE/Ref-1 and AP-1/Jun D reversed the induced stress by nitric oxide. Their studies also revealed that resveratrol bound to APE/Ref-1 protein resulted in decreased protein levels of downstream targets, leading to decreased melanoma progression. Additionally, NO. induced apoptosis in A375 human melanoma cells treated with a combination of capsaicin and resveratrol *via* activation of the mitochondrial apoptotic and death receptor cascades [50].

8.2.5 Binding Site of Resveratrol

Resveratrol has been shown to exert protective actions through specific binding sites in the epidermis. Bastianetto *et al.* [51] identified and characterized the presence of specific binding sites for (3H)-resveratrol in a human keratinocyte cell line (HaCaT) treated with a nitric oxide free radical donor, sodium nitroprusside, which resulted in cell death. On the contrary, the same resveratrol concentration protected the cell line by reducing mitochondrial apoptotic events triggered by sodium nitroprusside, thereby reducing the number of apoptotic cells in the HaCaT cell line.

8.2.6 Protection from UV-B Radiation

Resveratrol on topical application offered substantial protection against skin damage from UV-B radiation. Hung *et al.* [52] checked the feasibility of topical/transdermal delivery of resveratrol.

Delivery of resveratrol *via* skin has been an attractive alternative to oral dosing for topical and transdermal applications. The effect of vehicles on the *in-vitro* permeation and skin deposition from saturated solutions of aqueous buffers and soybean oil has been analyzed. They also found that viable epidermis acted as a barrier to a non-ionic solution, whereas stratum corneum and viable skin acted as a barrier to an anionic solution of resveratrol. Further, the viscosity of hydrogel form of topical and transdermal treatment with resveratrol controlled its permeation and deposition. The two significant contributions of the above study were that the hydrogel form of resveratrol caused more stratum corneum disruption or skin erythema (which ensures the safety profile of resveratrol to be used in hydrogel form in topical and transdermal application) and that they reported and established the permeation profile for topically applied resveratrol [52].

8.2.7 Antiangiogenic Effects of Resveratrol

Trapp *et al.* [53] found that resveratrol exhibits an antiangiogenic effect. They postulated that variation in the release of thrombospondin-1 (TSP1) and vascular endothelial growth factor (VEGF) by tumor cells into the extracellular matrix (ECM) leads to vascular endothelial cell (VEC) apoptosis. Evaluation of melanoma cell lines (YUZAZ6 and M14) co-cultured with VEC in monolayer and 3-D spheroids showed consistent proliferation in the case of isolated VEC but inhibition of VEC grown with melanoma cells.

This result can be related to the tumor suppressor protein p53 and matrix protein TSP1, which showed an increase in expression of melanoma cells, whereas expression of hypoxia-inducible factor-1α (HIF1α) decreased in addition to inhibition of VEGF production. Resveratrol also showed antiangiogenic properties in preclinical tumor models [54].

8.2.8 Aging and Resveratrol

Aging is a process that affects all living organisms and is a time-dependent decline in the functions of a living system [55]. According to Carlos López-Otín *et al.* [55], there are nine hallmarks of aging which can be described as follows: altered intercellular communication, telomere attrition, cellular senescence, genomic instability, deregulated nutrient sensing, mitochondrial dysfunction, loss of proteostasis, epigenetic alterations and stem cell exhaustion [55]. There has been a demographic shift with a more significant number of older people as human life expectancy has increased from about 50 years in the early 1900s to about 80 years in the present time. Although there has been an increase in the expected life expectancy, studies have suggested a limit to the maximum lifespan of human beings. There have been various reasons for this increase in life expectancy for older people. These include environmental factors such as lifestyle changes, reduced infant mortality rates, better sanitation facilities, and access to better medical facilities [56–58].

This demographic shift can be very detrimental as with aging come various economic and health implications. Aging causes many diseases such as cancers, neurodegenerative diseases, cardiovascular diseases, type 2 diabetes, and Alzheimer's [56, 59]. The only way to defer age-related issues is by delaying aging. The free radical theory of aging, given in 1955, suggested that the detrimental effects on health, including aging, are predominantly caused by reactive oxygen species [ROS] [60].

Harran *et al.* [61] suggested that the detrimental effects on health, including aging, are predominantly caused by ROS. Other factors include environmental pollution, degradation from the UV-B rays, and various other stress mediators. Various studies have shown that many polyphenolic compounds found naturally can reduce the detrimental effects of ROS [62]. *Polyphenols* are compounds that have been known for their various properties, such as antiaging, antioxidation, and anti-inflammation. One such polyphenolic compound is resveratrol. *Resveratrol* is an antioxidant used to treat cancers such as melanoma, neurodegenerative diseases, cardiovascular disease, infertility, Type-2 diabetes, sarcopenia and osteoporesis [60, 63] (Figure 8.2).

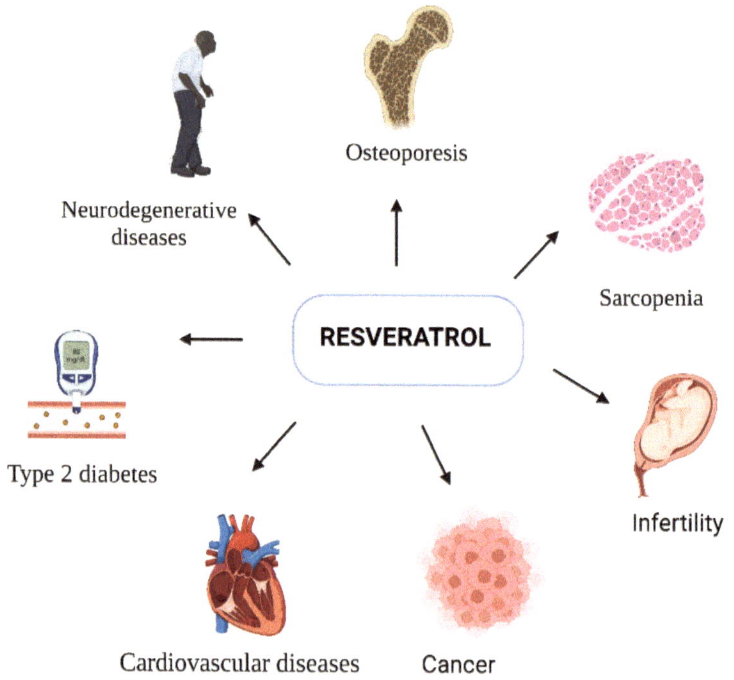

FIGURE 8.2 The age-related diseases on which resveratrol can have an effect.

8.2.9 Effects of Resveratrol on the Changes Caused by Reactive Oxygen Species

ROS have been involved in regulating three different activities, namely, metabolic functions, signaling pathways, and transcription factors (Figure 8.3). ROS can either directly damage or modulate redox-sensitive signaling pathways, depending on the molecular concentrations at which they are present [64]. Some antioxidant enzymes such as Cu-Zn-superoxide dismutase or oxidoreductase and transcription factors like AP-1 are redox-sensitive, so the action of ROS brings about changes in the regular cellular activities, thereby facilitating tumor formation.

The changes in the level of specific essential components that differentiate normal melanocytes from metastatic melanoma are illustrated in Table 8.2.

It shows upregulation and shows downregulation.

The wide range of changes induced in melanoma cells by resveratrol treatment is summarized in Table 8.3.

TABLE 8.2
Comparison of Various Components in Normal and Metastatic Melanoma

	Normal Melanocytes	Metastatic Melanocytes
Redox status	▼	▲
Intracellular ROS	▼	▲
AP-1	▼	▲
Catalase	▲	▼
Glutathione	▲	▼
Alpha-tocopherol	▲	▼

TABLE 8.3
Effects of Resveratrol on Melanoma Cells

Parameter	Effect of Resveratrol on Melanoma Cell
Anchorage dependency	↓
AP-1 binding	↓
Transcriptional activity of AP-1 gene	↓
Intracellular oxidative stress	↓
MHC class I antigen	↑
Fas expression	↑
Fra-2 expression	↑
C-jun	?

Note: Arrows indicate an increase [↑] or decrease [↓] in the levels of activation, expression, or unknown [?] effects of resveratrol on melanocytes.

Augmented ROS and high glycolytic flux are general features of malignant cells. Jung *et al.* [65] estimated the impact of resveratrol on cancer cell glucose metabolism and found the role of ROS. Resveratrol reduced cell content and more pronouncedly inhibited 18F Fludeoxyglucose (18F-FDG) uptake in Lewis lung carcinoma, HT-29 colon, and T47D breast cancer cells. After that, 18F-FDG uptake normalized to cell content was reduced to less than half of controls by 24-h exposure to resveratrol. The reduction was characterized by reduced glycolytic flux and GLUT-1 expression. Resveratrol also decreased intracellular ROS in patterns that closely paralleled 18F-FDG uptake. Scavenging ROS with N-acetyl cysteine, but not inhibition of nicotinamide adenine dinucleotide phosphate (NADP) oxidase, was sufficient to suppress 18F-FDG uptake.

Conversely, ROS inducers effectively reversed the metabolic response of resveratrol. HIF-1α protein was markedly reduced by resveratrol. Inhibited HIF-1α expression with cycloheximide or specific siRNA suppressed 18F-FDG uptake. The proteasomal inhibitor MG132 partly restored HIF-1α level and 18 F-FDG uptakes in resveratrol-treated cells. Resveratrol also inhibited Akt activation; inhibitors and siRNAs against phosphoinositide-3-kinase decreased 18F-FDG uptake. Resveratrol suppressed cancer cell 18F-FDG uptake and glycolytic metabolism in a manner that depends on the capacity of resveratrol to inhibit intracellular ROS, which downregulated HIF-1α accumulation.

8.3 THERAPEUTIC EFFECTS OF RESVERATROL THROUGH *IN VITRO* AND *IN VIVO* MECHANISMS

For *in vitro* and *in vivo* studies of resveratrol effects, several cell lines and mouse models of skin cancer have been described along with their mechanisms. Table 8.4 illustrates the chemo-preventive effects of resveratrol in a dose-specific manner through *in vitro* and *in vivo* mechanisms.

8.4 CONCLUSION

Natural agents have been believed to possess the potential to suppress and prevent carcinogenesis. In this aspect, the natural agent 'resveratrol' (3,4′,5-trihydroxy-trans-stilbene) has been demonstrated to prevent or slow down the progression of a wide variety of age-related illnesses such as cancer, diabetes, arthritis, coronary heart disease, neurodegenerative and pulmonary diseases.

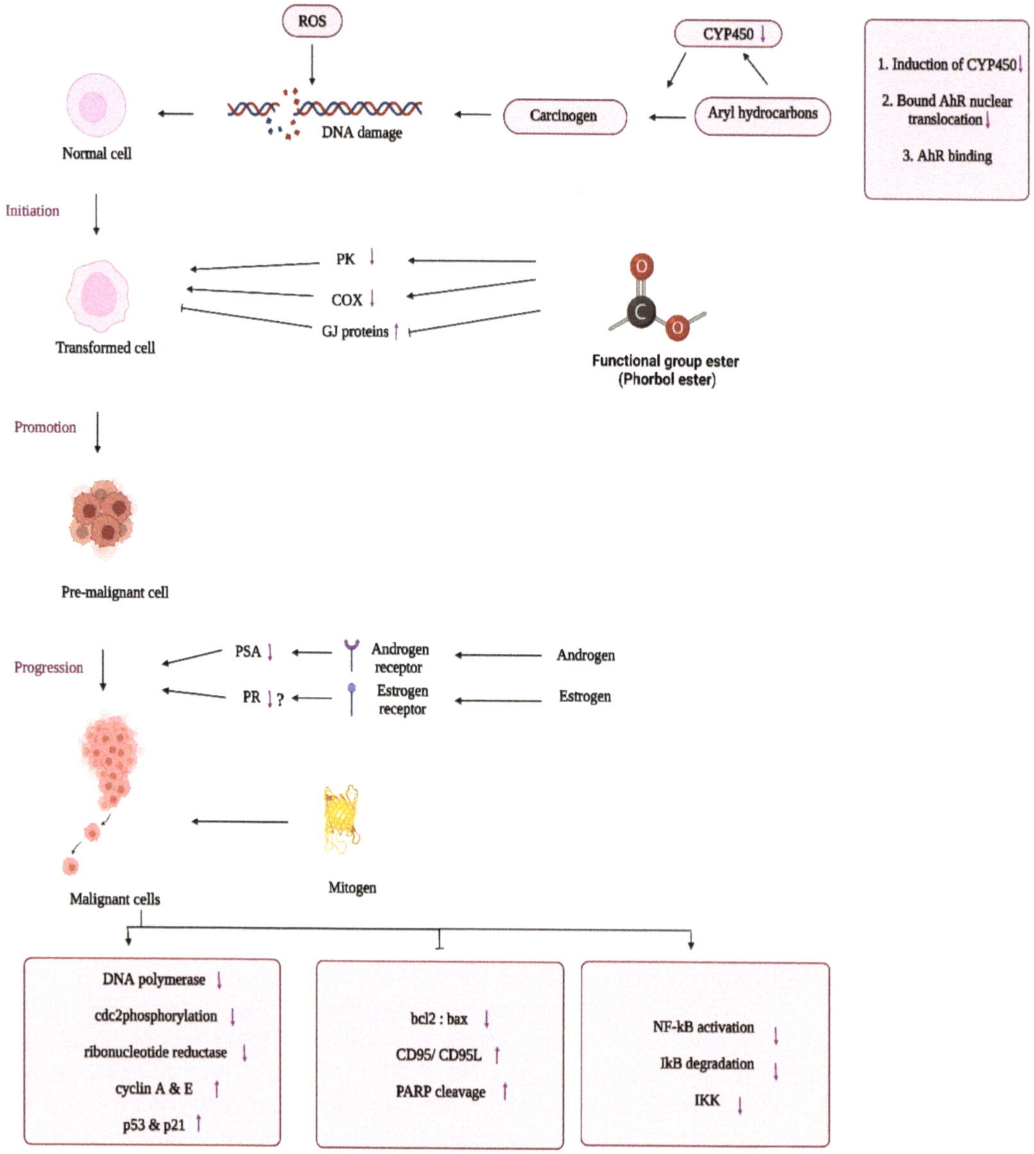

FIGURE 8.3 Effects of resveratrol on various pathways of carcinogenesis and cell proliferation.

Although an extensive study has been done for the past decade on this 'magic molecule,' it still fascinates researchers by revealing its secrets from experiment to experiment conducted so far.

Although melanoma was considered rare a century ago, the average lifetime risk for a person developing this cancer has increased its incidence from one in fifty in many Western populations. This review reveals various studies targeting the molecular factors associated with melanoma progression and resveratrol chemo-preventive effects. Although resveratrol is an old molecule, it kept its promise of fighting against melanoma by inhibiting uveal melanoma in animal models, causing tumor cell death via activating mitochondrial pathway, thus leading to melanoma regression. Resveratrol has been proven to act on specific molecular targets such as COX-2, IKK signaling, NF-κB, p53, MAPK signaling, and APE/Ref1, which are central molecular targets for resveratrol in controlling melanoma. It also targets mitochondrial signaling by reducing caspase3 activity, thus

TABLE 8.4
***In Vitro* and *In Vivo* Mechanism of Resveratrol**

	In Vitro		*In Vivo*				
Cell Line	**Dose**	**Mechanisms**	**Animal Model**	**Dose**	**Route of Administration**	**Mechanisms**	**References**
Induces apoptosis in melanoma [A375 and SK-mel28] cells.	50 μM for 48 hour	↑Phosphorylates ERK1/2	Mouse skin cancer model	1-25 μM, 2/week for 18 weeks,	Topical	Inhibits DMBA-induced preneoplastic lesions	[66, 67]
Induces apoptosis in epidermoid carcinoma [A431]	1-50 μM for 24 hour	↑p21Cip1/WAF1; ⊥ G1-phase	Mouse skin cancer model	1-25 μM, 30 min	Topical	Inhibits DMBA-induced mouse mammary] cell growth and PMA-promoted mouse skin tumors	[68, 69]
Inhibits prolifera-tion of epidermoid carcinoma [A431] cells	1,5,10&25 μM for 24 hours	↓Hyperphosphorylated Rb; ⊥ G0/G1-phase	Mouse skin cancer model	85 nM/day, daily	Topical	Reduces papillomas in DMBA-initiated [334] and PMA-promoted mouse skin two-stage cancer	[70, 71]
Induces apoptosis in JB6 P+ mouse epidermal cell line C1 41	1,5,10&25 μM for 24 hours	↑ p53-dependent apoptosis pathway	Mouse skin cancer model	1-25 μM, 2/week for 18 weeks	Topical	Inhibits DMBA-induced tumor incidence and tumor burden in CD-1	[70, 72]
Induces apoptosis of SK-Mel-28	100 μM for 48 hours	⊥ S-phase ↑cyclins A, E, and B1	Mouse skin cancer model		Topical	Protects against the damage caused by short term UVB exposed-SKH-1 hairless mouse skin	[73, 74]

Note: Arrows indicate an increase [↑] or decrease [↓] in the levels of protein expression/activation, genetic and epigenetic expression, or [⊥] inhibition of the particular phase of the cell cycle.

leading to apoptosis. The phenomenon of oxidative stress has become the central dogma for studying cancer progression. Concerning the above statement, resveratrol has been reported to balance intracellular oxidative stress by scavenging free radicals of nitric oxide.

Since skin plays a significant role in offering the first degree of protection to the human body from the external environment, it is also subjected to more significant stress caused by pollution, free radicals, and UV radiation. WHO recommends using sunscreen lotions to protect the skin from the stress induced by UV-B radiations, which may lead to several grades of sunburn, gradually leading to melanoma? Considering the statement, resveratrol is very promising and is used in sunscreen lotions as an anti-UVB and as an antioxidizing agent. Considering the bioavailability of resveratrol, a study proves that the usage of a non-ionic solution of resveratrol in a hydrogel form enhances skin permeation, thus improving the bioavailability. Thus, this paper suggests insights into the chemo-preventive effects of resveratrol in various molecular targets of melanoma.

8.5 FUTURE DIRECTIONS

Resveratrol is a potent polyphenol that can be used to treat melanoma. There are various applications of resveratrol as antiaging, anticancer, and antioxidant, which are of great importance for humankind. However, the problem lies with the bioavailability of resveratrol. Naturally occurring resveratrol present in grapes, blueberries, and other food products, has a very concentration of resveratrol and is quickly metabolized. Many studies have been done to increase the effect of resveratrol to be proven as an efficient treatment for melanoma. A combination of various drugs with resveratrol, such as quercetin, can also be used to increase the efficacy of the drugs that would target different molecular mechanisms [75]. Using a combination of the polyphenols curcumin, resveratrol, and solid lipid nanoparticles can increase the penetration and feasibility of the topical drug delivery system [76]. The permeability and retention of resveratrol can also be increased by using nano-emulsion gel which could be an efficient way for topical delivery of resveratrol to prevent UV-induced skin damage [77]. Studies have shown that when used along with IL-2, resveratrol has increased NKp30 [78]. Natural killer [NK] cells help kill malignant cells and control tumor metastasis [79]. Recent studies have been carried out which show that resveratrol can induce autophagy in melanoma cells. RSV-induced autophagy might function as a resistance mechanism against apoptotic cell death. Inhibition of autophagy could be a novel strategy for the adjuvant chemotherapy of melanoma cancer [80]. Recent research has been carried out, which unveils that resveratrol through upregulation of p53 can induce cell apoptosis and inhibit the proliferation of melanoma cells. Recent reports showed that antiapoptotic factors from intrinsic apoptosis pathways become potential targets for overcoming the apoptotic resistance of melanoma [81]. A bcl-2-mediated apoptotic pathway can be another possible association with resveratrol-induced apoptosis in melanoma cells except for the p53-related apoptotic pathway. Targeting the PD-L1-mediated tumor immune evasion represents a promising approach for immune checkpoint blockade therapies [82]. RSV alters N-glycosylation to disrupt PD-1/PD-L1 interaction. RSV interacts with and blocks α-glucosidase/α-mannosidase in tumoral cells and prevents glyco-PD-L1-processing, which, in turn, promotes the retention of the abnormally glycosylated form of PD-L1 in the endoplasmic reticulum [83].

ACKNOWLEDGMENT

The authors are grateful to Dr. Ashok K. Chauhan, Founder President, and Dr. Atul Chauhan, Chancellor of Amity University Uttar Pradesh, for their valuable guidance, kind support, and research facilities.

FINANCIAL SUPPORT AND SPONSORSHIP

Nil.

REFERENCES

1. Sung H, Ferlay J, Siegel RL, Laversanne M, Soerjomataram I, Jemal A, Bray F. Global cancer statistics 2020: GLOBOCAN estimates of incidence and mortality worldwide for 36 cancers in 185 countries. CA: a Cancer Journal for Clinicians. 2021 May;71(3):209–49.
2. Boyle P, Levin B. World cancer report 2008. IARC Press, International Agency for Research on Cancer; 2008.
3. Strickland LR, Pal HC, Elmets CA, Afaq F. Targeting drivers of melanoma with synthetic small molecules and phytochemicals. Cancer Letters. 2015 Apr 1;359(1):20–35.
4. Pal HC, Baxter RD, Hunt KM, Agarwal J, Elmets CA, Athar M, Afaq F. Fisetin, a phytochemical, potentiates sorafenib-induced apoptosis and abrogates tumor growth in athymic nude mice implanted with BRAF-mutated melanoma cells. Oncotarget. 2015 Sep 29;6(29):28296.
5. Ascierto PA, Kirkwood JM, Grob JJ, Simeone E, Grimaldi AM, Maio M, Palmieri G, Testori A, Marincola FM, Mozzillo N. The role of BRAF V600 mutation in melanoma. Journal of Translational Medicine. 2012 Dec;10(1):1–9.
6. Davies H, Bignell GR, Cox C, Stephens P, Edkins S, Clegg S, Teague J, Woffendin H, Garnett MJ, Bottomley W, Davis N. Mutations of the BRAF gene in human cancer. Nature. 2002 Jun;417(6892):949–54.
7. Greene VR, Johnson MM, Grimm EA, Ellerhorst JA. Frequencies of NRAS and BRAF mutations increase from the radial to the vertical growth phase in cutaneous melanoma. Journal of Investigative Dermatology. 2009 Jun 1;129(6):1483–8.
8. Johnson DB, Lovly CM, Flavin M, Panageas KS, Ayers GD, Zhao Z, Iams WT, Colgan M, DeNoble S, Terry CR, Berry EG. Impact of NRAS mutations for patients with advanced melanoma treated with immune therapies. Cancer Immunology Research. 2015 Mar 1;3(3):288–95.
9. Masaki T, Wang Y, DiGiovanna JJ, Khan SG, Raffeld M, Beltaifa S, Hornyak TJ, Darling TN, Lee CC, Kraemer KH. High frequency of PTEN mutations in nevi and melanomas from xeroderma pigmentosum patients. Pigment Cell & Melanoma Research. 2014 May;27(3):454–64.
10. Shull AY, Latham-Schwark A, Ramasamy P, Leskoske K, Oroian D, Birtwistle MRBuckhaults PJ. Novel somatic mutations to PI3K pathway genes in metastatic melanoma.
11. Tsao H, Chin L, Garraway LA, Fisher DE. Melanoma: From mutations to medicine. Genes & Development. 2012 Jun 1;26(11):1131–55.
12. Aguissa-Touré AH, Li G. Genetic alterations of PTEN in human melanoma. Cellular and Molecular Life Sciences. 2012 May;69(9):1475–91.
13. Houben R, Hesbacher S, Schmid CP, Kauczok CS, Flohr U, Haferkamp S, Müller CS, Schrama D, Wischhusen J, Becker JC. High-level expression of wild-type p53 in melanoma cells is frequently associated with inactivity in p53 reporter gene assays. PloS One. 2011 Jul 8;6(7):e22096.
14. Ji Z, Njauw CN, Taylor M, Neel V, Flaherty KT, Tsao H. p53 rescue through HDM2 antagonism suppresses melanoma growth and potentiates MEK inhibition. Journal of Investigative Dermatology. 2012 Feb 1;132(2):356–64.
15. Lu M, Miller P, Lu X. Restoring the tumour suppressive function of p53 as a parallel strategy in melanoma therapy. FEBS Letters. 2014 Aug 19;588(160:2616–21.
16. Ward KA, Lazovich D, Hordinsky MK. Germline melanoma susceptibility and prognostic genes: A review of the literature. Journal of the American Academy of Dermatology. 2012 Nov 1;67(5):1055–67.
17. Mumford BS, Robertson GP. Circulating melanoma cells in the diagnosis And monitoring of melanoma: An appraisal of clinical potential. Molecular Diagnosis & Therapy. 2014 Apr;18(2):175–83.
18. Garmpis N, Damaskos C, Garmpi A, Dimitroulis D, Spartalis E, Margonis GA, Schizas D, Deskou I, Doula C, Magkouti E, Andreatos N. Targeting histone deacetylases in malignant melanoma: A future therapeutic agent or just great expectations? Anticancer Research. 2017 Oct 1;37(10):5355–62.
19. Millward M, Price T, Townsend A, Sweeney C, Spencer A, Sukumaran S, Longenecker A, Lee L, Lay A, Sharma G, Gemmill RM. Phase 1 clinical trial of the novel proteasome inhibitor marizomib with the histone deacetylase inhibitor vorinostat in patients with melanoma, pancreatic and lung cancer based on in vitro assessments of the combination. Investigational New Drugs. 2012 Dec;30(6):2303–17.
20. Pal HC, Sharma S, Strickland LR, Katiyar SK, Ballestas ME, Athar M, Elmets CA, Afaq F. Fisetin inhibits human melanoma cell invasion through promotion of mesenchymal to epithelial transition and by targeting MAPK and nfκb signaling pathways. PloS One. 2014 Jan 23;9(1):e86338.
21. Ravindranath MH, Ramasamy V, Moon S, Ruiz C, Muthugounder S. Differential growth suppression of human melanoma cells by tea [Camellia sinensis] epicatechins [ECG, EGC and EGCG]. Evidence-Based Complementary and Alternative Medicine. 2009 Dec 1;6(4):523–30.

22. Sharma RA, Gescher AJ, Steward WP. Curcumin: The story so far. European Journal of Cancer. 2005 Sep 1;41(13):1955–68.
23. Shin DH, Kim OH, Jun HS, Kang MK. Inhibitory effect of capsaicin on B16-F10 melanoma cell migration via the phosphatidylinositol 3-kinase/Akt/Rac1 signal pathway. Experimental & Molecular Medicine. 2008 Oct;40(5):486–94.
24. Ji C, Yang YL, He L, Gu B, Xia JP, Sun WL, Su ZL, Chen B, Bi ZG. Increasing ceramides sensitizes genistein-induced melanoma cell apoptosis and growth inhibition. Biochemical and Biophysical Research Communications. 2012 May 11;421(3):462–7.
25. Yang Y, Wang H, Wang J, Li X, Ma M, Yang W. Effect of livin gene suppression by genistein on apoptosis, cell cycle and proliferation of malignant melanoma LiBr cells. Journal of Southern Medical University. 2012 Aug 1;32(8):1163–7.
26. Aronchik I, Kundu A, Quirit JG, Firestone GL. The antiproliferative response of indole-3-carbinol in human melanoma cells is triggered by an interaction with NEDD4-1 and disruption of wild-type PTEN degradation. Molecular Cancer Research. 2014 Nov 1;12(11):1621–34.
27. Nakashima S, Matsuda H, Oda Y, Nakamura S, Xu F, Yoshikawa M. Melanogenesis inhibitors from the desert plant Anastatica hierochuntica in B16 melanoma cells. Bioorganic & Medicinal Chemistry. 2010 Mar 15;18(6):2337–45.
28. Gupta C, Sharma G, Chan D. Resveratrol: A chemo-preventative agent with diverse applications. Prakash D. Sharma G. (Eds.), Phytochemicals of Nutraceutical Importance. 2014 Feb 28:47–60.
29. Jiang Q, Yang M, Qu Z, Zhou J, Zhang Q. Resveratrol enhances anticancer effects of paclitaxel in HepG2 human liver cancer cells. BMC Complementary and Alternative Medicine. 2017 Dec;17(1):1–2.
30. Chen L, Liu T, Wang Q, Liu J. Anti-inflammatory effect of combined tetramethylpyrazine, resveratrol and curcumin in vivo. BMC Complementary and Alternative Medicine. 2017 Dec;17(1):1–0.
31. Perrone D, Ardito F, Giannatempo G, Dioguardi M, Troiano G, Lo Russo L, De Lillo A, Laino L, Lo Muzio L. Biological and therapeutic activities, and anticancer properties of curcumin. Experimental and Therapeutic Medicine. 2015 Nov 1;10(5):1615–23.
32. Maioli E, Torricelli C, Valacchi G. Rottlerin and cancer: Novel evidence and mechanisms. The Scientific World Journal. 2012 Jan 1;2012.
33. Russo M, Russo GL, Daglia M, Kasi PD, Ravi S, Nabavi SF, Nabavi SM. Understanding genistein in cancer: The ‘good’ and the ‘bad’ effects: A review. Food Chemistry. 2016 Apr 1;196:589–600.
34. Filipa Brito A, Ribeiro M, Margarida Abrantes A, Salome Pires A, Jorge Teixo R, Guilherme Tralhao J, Filomena Botelho M. Quercetin in cancer treatment, alone or in combination with conventional therapeutics? Current Medicinal Chemistry. 2015 Sep 1;22(26):3025–39.
35. Park EJ, Pezzuto JM. The pharmacology of resveratrol in animals and humans. Biochimica et Biophysica Acta [BBA]-Molecular Basis of Disease. 2015 Jun 1;1852(6):1071–113.
36. Bonferoni MC, Rossi S, Sandri G, Ferrari F. Nanoparticle formulations to enhance tumor targeting of poorly soluble polyphenols with potential anticancer properties. In: Seminars in cancer biology 2017 Oct 1 (Vol. 46, pp. 205–214). Academic Press.
37. Platella C, Guida S, Bonmassar L, Aquino A, Bonmassar E, Ravagnan G, Montesarchio D, Roviello GN, Musumeci D, Fuggetta MP. Antitumour activity of resveratrol on human melanoma cells: A possible mechanism related to its interaction with malignant cell telomerase. Biochimica et Biophysica Acta [BBA]-General Subjects. 2017 Nov 1;1861(11):2843–51.
38. Luo H, Umebayashi M, Doi K, Morisaki T, Shirasawa S, Tsunoda T. Resveratrol overcomes cellular resistance to vemurafenib through dephosphorylation of akt in BRAF-mutated melanoma cells. Anticancer Research. 2016 Jul 1;36(7):3585–9.
39. Fuggetta MP, D’Atri S, Lanzilli G, Tricarico M, Cannavò E, Zambruno G, Falchetti R, Ravagnan G. In vitro antitumour activity of resveratrol in human melanoma cells sensitive or resistant to temozolomide. Melanoma Research. 2004 Jun 1;14(3):189–96.
40. Ivanov VN, Partridge MA, Johnson GE, Huang SX, Zhou H, Hei TK. Resveratrol sensitizes melanomas to TRAIL through modulation of antiapoptotic gene expression. Experimental Cell Research. 2008 Mar 10;314(5):1163–76.
41. Mukherjee S, Dudley JI, Das DK. Dose-dependency of resveratrol in providing health benefits. Dose-Response. 2010 Oct 1;8(4).
42. Van Ginkel PR, Darjatmoko SR, Sareen D, Subramanian L, Bhattacharya S, Lindstrom MJ, Albert DM, Polans AS. Resveratrol inhibits uveal melanoma tumor growth via early mitochondrial dysfunction. Investigative Ophthalmology & Visual Science. 2008 Apr 1;49(4):1299–306.
43. Kundu JK, Shin YK, Kim SH, Surh YJ. Resveratrol inhibits phorbol ester-induced expression of COX-2 and activation of NF-κB in mouse skin by blocking iκb kinase activity. Carcinogenesis. 2006 Jul 1;27(7):1465–74.

44. Ren Z, Wang L, Cui J, Huoc Z, Xue J, Cui H, Mao Q, Yang R. Resveratrol inhibits NF-κB signaling through suppression of p65 and IB kinase activities. Die Pharmazie-An International Journal of Pharmaceutical Sciences. 2013 Aug 1;68(8):689–94.
45. Mokhamatam RB, Sahoo BK, Manna SK. Suppression of microphthalmia-associated transcription factor, but not NF-kappa b sensitizes melanoma specific cell death. Apoptosis. 2016 Aug;21(8):928–40.
46. Thakur S, Sarkar B, Cholia RP, Gautam N, Dhiman M, Mantha AK. APE1/Ref-1 as an emerging therapeutic target for various human diseases: Phytochemical modulation of its functions. Experimental & Molecular Medicine. 2014 Jul;46(7):e106.
47. Raffoul JJ, Heydari AR, Hillman GG. DNA repair and cancer therapy: Targeting APE1/Ref-1 using dietary agents. Journal of Oncology. 2012 Jan 1;2012.
48. Yang S, Irani K, Heffron SE, Jurnak F, Meyskens FL Jr. Alterations in the expression of the apurinic/apyrimidinic endonuclease-1/redox factor-1 [APE/Ref-1] in human melanoma and identification of the therapeutic potential of resveratrol as an APE/Ref-1 inhibitor. Molecular Cancer Therapeutics. 2005 Dec;4(12):1923–35.
49. Lala PK, Chakraborty C. Role of nitric oxide in carcinogenesis and tumour progression. The Lancet Oncology. 2001 Mar 1;2(3):149–56.
50. Yang Z, Yang S, Misner BJ, Chiu R, Liu F, Meyskens FL. Nitric oxide initiates progression of human melanoma via a feedback loop mediated by apurinic/apyrimidinic endonuclease-1/redox factor-1, which is inhibited by resveratrol. Molecular Cancer Therapeutics. 2008 Dec 1;7(12):3751–60.
51. Kim MY. Nitric oxide triggers apoptosis in A375 human melanoma cells treated with capsaicin and resveratrol. Molecular Medicine Reports. 2012 Feb 1;5(2):585–91.
52. Bastianetto S, Dumont Y, Duranton A, Vercauteren F, Breton L, Quirion R. Protective action of resveratrol in human skin: Possible involvement of specific receptor binding sites. Plos One. 2010 Sep 23;5(9):e12935.
53. Hung CF, Lin YK, Huang ZR, Fang JY. Delivery of resveratrol, a red wine polyphenol, from solutions and hydrogels via the skin. Biological and Pharmaceutical Bulletin. 2008 May 1;31(5):955–62.
54. Trapp V, Parmakhtiar B, Papazian V, Willmott L, Fruehauf JP. Anti-angiogenic effects of resveratrol mediated by decreased VEGF and increased TSP1 expression in melanoma-endothelial cell co-culture. Angiogenesis. 2010 Dec;13(4):305–15.
55. López-Otín C, Blasco MA, Partridge L, Serrano M, Kroemer G. The hallmarks of aging. Cell. 2013 Jun 6;153(6):1194–217.
56. Omodei D, Fontana L. Calorie restriction and prevention of age-associated chronic disease. FEBS Letters. 2011 Jun 6;585(11):1537–42.
57. Niccoli T, Partridge L. Ageing as a risk factor for disease. Current Biology. 2012 Sep 11;22(17):R741–52.
58. Zhang W, Qu J, Liu GH, Belmonte JC. The ageing epigenome and its rejuvenation. Nature Reviews Molecular Cell Biology. 2020 Mar;21(3):137–50.
59. Crimmins EM. Lifespan and healthspan: Past, present, and promise. The Gerontologist. 2015 Dec 1;55(6): 901–11.
60. Rolt A, Cox LS. Structural basis of the anti-ageing effects of polyphenolics: Mitigation of oxidative stress. BMC Chemistry. 2020 Dec;14(1):1–3.
61. Harraan D. Aging: a theory based on free radical and radiation chemistry.
62. Liu M, Chen S, Zhang Z, Li H, Sun G, Yin N, Wen J. Anti-ageing peptides and proteins for topical applications: A review. Pharmaceutical Development and Technology. 2022 Jan 2;27(1):108–25.
63. Zhou DD, Luo M, Huang SY, Saimaiti A, Shang A, Gan RY, Li HB. Effects and mechanisms of resveratrol on aging and age-related diseases. Oxidative Medicine and Cellular Longevity. 2021 Jul 11;2021.
64. Sethi G, Shanmugam MK, Warrier S, Merarchi M, Arfuso F, Kumar AP, Bishayee A. Pro-apoptotic and anti-cancer properties of diosgenin: A comprehensive and critical review. Nutrients. 2018 May;10(5):645.
65. Ogrunc M. Reactive oxygen species: The good, The bad, and The enigma. Molecular & Cellular Oncology. 2014 Sep 15;1(3):e964033.
66. Niles RM, McFarland M, Weimer MB, Redkar A, Fu YM, Meadows GG. Resveratrol is a potent inducer of apoptosis in human melanoma cells. Cancer Letters. 2003 Feb 20;190(2):157–63.
67. Jang M, Cai L, Udeani GO, Slowing KV, Thomas CF, Beecher CW, Fong HH, Farnsworth NR, Kinghorn AD, Mehta RG, Moon RC. Cancer chemopreventive activity of resveratrol, a natural product derived from grapes. Science. 1997 Jan 10;275(5297):218–20.
68. Ahmad N, Adhami VM, Afaq F, Feyes DK, Mukhtar H. Resveratrol causes WAF-1/p21-mediated G1-phase arrest of cell cycle and induction of apoptosis in human epidermoid carcinoma A431 cells. Clinical Cancer Research. 2001 May 1;7(5):1466–73.
69. Jang M, Pezzuto JM. Effects of resveratrol on 12-o-tetradecanoylphorbol-13-acetate-induced oxidative events and Gene expression in mouse skin. Cancer Letters. 1998 Dec 11;134(1):81–9.

70. Adhami VM, Afaq F, Ahmad N. Involvement of the retinoblastoma [pRb]–E2F/DP pathway during antiproliferative effects of resveratrol in human epidermoid carcinoma [A431] cells. Biochemical and Biophysical Research Communications. 2001 Nov 2;288(3):579–85.
71. Kapadia GJ, Azuine MA, Tokuda H, Takasaki M, Mukainaka T, Konoshima T, Nishino H. Chemopreventive effect of resveratrol, sesamol, sesame oil and sunflower oil in the Epstein–Barr virus early antigen activation assay and the mouse skin two-stage carcinogenesis. Pharmacological Research. 2002 Jun 1;45(6):499–505.
72. Soleas GJ, Grass L, Josephy PD, Goldberg DM, Diamandis EP. A comparison of the anticarcinogenic properties of four red wine polyphenols. Clinical Biochemistry. 2002 Mar 1;35(2):119–24.
73. Larrosa M, Tomás-Barberán FA, Espín JC. Grape polyphenol resveratrol and the related molecule 4-hydroxystilbene induce growth inhibition, apoptosis, s-phase arrest, and upregulation of cyclins a, e, and B1 in human SK-mel-28 melanoma cells. Journal of Agricultural and Food Chemistry. 2003 Jul 30;51(16):4576–84.
74. Afaq F, Adhami VM, Ahmad N, Mukhtar H. Botanical antioxidants for chemoprevention of photocarcinogenesis. Frontiers in Biosciences. 2002 Apr 1;7:d784–92.
75. Imran M, Iqubal MK, Imtiyaz K, Saleem S, Mittal S, Rizvi MM, Ali J, Baboota S. Topical nanostructured lipid carrier gel of quercetin and resveratrol: Formulation, optimization, in vitro and ex vivo study for the treatment of skin cancer. International Journal of Pharmaceutics. 2020 Sep 25;587:119705.
76. Palliyage GH, Hussein N, Mimlitz M, Weeder C, Alnasser MH, Singh S, Ekpenyong A, Tiwari AK, Chauhan H. Novel curcumin-resveratrol solid nanoparticles synergistically inhibit proliferation of melanoma cells. Pharmaceutical Research. 2021 May;38(5):851–71.
77. Sharma B, Iqbal B, Kumar S, Ali J, Baboota S. Resveratrol-loaded nanoemulsion gel system to ameliorate UV-induced oxidative skin damage: From in vitro to in vivo investigation of antioxidant activity enhancement. Archives of Dermatological Research. 2019 Dec;311(10):773–93.
78. Lee Y, Shin H, Kim J. *In vivo* anti-cancer effects of resveratrol mediated by NK cell activation. Journal of Innate Immunity. 2021;13(2):94–106.
79. Kim HR, Lee KH, Park SJ, Kim SY, Yang YK, Tae J, Kim J. Anti-cancer activity and mechanistic features of a NK cell activating molecule. Cancer Immunology, Immunotherapy. 2009 Oct;58(10):1691–700.
80. Wang M, Yu T, Zhu C, Sun H, Qiu Y, Zhu X, Li J. Resveratrol triggers protective autophagy through the ceramide/Akt/mTOR pathway in melanoma B16 cells. Nutrition and Cancer. 2014 Apr 3;66(3):435–40.
81. Zhao H, Han L, Jian Y, Ma Y, Yan W, Chen X, Xu H, Li L. Resveratrol induces apoptosis in human melanoma cell through negatively regulating Erk/PKM2/Bcl-2 axis. OncoTargets and Therapy. 2018;11:8995.
82. Leung M, Pipchuk A, Yang X. Current progress in cancer immunotherapies using small molecules targeting PD-L1 stability. OBM Genetics. 2021 Mar;5(1):1.
83. Delmas D, Hermetet F, Aires V. PD-1/PD-L1 checkpoints and resveratrol: A controversial new way for a therapeutic strategy. Cancers. 2021 Jan;13(18):4509.

9 Effects of Plant-Derived Compounds against Cancer Stem Cells

Rose Mary Stiffin[1], *Madhumita Das*[2,3], *and Jayanta Kumar Das*[1]
[1]Florida Memorial University
Miami Gardens, FL, United States
[2]Miami Dade College
Miami, FL, United States
[3]Palm Beach State College
Lake Worth, FL, United States

9.1 INTRODUCTION

Cancer is a disorder that involves the abnormal growth and proliferation of cells that can exit their primary site, travel to other parts of the body and infiltrate there, resulting in the development of a secondary site for cancer (motility and metastasis are common traits for several malignancies) (Das, 2015). Currently, the three treatment modalities for cancer treatment are excision of tissue (surgery), chemotherapy, and radiotherapy (use of ionizing radiation to target cancer cells) (Das, et al., 2021). However, although many cancers are treatable with either of these modes of treatment, many other malignancies remain refractory, with the cancers having myriad complications and side effects. Therefore, this dearth of suitable and relatively safe chemotherapeutics necessitates the development or discovery ofnovel compounds that do not possess these traits. The development of novel small molecules with minimal toxicity to non-cancerous tissues but possessing high efficacy against cancerous cells is important for advancing cancer therapy (Das et al., 2022). Several studies have shown that natural products isolated from plants, or their extracts (phytochemicals) possess potential anticancer properties. Because these compounds occur naturally with little to no chemical modification, they are expected to serve as effective anticancer chemotherapies with minimal toxicity to patients (Das et al., 2014).

9.2 PLANT-DERIVED COMPOUNDS TARGETING CANCER STEM CELLS SIGNALING PATHWAYS

Recent scientific studies mentioned the importance of epigenetically clinical targets for bivalent histone modifications namely H3K4me3 as well as H3K27me3 at transcript start sequences (TSS) for target genes with the heterogenic expression of several cancer stem cells (CSCs) biomarkers namely CD49f, ALDH1A3, CD133, CD44, CD15 CD24, and CXCR4) associated with glioblastomas, oral, and colorectal cancers (CRC) (Farhana L, et al., 2012). Therefore, plant-resultant bio-effective compounds could be promising as likely new clinical therapies targeting heterogeneous CSCs as well as metastases (Gupta et al., 2021). Yang et al. (2020) reported that the metabolites from several plants have been proven to imply anticancer abilities in preclinical investigations. The different metabolites from different plants may be effective on different CSCs signaling pathways including in CSC initiation, apoptosis, and reprogramming of differentiation which would possibly daze the drug resistance of CSCs and would decrease the tumor relapse as well as metastasis. Scientific studies (Ramasamy et al., 2015; Wang et al., 2018; Namiki et al., 2020; Qin et al., 2020; Liao et al., 2023;

DOI: 10.1201/9781003486237-9

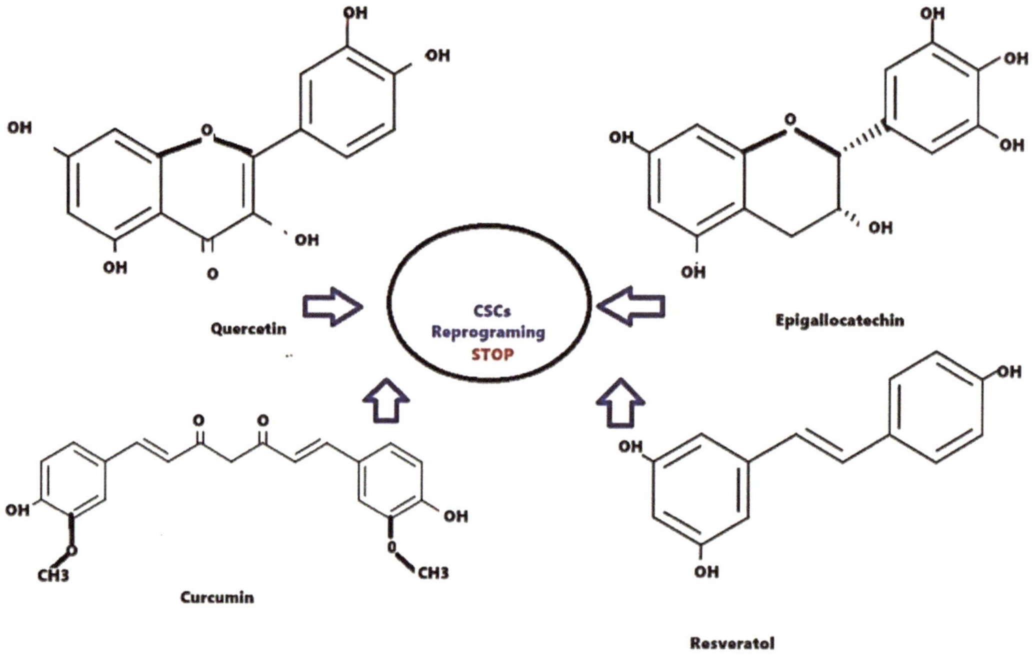

FIGURE 9.1 Effects of the different plant metabolites on CSCs reprogramming.

Liju et al., 2023) mentioned that the different plant metabolites namely epigallocatechin gallate of green tea, quercetin of fruits and vegetables, curcumin of turmeric, and resveratrol of grapes, the anticancer effects on CSCs (Figure 9.1) of in vivo mouse models aiming CSCs as well as CSCs signaling pathways.

Therefore, both natural as well as synthetic compounds are considered for chemoprevention which revokes carcinogenesis, tumor initiation, and metastasis, through different CSC signaling pathways, some of which have already been discovered by (Das *et al.,* 2018; Solomon et al., 2022). The epigenetic pathways may imply the triggering of free radical scavenging enzymes, inhibition of specific signaling pathways, and the control of chronic inflammation are correlated with the chemo-preventive capability of many plant-derived compounds, namely sulforaphane, green tea resulting compounds, resveratrol, isoflavones, and others (Das et al., 2024; Das et al., 2020). Although many chemopreventive compounds work in more than one molecular signaling pathways, it is proven that green tea compounds, curcumin, and polyunsaturated fatty acids, reduce inflammation, the resveratrol as well as sulforaphane reduces reactive oxygen species (ROS), piperlonguminine, resveratrol, apigenin, fisetin, curcumin, quercetin, [6]-gingerol promote autophagy, apoptosis as well as inflammation (Dash et al., 2023). Scientific studies showed that curcumin, silibinin, [6]-gingerol, green tea compounds, and resveratrol increase E-cadherin expression as well as decrease MMP-9 expression, inhibiting cell metastasis, and epithelial-mesenchymal transition (EMT) (Das et al., 2021) (Figure 9.2).

9.3 ANTICANCER EFFECT OF TWO PLANT ETHANOLIC EXTRACTS ON ORAL CANCER

AH HP and Nachiammai N (2023) obtained the leaf extract (of the plant *Plectranthus amboinicus*) and the root extract (of the plant *Glycyrrhiza glabra*) to test as anticancer agents for the treatment of oral cancer using the KB cell line. The plant specimens were procured from the medicinal plant garden in Chennai, India. Each extract was prepared by maceration of aqueous-sterilized leaves and roots, followed by drying and ground to powder form, respectively. The powders from each plant were solubilized in 0.1% dimethyl sulfoxide (DMSO) to the final concentrations for each extract of

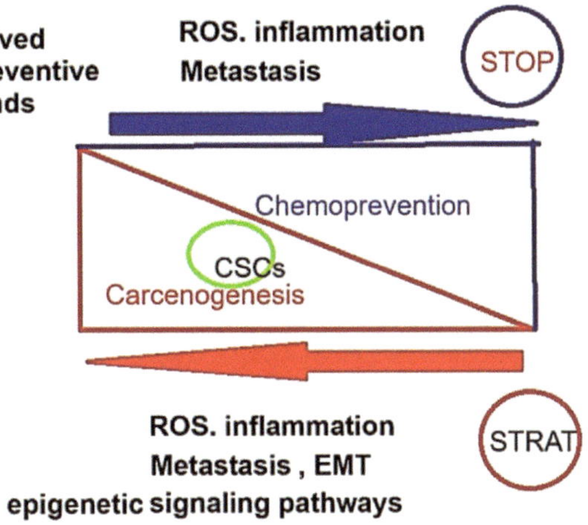

FIGURE 9.2 The plant-derived chemo-preventive compounds stop the signal for CSCs reprogramming whereas ROS, inflammation, EMT, and epigenetic signaling pathways start the CSCs reprogramming for carcinogenesis.

15.6 μg/ml, 31.2 μg/ml, 62.5 μg/ml, 125 μg/ml, 250 μg/ml, and 500 μg/ml. Cells were cultured in Minimal Essential Medium (MEM) under appropriate and standard conditions (a cocktail of antibiotics, 10% FBS for food, L-glutamine, 37°C, and 5% CO_2 at pH of ~7.3). To determine the anticancer effect of each extract (Sample 1 is extracts from *Plectranthus amboinicus* and Sample 2 is extracts from *Glycyrrhiza glabra*), 1 × 10^5 KB cells were seeded into 96-well plates and subsequently treated with the increasing concentrations of Samples 1 and 2, respectively. The determine the cell viability, the MTT assay was done 245 hours after treating cells with the extracts. The percent cell viability was determined using the following formula: %Cell viability = (absorbance$_{540nm}$ for treated cells/Absorbance$_{540nm}$ for control cells) × 100. The IC_{50} for each extract was determined graphically. Sample 2 (root extracts from *Glycyrrhiza glabra*) was more effective than sample 1 (leaf extracts from *Plectranthus amboinicus*). The IC_{50} value, defined as the half maximal inhibitory concentration from Sample 1 was 53 μg/ml, whereas Sample 2 had an IC_{50} value of 43.6 μg/ml. Further, at the highest concentration of each extract used (500 μg/ml), Sample 2 again outperformed Sample 1 with 1.6% cell viability *versus* 2.5% for Sample 1. Although the data are similar, the results do suggest that there are components in the ethanolic root extract of *Glycyrrhiza glabra* that possess anticancer properties. Further studies should be done to determine the compounds in each extract, as both samples showed activity against KB oral cancer cells (AH HP et al., 2023).

9.4 THE DIFFERENTIAL EXTRACTS OF *JUGLANS REGIA* SHELLS SHOWED EFFICACY AGAINST MDA-MB-231 AND AS49 CANCER CELL LINES

As stated previously, three standard forms of treatment for cancer involve chemotherapies, radiotherapies, and excision (tumor) of malignancies. Several chemotherapeutic agents, particularly alkylating agents, are cytotoxic to host and result in the death of nearby normal cells. The consequence of using such agents has resulted in chemotherapy-induced alopecia (CIA), gastrointestinal problems, and loss of bone reformation (Dunnill *et al.*, 2018). Although rare, many antitumor agents can in and of themselves induce carcinogenesis. Therefore, the development or isolation of novel, better-tolerated, less toxic chemotherapies is of utmost importance.

Dalkiliç *et al.* (2023) did differential extraction of the shell of the English walnut (*Juglans regia*) using methanol, ethanol, and hexanes as extractants. Given the decrease in polarity of the three solvents (methanol>ethanol≫hexane, respectively), their anticipation was that each extract would yield a plethora of different compounds from the powdered shell. After each separate extraction, 1 g of the dried, powdered extract was reconstituted in 10 ml DMSO for a 100 mg/ml stock solution.

Two different cancer cells were used for these *in vitro* studies, MDA-MB-231, a metastasizing breast cancer cell line, and A549, human alveolar basal epithelial cells (lung cancer). The cells were seeded in 96-well plates and incubated at 37°C, 5% CO_2 for 72 hours with three concentrations of each extract. The topoisomerase II (Topo II) inhibitor Doxorubicin, at 2.5 µg/ml, was used as the positive control, and the medium DMEM was used as the negative control.

After incubation, the cells were treated with MTT to assess cell viability, proliferation, and cytotoxicity. The cells were placed in a plate reader measuring the absorbance at 570 nm. The absorbance values (due to the mitochondrial reduction of yellow MTT to the purple-colored formazan) obtained from the plant extracts were compared to those from the control absorbances (positive and negative) and cell viability was calculated.

The results clearly indicate that the ethanol extract did not provide any inhibition of cell proliferation for the MDA-MB-231 breast cancer cells, with each concentration having approximately 100% cell viability. At the highest concentration (1/1000 µg/ml) the hexane extract showed approximately 40% cytotoxicity. At lower concentrations (1/2000 µg/ml and 1/4000 µg/ml), the hexane extract showed less efficacy. It is of interest to note that at 1/4000 µg/ml, the hexane extract increased its cytotoxicity from ~0% to ~20%. The biphasicity of this extract was not discussed. Methanol being more polar than ethanol showed a concentration-dependent cytotoxicity, with the highest concentration displaying ~40% cytotoxicity. At 1/4000 µg/ml, this extract was almost as effective as the negative control (~10% cell death).

For the lung cancer cell line A549, cells treated with the two alcoholic extracts from the bark of the English walnut tree at a concentration of 2 µg/ml were stained with Hoechst 33258 dye to visualize cell death. Methanol and ethanol extract applied in the A549 cell line induced apoptosis in cells. However, when the two extracts were compared, it was observed that the methanol extract showed higher apoptotic activity than the ethanol extract, as seen with the breast cancer cells using the shell extract (Kadıoğlu Dalkiliç et al., 2023).

Although no compounds were identified from each extract, it has been reported by many authors that *J. regia* is one of the nuts with the highest antioxidant activity (Halvorsen *et al.*, 2002; Kornsteiner *et al.*, 2006; Mishra *et al.*, 2010). Among various dried fruits, the methanolic extract of *J. regia* has been found to have very high antioxidant activity (as demonstrated in this study, the methanolic extract of the shells showed that antioxidant was concentration dependent, with the 2 mg/ml solution having ~50% scavenging activity. The antioxidant activity was assessed using the compound DPPH.

Other studies have suggested that the shell or seed coat surrounding the kernel has high antioxidant activity (Jahanban-Esfahlan *et al.*, 2019). Some studies in the literature have shown that *J. regia* green bark has an antioxidant effect in the presence of polyphenolic compounds (Oliveira *et al.*, 2008; Zhou *et al.*, 2015). The results obtained from this study emphasized that the *J. regia* tree (shell, flowers, and bark) constitutes a very good source of effective natural antioxidants that can act as anticancer agents. It is also suggested that *J. regia* bark extract be considered as an effective agent in the fight against cancer.

9.5 ANTICANCER ACTIVITY OF INDIAN MEDICINAL PLANTS

Historically, before the advent of so-called 'modern medicine', plants and their parts were used as sources to obtain medicines that could treat or improve a given disease or condition, such as various forms of cancer. Cancer is a multicellular phenomenon characterized by uncontrolled cell division, where the signals for apoptosis are dysfunctional. Because many classically used chemotherapeutics can be toxic, poorly tolerated, or possess other, harmful side effects, it has become imperative that newer, more effective compounds be developed or isolated that do not have these drawbacks.

There are several thousand plan species that possess bioactivity. Of this several thousand, at least a thousand plants have been identified as possessing compounds that have anticancer properties. Thus, plant-based cancer therapies are continuously being discovered and researched.

Recently, Lanchhana *et al.* (2023) examined three such plants indigenous to India, namely *Croton bonplandiaman* Baill, *Heliotropium indicum* L., and *Quisqualis indica* L. The parts of the plant used for each were the whole plant for the first two, and only the leaves and flowers for the third. The various parts of the dried plant parts were subjected to ethanol extraction. The filtrates were then collected, concentrated using a rotary evaporator, and stored at 4°C until use. Before while-cell studies, each extract was subjected to phytochemical screening so as to determine the components of each extract from each plant (plant part). All extracts from each plant (part) contained alkaloids, flavonoids, and tannins. Studies have shown that this family of organic small molecules possess biological activity, including anticancer.

The purpose of this study was to test the different ethanolic extracts for possessing anticancer activity against several human cancer cell lines.

The cancer cell lines utilized in this study were MDA-MB-231 (metastasizing breast cancer), A-549 (human lung cancer of epithelial cells), PC-3 (prostate cancer), and Hep-G2 (hepatoma or liver cancer) cells.

Cells underwent standard culture conditions before the addition of each extract. Increasing concentrations of extracts were added to the cells (0–80 g/ml) and the cells were incubated for 48 hours at 37°C, 5% CO_2, and 90% relative humidity.

To determine cell viability, the sulforhodamine assay (SRB) was done. After the addition of this compound, if the cells are viable, a pink color develops and the number of live cells are quantified by reading the absorbance at 560 nm. The percent growth inhibition (GI) was calculated using the formula: $GI = 100 - [OD_{(test\ sample)} - OD_{(blank)}/OD_{(control)} - OD_{(blank)}]$.

The results of the study showed that extracts from the whole plant *C. bonplandiaman* Baill worked best against A549 and PC-3 cells, but at maximal concentration administered (80 mg/ml). Ethanolic extracts of the whole plant *Heliotropium indicum* L. performed well against all four cancer cell lines; however, only MDA-MB-231 and A-549 cells showed a concentration-dependent inhibition. Inhibition of PC-3 and Hep-G2 cells only occurred at the higher concentrations (40 and 80 mg/ml). When the extract (leaves and flowers of *Quisqualis indica* L. was used, inhibition occurred only for one cell line, A-549, and at the highest dosage, 80 mg/ml.

Taken together, these results suggest that plant-based phytochemicals do possess varying degrees of anticancer activity against disparate cancer cell lines. Further, additional research should be done as a way to identify, isolate, and characterize the small molecules that have anticancer activity, as well as determine the mechanism of action for each.

9.6 THE FRUITS OF THE AMERICAN SWEETGUM TREE (*LIQUIDAMBAR STYRACIFLUA* L.) POSSESS POTENTIAL ANTICANCER DRUGS

Although the aromatic, dicotyledonous, hardwood tree commonly known as sweet gum (*Liquidambar styraciflua* L.) is indigenous to various parts of the United States, Mexico, and Central America, it is also found in China. In traditional Chinese medicine, the extracts from this tree have been used to treat diarrhea, dysentery, coughs, and skin sores. Further, other research has shown that its alcoholic extracts from several parts of the plant, including bark, stems, and leaves of the tree have anti-inflammatory properties and potentially could be used as anticancer drugs.

In the present study, Pozzobon *et al.* (2023) studied the effect of the fruits of this tree, as the fruits had not been studied as much as the other parts of the plant. To that end, the authors obtained the fruits, ground and dried them, and then did an aqueous extraction for the main isolate of any metabolite or macromolecule. Then the aqueous extract's volume was reduced under low pressure and the resultant residue was extracted with ethanol. From the ethanol extract, a precipitate and a supernatant were obtained. The precipitate (polysaccharide extract) was eventually modified to alditol acetates and the identity of these compounds was determined using GC-MS. The supernatant contained soluble small molecules, whose structures were eventually examined using GC-MS.

Using the Folin-Ciocâlteu method and gallic acid as a standard, assay, it was determined that the supernatant fraction (S-EA) had polyphenolic compounds present to act as radical scavengers. The polysaccharide and small molecule fractions were subsequently tested for anticancer activity at various concentrations (0–100 μg/ml, using ultrapure water or 1% DMSO as vehicle control). The cells tested were human breast cancer (MCF7, a non-metastasizing lineage), normal human breast epithelial (MCF10A), and human hepatocellular carcinoma (HepG2).

After testing against the extracts (polysaccharide and small molecule), the cell viability was determined using the well-established MTT assay. Although the crude extract (from aqueous extract and concentration to low volume) showed efficacy against MCF7 cells in a concentration- and time-dependent manner, it is noteworthy to mention that the supernatant fraction obtained after ethanolic extraction that contained small molecules also showed efficacy in a concentration- and time-dependent manner. The IC_{50} values after 72-hour treatment using the crude extract, the ethanolic extract containing the chemically modified saccharides, and the supernatant from the ethanolic extract (small molecules and soluble macromolecules) were 328 μg/ml, 548 μg/ml, and 502 μg/ml, respectively.

When the normal breast epithelial cells were treated with the three extracts, the crude and the supernatant from the ethanolic extract proved to be toxic at high concentrations (0.5–1.0 mg/ml). However, the polysaccharide extract was consistently nontoxic, even at the highest concentration and longest exposure (1 mg/ml and 72 hours, respectively).

For the HepG2 carcinoma cell line, the supernatant of the ethanolic extract (small molecules and soluble macromolecules) was toxic only at the high dose (1000 μg/ml at 24- and 48-hour exposure times). The other two extracts were ineffective at cell killing, as a function of concentration and time of exposure.

When each extract was tested for antioxidant activity, using the DPPH assay, the supernatant (labeled S-EA in this paper) ethanol extract performed the best, with an IC_{50} value of 3.67 μg/ml *versus* 16.5 μg/ml for the polysaccharide ethanol extract (P-EA). The crude extract (EA), which would contain both polyphenols and polysaccharides, had an IC_{50} of 4.6 μg/ml.

When taken together, the results from this study show that the ethanolic extract of the fruits of the sweet gum tree possesses antioxidant activity, especially in the polyphenolic (S-EA) extract. Therefore, it is hypothesized that the fruits of this tree may have great potential for the treatment of pathologies that are directly related to the imbalance or damage to the organism/cell/organelle caused by the uncontrolled generation of reactive oxygen species (ROS).

As for the anticancer activity, as determined by the MTT assay, the polysaccharide extract (P-EA) worked best against MCF7 cells. The observation that S-EA extract was toxic against normal cells suggests that the method of extraction of each plant part be scrupulously studied. The P-EA extract, moderately effective against tumor cells and nontoxic to normal cells after 72 hours of exposure suggests that this extract has potential as a promising candidate for studies against cancer and may mark the advent of a novel set of compounds with such activity.

9.7 SEVENTY-SIX PLANT SPECIES FROM THE ANDALUSIA AREA OF SPAIN WAS EVALUATED FOR THEIR ANTICANCER ACTIVITY AGAINST LUNG CANCER CELLS

Because of the aging population around the world, the incidence of cancer will become the leading cause of death by 2030. According to the World Health Organization, in 2020, there were 19.3 million new cancer cases and 10 million deaths due to cancer. When traditional chemotherapeutics are used, chemotoxicity may occur in which normal cells are also targeted for death, which causes numerous adverse effects that may require lower doses of the chemotherapeutic, resulting in lower efficacy. Further, detection of cancer at later stages may result in poorer treatment; some cancer drugs are also cost-prohibitive; and some treatments may work for only certain forms of cancer.

Therefore, it becomes of the utmost importance to find new anticancer drugs with high selectivity against cancer and to develop more effective, affordable treatments for patients with various types of cancers. One reservoir of putative, novel anticancer drugs is in the plant kingdom.

Jiménez-González *et al.* (2023) suggested that the random screening of plants from areas high in plant diversity could lead to the discovery of novel compounds to test as anticancer agents, given that biodiversity could generate a variety of natural compounds with pharmacological activities. To that end, his group examined the plant diversity in the Andalusian region of Spain. That region of Europe provided plants in 4437 taxa. Many of these plants had not been evaluated for putative anticancer properties. Thus, the authors screened and evaluated the anticancer activity of 82 extracts from 76 plant species that were collected in several regions of Andalusia. Various parts of the plants (leaves, fruit, flowers) were obtained and the extraction was done using a mixture of ethanol/ethyl acetate/water (1:1:1) as the extractant.

The authors tested the 82 extracts against A549 lung cancer cells (this cell line was chosen because lung cancer is the second most common cancer worldwide, and when discovered, it is in advanced stages of Stage III and Stage IV). The extracts were also tested against HaCaT cells, which are non-malignant cells.

Cell viability was determined using the resazurin assay, in which live cells convert the blue resazurin to a pink, fluorescent compound called resorufin, which is soluble in aqueous media. Extracts 8, 30, 35, 43, 46, 47, 63, and 81 worked well against A549 cells in a dose-dependent manner (0–1000 μM). Only at high concentrations (>100 μM) did they show lethality to the non-malignant cells. It must be noted that cisplatin and 5-Fluorouracil, standard anticancer drugs, showed cytotoxicity behavior for both cell lines in a concentration-dependent manner at lower concentrations.

Although several plant extracts showed selective toxicity against A549 cells, extract 81 from the plant *Xiphion xiphium* demonstrated the most selectivity against A549 cells. Therefore, it was evaluated against an additional three cancer cell lines and showed a cytotoxic effect against T24 urinary bladder cancer cells.

This study indicates that various plants are potential sources for the isolation and development of novel anticancer drugs with selective activity toward different cancer cell types. However, more studies must be conducted to understand the identity, nature, and complexity of the compounds and their anticancer activity, as well as their mechanism of action.

9.8 CONCLUSIONS

Prevention of cancer development will be a good strategy using several plant-derived bioactive compounds for everyone's daily food chart. Several cancer treatments have failed due to CSC drug registrant where scientific studies showed particularly in secondary care of cancer patients, due to the capability of flavonoids to contest drug resistance as well as offer more beneficial therapeutic outcomes for many cancer patients. Future research is expected to successfully translate plant-derived biomolecules into clinical testing by referring to trials on formulation, safety, bioavailability, as well as consistency of plant-derived biomolecules as phytomedicines for cancer treatment.

ACKNOWLEDGMENTS

This work was partly used the financial support from the National Science Foundation (NSF) grant awarded to Co-Principal Investigator, Jayanta Kumar Das (Award Number FAIN: 2332021), and by the FMU STAR2 award to Principal Investigator, Jayanta Kumar Das, for his SEGTBACR Lab.

ABBREVIATION

CSCs cancer stem cells

REFERENCES

Caroline M, L., Muthukumar R, S., A H, H. P., & N, N. (2023). Anticancer Effect of Plectranthus Amboinicus and Glycyrrhiza Glabra on Oral Cancer Cell Line: An Invitro Experimental Study. Asian Pacific journal of cancer prevention : APJCP, 24(3), 881–887. https://doi.org/10.31557/APJCP.2023.24.3.881

Das J, Das M, Doke M, Wnuk S, Stiffin RM, Ruiz M, et al. A small molecule inhibits pancreatic cancer stem cells. Int J Exp Res Rev 2021;26(2021):1–15. https://doi.org/10.52756/ijerr.2021.v26.001

Das JK. 3-D spheroid model as a new tool for toxicity testing. In: ICCVAM public forum. NIEHS; 2015. p. 1. https://ntp.niehs.nih.gov/iccvam/meetings/iccvam-forum-2015/das-fiu-508.pdf.

Das JK, Deoraj A, Roy D, Felty Q. Brain infiltration of breast cancer stem cells is facilitated by paracrine signaling by inhibitor of differentiation 3 to nuclear respiratory factor 1. J Cancer Res Clin Oncol 2022;148:288191. https://doi.org/10.1007/s00432-022-04026-w.

Das JK, Felty Q. PCB153-induced overexpression of ID3 contributes to the development of microvascular lesions. PLoS One 2014;9:e104159. https://doi.org/10.1371/journal.pone.0104159.

Das JK, Felty Q, Poppiti R, Jackson RM, Roy D. Nuclear respiratory factor 1 acting as an oncoprotein drives estrogen-induced breast carcinogenesis. Cells 2018;7:234. https://doi.org/10.3390/cells 7120234

Das JK, Stella JH, Saunders M, et al. 2024. In: *Cancer Stem Cells and Signaling Pathways*, Editor(s): S Pathak, A Banerjee, Chapter 22: Epigenetic modulation of cancer stem cells in vitro 3D models, Academic Press 2024; pp 381-394. https://doi.org/10.1016/B978-0-443-13212-4.00035-0

Das JK, Venkatesan T, Natarajan U, Algahtani M, Mathuram TL, Dondapati P, et al. 3D in vitro spheroid: a preclinical model for drug discovery against different cancer, In: Health professions division reach day, Nova Southeastern University, Davie, FL; 2020. p. 15. https://nsuworks.nova.edu/hpdrd/2020/2020programbook/15/

Das M, Santana MC, Barraque S, et al. 2021. 3D spheroid: a rapid drug screening model for epigenetic clinical targets against heterogenic cancer stem cells. In: *Proceedings of the American Association for Cancer Research Annual Meeting 2021, AACR, Philadelphia. Cancer Res* 2021;81. https://doi.org/10.1158/1538-7445.AM2021-2104 Abstract nr 2104.

Dash SL, Mohapater R, Mishra SK, Anticancer activity study of some selected Indian medicinal plants used traditionally, J Pharm Negative Results 2023;14(1):125–132.

Dunnill CJ, Al-Tameemi W, Collett A, Haslam IS, Georgopoulos NT. A Clinical and Biological Guide for Understanding Chemotherapy-Induced Alopecia and Its Prevention. Oncologist. 2018 Jan;23(1):84–96. https://doi.org/10.1634/theoncologist.2017-0263. Epub 2017 Sep 26. PMID: 28951499; PMCID: PMC5759815.

Dalkiliç LK, Dalkılıç S, Uygur L. Investigation of apoptotic, cytotoxic, and antioxidant effects of Juglans regia against MDA-MB-231 and A549 cell lines. International Journal of Plant Based Pharmaceuticals. 2023 Jan 31;3(1):62–7.

Farhana L, Dawson MI, Das JK, Murshed F, Xia Z, Hadden TJ, et al. Adamantyl retinoid-related molecules induce apoptosis in pancreatic cancer cells by inhibiting IGF-1R and Wnt/β-catenin pathways. J Oncol 2012;14. https://doi.org/10.1155/2012/796729.

Gupta PK, Saraff M, Gahtori, et al. 2021. Phytomedicines targeting cancer stem cells: therapeutic opportunities and prospects for pharmaceutical development. *Pharmaceuticals* 14 (7), 676. https://doi.org/10.3390/ph14070676

Halvorsen, B., Holte, K., Myhrstad, M. C., Barikmo, I., Hvattum, E., Remberg, S. F., Wold, A. B., & Haffner, K. (2002). A Systematic Screening of Total Antioxidants in Dietary Plants. Journal of Nutrition, 132, 461–471.

Jahanban-Esfahlan, A., Ostadrahimi, A., Tabibiazar, M., & Amarowicz, R. (2019). A comparative review on the extraction, antioxidant content and antioxidant potential of different parts of walnut (Juglans regia L.) fruit and tree. Molecules, 24(11), 2133.

Jiménez-González V, Benítez G, Pastor JE, López-Lázaro M, Calderón-Montaño JM. Evaluation of Anticancer Activity of 76 Plant Species Collected in Andalusia (Spain) against Lung Cancer Cells. Plants (Basel). 2023 Sep 15;12(18):3275. https://doi.org/10.3390/plants12183275

Kadıoğlu Dalkılıç, L., Dalkılıç, S., & Uygur, L. (2023). Investigation of apoptotic, cytotoxic, and antioxidant effects of Juglans regia against MDA-MB-231 and A549 cell lines. International Journal of Plant Based Pharmaceuticals, 3(1), 62–67. https://doi.org/10.29228/ijpbp.17

Kornsteiner, M., Wagner, K. H., & Elmadfa, I. (2006). Tocopherols and total phenolics in 10 different nut types. Food Chemistry, 98(2), 381–387.

Lanchhana DS, Ranjit M, Kumar MS. Anticancer activity study of some selected Indian medicinal plants used traditionally. Journal of Pharmaceutical Negative Results. 2023 Jan 1:1123–32.

Liao, W., Zhang, L., Chen, X., Xiang, J., Zheng, Q., Chen, N., et al. (2023). Targeting cancer stem cells and signalling pathways through phytochemicals: a promising approach against colorectal cancer. *Phytomedicine* 108, 154524. https://doi.org/10.1016/j.phymed.2022.154524

Liju, V. B., Martin, S., Nath, L. R., Gopinadhan Nair, G. K., & Thayele Purayil, H. (2023). Editorial: Plant-based drugs: the potential novel therapeutic intervention against cancer stemness and metastasis. Frontiers in pharmacology, 14, 1283694. https://doi.org/10.3389/fphar.2023.1283694

Mishra, N., Dubey, A., Mishra, R., & Barik, N. (2010). Study on antioxidant activity of common dry fruits. Food and Chemical Toxicology, 48(12), 3316–3320.

Namiki, K., Wongsirisin, P., Yokoyama, S., Sato, M., Rawangkan, A., Sakai, R., et al. (2020). (-)-Epigallocatechin gallate inhibits stemness and tumourigenicity stimulated by AXL receptor tyrosine kinase in human lung cancer cells. Sci. Rep. 10 (1), 2444. https://doi.org/10.1038/s41598-020-59281-z

Oliveira, I., Sousa, A., Ferreira, I. C., Bento, A., Estevinho, L., & Pereira, J. A. (2008). Total phenols, antioxidant potential and antimicrobial activity of walnut (Juglans regia L.) green husks. Food and Chemical Toxicology, 46(7), 2326–2331.

Pozzobon, R.G.; Rutckeviski, R.; Carlotto, J.; Schneider, V.S.; Cordeiro, L.M.C.; Mancarz, G.F.F.; Souza, L.M.d.; Mello, R.G.; Smiderle, F.R. Chemical Evaluation of Liquidambar styraciflua L. Fruits Extracts and Their Potential as Anticancer Drugs. Molecules 2023, 28, 360. https://doi.org/10.3390/molecules28010360

Qin, T., Cheng, L., Xiao, Y., Qian, W., Li, J., Wu, Z., et al. (2020). NAF-1 inhibition by resveratrol suppresses cancer stem cell-like properties and the invasion of pancreatic cancer. *Front. Oncol.* 10, 1038. https://doi.org/10.3389/fonc.2020.01038

Ramasamy, T. S., Ayob, A. Z., Myint, H. H. L., Thiagarajah, S., and Amini, F. (2015). Targeting colorectal cancer stem cells using curcumin and curcumin analogues: insights into the mechanism of the therapeutic efficacy. Cancer Cell Int. 15 (1), 96–15. https://doi.org/10.1186/s12935-015-0241-x

Solomon B, Sturrup R, Dixon J, et al. 2022. Carcinogenic effects of metal and metalloids through epigenetic reprogramming of human bone marrow stem cells: the bioinformatics database analysis In: *Proceedings of the American Association for Cancer Research Annual Meeting, Apr 8-13. Philadelphia (PA): AACR. Cancer Res* 2022;82. https://doi.org/10.1158/1538-7445.AM2022-5830 Abstract nr 5830

Wang, R., Yang, L., Li, S., Ye, D., Yang, L., Liu, Q., et al. (2018). Quercetin inhibits breast cancer stem cells via downregulation of aldehyde dehydrogenase 1A1 (ALDH1A1), chemokine receptor type 4 (CXCR4), mucin 1 (MUC1), and epithelial cell adhesion molecule (EpCAM). *Med. Sci. Monit.* 24, 412–420. https://doi.org/10.12659/msm.908022

Yang, L., Shi, P., Zhao, G., et al. 2020. Targeting cancer stem cell pathways for cancer therapy. *Signal Transduct. Target Ther.* 5 (1), 8. https://doi.org/10.1038/s41392-020-0110-5

Zhou, X., Cai, J., Zhu, W., Zhao, H., Wang, K., & Zhang, X. (2015). Boswellic acid attenuates asthma phenotype by downregulation of GATA3 via inhibition of PSTAT6. Genetics and Molecular Research, 14(3), 7463–7468.

10 An Overview of Cardiovascular Disease Management with Plant-Derived Bioactive Compounds

Diptimayee Das[1], *Keerthi N*[1], *Antara Banerjee*[1], *Asim K Duttaroy*[2], *and Surajit Pathak*[1]

[1]Faculty of Allied Health Sciences, Chettinad Academy of Research and Education (CARE), Chettinad Hospital and Research Institute (CHRI) Chennai, India

[2]Department of Nutrition, Institute of Medical Sciences, Faculty of Medicine, University of Oslo Oslo, Norway

10.1 INTRODUCTION

Cardiovascular diseases (CVDs) represent a prominent and significant contributor to global mortality rates on an annual basis. The phenomenon under consideration is typically observed in developed nations, where healthcare expenditures show a notable upward trend, indicating a significant growth in economic investment in this sector [1]. According to the report published by the World Health Organization (WHO), it has been observed that approximately 7.4 million deaths are attributed to coronary heart disease (CHD). In comparison, stroke is responsible for approximately 6.7 million deaths. The prevailing modifiable risk factors associated with CVDs encompass a notable presence of psychosocial factors, excessive alcohol consumption, elevated levels of low-density lipoprotein (LDL) and cholesterol, hypertension, inadequate intake of fruits and vegetables, diabetes mellitus, abdominal obesity, smoking, and sedentary lifestyle. According to the research conducted by Stewart J., 2017 [2], it can be observed that the confluence of cholesterol, lipids, calcium, and various harmful compounds can give rise to the development of plaques, thereby instigating the pathological condition known as atherosclerosis or arterial calcification [2].

The primary drivers behind contemporary consumer food choices are the pursuit of health maintenance and disease prevention. Consequently, a growing demand for food products perceived as "healthier" or possessing functional food exists. In the context of this matter, it has been proposed that lifestyle alterations, particularly dietary patterns, may have the potential to effectively alter numerous risk factors associated with diseases and positively influence disease progression [3]. The abovementioned idea aligns with the contemporary understanding of food, which has evolved beyond its conventional function of providing essential nutrients and encompass its potential in preventing and managing various diseases [4]. The persistent consumer demand for "functional foods" has prompted processors to actively seek edible crops with bioactive compounds that bestow various health benefits. As exemplified, fruits with a notable concentration of bioactive compounds demonstrate various biological activities, including antioxidant, anti-microbial, anti-inflammatory, and anti-tumorigenic properties [5]. The present nutritional methodologies are increasingly aligning

 DOI: 10.1201/9781003486237-10

with a paradigm shift in our comprehension of well-being. The expansion of scientific understanding of the influence of dietary factors on genetic and molecular regulation has prompted a paradigm shift in our perception of the significance of nutrition.

Consequently, novel dietary approaches are being developed. The role of functional food extends beyond fulfilling metabolic needs by potentially positively impacting human health [6]. Therefore, it is imperative to identify and establish extracts of plants or bioactive compounds that are hypothesized to have potential health benefits for humans. These extracts or compounds can then be introduced to enhance dietary supplements in CVDs.

10.2 THE DEVELOPMENT AND PROGRESSION OF CARDIOVASCULAR DISEASES

CVDs encompass various clinical conditions that impact the functioning of the blood vessels and heart. This includes CHD, dyslipidemia, hypertension, myocardial infarction, peripheral vascular diseases, stroke, heart failure, and cardiomyopathies, among other related disorders [7]. The development of CVD is a complex and multifactorial process, encompassing many genetic, environmental, and lifestyle factors that interact in intricate ways.

Individuals possessing a familial predisposition to CVD exhibit an elevated susceptibility to such conditions. The persistent increase in blood pressure can potentially induce detrimental effects on the arterial system, rendering it more vulnerable to the development of atherosclerosis [8]. This pathological process significantly contributes to the onset and progression of various CVDs. The presence of elevated levels of LDL cholesterol and reduced levels of high-density lipoprotein (HDL) cholesterol is recognized as a significant factor in the progression of atherosclerosis. Atherosclerosis is a chronic and persistent disease characterized by the gradual build-up of lipid deposits within the luminal surface of arterial walls, including fatty substances, cholesterol, and other substances [9]. The occurrence of this phenomenon can give rise to the development of atherosclerotic plaques, which possess the propensity to undergo rupture, thereby instigating the initiation of thrombus formation and conceivably culminating in the manifestation of myocardial infarction or cerebrovascular accident [10], as shown in Figure 10.1. The inhalation of tobacco smoke exposes individuals to a multitude of detrimental chemicals, which have been found to possess the capacity to inflict harm upon blood vessels. This harm manifests in impaired functionality, including reduced oxygen supply to the heart [11]. The presence of diabetes is correlated with an increased susceptibility to developing CVDs. Elevated levels of glucose in the bloodstream have the potential to inflict harm upon the structures of blood vessels and nerves, thereby precipitating a range of complications, including atherosclerosis and peripheral artery disease. Excessive body weight, particularly the accumulation of fat in the abdominal region, and the absence of consistent engagement in physical activity significantly increase the probability of developing CVDs [12]. Obesity frequently exhibits comorbidities, including hypertension, dyslipidemia, and insulin resistance. Regular physical activity has been shown to play a crucial role in the maintenance of a healthy body weight, as well as in the reduction of blood pressure levels. Furthermore, engaging in exercise has been associated with notable improvements in cardiovascular health, encompassing various aspects of heart function and overall well-being [13].

A diet that is high in saturated and trans fats, cholesterol, and sodium has been linked to a greater chance of developing CVDs. On the other hand, adopting a diet rich in whole grains, vegetables, lean proteins, and fruits has been found to provide a protective effect [14]. The involvement of chronic inflammation in the pathogenesis and advancement of atherosclerosis has been well-documented in research. Elevated levels of inflammatory markers have been frequently observed in individuals diagnosed with CVDs. The presence of chronic stress, depression, and social isolation has been observed to potentially contribute to the onset and progression of CVDs, thereby exerting a negative impact on overall cardiovascular well-being [15].

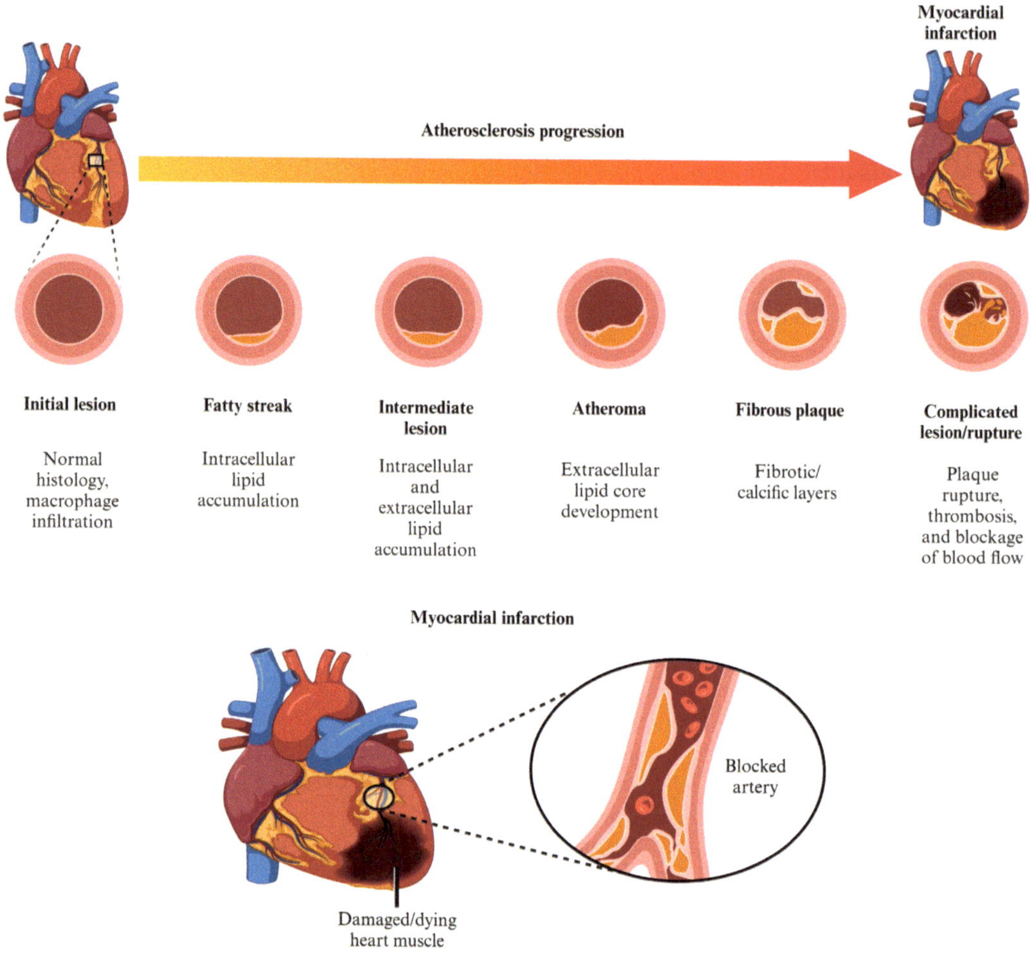

FIGURE 10.1 A plaque fatty deposit in the arteries turns a healthy artery into an unhealthy one, causing a complex atherosclerotic lesion. Cholesterol, fatty substances, cellular waste, calcium, and fibrin form artery deposits. Deposits may cause myocardial infarction.

10.3 MOLECULAR MECHANISM OF CARDIOVASCULAR DISEASE PROGRESSION

Understanding the progression of CVD involves delving into the complex molecular mechanisms that affect the blood vessels and heart, ultimately influencing their structure and function. Several critical mechanisms are involved in the onset and advancement of CVD. These processes include atherosclerosis, endothelial dysfunction, inflammation, platelet hyperactivity, oxidative stress, and vascular remodeling, as shown in Figure 10.2.

10.3.1 Atherosclerosis

The accumulation of plaques within the arteries is a critical factor in the occurrence of CVD. The process starts when the endothelial cells lining the arteries get damaged, caused by factors like high blood pressure, smoking, or high cholesterol levels. When there is damage to the

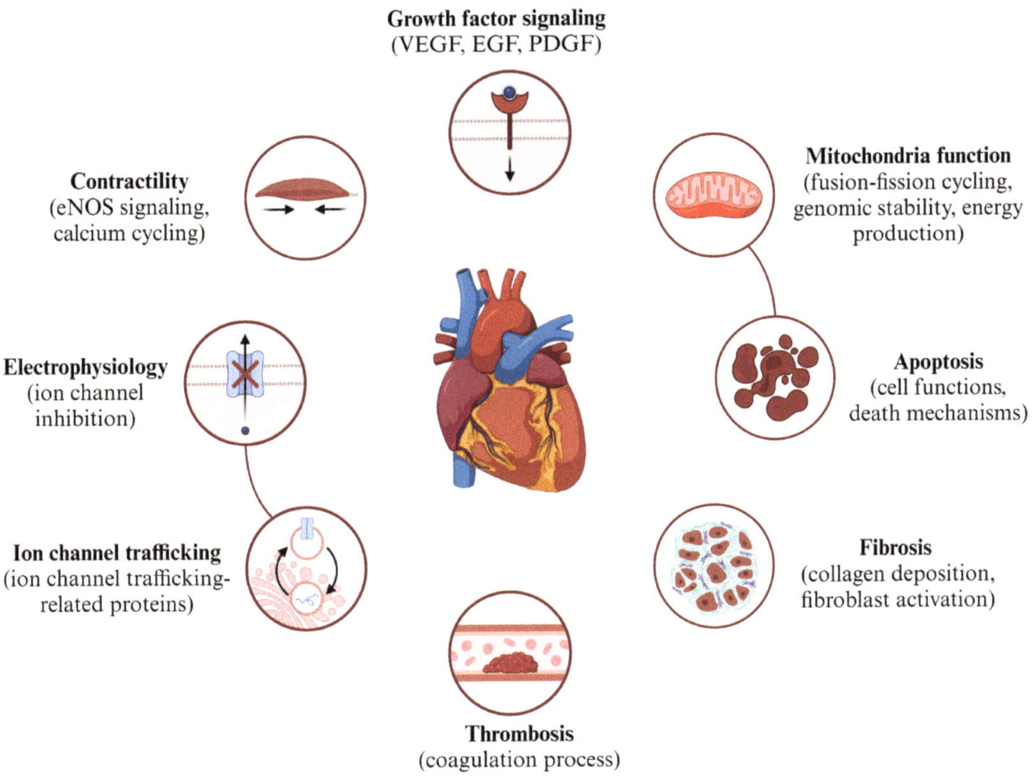

FIGURE 10.2 Mechanism of cardiovascular disease: Alteration of contractility, molecular processes, and vasculature, leading to cardiac dysfunction. Cardiotoxicity can result from interactions between medications and cardiac receptors, channels, or proteins, which may inhibit enzymatic activity, vesicle trafficking, or disrupt signaling pathways.

endothelium, inflammatory cells, specifically monocytes, migrate into the arterial wall. These cells then proceed to absorb oxidized LDL cholesterol, resulting in the formation of foam cells. Foam cells, smooth muscle cells, and other cellular components play a role in developing atherosclerotic plaques [9].

Platelet hyperactivity and systemic inflammation play a significant role in the development and progression of atherosclerosis. Platelet hyperactivity and systemic inflammation contribute to an adverse cycle, enhancing one another and accelerating the development of atherosclerosis. Platelets are activated by inflammatory mediators, leading to their aggregation. Platelets in an activated state contribute to the release of pro-inflammatory factors, further intensifying the body's overall inflammatory response. This continuous cycle results in gradual damage to the endothelial lining, the development of plaque, and, ultimately, a heightened susceptibility to heart attacks, strokes, and peripheral artery disease [16].

10.3.2 Platelet Hyperactivity

Platelets play a crucial role in the body by forming clots to prevent bleeding following an injury. In the bloodstream of healthy individuals, platelets remain inactive until they are required. However, several factors can cause an increase in platelet activation, resulting in the formation of microthrombi that can potentially obstruct blood vessels. Platelets adhere to damaged endothelium, leading to an inflammatory response and plaque formation. Platelets release molecules such as thromboxane and platelet-derived growth factor (PDGF), contributing to inflammation and attracting additional immune cells. Several factors, such as high cholesterol, diabetes, smoking, obesity, chronic kidney disease, and genetic predispositions, can contribute to a condition associated with platelet hyperactivity [17].

10.3.3 Systemic Inflammation

There is a strong connection between chronic, low-grade inflammation throughout the body and atherosclerosis, along with other diseases. This inflammation can be caused by various factors such as overconsumption of food, which is important to be mindful of the types of fats and sugars we consume, as well as the quality of carbohydrates we choose and the impact of infections and chronic stress where they increase levels of inflammatory markers such as C-reactive protein (CRP) and interleukin-6 (IL-6). Autoimmune diseases are a complex and fascinating area of study. Conditions such as lupus and rheumatoid arthritis can result in widespread inflammation throughout the body. The chronic inflammatory state can harm the endothelium, increasing its vulnerability to the adhesion of monocytes and leukocytes which contributes to the development of fatty streaks and raises the likelihood of rupture and cardiovascular events [18].

10.3.4 Oxidative Stress

Reactive oxygen species (ROS) and oxidative stress play a role in endothelial dysfunction and damage. Oxidative stress plays a role in the oxidation of LDL cholesterol, increasing its potential to contribute to the development of atherosclerosis. Oxidative stress can be worsened by mitochondrial dysfunction and a compromised antioxidant defense system [19].

10.3.5 Vascular Remodeling

The gradual alterations influence the development of CVD in the structure and function of blood vessels. Vascular smooth muscle cells (VSMCs) can undergo hypertrophy and hyperplasia, resulting in the vessel wall's thickening. Matrix metalloproteinases (MMPs) are a group of enzymes that play a crucial role in remodeling the extracellular matrix. When these enzymes are not adequately regulated, it can lead to instability in the vessel wall [20].

10.3.6 Endothelial Dysfunction

The endothelium, a single layer of cells lining the inner surface of blood vessels, plays a vital role in cardiovascular health. It regulates blood flow, prevents inflammation, and inhibits blood clots when functioning optimally. However, endothelial dysfunction, characterized by impaired endothelium function, emerges as a key player in the development and progression of various CVDs. It disrupts the maintenance of vascular homeostasis [21].

Endothelial dysfunction causes multiple factors that contribute to high blood pressure, high cholesterol, diabetes, smoking, obesity, physical inactivity, autoimmune diseases, and chronic infections. It disrupts essential functions, leading to Impaired vasodilation, where blood vessels become stiff and less responsive to relaxation signals, leading to higher blood pressure. Increased inflammation in which endothelial cells produce inflammatory molecules attracts immune cells and promotes plaque formation and enhanced platelet aggregation, where blood platelets become more prone to clumping, increasing the risk of blood clots [22]. Endothelial dysfunction is a cornerstone of atherosclerosis, causes a build-up of fatty plaques in blood vessels, restricts blood flow, and increases the risk of heart attacks, strokes, and peripheral artery disease. Impaired endothelium function contributes to heart failure by affecting contractility, relaxation, and overall heart function. It can disrupt electrical signals in the heart, leading to irregular heartbeats [23]. Endothelial dysfunction is marked by a reduction in nitric oxide (NO), a chemical recognized for its ability to widen blood vessels and reduce inflammation [24].

10.3.7 Role of Endothelial Adhesion Factors in Atherosclerosis

Endothelial adhesion factors (EAFs) play a pivotal role in initiating and driving this process by acting as molecular "traffic signals" that guide white blood cells, particularly monocytes, into the vessel wall. Normally, EAFs are present on the endothelium in low levels. However, risk factors like high cholesterol, smoking, and inflammation trigger their upregulation. Elevated EAFs act as "sticky pads" for circulating monocytes. These monocytes bind to EAFs and then squeeze through the endothelium, entering the sub-intimal space of the artery wall. Once inside the vessel wall, monocytes transform into macrophages, which engulf oxidized LDL cholesterol and become foam cells [25]. Accumulation of foam cells leads to the formation of fatty streaks, the initial stage of atherosclerotic plaque. Key EAFs in atherosclerosis are Vascular cell adhesion protein 1 VCAM-1, which is responsible for monocyte adhesion; intercellular adhesion molecule-1 (ICAM-1), which contributes to both monocyte and leukocyte adhesion; and E-selectin and P-selectin, which play a role in the initial tethering of white blood cells to the endothelium [26].

10.4 THROMBOSIS

The rupture of plaque can expose the blood to substances that promote blood clotting, which can result in the formation of blood clots. Thrombosis is a condition that can lead to the blockage of blood flow, which in turn can cause heart attacks or strokes [27].

10.5 CARDIOPROTECTIVE EFFECT OF NATURAL-DERIVED BIOACTIVE COMPOUNDS

Recently, according to a study, plant-based dietary supplements are beneficial and effective in reducing the risk of CVDs [28]. Food provides a wide range of nutrients vital for our well-being, along with various bioactive substances that promote good health and prevent CVDs. A wide range of bioactive compounds are crucial for our health and can be found in various natural sources, forming an integral part of the food chain. Minerals found in plants are essential for human nutrition. Furthermore, they include several metabolites, both primary and secondary, that have the potential to impact human health and nutrition significantly [29]. The influence of secondary metabolites on plant development and function is minimal. Vegetables and fruits are excellent sources of several critical micronutrients, including potassium, magnesium, and calcium. Their composition includes natural compounds like phytochemicals, plant sterols, anthocyanins, betalains, glucosinolates, tannins, polyphenols, flavonoids, dietary fiber, carotenoids, and vitamins have been found to influence human health significantly. These compounds, referred to as nutraceuticals, have gained recognition for their occurrence in the human diet and biological activity [30].

Natural antioxidants, known as polyphenols, are primarily obtained from various sources such as medicinal herbs, medicinal plants, vegetables, beverages, fruits, spices, and cereals. Plant-derived natural antioxidants are phenolic acids, flavonoids, anthocyanins, and other compounds, which fall under polyphenols [31]. Scientists have discovered that certain natural compounds, like carotenoids and polyphenols, have a range of beneficial effects on the body. These compounds have been found to have anti-diabetic, anti-cancer, anti-aging, anti-inflammatory, and cardioprotective properties [32]. There is a growing recognition of the potential of bioactive compounds in addressing various chronic diseases of the 21st century, particularly those related to cardiovascular health. These compounds are considered promising strategies for prevention, adjuvant therapy, and even

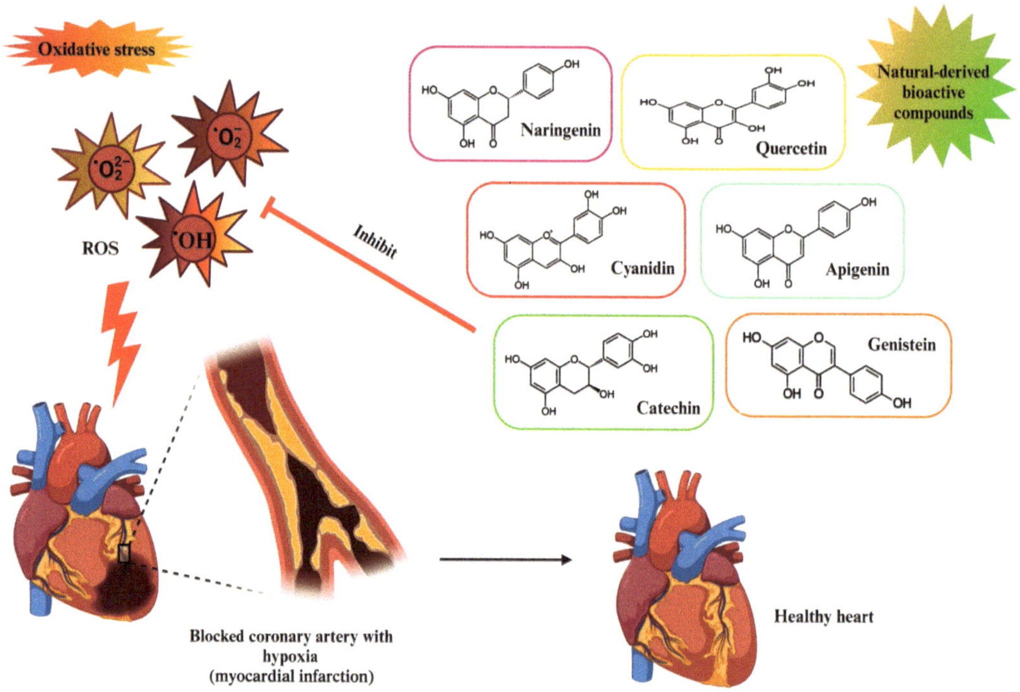

FIGURE 10.3 Natural bioactive compound in the management of cardiovascular disease: Oxidative stress produces superoxide, peroxide ions, and hydroxyl radicals, which combine to form reactive oxygen species that cause an increase in fat lipids in the blood vessels, resulting in an artery blockage known as myocardial infarction. Natural compounds that suppress oxidative stress and increase antioxidant and anti-inflammatory levels have a cardioprotective impact and help to restore a healthy heart.

potential cures. The diverse characteristics of their dietary lifestyles have been found to have beneficial effects on managing and preventing CVD [33], as shown in Figure 10.3.

10.6 FLAVONOIDS

Flavonoids, a heterogeneous class of polyphenolic compounds, are abundantly present in various plant species. These compounds have gained attention due to their notable antioxidant and anti-inflammatory properties. These sources primarily include fruits, vegetables, tea, red wine, and cocoa. Flavonoid subclasses encompass a variety of compounds, such as flavonols (e.g., myricetin, quercetin, isorhamnetin, and kaempferol), flavones (e.g., apigenin, luteolin, tangeretin, and baicalein), anthocyanins (e.g., cyanidin, delphinidin, malvidin, and peonidin), isoflavones (e.g., daidzein, genistein, glycitein, and biochanin A), flavanones (e.g., eriodictyol, naringenin, and hesperetin), and flavanols (e.g., catechin, gallocatechin, epicatechin, and epigallocatechin-3-gallate [EGCG]). This indicates potential cardiovascular advantages associated with flavonoids [34].

Flavonoids are potent antioxidants that can neutralize ROS and reduce oxidative stress. Oxidative stress contributes to CVDs like atherosclerosis. Flavonoids reduce inflammation in the vascular system. Chronic inflammation contributes to cardiovascular conditions. Flavonoids may improve blood flow by promoting NO production, which relaxes blood vessels. Better endothelial function lowers the risk of atherosclerosis and heart events [35]. Certain flavonoids may lower blood pressure, which is essential for preventing and managing hypertension, a significant risk factor for CVD. Flavonoids in fruits and vegetables can reduce blood clotting. This activity prevents thrombosis and related cardiovascular events. Flavonoids can affect lipid metabolism, including cholesterol regulation. Flavonoids are linked to lower LDL cholesterol. Flavonoids can relax blood vessels and improve blood flow. This effect is partly due to an increase in NO production. Flavonoids protect against ischemia-reperfusion injury, which causes tissue damage due to temporary interruption and restoration of blood supply [36].

10.6.1 Quercetin

In the past few decades, there has been a growing interest in the potential therapeutic attributes associated with quercetin, which belongs to the class of flavonols and has gained attention as a potential therapeutic agent in CVDs. Quercetin has various beneficial effects, including reducing inflammation, dilating blood vessels, lowering blood pressure, preventing irregular heart rhythms, and protecting against cell death. It can potentially lower LDL and prevent cholesterol oxidation, which can help protect against endothelial dysfunction in CVDs [37]. Studies have shown that quercetin protects against oxidative stress-induced damage in the heart, kidney, and vessel cells [38, 39]. According to recent studies, certain factors can lead to a reduction in inflammatory markers like tumour necrosis factor alpha (TNF-α), interleukin-1 beta (IL-1β), and interleukin-17 (IL-17) [40]. Regular consumption of quercetin reduces VSMC counts, which slows CVD development. In addition, quercetin has been found to reduce the levels of nicotinamide adenine dinucleotide phosphate (NADPH) oxidase, ROS, superoxide anion, and free radicals, which have been shown to possess anti-inflammatory and anti-hypertensive effects. It has been observed that the level of MMP-2 tends to increase in cases of hypertension, primarily due to stress oxidation. Quercetin has been found to reduce the elevated MMP-2 activity in rat models with hypertension, leading to improvements in vascular remodeling [41]. Quercetin can remarkably shield the myocardium from damage caused by cardiac ischemia/reperfusion. Its primary mechanism of action is to inhibit apoptosis and the inflammatory cascade via the phosphatidylinositol 3-kinase/protein kinase B (PI3K/Akt) pathway [42]. Quercetin has been found to protect cardiomyocytes, reducing the risk of apoptosis and minimizing myocardial injury caused by mechanical trauma. Additionally, it has shown potential in improving cardiac dysfunction by regulating intracellular Ca^{2+} levels and reducing the presence of ROS [43].

10.6.2 Apigenin

Apigenin is a natural flavonoid in various vegetables, fruits, and plant-based beverages [44]. It has a wide array of biological properties exhibited by this substance, encompassing its anti-inflammatory, anti-bacterial, antioxidant, and anti-cancer effects. The reported effects of apigenin are helpful in a range of medical conditions, including diabetic cardiomyopathy, hypertension, atherosclerosis, stroke, and ischemia/reperfusion-induced myocardial injury. In addition, it provides a defense against the harmful effects of drugs on the heart [45]. The prevention of CVD relies on activating anti-apoptotic and antioxidant mechanisms in cardiomyocytes and vascular endothelial cells. Vascular endothelial dysfunction, caused by oxidative stress and abnormal glucose metabolism, significantly affects CVD [46]. The use of apigenin to treat endothelial dysfunction has been the subject of several investigations. Firstly, apigenin has been found to effectively enhance apelin expression, which can help address endothelial dysfunction. Apelin is a molecule that occurs naturally and binds to the apelin (APJ) receptor, which is present in the cardiovascular system. The study highlights the beneficial impact of the treatment on glucose uptake and superoxide dismutase (SOD) activity in human endothelial cells. This leads to restoring impaired glucose metabolism and homeostasis and decreasing oxidative stress [47]. Researchers have found that apigenin can hinder nuclear factor kappa B (NF-kB)-associated signaling pathways and reduce the expression of ICAM-1. ICAM-1 facilitates communication between keratinocytes and leukocytes by interacting with lymphocyte function-associated antigen 1 (LFA-1) [48]. ICAM-1 regulates migration, adhesion, and abnormal inflammatory infiltration [49].

Furthermore, apigenin can deactivate the PI3K/Akt axis in human umbilical vein endothelial cells (HUVECs), essential for preventing vascular endothelial dysfunction. The PI3K/Akt axis plays a crucial role in the development of vascular complications in individuals with type-2 diabetes mellitus (T2DM) [50]. Apigenin therapy effectively averted vascular endothelial dysfunction by reducing the phosphorylation of Akt residues serine 473 (Ser473) and threonine 308 (Thr308) [51]. A study found that apigenin was effective in reducing oxidative stress and increased NO levels. As a result, this contributed to reducing vascular endothelial dysfunction caused by mitochondria-dependent apoptosis. This was accomplished by blocking the phosphorylation of protein kinase C βII (PKCβII). In vascular endothelial dysfunction, PKCβII is involved in oxidative stress, ROS production, and the activation of mitochondria-dependent apoptosis [52]. While apigenin therapy increased the expression of B-cell lymphoma-2 (Bcl-2), an anti-apoptotic gene, it decreased the expression of Bcl-2 associated X (Bax), a pro-apoptotic gene. This led to endothelial cells exhibiting less mitochondria-dependent apoptosis [53]).

10.6.3 Cyanidin

Cyanidin belongs to the class of flavonoids known as anthocyanins. Anthocyanins are phenolic pigments in berries, fruits, and cereal crops. Anthocyanins are pigments that give fruits and vegetables vibrant red, orange, purple, and blue colors. Anthocyanins are crucial in plant metabolism, as phytoprotective agents [54]. As a member of the phenolic compounds, anthocyanin has excellent potential as an antioxidant. Research has shown that anthocyanin can effectively neutralize harmful free radicals and significantly protect the heart. Their potent anti-inflammatory activity is widely recognized, as it explains these substances' numerous reported biological effects. These effects include their potential as anti-diabetic and anti-cancer agents, as well as their role in preventing CVD [55]. Studies have demonstrated the potential benefits of cyanidin in enhancing hepatic glucose uptake, insulin production, and insulin resistance in individuals with T2DM [56]. Flavanols have gained recognition for reducing blood pressure and protecting the endothelium by lowering triglycerides and preventing harmful lipid build-up. Research has shown that specific treatments may enhance mitochondrial bioenergetics, reduce oxidative stress in cardiac cells, and maintain the proper functioning of complex I [57]. Research has shown that cyanidin supplements may prevent LDL cholesterol oxidation [58]. This is evidenced by the decrease in levels of serum anti-oxLDLs and markers of lipid peroxidation, including F_2-isoprostane and malondialdehyde (MDA) [59].

10.6.4 Genistein

Genistein is a phytochemical that falls under the category of isoflavones. It is a prominent isoflavone commonly consumed in the daily diet of Asian populations [60]. In 1899, researchers successfully isolated genistein from the plant *Genista tinctoria* L. This compound is commonly found in soy-based foods, including soy cheese and soy drinks [61]. Genistein is widely recognized as a prominent phytoestrogen. It binds to the estrogen receptor and acts similarly to estrogen [62]. Multiple studies have demonstrated the cardioprotective effects of genistein [63]. Studies have shown that certain compounds found in plants can reduce symptoms experienced during postmenopausal. These compounds have also been found to lower the risk of heart attack, reduce blood cholesterol levels, prevent blood clotting, and promote blood vessel growth and relaxation [64]. It has been suggested that genistein exhibits estrogenic effects at lower doses and anti-estrogenic effects at higher doses [62]. In lower doses, genistein has been found to reduce ventricular arrhythmia and infarct sizes in female rats that have undergone ovariectomy [65]. Genistein has shown various biological activities, including the inhibition of protein tyrosine kinase (PTK) [66], as well as anti-apoptosis, anti-cancer, anti-oxidation, and anti-inflammation activities [67]. In a study conducted on type 2 diabetic mice, genistein was found to have a positive impact on various health markers [68]. Researchers found that it reduced inflammation, lowered fasting blood glucose levels, improved insulin resistance, and addressed liver abnormalities. Researchers found that these effects were caused by alterations in the gut flora [69]. Research in diabetic rats has shown that genistein may preserve the kidneys via processes related to the nuclear factor erythroid 2-related factor 2- heme oxygenase-1/ NAD(P)H:quinone-oxidoreductase-1 (Nrf2-HO-1/NQO1) pathway [70]. In rats, genistein efficiently protects against streptozotocin-induced type 1 diabetes-induced cardiac damage. Its capacity to decrease oxidative stress and inflammation is responsible for this protective effect [71]. Recently, there has been increased focus on the positive impact of genistein on the cardiovascular system. Studies have demonstrated that genistein can effectively prevent thrombotic occlusion in mouse femoral arteries and inhibit platelet aggregation in whole blood [72]. According to studies, Japanese women, especially those who have gone through menopause, have a reduced risk of cerebral and myocardial infarction if they consume a diet high in soy and isoflavones [73].

10.6.5 Naringenin

The cardiovascular and cardiometabolic benefits of citrus flavonoids are widely recognized [74]. The citrus flavonoids naringenin and naringin, and in particular naringin's aglycone, are quite promising in managing cardiovascular complications. Research has shown that they have a beneficial effect on the metabolic profile and on systolic blood pressure levels [75]. Naringenin is a naturally occurring flavanone found in vegetables, citrus fruits, tomatoes, and grapes [76]. The liver enzyme naringinase breaks down naringin through hydrolysis, producing naringenin and rhamnose [77]. The therapeutic effects of naringenin have been extensively studied and have shown positive results in various diseases. Its anti-inflammatory, anti-tumor, antioxidant, and anti-apoptotic activities have been observed in conditions such as liver disease, chronic kidney disease, neurological disorders, and cancer [78]. The left ventricle's systolic function was markedly improved in rats before ischemia-reperfusion damage by preconditioning with naringenin, which activated particular membrane channels [79]. A study found that naringenin has potential benefits in reducing diabetic nephropathy and endothelial dysfunction caused by a high-cholesterol diet. This is achieved by inhibiting the activity of NADPH oxidase 4&2 (NOX4 and NOX2) enzymes [80]. Researchers have shown that naringenin might prevent hyperglycemia-induced cardiac fibrosis, protect against oxidative stress, and decrease NADPH oxidase activity in patients with cardiovascular disorders.

Furthermore, by blocking the PI3K/Akt, extra-cellular signal regulated kinase (ERK), and Jun N-terminal kinase (JNK) signaling pathways, naringenin therapy aids in reducing pressure overload-induced ventricular hypertrophy [81]. Studies have found that naringenin positively affects cholesterol levels, inflammation, and the development of atherosclerosis. It has been shown to reduce LDL, increase HDL, and inhibit the formation of foam cells.

Additionally, it down-regulates the expression of genes associated with atherosclerosis, indicating its potential as an effective anti-atherosclerosis agent [82]. Research has shown that naringenin may suppress mitochondrial oxidative stress damage, reducing heart damage from ischemia-reperfusion injury [83]. By successfully blocking NF-kB-dependent cardiac inflammation, naringenin protected against septic cardiac dysfunction in a mouse sepsis model [84]. Naringenin also reduced ischemia-reperfusion injury-induced cardiac cell death in rat models [85].

10.6.6 Catechin

Catechin is a type of flavonoid found in the flavanols family. Plant-based foods, such as cocoa, tea, fruits, and legumes, naturally contain catechins [86]. Its remarkable pharmacological and biological properties have made it prominent. Green tea has been widely studied for its health benefits, precisely due to the presence of phenolic compounds that have strong antioxidant properties and aid in removing free iron. Catechin stands out as the most abundant and potent among the various catechins present in green tea, such as EGCG, (–)-epigallocatechin, and (–)-epicatechin [87]. After consumption, catechins are absorbed by the small intestine and undergo various modifications in the liver, such as glucuronidation, sulphation, and methylation [86]. Multiple studies, including both epidemiological and experimental research, have consistently demonstrated a clear correlation between the consumption of green tea and the promotion of a healthy cardiovascular system [88].

Some of the conditions that are being studied are congestive heart failure, ischemic heart diseases, endothelial dysfunction, cardiac hypertrophy, hypertension, cardiomyopathy, and atherosclerosis. Catechins achieve these effects through the reduction of oxidative stress, prevention of inflammatory events, reduction of platelet aggregation, and inhibition of the proliferation of VSMCs [89]. Studies have shown that the catechins found in green tea can positively impact blood vessels. These effects encompass a range of health benefits, such as combating oxidative stress, regulating blood pressure, alleviating inflammation, impeding cell proliferation, averting blood clot formation, and managing lipid levels [90]. Tea catechins demonstrate antioxidant activity through a range of mechanisms. These include the ability to scavenge free radicals, bind to transition-metal ions, inhibit transcription factors related to oxidative stress, suppress enzymes that promote oxidation, and enhance the activity of antioxidant enzymes [91]. Tea catechins possess the remarkable capability to hinder the vital enzymes that play a role in lipid biosynthesis, as well as aid in diminishing the absorption of lipids in the intestines. As a result, it has a positive impact on the blood lipid profile [92]. Catechins have been found to effectively prevent vascular inflammation, a crucial factor in the formation of atherosclerotic lesions. The ability of catechins to inhibit leukocyte adhesion to endothelium and prevent their migration may contribute to their anti-inflammatory activities. This is accomplished by reducing the production of cytokines and adhesion molecules by inhibiting the transcriptional factor NF-*kB* in both endothelial cells and inflammatory cells [93]. Catechins have been found to inhibit the proliferation of VSMCs by disrupting the activity of vascular cell growth factors that are linked to atherogenesis [94]. Catechins have been found to effectively inhibit thrombogenesis by suppressing platelet adhesion. In addition, various studies have shown that EGCG can boost the expression of specific proteins, resulting in the programed cell death of VSMCs and effectively halting the advancement of atherosclerosis [95]. In addition, the interaction between catechins and free radicals in the body can aid in lowering the build-up of cholesterol and its oxidation products in the walls of arteries.

As a result, blood circulation may be enhanced, potentially aiding in the prevention of atherosclerosis [96]. Studies have indicated that different types of catechins can improve total cholesterol levels, apolipoprotein B levels, serum triglyceride levels, and LDL cholesterol levels. There is evidence suggesting that this may have a positive impact on reducing fat build-up in the blood [97]. Furthermore, catechins can catalyze the activities of SOD in both the liver and serum. In addition, catechins have shown resistance to oxidation by forming half-quinone radicals [98]. Studies have indicated that EGCG can effectively inhibit the oxidation of LDL induced by Cu^{2+} and enhance its antioxidant capabilities within the body. There is evidence suggesting that this may reduce the risk of CVD [98].

10.7 ENDOTHELIAL ADHESION FACTORS AND POLYPHENOLS: A BALANCING ACT FOR VASCULAR HEALTH

EAFs and polyphenols play crucial roles in maintaining healthy blood vessels. Understanding their interplay is essential for preventing and managing CVDs. EAFs are molecules expressed on the surface of endothelial cells, lining the inner layer of blood vessels. It facilitates the adhesion of various blood cells, including leukocytes and platelets, to the endothelium. Play a key role in immune responses, inflammation, and blood clotting. Elevated levels of certain adhesion factors, like VCAM-1 and ICAM-1, are associated with increased risk of CVDs [24].

Polyphenols are natural bioactive compounds found in fruits, vegetables, tea, and other plant-based foods. It possesses diverse biological activities, including antioxidant, anti-inflammatory, and anti-proliferative properties. It can modulate the expression and function of EAFs. Polyphenols reduce the expression of adhesion molecules by scavenging free radicals, preventing oxidative stress-induced upregulation of adhesion factors. It inhibits the activation of signaling pathways where they interfere with pathways leading to adhesion molecule expression and leukocyte adhesion [99]. It modulates inflammatory responses by suppressing the release of pro-inflammatory cytokines that further activate adhesion molecules. By downregulating adhesion factors, polyphenols can prevent leukocyte infiltration into the vessel wall by reducing inflammation and the formation of atherosclerotic plaques. It inhibits platelet aggregation by preventing blood clots and thrombotic events. It promotes endothelial repair and regeneration by maintaining vascular integrity and function [100].

Some polyphenols with beneficial effects on adhesion factors include resveratrol which inhibits NF-kB activation, reducing VCAM-1 and ICAM-1 expression, curcumin which downregulates VCAM-1 and E-selectin expression, suppressing leukocyte adhesion and quercetin which modulates signaling pathways involved in adhesion molecule activation [101].

10.8 ADVANTAGES OF PLANT-DERIVED COMPOUNDS IN CARDIOVASCULAR DISEASE

The effectiveness of phenolic components in inhibiting free radicals by donating hydrogen atoms or electrons has been widely recognized [102]. Research has indicated that incorporating fruits and vegetables in a daily diet can benefit a range of health conditions, such as arteriosclerosis, heart disorders, neurodegenerative diseases, arthritis, brain dysfunction, and cancer [103]. Several naturally derived bioactive peptides from soy, rice, oats, cereals, barley, and wheat have been discovered to possess anti-hypertensive properties [104]. Food protein hydrolysis with the help of protease can also result in the release of peptide sequences that can positively impact lipid and cholesterol levels [105]. Research has indicated that including vegetables such as garlic, leek, and onion in one's regular diet can protect against CVDs, diabetes, and various metabolic conditions [106]. Extensive research is being conducted on the potential health benefits of plant-based bioactive compounds. Researchers have conducted several studies that have revealed promising findings about various compounds, including flavonoids, salicylates, polyphenols, glucosinates, and plant sterols [107]. Studies have shown that bioactive compounds obtained from plants have a comprehensive array of biological functions; these activities encompass anti-diabetic and anti-cancer effects, enhance blood lipid profile, regulate gastrointestinal health, improve cardiovascular health, maintain oxidoreductive balance, and promote brain health [30].

10.9 EPIDEMIOLOGICAL AND INTERVENTIONAL TRIALS INVOLVING PLANT PRODUCTS AND CVD

Research exploring the role of plant products in CVD prevention and management has flourished in recent years. These trials aim to explore the potential benefits of various plant-derived compounds, such as polyphenols, flavonoids, and other bioactive substances, in reducing the risk of CVD and improving cardiovascular health. Research has suggested that diets rich in fruits, vegetables, whole grains, nuts, and seeds may lower the risk of developing CVD due to their high content of antioxidants, fiber, vitamins, and minerals [108]. While promising, further research is needed to better understand the mechanisms of action, optimal dosages, and long-term effects of plant-based interventions on cardiovascular health, as shown in Table 10.1. Additionally, lifestyle factors such as exercise, smoking cessation, and stress management should be considered alongside dietary interventions for comprehensive cardiovascular risk reduction.

TABLE 10.1
Epidemiological and Interventional Trials of Plant Products in Cardiovascular Disease

Study Type	Plant Product (Flavonoids)	CVD Focus	Dosage	Study Duration	Population Characteristics	Key Findings	Limitations	References
Epidemiological	Fruits and vegetables	Overall CVD risk	Varied intake estimates	Ongoing; long-term observations	Diverse	Higher intake generally associated with lower CVD risk; specific fruits/vegetables show varying associations	Observational design, potential confounding factors	[109]
Epidemiological	Green tea	Coronary heart disease	2–4 cups/day	Varied (up to 20+ years)	Diverse	Moderate intake potentially reduces CHD risk	Observational design, potential confounding factors, variable quality of green tea	[110]
Epidemiological	Nuts	Coronary heart disease	1–2 servings/day	Varied (up to 20+ years)	Diverse	Moderate intake potentially reduces CHD risk	Observational design, potential confounding factors	[111]
Epidemiological	Soy protein	Cardiovascular risk factors	25–50 g/day	Varied (up to 10+ years)	Diverse	May improve lipid profile and blood pressure	Observational design, potential confounding factors	[112]
Interventional	Garlic	Hypertension	400–1800 mg/day	4–24 weeks	Overweight/obese, hypertensive adults	May modestly reduce blood pressure in some studies	Limited intervention duration, variable quality of garlic preparations	[113]
Interventional	Cocoa flavanols	Endothelial function	200–500 mg/day	4–24 weeks	Healthy adults, high CVD risk individuals	May improve endothelial function in some studies	Limited study duration, small sample sizes	[114]
Interventional	Red wine (resveratrol)	Atherosclerosis	1–2 glasses/day	3–12 months	Healthy adults, high CVD risk individuals	Evidence remains inconclusive regarding clinical benefits	High alcohol intake poses health risks, limited high-quality studies	[115]
Interventional	Curcumin	Inflammatory markers	500–2000 mg/day	4–12 weeks	Overweight/obese adults, individuals with metabolic syndrome	May reduce inflammation markers but clinical CVD benefits unconfirmed	Limited study duration, variable dosages and formulations	[116]

10.10 LIMITATIONS OF USING PLANT-DERIVED COMPOUNDS IN CARDIOVASCULAR DISEASE

While plant-derived bioactive compounds may be easily accessible and intriguing for further investigation, it is essential to consider their various associated obstacles. Nevertheless, a small number of clinical trials have supported the evidence of their health-promoting potential. Therefore, before evaluating these bioactive components for usage in human health-related applications, large-scale, long-term human investigations are necessary to ascertain their potential [30]. The primary challenge is securing approval for the drug's commercialization due to limited resources. Due to the constraints in synthesizing specific compounds using semi-synthetic methods or artificial plant growth, the product will inevitably rely more on natural resources. The low bioavailability of chemicals obtained from plants is a significant issue. The solubility of these chemicals in the systemic fluid also raises concerns about their possible potency. In addition, it's important to mention that some phytochemicals may not stay stable in the body, which can affect how well they treat CVD [117]. There is a noticeable chemical difference between pharmaceuticals made chemically and those from natural sources. Natural product compounds have unique features such as multiple chiral centers, complex steric structures, eased molecular weight, a variety of ring systems, a high concentration of oxygen atoms, and a lack of hydrogen bond donors [118]. Natural compounds may have to change some of their properties to pass the drug-likeness test. With a low molecular weight, a stable and flexible structure, few hydrogen bond acceptors, and a limited quantity of hydrogen bond donors, an ideal medicine would have these chemical properties. The bioavailability of the chemical and the advancement of its analog are affected by several circumstances. Modifying the criteria for draggability and improving the efficacy of natural compounds while maintaining their inherent benefits can pose significant challenges [119]. Many natural pro-exhibits show excellent biological efficiency play impressive characteristics during their metabolism, absorption, excretion, distribution, and toxicity testing. However, if they do not fulfil the drug-likeness requirements, these medications will not go to the second phase of a clinical study [120]. This drawback needs to be addressed as it significantly hinders the study and advancement of naturally derived bioactive compounds. Phytochemicals can have varying effects on CVD prevention.

Additionally, certain phytochemicals may have the ability to act as carcinogens [121]. More research is needed to fully understand the toxicity index of phytochemicals in a clinical setting. Therefore, the clinical application of phytochemicals for treating CVD has not yet been approved. Some phytochemicals can harm the human body and may interfere with absorbing essential compounds, acting as anti-nutrients. Furthermore, phytochemical research often suffers from methodological flaws, including small sample sizes, limited trial durations, and inadequate control during clinical trials. Caution is necessary when combining traditional plants with other medicines, as it can result in harmful drug interactions that could be life-threatening [122].

10.11 CONCLUSION

The study of naturally derived bioactive compounds from plants for managing CVD presents a fascinating new field of research with potential advantages and notable limitations. The advantages of utilizing these compounds arise from their natural origins, which frequently lead to fewer side effects than synthetic drugs. Furthermore, the different mechanisms by which they work, including their capacity to combat oxidation, alleviate inflammation, and expand blood vessels, all contribute to their potential benefits for the cardiovascular system. However, it's crucial to acknowledge the limitations associated with this field of research. Establishing standardized treatments can be challenging due to the diverse composition of plant extracts and factors such as bioavailability, dosage, and individual patient reactions.

Furthermore, it is crucial to be cautious when considering the widespread use of these compounds because of the limited number of well-designed clinical trials and the absence of long-term safety and effectiveness data. It is essential to investigate how each bioactive compound works, improve formulations for better absorption, and conduct comprehensive clinical trials to determine their effectiveness and safety. In addition, it is crucial to fully understand the potential interactions with existing medications to successfully integrate plant-derived compounds into conventional CVD treatment. Understanding plant-derived bioactive compounds requires a comprehensive approach that combines traditional and evidence-based medicine. Evaluation of these drugs' use in clinical practice must be based on reliable scientific data. This will ensure the effectiveness and safety of the treatment for patients with CVD.

CONSENT FOR PUBLICATION

Not applicable.

FUNDING

Not applicable.

ETHICS APPROVAL

Not applicable. This research does not involve human/animal participants, human/animal material, or humans.

CONFLICT OF INTEREST

The authors report no conflict of interest.

ACKNOWLEDGMENTS

The authors thank the Chettinad Academy of Research and Education (CARE) for providing the infrastructural and financial support to complete this work. We would also like to thank the Provexis, PLC, UK project grants-in-aid for giving the grant (Grant number: Ref. No. 078/Regr./AR-Research/2023) to Prof. Surajit Pathak.

AUTHOR'S CONTRIBUTIONS

SP, AKD, and DD conceptualized the chapter, and DD wrote it. DD was involved in designing the images and table. SP, AKD, and AB critically reviewed the chapter. All authors read, reviewed, and approved the final chapter.

REFERENCES

1. Mahmood, Syed S., Daniel Levy, Ramachandran S. Vasan, and Thomas J. Wang. "The Framingham Heart Study and the epidemiology of cardiovascular disease: A historical perspective." *The Lancet* 383, no. 9921 (2014): 999–1008.
2. Stewart, Jack, Gavin Manmathan, and Peter Wilkinson. "Primary prevention of cardiovascular disease: A review of contemporary guidance and literature." *JRSM cardiovascular disease* 6 (2017): 2048004016687211.
3. Adefegha, Stephen Adeniyi. "Functional foods and nutraceuticals as dietary intervention in chronic diseases; novel perspectives for health promotion and disease prevention." *Journal of dietary supplements* 15, no. 6 (2018): 977–1009.
4. Gul, Khalid, A. K. Singh, and Rifat Jabeen. "Nutraceuticals and functional foods: The foods for the future world." *Critical reviews in food science and nutrition* 56, no. 16 (2016): 2617–2627.

5. Ono, Kenjiro, Daisy Zhao, Qingli Wu, James Simon, Jun Wang, Aurelian Radu, and Giulio Maria Pasinetti. "Pine bark polyphenolic extract attenuates amyloid-β and tau misfolding in a model system of Alzheimer's disease neuropathology." *Journal of Alzheimer's disease* 73, no. 4 (2020): 1597–1606.
6. Biesalski, Hans-Konrad, Lars Ove Dragsted, Ibrahim Elmadfa, Rolf Grossklaus, Michael Müller, Dieter Schrenk, Paul Walter, and Peter Weber. "Bioactive compounds: Definition and assessment of activity." *Nutrition* 25, no. 11–12 (2009): 1202–1205.
7. Arnett, Donna K., Roger S. Blumenthal, Michelle A. Albert, Andrew B. Buroker, Zachary D. Goldberger, Ellen J. Hahn, Cheryl Dennison Himmelfarb, et al. "2019 ACC/AHA guideline on the primary prevention of cardiovascular disease: A report of the American College of Cardiology/American Heart Association task force on clinical practice guidelines." *Circulation* 140, no. 11 (2019): e596–e646.
8. Fuchs, Flávio D., and Paul K. Whelton. "High blood pressure and cardiovascular disease." *Hypertension* 75, no. 2 (2020): 285–292.
9. Rafieian-Kopaei, Mahmoud, Mahbubeh Setorki, Monir Doudi, Azar Baradaran, and Hamid Nasri. "Atherosclerosis: Process, indicators, risk factors and new hopes." *International journal of preventive medicine* 5, no. 8 (2014): 927.
10. Badimon, Lina, Teresa Padró, and Gemma Vilahur. "Atherosclerosis, platelets and thrombosis in acute ischaemic heart disease." *European heart journal: Acute cardiovascular care* 1, no. 1 (2012): 60–74.
11. Benjamin, Regina M. "Exposure to tobacco smoke causes immediate damage: A report of the surgeon general." *Public health reports* 126, no. 2 (2011): 158–159.
12. Rask-Madsen, Christian, and George L. King. "Vascular complications of diabetes: Mechanisms of injury and protective factors." *Cell metabolism* 17, no. 1 (2013): 20–33.
13. Jehan, Shazia, Ferdinand Zizi, Seithikurippu R. Pandi-Perumal, Samy I. McFarlane, Girardin Jean-Louis, and Alyson K. Myers. "Energy imbalance: Obesity, associated comorbidities, prevention, management and public health implications." *Advances in obesity, weight management & control* 10, no. 5 (2020): 146.
14. Wali, Jibran A., Natalia Jarzebska, David Raubenheimer, Stephen J. Simpson, Roman N. Rodionov, and John F. O'Sullivan. "Cardio-metabolic effects of high-fat diets and their underlying mechanisms—a narrative review." *Nutrients* 12, no. 5 (2020): 1505.
15. Raggi, Paolo, Jacques Genest, Jon T. Giles, Katey J. Rayner, Girish Dwivedi, Robert S. Beanlands, and Milan Gupta. "Role of inflammation in the pathogenesis of atherosclerosis and therapeutic interventions." *Atherosclerosis* 276 (2018): 98–108.
16. Johny, Ebin, Pathoori Bhaskar, Md Jahangir Alam, Indra Kuladhipati, Rupam Das, and Ramu Adela. "Platelet mediated inflammation in coronary artery disease with type 2 diabetes patients." *Journal of inflammation research* 14 (2021): 5131.
17. Holinstat, Michael. "Normal platelet function." *Cancer and metastasis reviews* 36 (2017): 195–198.
18. Chen, Linlin, Huidan Deng, Hengmin Cui, Jing Fang, Zhicai Zuo, Junliang Deng, Yinglun Li, Xun Wang, and Ling Zhao. "Inflammatory responses and inflammation-associated diseases in organs." *Oncotarget* 9, no. 6 (2018): 7204.
19. Incalza, Maria Angela, Rossella D'Oria, Annalisa Natalicchio, Sebastio Perrini, Luigi Laviola, and Francesco Giorgino. "Oxidative stress and reactive oxygen species in endothelial dysfunction associated with cardiovascular and metabolic diseases." *Vascular pharmacology* 100 (2018): 1–19.
20. Brozovich, F. V., C. J. Nicholson, C. V. Degen, Yuan Z. Gao, M. Aggarwal, and Kathleen G. Morgan. "Mechanisms of vascular smooth muscle contraction and the basis for pharmacologic treatment of smooth muscle disorders." *Pharmacological reviews* 68, no. 2 (2016): 476–532.
21. Favero G, Paganelli C, Buffoli B, Rodella LF, Rezzani R. Endothelium and its alterations in cardiovascular diseases: life style intervention. Biomed Res Int. 2014;2014:801896. doi:10.1155/2014/801896.
22. Kolluru GK, Bir SC, Kevil CG. Endothelial dysfunction and diabetes: effects on angiogenesis, vascular remodeling, and wound healing. Int J Vasc Med. 2012;2012:918267. doi:10.1155/2012/918267.
23. Hadi, Hadi A. R., Cornelia S. Carr, and Jassim Al Suwaidi. "Endothelial dysfunction: Cardiovascular risk factors, therapy, and outcome." *Vascular health and risk management* 1, no. 3 (2005): 183.
24. Sun, Hai-Jian, Zhi-Yuan Wu, Xiao-Wei Nie, and Jin-Song Bian. "Role of endothelial dysfunction in cardiovascular diseases: The link between inflammation and hydrogen sulfide." *Frontiers in pharmacology* 10 (2020): 1568.
25. Radeke, Heinfried H., Ralf J. Ludwig, and Wolf-Henning Boehncke. "Experimental approaches to lymphocyte migration in dermatology in vitro and in vivo." *Experimental dermatology* 14, no. 9 (2005): 641–666.

26. Gui, Yuzhou, Hongchao Zheng, and Richard Y. Cao. "Foam cells in atherosclerosis: Novel insights into its origins, consequences, and molecular mechanisms." *Frontiers in cardiovascular medicine* 9 (2022): 845942.
27. Bentzon, Jacob Fog, Fumiyuki Otsuka, Renu Virmani, and Erling Falk. "Mechanisms of plaque formation and rupture." *Circulation research* 114, no. 12 (2014): 1852–1866.
28. Pashikanti, Srinath, David R. de Alba, Gilbert A. Boissonneault, and Daniel Cervantes-Laurean. "Rutin metabolites: Novel inhibitors of nonoxidative advanced glycation end products." *Free radical biology and medicine* 48, no. 5 (2010): 656–663.
29. Gürbüz, Nergiz, Selman Uluişik, Anne Frary, Amy Frary, and Sami Doğanlar. "Health benefits and bioactive compounds of eggplant." *Food chemistry* 268 (2018): 602–610.
30. Samtiya, Mrinal, Rotimi E. Aluko, Tejpal Dhewa, and José Manuel Moreno-Rojas. "Potential health benefits of plant food-derived bioactive components: An overview." *Foods* 10, no. 4 (2021): 839.
31. Pandey, Kanti Bhooshan, and Syed Ibrahim Rizvi. "Plant polyphenols as dietary antioxidants in human health and disease." *Oxidative medicine and cellular longevity* 2 (2009): 270–278.
32. Dhalaria, Rajni, Rachna Verma, Dinesh Kumar, Sunil Puri, Ashwani Tapwal, Vinod Kumar, Eugenie Nepovimova, and Kamil Kuca. "Bioactive compounds of edible fruits with their anti-aging properties: A comprehensive review to prolong human life." *Antioxidants* 9, no. 11 (2020): 1123.
33. Alfahel, Raz, Tomasz Sawicki, Monika Jabłońska, and Katarzyna E. Przybyłowicz. "Anti-hyperglycemic effects of bioactive compounds in the context of the prevention of diet-related diseases." *Foods* 12, no. 19 (2023): 3698.
34. Diwan, A. D., A. S. Ninawe, and S. N. Harke. "Gene editing (CRISPR-Cas) technology and fisheries sector." *Canadian journal of biotechnology* 1, no. 2 (2017): 65–72.
35. Ullah, Asad, Sidra Munir, Syed Lal Badshah, Noreen Khan, Lubna Ghani, Benjamin Gabriel Poulson, Abdul-Hamid Emwas, and Mariusz Jaremko. "Important flavonoids and their role as a therapeutic agent." *Molecules* 25, no. 22 (2020): 5243.
36. Ciumărnean, Lorena, Mircea Vasile Milaciu, Octavia Runcan, Ştefan Cristian Vesa, Andreea Liana Răchişan, Vasile Negrean, Mirela-Georgiana Perné, et al. "The effects of flavonoids in cardiovascular diseases." *Molecules* 25, no. 18 (2020): 4320.
37. Papakyriakopoulou, Paraskevi, Nikolaos Velidakis, Elina Khattab, Georgia Valsami, Ioannis Korakianitis, and Nikolaos P. E. Kadoglou. "Potential pharmaceutical applications of quercetin in cardiovascular diseases." *Pharmaceuticals* 15, no. 8 (2022): 1019.
38. Abarikwu, Sunny O. "Protective effect of quercetin on atrazine-induced oxidative stress in the liver, kidney, brain, and heart of adult Wistar rats." *Toxicology international* 21, no. 2 (2014): 148.
39. Renugadevi, J., and S. Milton Prabu. "Quercetin protects against oxidative stress-related renal dysfunction by cadmium in rats." *Experimental and toxicologic pathology* 62, no. 5 (2010): 471–481.
40. Ożarowski M, Mikołajczak PŁ, Kujawski R, et al. Pharmacological Effect of Quercetin in Hypertension and Its Potential Application in Pregnancy-Induced Hypertension: Review of In Vitro, In Vivo, and Clinical Studies. *Evid Based Complement Alternat Med.* 2018;2018:7421489. Published 2018 Dec 2. doi:10.1155/2018/7421489.
41. Pereira, Sherliane C., Juliana M. Parente, Vanessa A. Belo, Atlante S. Mendes, Natália A. Gonzaga, Gabriel T. do Vale, Carla S. Ceron, José Eduardo Tanus-Santos, Carlos R. Tirapelli, and Michele M. Castro. "Quercetin decreases the activity of matrix metalloproteinase-2 and ameliorates vascular remodeling in renovascular hypertension." *Atherosclerosis* 270 (2018): 146–153.
42. Wang, Y., Z. Z. Zhang, Y. Wu, J. J. Ke, X. H. He, and Y. L. Wang. "Quercetin postconditioning attenuates myocardial ischemia/reperfusion injury in rats through the PI3K/Akt pathway." *Brazilian journal of medical and biological research* 46 (2013): 861–867.
43. Dias, Patrícia, Jana Pourová, Marie Vopršalová, Iveta Nejmanová, and Přemysl Mladěnka. "3-Hydroxyphenylacetic acid: A blood pressure-reducing flavonoid metabolite." *Nutrients* 14, no. 2 (2022): 328.
44. Xu, Kun, Yao Yang, Ming Lan, Jiannan Wang, Bing Liu, Mingjing Yan, Hua Wang, et al. "Apigenin alleviates oxidative stress-induced myocardial injury by regulating SIRT1 signaling pathway." *European journal of pharmacology* 944 (2023): 175584.
45. Thomas, Shilu Deepa, Niraj Kumar Jha, Saurabh Kumar Jha, Bassem Sadek, and Shreesh Ojha. "Pharmacological and molecular insight on the cardioprotective role of apigenin." *Nutrients* 15, no. 2 (2023): 385.
46. Knapp, Maura, Xin Tu, and Rongxue Wu. "Vascular endothelial dysfunction, a major mediator in diabetic cardiomyopathy." *Acta Pharmacologica Sinica* 40, no. 1 (2019): 1–8.

47. Zhao, Yuewu, Yong Li, Zhengzhang Li, Bing Xu, Peng Chen, and Xiangjun Yang. "Superoxide anions modulate the performance of apelin in the paraventricular nucleus on sympathetic activity and blood pressure in spontaneously hypertensive rats." *Peptides* 121 (2019): 170051.
48. Meng D, Qin Y, Lu N, et al. Kupffer Cells Promote the Differentiation of Adult Liver Hematopoietic Stem and Progenitor Cells into Lymphocytes via ICAM-1 and LFA-1 Interaction. Stem Cells Int. 2019;2019:4848279. Published 2019 Jul 1. doi:10.1155/2019/4848279."
49. Ueki, Shigeharu, Junko Nishikawa, Mineyo Fukuchi, Yasunori Konno, Masahide Takeda, Yuki Moritoki, Junichi Chihara, Ayumi Omokawa, Tomoo Saga, and Makoto Hirokawa. "ICAM-1 upregulation is not required for retinoic acid-induced human eosinophil survival." *Immunology letters* 196 (2018): 68–73.
50. Yu, Wei, Huirong Sun, Wenliang Zha, Weili Cui, Ling Xu, Qing Min, and Jiliang Wu. "Apigenin attenuates adriamycin-induced cardiomyocyte apoptosis via the PI3K/AKT/mTOR pathway." *Evidence-based complementary and alternative medicine* 2017 (2017).
51. Dirimanov, Stoyan, and Petra Högger. "Screening of inhibitory effects of polyphenols on Akt-phosphorylation in endothelial cells and determination of structure-activity features." *Biomolecules* 9, no. 6 (2019): 219.
52. Zhu X, Liu Y, Huang H, et al. PKCβII-induced upregulation of PGP9.5 and VEGF in postoperative persistent pain in rats. *Journal of pain research* 11 (2018):2095–2106. doi:10.2147/JPR.S144852
53. Qin, Weiwei, Bei Ren, Shanshan Wang, Shujun Liang, Baiqiu He, Xiaoji Shi, Liying Wang, Jingyu Liang, and Feihua Wu. "Apigenin and naringenin ameliorate PKCβII-associated endothelial dysfunction via regulating ROS/caspase-3 and NO pathway in endothelial cells exposed to high glucose." *Vascular pharmacology* 85 (2016): 39–49.
54. Zafra-Stone, Shirley, Taharat Yasmin, Manashi Bagchi, Archana Chatterjee, Joe A. Vinson, and Debasis Bagchi. "Berry anthocyanins as novel antioxidants in human health and disease prevention." *Molecular nutrition & food research* 51, no. 6 (2007): 675–683.
55. Salehi, Bahare, Javad Sharifi-Rad, Francesca Cappellini, Željko Reiner, Debora Zorzan, Muhammad Imran, Bilge Sener, et al. "The therapeutic potential of anthocyanins: Current approaches based on their molecular mechanism of action." *Frontiers in pharmacology* 11 (2020): 1300.
56. Siow, Richard C. M., and Giovanni E. Mann. "Dietary isoflavones and vascular protection: Activation of cellular antioxidant defenses by SERMs or hormesis?" *Molecular aspects of medicine* 31, no. 6 (2010): 468–477.
57. Liobikas, Julius, Kristina Skemiene, Sonata Trumbeckaite, and Vilmante Borutaite. "Anthocyanins in cardioprotection: A path through mitochondria." *Pharmacological research* 113 (2016): 808–815.
58. Xia, Min, Wen Hua Ling, Jing Ma, David D. Kitts, and Jerzy Zawistowski. "Supplementation of diets with the black rice pigment fraction attenuates atherosclerotic plaque formation in apolipoprotein E deficient mice." *The journal of nutrition* 133, no. 3 (2003): 744–751.
59. Jiang, Yan, Min Dai, Wen-Jing Nie, Xiao-Rong Yang, and Xian-Chun Zeng. "Effects of the ethanol extract of black mulberry (*Morus nigra* L.) Fruit on experimental atherosclerosis in rats." *Journal of ethnopharmacology* 200 (2017): 228–235.
60. Spagnuolo, Carmela, Gian Luigi Russo, Ilkay Erdogan Orhan, Solomon Habtemariam, Maria Daglia, Antoni Sureda, Seyed Fazel Nabavi, et al. "Genistein and cancer: Current status, challenges, and future directions." *Advances in nutrition* 6, no. 4 (2015): 408–419.
61. Tuli, Hardeep Singh, Muobarak Jaber Tuorkey, Falak Thakral, Katrin Sak, Manoj Kumar, Anil Kumar Sharma, Uttam Sharma, Aklank Jain, Vaishali Aggarwal, and Anupam Bishayee. "Molecular mechanisms of action of genistein in cancer: Recent advances." *Frontiers in pharmacology* 10 (2019): 1336.
62. Suetsugi, Masatomo, Leila Su, Kimberly Karlsberg, Yate-Ching Yuan, and Shiuan Chen. "Flavone and isoflavone phytoestrogens are agonists of estrogen-related receptors." *Molecular cancer research* 1, no. 13 (2003): 981–991.
63. Carlson, Scott, Ning Peng, Jeevan K. Prasain, and J. Michael Wyss. "Effects of botanical dietary supplements on cardiovascular, cognitive, and metabolic function in males and females." *Gender medicine* 5 (2008): S76–S90.
64. Wroblewski Lissin, Lynette, and John P. Cooke. "Phytoestrogens and cardiovascular health." *Journal of the American College of Cardiology* 35, no. 6 (2000): 1403–1410.
65. Hayward, Christopher S., Raymond P. Kelly, and Peter Collins. "The roles of gender, the menopause and hormone replacement on cardiovascular function." *Cardiovascular research* 46, no. 1 (2000): 28–49.
66. Ruiz-Larrea, M. Begona, Aarthi R. Mohan, George Paganga, Nicholas J. Miller, G. Paul Bolwell, and Catherine A. Rice-Evans. "Antioxidant activity of phytoestrogenic isoflavones." *Free radical research* 26, no. 1 (1997): 63–70.

67. Li, D-Y., L. Tao, H. Liu, T. A. Christopher, B. L. Lopez, and X. L. Ma. "Role of ERK1/2 in the anti-apoptotic and cardioprotective effects of nitric oxide after myocardial ischemia and reperfusion." *Apoptosis* 11 (2006): 923–930.
68. Yang, Rui, Qiang Jia, Shomaila Mehmood, Shanfeng Ma, and Xiaofen Liu. "Genistein ameliorates inflammation and insulin resistance through mediation of gut microbiota composition in type 2 diabetic mice." *European journal of nutrition* 60 (2021): 2155–2168.
69. Jia, Qiang, Rui Yang, Xiao-Fen Liu, Shan-Feng Ma, and Lei Wang. "Genistein attenuates renal fibrosis in streptozotocin-induced diabetic rats." *Molecular medicine reports* 19, no. 1 (2019): 423–431.
70. Li, Zhao-Hui, Ying Zhang, Yan Lu, Qing-Ming Lu, and Xiao-Hua Xie. "Effect of fracture of lower limbs with hemorrhage on myocardial injury and its mechanism in rats." *Zhongguo Ying Yong Sheng li xue za zhi= Zhongguo Yingyong Shenglixue Zazhi= Chinese journal of applied physiology* 34, no. 2 (2018): 115–119.
71. Kondo, Kazunao, Yasuhiro Suzuki, Yasuhiko Ikeda, and Kazuo Umemura. "Genistein, an isoflavone included in soy, inhibits thrombotic vessel occlusion in the mouse femoral artery and in vitro platelet aggregation." *European journal of pharmacology* 455, no. 1 (2002): 53–57.
72. Kokubo, Yoshihiro, Hiroyasu Iso, Junko Ishihara, Katsutoshi Okada, Manami Inoue, and Shoichiro Tsugane. "Association of dietary intake of soy, beans, and isoflavones with risk of cerebral and myocardial infarctions in Japanese populations: The Japan Public Health Center–based (JPHC) study cohort I." *Circulation* 116, no. 22 (2007): 2553–2562.
73. Testai, Lara, and Vincenzo Calderone. "Nutraceutical value of citrus flavanones and their implications in cardiovascular disease." *Nutrients* 9, no. 5 (2017): 502.
74. Naeini, Fatemeh, Zahra Namkhah, Helda Tutunchi, Seyed Mahdi Rezayat, Siavash Mansouri, Mehdi Yaseri, and Mohammad Javad Hosseinzadeh-Attar. "Effects of naringenin supplementation on cardiovascular risk factors in overweight/obese patients with nonalcoholic fatty liver disease: A pilot double-blind, placebo-controlled, randomized clinical trial." *European journal of gastroenterology & hepatology* 34, no. 3 (2022): 345–353.
75. Alam, Mahboob, Farogh Ahsan, Tarique Mahmood, Arshiya Shamim, Saba Parveen, Mohammad Shariq, and Vaseem Ahmad Ansari. "Meticulous parade on naringin respecting its pharmacological activities and novel formulations." *Avicenna journal of phytomedicine* 12, no. 5 (2022): 457.
76. Zeng, Xuan, Yuying Zheng, Yan He, Jiashuo Zhang, Wei Peng, and Weiwei Su. "Microbial metabolism of naringin and the impact on antioxidant capacity." *Nutrients* 14, no. 18 (2022): 3765.
77. Hamsalakshmi, Alex AM, Arehally Marappa M, Joghee S, Chidambaram SB. Therapeutic benefits of flavonoids against neuroinflammation: A systematic review. *Inflammopharmacology* 30, no. 1 (2022): 111–136. doi:10.1007/s10787-021-00895-8
78. Meng, Li-Min, Hui-Jie Ma, Hui Guo, Qian-Qian Kong, and Yi Zhang. "The cardioprotective effect of naringenin against ischemia–reperfusion injury through activation of ATP-sensitive potassium channel in rat." *Canadian journal of physiology and pharmacology* 94, no. 9 (2016): 973–978.
79. Jin Q, Liu T, Qiao Y, et al. Oxidative stress and inflammation in diabetic nephropathy: role of polyphenols. Front Immunol. 2023;14:1185317. Published 2023 Jul 21. doi:10.3389/fimmu.2023.1185317.
80. Zhang, Ning, Zheng Yang, Yuan Yongle, Fangfang Li, Yuan Liu, Zhenguo Ma, Haihan Liao, et al. "Naringenin attenuates pressure overload-induced cardiac hypertrophy." *Experimental and therapeutic medicine* 10, no. 6 (2015): 2206–2212.
81. E. Orhan, Ilkay, Seyed F. Nabavi, Maria Daglia, Gian C. Tenore, Kowsar Mansouri, and Seyed M. Nabavi. "Naringenin and atherosclerosis: A review of literature." *Current pharmaceutical biotechnology* 16, no. 3 (2015): 245–251.
82. Yu, Li-Ming, Xue Dong, Xiao-Dong Xue, Jian Zhang, Zhi Li, Hong-Jiang Wu, Zhong-Lu Yang, Yang Haiyan, and Hui-Shan Wang. "Naringenin improves mitochondrial function and reduces cardiac damage following ischemia-reperfusion injury: The role of the AMPK-SIRT3 signaling pathway." *Food & function* 10, no. 5 (2019): 2752–2765.
83. Ye, Gang, Minna Wang, Delin Liu, Li Cheng, Xiaoxue Yin, Qifeng Zhang, and Wei Liu. "Mechanism of naringenin blocking the protection of LTB4/BLT1 receptor against septic cardiac dysfunction." *Annals of clinical & laboratory science* 50, no. 6 (2020): 769–774.
84. Xu, Shujun, Bingxin Wu, Biying Zhong, Luoqi Lin, Yining Ding, Xiao Jin, Zhiwei Huang, Miaoyang Lin, Huanlin Wu, and Danping Xu. "Naringenin alleviates myocardial ischemia/reperfusion injury by regulating the nuclear factor-erythroid factor 2-related factor 2 (Nrf2)/System xc-/glutathione peroxidase 4 (GPX4) axis to inhibit ferroptosis." *Bioengineered* 12, no. 2 (2021): 10924–10934.

85. Ottaviani, Javier I., Tony Y. Momma, Gunter K. Kuhnle, Carl L. Keen, and Hagen Schroeter. "Structurally related (−)-epicatechin metabolites in humans: Assessment using de novo chemically synthesized authentic standards." *Free radical biology and medicine* 52, no. 8 (2012): 1403–1412.
86. Mak, Judith C. W. "Potential role of green tea catechins in various disease therapies: Progress and promise." *Clinical and experimental pharmacology and physiology* 39, no. 3 (2012): 265–273.
87. Anandh Babu, V. Pon, and Dongmin Liu. "Green tea catechins and cardiovascular health: An update." *Current medicinal chemistry* 15, no. 18 (2008): 1840–1850.
88. Deka, Apranta, and Joseph A. Vita. "Tea and cardiovascular disease." *Pharmacological research* 64, no. 2 (2011): 136–145.
89. Balentine, Douglas A., Sheila A. Wiseman, and Liesbeth C. M. Bouwens. "The chemistry of tea flavonoids." *Critical reviews in food science & nutrition* 37, no. 8 (1997): 693–704.
90. Dufresne, Christiane J., and Edward R. Farnworth. "A review of latest research findings on the health promotion properties of tea." *The journal of nutritional biochemistry* 12, no. 7 (2001): 404–421.
91. Higdon JV, Frei B. Tea catechins and polyphenols: Health effects, metabolism, and antioxidant functions. *Critical reviews in food science and nutrition* 43, no. 1(2003):89–143. doi:10.1080/10408690390826464
92. Khan, Naghma, and Hasan Mukhtar. "Tea polyphenols for health promotion." *Life sciences* 81, no. 7 (2007): 519–533.
93. Rice-Evans, Catherine A., Nicholas J. Miller, and George Paganga. "Structure-antioxidant activity relationships of flavonoids and phenolic acids." *Free radical biology and medicine* 20, no. 7 (1996): 933–956.
94. Hofmann, Claudia S., and Gail E. Sonenshein. "Green tea polyphenol epigallocatechin-3 gallate induces apoptosis of proliferating vascular smooth muscle cells via activation of p53." *The FASEB journal* 17, no. 6 (2003): 702–704.
95. Liu, Litong, Izumi Nagai, Ying Gao, Yoshibumi Matsushima, Yoshichika Kawai, and Kazutoshi Sayama. "Effects of catechins and caffeine on the development of atherosclerosis in mice." *Bioscience, biotechnology, and biochemistry* 81, no. 10 (2017): 1948–1955.
96. Zhang, Jianyong, Rixin Liang, Lan Wang, Renyi Yan, Rui Hou, Shuangrong Gao, and Bin Yang. "Effects of an aqueous extract of *Crataegus pinnatifida* Bge. var. major NE Br. fruit on experimental atherosclerosis in rats." *Journal of ethnopharmacology* 148, no. 2 (2013): 563–569.
97. Arts, Ilja C. W., Peter C. H. Hollman, Edith J. M. Feskens, H. Bas Bueno de Mesquita, and Daan Kromhout. "Catechin intake might explain the inverse relation between tea consumption and ischemic heart disease: The Zutphen Elderly Study." *The American journal of clinical nutrition* 74, no. 2 (2001): 227–232.
98. Miura, Yukiko, Tsuyoshi Chiba, Shinji Miura, Isao Tomita, Keizo Umegaki, Masahiko Ikeda, and Takako Tomita. "Green tea polyphenols (flavan 3-ols) prevent oxidative modification of low density lipoproteins: An ex vivo study in humans." *The journal of nutritional biochemistry* 11, no. 4 (2000): 216–222.
99. Khurana, Sandhya, Krishnan Venkataraman, Amanda Hollingsworth, Matthew Piche, and T. C. Tai. "Polyphenols: Benefits to the cardiovascular system in health and in aging." *Nutrients* 5, no. 10 (2013): 3779–3827.
100. Ziółkiewicz, Agnieszka, Kamila Kasprzak-Drozd, Robert Rusinek, Ewa Markut-Miotła, and Anna Oniszczuk. "The influence of polyphenols on atherosclerosis development." *International journal of molecular sciences* 24, no. 8 (2023): 7146.
101. Han, Xiuzhen, Tao Shen, and Hongxiang Lou. "Dietary polyphenols and their biological significance." *International journal of molecular sciences* 8, no. 9 (2007): 950–988.
102. Singh, Balwinder, Jatinder Pal Singh, Amritpal Kaur, and Narpinder Singh. "Bioactive compounds in banana and their associated health benefits—a review." *Food chemistry* 206 (2016): 1–11.
103. Carrillo, Juan Ángel, M. Pilar Zafrilla, and Javier Marhuenda. "Cognitive function and consumption of fruit and vegetable polyphenols in a young population: Is there a relationship?" *Foods* 8, no. 10 (2019): 507.
104. Majid, Abdul, and Poornima Priyadarshini. "Millet derived bioactive peptides: A review on their functional properties and health benefits." *Critical reviews in food science and nutrition* 60, no. 19 (2020): 3342–3351.
105. Udenigwe, Chibuike C., and Rotimi E. Aluko. "Food protein-derived bioactive peptides: Production, processing, and potential health benefits." *Journal of food science* 77, no. 1 (2012): R11–R24.
106. Zhao, Xin-Xin, Fang-Jun Lin, Hang Li, Hua-Bin Li, Ding-Tao Wu, Fang Geng, Wei Ma, Yu Wang, Bao-He Miao, and Ren-You Gan. "Recent advances in bioactive compounds, health functions, and safety concerns of onion (*Allium cepa* L.)." *Frontiers in nutrition* 8 (2021): 669805.

107. Aziz, Ejaz, Riffat Batool, Muhammad Usman Khan, Abdur Rauf, Wasim Akhtar, Mojtaba Heydari, Shazia Rehman, et al. "An overview on red algae bioactive compounds and their pharmaceutical applications." *Journal of complementary and integrative medicine* 17, no. 4 (2020): 20190203.
108. Iqbal, Iram, Polrat Wilairatana, Fatima Saqib, Bushra Nasir, Muqeet Wahid, Muhammad Farhaj Latif, Ahmar Iqbal, Rabia Naz, and Mohammad S. Mubarak. "Plant polyphenols and their potential benefits on cardiovascular health: A review." *Molecules* 28, no. 17 (2023): 6403.
109. Bhupathiraju, Shilpa N., and Katherine L. Tucker. "Coronary heart disease prevention: Nutrients, foods, and dietary patterns." *Clinica chimica acta* 412, no. 17–18 (2011): 1493–1514.
110. Wolfram, Swen. "Effects of green tea and EGCG on cardiovascular and metabolic health." *Journal of the American College of Nutrition* 26, no. 4 (2007): 373S–388S.
111. Zhou, Donghao, Haibing Yu, Fang He, Kathleen Heather Reilly, Jingling Zhang, Shuangshuang Li, Tao Zhang, Baozhen Wang, Yuanlin Ding, and Bo Xi. "Nut consumption in relation to cardiovascular disease risk and type 2 diabetes: A systematic review and meta-analysis of prospective studies." *The American journal of clinical nutrition* 100, no. 1 (2014): 270–277.
112. Nachvak, Seyed Mostafa, Shima Moradi, Javad Anjom-Shoae, Jamal Rahmani, Morteza Nasiri, Vahid Maleki, and Omid Sadeghi. "Soy, soy isoflavones, and protein intake in relation to mortality from all causes, cancers, and cardiovascular diseases: A systematic review and dose–response meta-analysis of prospective cohort studies." *Journal of the academy of nutrition and dietetics* 119, no. 9 (2019): 1483–1500.
113. Stevinson, Clare, Max H. Pittler, and Edzard Ernst. "Garlic for treating hypercholesterolemia: A meta-analysis of randomized clinical trials." *Annals of internal medicine* 133, no. 6 (2000): 420–429.
114. Ried, Karin, Thomas Sullivan, Peter Fakler, Oliver R. Frank, and Nigel P. Stocks. "Does chocolate reduce blood pressure? A meta-analysis." *BMC medicine* 8 (2010): 1–11.
115. Bonnefont-Rousselot, Dominique. "Resveratrol and cardiovascular diseases." *Nutrients* 8, no. 5 (2016): 250.
116. Pourbagher-Shahri, Ali Mohammad, Tahereh Farkhondeh, Milad Ashrafizadeh, Marjan Talebi, and Saeed Samargahndian. "Curcumin and cardiovascular diseases: Focus on cellular targets and cascades." *Biomedicine & pharmacotherapy* 136 (2021): 111214.
117. Koehn, Frank E., and Guy T. Carter. "The evolving role of natural products in drug discovery." *Nature reviews drug discovery* 4, no. 3 (2005): 206–220.
118. Choi, Yuni, Nicole Larson, Lyn M. Steffen, Pamela J. Schreiner, Daniel D. Gallaher, Daniel A. Duprez, James M. Shikany, Jamal S. Rana, and David R. Jacobs, Jr. "Plant-centered diet and risk of incident cardiovascular disease during young to middle adulthood." *Journal of the American Heart Association* 10, no. 16 (2021): e020718.
119. Chang X, Zhang T, Zhang W, Zhao Z, Sun J. Natural Drugs as a Treatment Strategy for Cardiovascular Disease through the Regulation of Oxidative Stress. Oxid Med Cell Longev. 2020;2020:5430407. Published 2020 Sep 27. doi:10.1155/2020/5430407.
120. Bahramsoltani R, Farzaei MH, Ram M, Nikfar S, Rahimi R. Bioactive Foods and Medicinal Plants for Cardiovascular Complications of Type II Diabetes: Current Clinical Evidence and Future Perspectives. Evid Based Complement Alternat Med. 2021;2021:6681540. Published 2021 Sep 16. doi:10.1155/2021/6681540.
121. Akram, Wasim, Mohd Rihan, Sakeel Ahmed, Swamita Arora, Sameer Ahmad, and Rahul Vashishth. "Marine-derived compounds applied in cardiovascular diseases: Submerged medicinal industry." *Marine drugs* 21, no. 3 (2023): 193.
122. Pagliaro B, Santolamazza C, Simonelli F, Rubattu S. Phytochemical Compounds and Protection from Cardiovascular Diseases: A State of the Art. Biomed Res Int. 2015;2015:918069. doi:10.1155/2015/918069.

11 Analysis of Phytochemicals
The Free Radical Warrior in Colostrum

Rachana R Rao[1], Sudipta Choudhury[1], Logesh S[2], Gopenath TS[2], and Kanthesh M Basalingappa[1]

[1]Division of Molecular Biology, School of Life Sciences, JSS AHER Mysore, India

[2]Department of Biotechnology and Bioinformatics, School of Life Sciences, JSS AHER Mysore, India

11.1 INTRODUCTION

Plant-based diets rich in a variety of foods can provide a plethora of phytochemicals and non-nutritive molecules that are found in plants and are proven to boost health. Different beneficial health effects are associated with phytochemicals, which are secondary plant metabolites (Sibeko, Johns, and Cordeiro 2021). The medicinal qualities of phytochemicals in the treatment of numerous illnesses and ailments are well recognized. It also has a lot of potential for the creation of novel medications for a range of human illnesses. In traditional medicine, many medicinal plants are said to have antibacterial, antioxidant, and wound-healing qualities (Rex *et al.*, 2019). The redox balance has been restored and excess oxidants have been quenched or eliminated in adult populations through the comprehensive study of antioxidant phytochemicals (Tan *et al.*, 2018). Oxidative stress is a component of all these disorders.

An array of valuable component groups, including fatty acids, glycosides, polyphenols, tannins, terpenes, steroids, flavonoids, and alkaloids, have been revealed by phytochemical analyses of historically used therapeutic plants. These phytochemical constituents are crucial in assisting in the quicker healing of wounds, combating infection, and scavenging free radicals. A more affordable alternative therapeutic agent for treating human illnesses can be phytochemicals (Rex *et al.*, 2019). As a living being, man has always engaged in a variety of tasks to guarantee his existence and welfare. By doing this, he has caused a variety of reactive chemicals, or free radicals, to be produced or released, which can be inhaled or swallowed. Free radicals or pollutants are also produced by the body during specific physiological processes. Because of their instability and lack of electrons, these free radicals, also known as reactive species, assault centres rich in electrons, such as lipid membranes, proteins, and nucleic acids, causing harm to the body's cells and tissues (Engwa 2018).

In accordance with the Pan American Health Organisation and the World Health Organization (WHO) (2009), breastfeeding is recommended for up to two years of life; however, during the first six months of life, human milk (HM) is the only diet that is suitably tailored to meet the nutritional requirements of newborns. According to Rex *et al.* (2019), along with providing all the nutrients needed for a baby's healthy growth and development, it also provides immunological protection. A growing child's changing demands are considered by the dynamic composition of milk, which changes over time. The mother's diet, health, and exposure to contaminants or heavy metals in the environment all have an impact on these variations, as does the breastfeeding stage (Lee and Kelleher 2016; Rex *et al.*, 2019).

DOI: 10.1201/9781003486237-11

In the neonatal period, oxidative stress has been associated with necrotizing enterocolitis (NEC), periventricular leukomalacia, and retinopathy. However, compared to infant formulae (IF), HM has been demonstrated to offer benefits. Accordingly, it has been demonstrated that HM protects newborns from common illnesses associated with oxidative stress (Rex *et al.,* 2019). The antioxidant qualities of phenolic compounds are well known. A minimum of one phenol group is present in them, which is an aromatic ring connected to a minimum of one hydroxyl functional group (Rex *et al.,* 2019). It is well recognized that these substances can help prevent a number of illnesses, including cancer, osteoporosis, and cardiovascular diseases (Poniedziałek *et al.,* 2018). But the human body is unable to produce them; instead, one must consume fruits, vegetables, and cereals.

Phenolic substances are often categorized as follows: tannins, lignans, stilbenes, coumarins, phenolic acids, and flavonoids. Flavonoids include anthocyanidins, isoflavonoids, flavonols, flavononols, flavones, and flavanones. The two categories of phenolic acids are hydroxycinnamic and hydroxybenzoic acid derivatives. The first group contains gallic, syringic, vanillic, p-hydroxybenzoic, protocatechuic, and ellagic acids; the second group contains caffeic, p-coumaric, ferulic, and chlorogenic acids (Rex *et al.,* 2019). It has been demonstrated that the mothers' food provides the phenolic compounds and carotenoids seen in human milk (Tsopmo 2018) (Figure 11.1).

It is the best diet for newborns in their first few days of life, and there are no artificial substitutes that can match it (Gila-Diaz *et al.*, 2019). Colostrum is a biologically complex fluid secreted by the breast gland that stimulates immunomodulation and supports the proper growing of the baby. Breastfed babies can benefit from optimal development as well as several advantages, such as a lower risk of developing allergies, a decreased chance of developing gastrointestinal disorders like celiac disease and inflammatory bowel disease, and a decreased risk of suffering from sudden infant death syndrome (SIDS) (Cacho and Lawrence 2017).

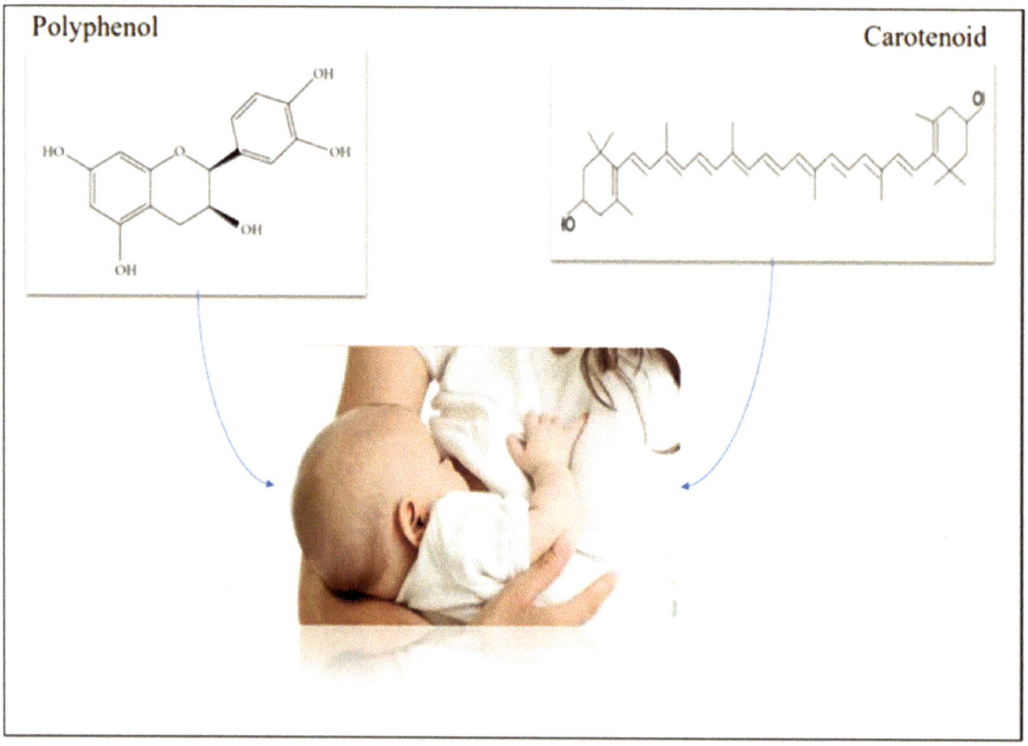

FIGURE 11.1 Primary constituents of breast milk.

(Created by PowerPoint, Courtesy: Google free licensed pictures.)

Human milk is recognized as a more effective free radical scavenger than infant formula because it contains bioactive components that the latter does not (Gila-Diaz *et al.*, 2019). As breastfeeding advances, the composition of HM constantly changes to accommodate the evolving needs of developing infants. The baby receives nitrogen, amino acids, and peptides from the proteins in HM. Moreover, HM proteins contain non-immunological defences and antioxidant peptides (Cacho and Lawrence 2017). Antioxidants are compounds that shield the target molecule from oxidative damage. Because of their ability to donate redox hydrogen and quench singlet oxygen, antioxidants can scavenge free radicals. Natural antioxidants have replaced synthetic antioxidants in recent years because they may be safer and have fewer adverse effects (G. *et al.*, 2020).

As a crucial oxidative fuel for enterocytes, the glutamate in HM can maintain the barrier's ability to operate in the digestive system. Oligosaccharides and polysaccharides in HM can restrict bacterial attachment to epithelial cells, helping to prevent infectious diseases in the infant in addition to lowering the risk of retinal and pulmonary damage in premature neonates with respiratory distress syndrome (Cacho and Lawrence 2017).

There are different stages for the chemical composition of HM, which varies during lactation. The first stage, called pre-colostrum, is a plasma exudate that begins to form approximately the 16th week of pregnancy. Then, five to seven days after birth, the colostrum is released. Thanks to its b-carotenes, it has a sticky consistency and a yellow colour. The nutritional, immunological, and growth-promoting qualities of colostrum distinguish it. The emulsification of lipids and the presence of calcium caseinate give the third stage, transition milk, its white colour between days 4 and 15 after delivery. Mature milk can be generated for more than 15 months and is thinner and more diluted than colostrum. It begins to be formed on the 15th postpartum day (Ríos, Valero-Jara and Thomas-Valdés 2022). Following colostrum feeding, there were increases in plasma glucose, lactate, total protein, immunoglobulins, triglycerides, cholesterol, Trolox equivalent antioxidative capacity, and thiobarbituric acid reactive species (Gruse *et al.*, 2016).

11.2 THE COMPONENTS OF HUMAN MILK

As per the recommendations of the WHO, infants should be breastfed exclusively for the first six months of life, followed by continued breastfeeding for at least two years or more, as human milk is considered to be the optimal source of nutrition for newborns' growth and development. Analysing the makeup of human milk is difficult since it is dynamic and changes during the day, depending on the lactation stage, even between and among populations during a single feeding (Eriksen *et al.*, 2018) (Figure 11.2).

11.2.1 Macronutrients

Variable levels of protein, fat, and carbohydrates can be found in human milk. Infants who were nursed experienced slower postnatal growth than those who were fed formula; this could be because human milk contains fewer macronutrients or has a lower overall calorie intake (Eriksen *et al.*, 2018).

11.2.1.1 Carbohydrates

The main carbohydrate in human milk is disaccharide lactose (Mosca and Gianni 2017). In the study, some researchers theorize that consuming more lactose from human milk might encourage the accumulation of fat and glycogen, which would foster growth. Low-fat or protein human milk may lead to a high lactose content, which may make babies feel less full and maybe overfeed. A prior study that discovered a positive correlation between lactose content and milk volume and an inverse association between fat concentration and milk volume supports this (Gates *et al.*, 2021).

Owing to the elevated nutritional needs associated with the human brain, the elevated level of human breast milk (6.7 g/100 mL) exceeds that of other animal species. Besides, lactose contains the majority of galactose, which is essential for supporting the development of the nervous system

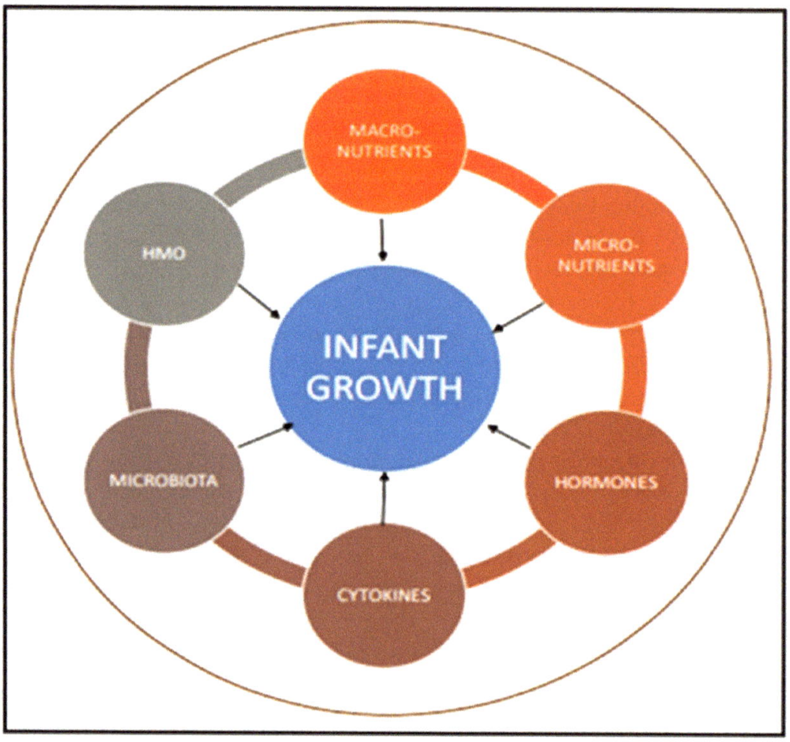

FIGURE 11.2 Infant growth and human milk components. This original chart illustrates several milk constituents that may have an impact on a newborn's development in terms of obesity and undernourishment (HMO, human milk oligosaccharides) (created with PowerPoint). (Adapted from Eriksen *et al.* 2018).

in adults. Other significant carbohydrates in human milk are called oligosaccharides (human milk oligosaccharides [HMOs]), and they vary in concentration from 1–10 g/L in the mature milk to 15–23 g/L in colostrum. Mature milk contains varying amounts of fucosylated, sialylated, and non-fucosylated neutral HMO (Mosca and Gianni 2017).

A small study done in America discovered a relationship between the fructose content of human milk and the physical weight, body mass index (BMI), fatty tissue, and mineral density of newborns at six months of age. Human milk has modest fructose levels (7 mg/mL) which may be caused by maternal use of fructose-containing goods, such as beverages sweetened with sugar. According to the study, toddler obesity may result from even very low quantities of sugar in human milk (Eriksen *et al.* 2018).

11.2.1.2 Protein

The third most prevalent solid in human milk is protein, which serves a variety of bioactive purposes in addition to being a source of nutrition. In fact, proteins are necessary for a baby's healthy growth because they act as carriers of other nutrients (such as beta-casein, haptocorrin, alfa-lactalbumin, growth factors, and lactoferrin [LF]), as well as because they stimulate the development of the gut and aid in the absorption of nutrients (like bile salt-stimulated lipase, amylase, and alpha-1 antitrypsin), as well as because LF, lysozyme, osteopontin, cytokines, secretory immunoglobulin A (IgA), and other immunological and antibacterial qualities are among them (Mosca and Gianni 2017).

Human milk's amino acid composition probably affects both overall protein concentration and growth. The hypothesis is that the levels of free amino acids (FAAs), particularly glutamic acid and glutamine concentrations, can regulate hunger. These FAAs are found in human milk in significantly higher amounts than other FAAs, and there is a significant variance between mothers.

Additionally, women who were tall or had large weights had greater glutamic acid concentrations (Eriksen *et al.*, 2018).

11.2.1.3 Fat

The primary energy sources are lipids, which provide for 44% of the total energy contained in human milk. Additionally, they are a significant source of vital nutrients such complex lipids, lipid-soluble vitamins, polyunsaturated fatty acids (PUFA), and bioactive substances. Lipoids are found in the form of dispersed fat globules, which are made up of an inner core of non-polar lipids (primarily triacylglycerols) produced in the membrane of the endoplasmic reticulum of mammary epithelium cells (Mosca and Gianni 2017).

Fat from human milk is essential to a baby's development. Increased human breast milk fat levels may be linked to lower infant BMI and adiposity at 3 to 12 months of age, in addition to decreased weight gain and adiposity during that time, according to a study. These results imply that a high proportion of fat in human milk may prevent premature growth and enhance body composition in later life (Eriksen *et al.*, 2018).

11.2.2 Micronutrients

Studies on the connections between human milk micronutrients and baby growth are noticeably limited. One reason for this study restriction could be the assumption that the amounts of micronutrients in human milk have little impact on the growth of newborns. Human milk has high concentrations of type I nutrients include calcium, riboflavin, thiamine, and vitamin B_{12}. On the other hand, type II nutrients, including phosphorus, zinc, essential amino acids, and energy, are also present in human milk but in smaller amounts. The reaction to a type I nutritional deficit is frequently defined by the clinical indication of deficiency rather than growth lag. On the other hand, nutrient deficit reactions are frequently associated with a reduction in either body size or body weight. However, more research needs to be done on how micronutrients in human milk affect a baby's growth (Eriksen *et al.*, 2018). This may be significant in environments with limited resources when breastfeeding mothers may be at risk of micronutrient shortages, which may then affect the amounts of micronutrients in human milk (Gates *et al.*, 2021).

11.2.3 Microbiota

It has been suggested that the development of the newborn gut microbiota has long-term consequences on child development and that human milk is essential for the passage of microorganisms from the breast into the gut. Although the exact origin of the bacteria detected in human milk is unknown, possible sources involve the mother's skin, the baby's mouth, or a conduit between the maternal digestive tract and mammary glands. Typically, bacteria like *Staphylococcus* and *Streptococcus* are found in human milk. The microbial makeup of human milk may be affected by a number of variables, including antibiotic use, gestational age, mother weight, and lactational stage (Eriksen *et al.*, 2018).

The way that the gut bacteria store and extract energy from meals may affect the growth of a newborn. Preclinical studies indicate that an underdeveloped gut microbiota may be a contributing factor to malnutrition and growth failure in children (Santhiravel *et al.*, 2022). The growth pattern of mice that received gut microbiota from undernourished children differed from that of germ-free mice that received microbiota from healthy children, even though they were given the same food. Moreover, manipulating the microbiota of human milk may one day be used to treat malnutrition by targeting the gut microbiota as a therapeutic target (Eriksen *et al.*, 2018).

11.2.4 Human Milk Oligosaccharides

Human milk contains bioactive substances called HMOs, which are specific to human milk, in addition to simple carbs. HMOs are made of lactose molecules that can be linked together with other disaccharides through various enzymatic processes. HMOs are believed to improve infant health by promoting gut microbiota growth and enhancing immune function, among other processes (Eriksen *et al.*, 2018). It has been suggested that HMOs function in the newborn gut as prebiotics by promoting gut microbiota development, particularly *Bifidobacterium longum* subsp. infants, as their gastrointestinal tract absorbs only 1% of the substance and they are not susceptible to the low pH of the stomach. HMOs are considered necessary functional substances vital in forming newborns' intestinal microbiota, which may affect human milk's growth and gut microbiome (Santhiravel *et al.*, 2022).

11.2.5 Other Bioactive Components

Human milk contains a very high concentration of microRNAs (miRNAs), which may be important for an infant's development and defence. Small, non-coding RNA molecules known as miRNAs control the expression of genes at the post-transcriptional level, influencing a variety of cellular processes including the cell cycle, proliferation, differentiation, apoptosis, and immune response (Alsaweed *et al.*, 2016). Since human milk contains a complex microbial population with over 200 phylotypes, it has also been determined to be the first probiotic diet given to babies. How these bacteria enter the mammary gland and eventually find their way into human milk, however, remains unclear (Mosca and Gianni 2017).

Many hormones and associated compounds, including growth factors and cytokines, are thought to be present in human milk and may have an impact on an infant's development. The growth and composition of an infant's body are thought to be significantly influenced by the hormones found in human milk. Given its involvement in the absorption and metabolism of glucose, cortisol may affect an infant's growth. It's thought that human milk cytokines may contribute to the development of the immune system, even if the precise mechanisms linking them to neonatal growth are not well understood (Eriksen *et al.*, 2018).

11.3 PHYTOCHEMICALS IN BREAST MILK

Phytochemicals are organic compounds produced by plants that help plants resist diseases and are extensively used in traditional medicine. Recent studies have mainly focused on the phytochemical content in breast milk, which to date is marked to be doing wonders for infants' growth and immunity. Polyphenols and carotenoids are a major group of phytochemicals found in colostrum (Poniedziałek *et al.*, 2018). The presence of these phytochemicals extensively depends on the diet consumed by the mother during her pregnancy and nursing stages, which becomes an important essential element for foetal and maternal health.

Polyphenols: A composed unit of phenols is a naturally derived organic compound or a secondary plant metabolite; is mainly found in nuts, berries, cocoa, red wine, etc.; and is also one of the substances present in colostrum (Sánchez-Hernández *et al.*, 2021). According to a few studies, it was proved that the polyphenolic content in berries exhibits a constructive effect in treating age-related disorders like Alzheimer's and Parkinson's. As these polyphenols are not completely absorbed by the intestine, they are not very active inside the body but their consumption through diet is prescribed for pregnant women and mothers as they are effective in the prevention of early borne disorders. Polyphenols are classified as stilbenes, flavonoids, lignans, and phenolic acids based on the quantity of phenol rings they contain. Polyphenols can be primarily classified into two groups: flavonoids and non-flavonoids. Catechins, flavones, isoflavones, flavanones, and anthocyanins are other subgroups of flavonoids. The biological activities of polyphenols, such as their anti-inflammatory and antioxidant properties, have been studied in a variety of systems (both in vitro and in vivo). A typical biological function of polyphenols is their capacity to act as antioxidants, which may shield human beings from oxidative stress and inflammation while also lowering the chance of chronic and degenerative diseases like macular degeneration, cancer, obesity, and diabetes, including cardiovascular diseases. Bearing in mind the importance of polyphenols, it is necessary to systematically study their levels in colostrum and the factors that influence them. Even though current knowledge in this field is limited, diverse methods have been adapted to determine the polyphenol content of the milk of humans and other species. Considering the role of polyphenols in oxidative stress, it is significant to evaluate their presence and role in a more detailed manner. Polyphenols like catechins, epicatechins, epigallocatechin's, etc. have also proven their significance as anticancer drugs (Sánchez-Hernández *et al.*, 2021).

Carotenoids: Another plant-derived organic compound that is primarily responsible for the characteristic pigmentation of plants is also the pigment or colour of breast milk too. Consumption of carotenoids through diet imparts a deep to pale yellow colour in breast milk. As the number of carotenoids in the maternal diet increases, so does the amount in breast milk (Tsopmo 2018). The concentration of carotenoids in colostrum decreases by the end of the third week postpartum and later the milk turns from a pale yellow to a lighter shade of white.

The human body cannot produce carotenoids, so they must be obtained through dietary intake and stored in adipose tissues. Carotenoids are transported to the epithelial surface by lipoproteins and are related to low-density lipoprotein (LDL). Carotenoids also possess an antioxidant property by reducing stress caused by free radicals. Depending on the presence of oxygen, carotenoids are divided into two classes: carotenes and xanthophylls. One such example of carotene present in breast milk is beta-carotene (β-carotene) which gets transformed into vitamin A in the body and lycopene. Referring to some research, the standard intake of β-carotene and lycopene through diet is suggested to be 10 to 15 mg on a daily basis. Through this standard quantity of food consumption, it was observed in a study that levels of lycopene and β-carotene rose by 130 to 200%. The assimilation of β-carotene is dependent on fat content and can be facilitated by drugs or supplements. Another carotenoid, namely lutein, whose content is rich in dark leafy vegetables and fruits, is another fundamental substance required for infant development (Schaefer *et al.*, 2020). Lutein and its derivative forms are the only carotenoids present in the neural retina that absorb the short wavelength of light and prevent it from damaging it, and they enhance and prevent certain conditions in both mothers and infants. The concentration of lutein is high in the breast milk of lactating women, while in pregnant women, lutein is transported to the foetus through the placenta. The presence of lutein in both pregnant and lactating women depends on their diet, and understanding its influence on it becomes more critical. As lutein is a supreme carotenoid present in the neural retina, it is also predominant in the neocortex, where cognitive functioning originates (Roy *et al.*, 2019). Other phytochemicals like allyl methyl sulphate and sulphide, sulphone, sulphoxide, theobromine, xanthine, etc. are also present in breast milk. Theobromine and xanthine are catabolic products of caffeine, whereas allyl methyl sulphates are majorly present in garlic and onions, which are consumed through diet. Collectively, all these phytochemicals quench hydroxyl free radicals and prevent or reduce oxidative stress caused by them. Few of the benefits of phytochemicals are restricted to the mother alone, and their role in infant growth is unknown and yet to be determined but low intake of phytochemicals exposes the infant to oxidative stress and its harmful effects.

11.4 miRNA: THE UNSUNG HERO IN COLOSTRUM

miRNA, or "microRNA," is a short, single-strand junk RNA molecule comprising 21–23 nucleotides found as free molecules or inside vesicles (the exosomes) (Planat *et al.*, 2023). miRNAs are also present as clusters and are expressed as polycistronic transcripts of RNA. They have promoter region and helps in initiation of transcription and have ability of individually transcribing into monocistronic primary transcripts. It is very important to eliminate the miRNA once its function is complete because a single miRNA has so many other mRNA targets and therefore can inadvertently cleave or silence other mRNAs as well. Imbalanced gene expression can be caused by undesired miRNA activity, requiring it to be destroyed once its function has ceased. Small RNA-degrading nucleases, exoribonucleases, and other endonucleases and exonucleases inactivate miRNA by cleaving it. Some nucleases have an activity from 5′ to 3′ (exoribonuclease), while others have an activity from 3′ to 5′ (SRDN [smaller RNA degrading nuclease] minor RNA degrading nuclease). In addition, dysregulation of miRNA activity also causes disease states, any mutation or lack of function of miRNAs results in down-regulation or up-regulation of a related protein, thus resulting in abnormal phenotypes. Obesity, certain nervous system disorders, kidney disease, heart diseases, cardiovascular disorders, and cancer are some of the most common conditions associated with abnormal miRNAs.

Because miRNA is a naturally occurring molecule, its normal function can be restored with a treatment called "miRNA replacement therapy." This method can be used to restore normal miRNA function in each tissue. More than 60% of human coding genes are regulated by miRNA-mediated signalling pathways. Approximately 40% of miRNAs are in intron regions. As already mentioned, some miRNAs are highly conserved in all mammals, these miRNAs represent on average over 400 different conserved target mRNAs. Additionally, about 16% of pre-miRNAs undergo RNA editing (which cannot function properly during gene silencing).

Approximately 1,400 miRNAs have been discovered in breast milk. The miRNA in breast milk was recently discovered in 2010, when scientists proposed two theories of its role in supporting the immune system of the newborn. The nutritional hypothesis proposed that miRNA is present in breast milk as a series of proteins in serum albumin, which is digested later in the gut. The other functional hypothesis proposes that miRNAs possess the role of a regulatory gene (Nguyen 2020) (Figure 11.3).

Exosomes protect these miRNAs from susceptible conditions and prevent them from being degraded by acidic conditions and other potential hazards that can be caused by RNases, gut microflora, pH, etc. Through these activities of exosomes, the integrity of miRNA is well preserved for its further actions too (Vélez-Ixta *et al.*, 2022). According to an experiment conducted by Bo Lönnerdal and his colleagues at the University of California in 2017, miRNA in exosomes were incubated along with human cells, and it was found that miRNA were transported to the cell nucleus, where they played a role in the regulation of genes switch on and off mechanisms. Though these miRNAs are active at low levels, their role and presence in colostrum are crucial. The miRNA-148a is expressed more in breast milk and has the activity of promoting the proliferation of tumour cells by down-regulating cell lines of organ-specific cancer, targeting single nucleotide proteins (SNPs) to affect the risk of carcinoma cells, hypermethylation of tumour tissues, and many more (Li *et al.*, 2016).

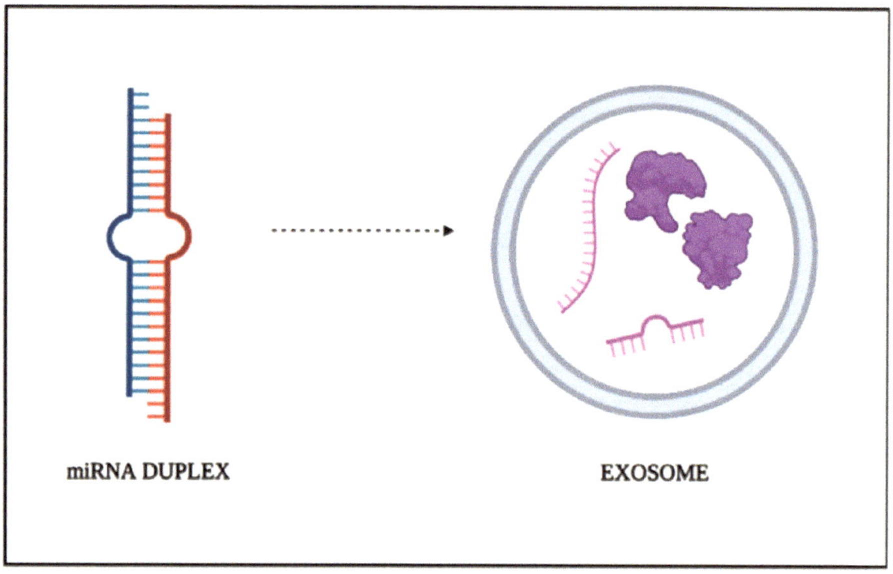

FIGURE 11.3 miRNA in exosomes.

(Created using BioRender website.)

miRNA-148a is involved in glioma apoptosis by inhibiting the B-cell lymphoma 2 gene (BCL2 gene), which promotes oncogenesis, and by suppressing BCL2 interacting mediator of cell death (BIM), a pro-apoptotic protein found in the BCL2 gene. Upregulation of miRNA-148a suppressed BIM, which inhibited apoptosis, but vice versa, it induced apoptosis (Yu *et al.*, 2022).

Research conducted by Steven Hicks at the University of Pennsylvania suggests that breast milk from mothers who gave birth prematurely contains significantly more miRNA compared to breast milk from mothers who gave birth to full-term babies. This research also proved that breast milk containing rich amounts of miRNA supported the growth of preterm babies to achieve the growth of full-term birthed ones. "All these studies have led to far more discoveries and ongoing research to introduce them in infant formulas," says Hicks.

Though breast milk is a rich source of miRNAs and its profile is widely studied, their interactions between mother and infant are not completely defined because of divergence in other research.

11.5 INTERACTION OF COLOSTRUM AS ANTIOXIDANTS IN NEWBORNS

Human milk is widely considered the most beneficial source of nourishment for newborns and infants during the first six months of life. This is due to its rich content of essential nutrients, such as vitamins, minerals, fatty acids, and amino acids. Human milk is a dynamic and ever-changing combination of nutrients and bioactive components that is produced by the human breast. As human milk evolves from colostrum to transitional milk to mature milk, it provides appropriate nourishment and protection that is suitable for the infant's time-dependent growth and development, just as the uterus, placenta, and amniotic fluid provide nourishment and protection to the foetus (Tsopmo 2018).

The mother-child dyad benefits from bonding that is caused by the release of oxytocin by proximity and contact with one another more immediately and shortly after childbirth (Moberg *et al.*, 2019). Oxytocin reduces blood loss following labour and contracts the uterus after delivery. In long-term, the endometrium, ovarian, ovarian epithelial, and breast cancers are less common in nursing moms (Ma *et al.*, 2018; Lorenzetti, Plösch and Teller 2021). Additionally, they are less likely to develop type 2 diabetes, hypertension, cardiovascular disease, metabolic syndrome, and hypertension (Nguyen, Jin and Ding 2017; Zachou, Armeni and Lambrinoudaki 2019). These benefits are positively correlated with the duration of breastfeeding (Lorenzetti *et al.*, 2021).

The foetus experiences significant stress during birth as it shifts from a hypoxic intrauterine environment to a normoxic extrauterine environment. Thus, environmental stress is present at birth for the newborn. This needs extensive antioxidant protection against reactive oxygen species (ROS)-induced lipid peroxidation (LPO) after delivery. Colostrum has a much higher total antioxidant capacity (TAC) than umbilical blood. Many enzymes found in human milk are either produced by the mammary gland or delivered by the blood to the gland (Yuksel *et al.*, 2015).

Numerous direct benefits are linked to the various bioactive compounds present in human milk, especially colostrum. They have probiotic, trophic, antioxidant, and anti-inflammatory qualities. They also help human milks' immunoprotective qualities and significantly lower death rates during the first five years of life. The necessity and length of hospital stay, the risk of (infectious) diarrhoea, gastroenteritis, respiratory and acute ear infections, and SIDS all contribute to this "survival through breastfeeding" (Victora *et al.*, 2016; Lorenzetti *et al.*, 2021).

Antioxidants found in human milk, including vitamin E, vitamin C, β-carotene, glutathione peroxidase (GPX), superoxide dismutase (SOD), and catalase (CAT), can be classified as enzymatic and non-enzymatic and may shield newborns from the formation of ROS in their early years of life. Radicals and peroxides, such as hydrogen peroxide, are inhibited by enzymes such as the enzyme GPX, the antioxidant SOD, and CAT. Additionally, LF is a metal-binding and anion-binding protein found in human milk (Yuksel *et al.*, 2015). It demonstrates antibacterial and antifungal activities that out-compete the effects of microbes that need iron for growth. Breastfed newborns are protected from peroxidation brought on by free radicals by the high antioxidant activity of breast milk (Tsopmo 2018).

Molecular entities that possess unpaired electrons in their atomic orbitals and are capable of independent existence are known as free radicals. Because of their high reactivity, these radicals can function as reductants or oxidants by either taking an electron from a molecule or giving one to another. Free radicals exhibit tremendous reactivity, although in biological systems, their half-lives are typically less than 10^{-6} s. Free radicals can be produced by some oxygen species, also referred to as ROS, even if they are not reactive in their native state (Engwa 2018). ROS, while destructive, can be harmful to cellular structures when present in excessive amounts. These compounds are highly cytotoxic due to their ability to react with and damage essential cellular molecules. The body's innate antioxidant defence mechanisms can either prevent ROS production or counteract its effects. Oxidative stress arises when the body's antioxidant defences are inadequate to remove ROS quickly. Oxidative stress is believed to be a factor in the pathogenesis of numerous neonatal diseases, including NEC, retinopathy of prematurity, and bronchopulmonary dysplasia (Tsopmo 2018).

The potential health effects of antioxidative bioactive compounds and/or their bioactive derivatives are still largely understood, as is the full spectrum of pollutants found in milk. In the intricate framework, that is milk, interactions between milk bioactive compounds and signalling pathways, transcription factors, and environmental and dietary pollutants need to be investigated further (Lorenzetti *et al.*, 2021).

11.6 DIFFERENTIAL STUDY – INFANT FORMULA VERSUS BREAST MILK

Infant formula is a manufactured powder food product intended for nourishing newborns during the initial stage of life, which typically uses cow's milk as a base that has been modified to mimic human breast milk. Animal-derived and plant-derived kinds of milk are also used if the infants are vulnerable to lactose or cow's milk (Logeshwaran, Selvasekaran and Chidambaram 2020). Infant formula is fed to the baby in cases where the mother has certain health conditions, like HIV or tuberculosis; if the baby is unable to be breastfed because the quantity of the mother's milk is inadequate; if the baby is on the verge of malnutrition in the absence of the mother; and many more (Chrustek, Dombrowska-Pali and Olszewska-Słonina 2021). Though infant formula has been used for all the above-mentioned causes, it comes with its own set of risks to health; hence, breastfeeding is given more importance. The health benefits of breast milk have more dominance over infant formula as it produces antibodies, stem cells, hormones, and prebiotics to support and enhance the immune system of the baby, whereas infant formula has fewer health benefits and differs in constituents as well (Iguacel *et al.*, 2019).

Colostrum comprises 87–88% water, carbohydrates, proteins, and fats in proportions of 7%, 1%, and 3.8%, respectively. They are rich in proteins and amino acids, more than 40 enzymes, several growth factors, vitamins, minerals, and billions of live cells like WBCs, stem cells, microbiota, etc. (Palmeira and Carneiro-Sampaio 2016). The WHO and the Codex Alimentarius Commission of the Food and Agriculture Organization of the United Nations (FAO) have established guidelines for the composition of infant formula, which state that it should be prepared according to the manufacturer's instructions and contain a minimum of 60 kcal per 100 mL, with a maximum energy content of 70 kcal, 3.8 g of fat, and specific amounts of sodium, calcium, and phosphorus in mmol (Vass *et al.*, 2021). Infant formula can contain optional components such as taurine, nucleotides, phospholipids, long-chain polyunsaturated fatty acids (LCPUFAs), and carrageenan. Both breast milk and infant formula contain proteins such as whey and casein, but breast milk has a higher proportion of whey protein that is more easily digestible for newborns, while infant formula has a higher concentration of casein protein, which is digested at a slower rate than whey (Palmeira and Carneiro-Sampaio 2016).

Long-term breastfeeding protects lactating mothers from possible future health risks like diabetes, high blood pressure, and breast and ovarian cancers and helps the mother overcome postpartum depression (Siraneh, Woldie and Birhanu 2021). The main challenges in formula feeding are lack of antibodies, expense, no comparison with the convolution of breast milk, chances of an infant being exposed to bowel syndrome, etc. Though infant formula has its own set of risks, it is widely used, but the importance of breast milk must be promoted, and knowledge drives to educate lactating and pregnant women must be prioritized (Zielinska *et al.*, 2019).

11.7 CONCLUSION

The ideal form of baby nourishment is breastmilk because it is a complex diet. It is not necessary to provide a newborn with an "enough" number of crucial nutrients and a protective level of bioactive substances while feeding them human breast milk. Factors that influence the development of an infant's digestive and immune systems can also affect their innate immunity. The findings from the current analysis show that bioactive substances can evolve with time and that colostrum typically contains the highest concentrations. Organ maturation and development are aided by macronutrients as well as the growth factors and adipokines found in breast milk. To control the infant's short- and long-term inflammatory responses, it is crucial that breastfeeding cytokines are readily available. Newborns receive vital nutrients and immunity from human milk, which helps them fight off infections in their upper respiratory, gastrointestinal, respiration, and digestive and urinary systems. The immunoglobulins in milk, including glycosylated components, oligosaccharides, and secretory IgA, inhibit microbial adherence. Additionally, other milk constituents foster a condition in the gastrointestinal system that is conducive to non-pathogenic microbes, hence averting infections. Additionally, some non-immunoglobulin components in milk possess antimicrobial properties that aid in protecting against infections.

In addition, anti-inflammatory components present in human milk provide protection to the infant by decreasing clinical symptoms of injury or infection and by reducing scarring after healing. Human milk microbiome has been studied using more advanced culture-dependent and independent approaches, which has led to the discovery of a complex ecosystem with significantly more variety than initially thought. Infant health, both short-term and long-term, as well as mammary health, may be affected by the milk microbiome's possible function. New directions in the study of lactation and pregnancy will be made possible by a deeper comprehension of the relationship between the milk microbiota and health advantages, as well as the possible factors that may have an impact on this association (Kim and Yi 2020). The WHO advises moms to breastfeed their children for a maximum of six months. Breastfeeding offers infants several advantages, such as bolstering the growth of a strong immune system and encouraging the best possible biological functions and nutrition. Additionally, nursing women can lower their chance of developing several illnesses, including osteoporosis, diabetes, and breast cancer (Figure 11.4).

Breast milk is given more importance in today's world and hence the donation of it holds great value too. Mothers lacking in the production of milk opt for acceptance of breast milk from a donor, while the excessive production of breast milk is safely stored, refrigerated, and transported for the receivers (Meneses, Oliveira and Boccolini 2017). Breast milk holds utmost value, especially in premature or medically fragile newborns as they are prone to illness. A programme to screen donors and pasteurized breast milk from recognized milk banks – which are approved by the FDA and AAP and guarantee the preservation of the milk's nutritional and immunological properties – has been started by the Human Milk Banking Association of North America (HMBANA). The milk donors undergo protocol screenings for HIV, hepatitis B and C viruses (Hep B and C), and other medical evaluations before being approved to donate breast milk. Milk donors also are provided specific instructions on how to collect milk, freeze it, and send it to the nearest milk bank (Haiden *et al.*, 2017). The received frozen milk is then thawed and undergoes screening, pasteurization, cooling, and freezing again. During the process of pasteurization and refreezing, minimal quantity of nutrition is lost but this does not affect the overall integrity of breast milk.

Hereby, this chapter concludes with the importance of breast milk, the major phytochemicals present, their impact on child health, and the factors influencing them.

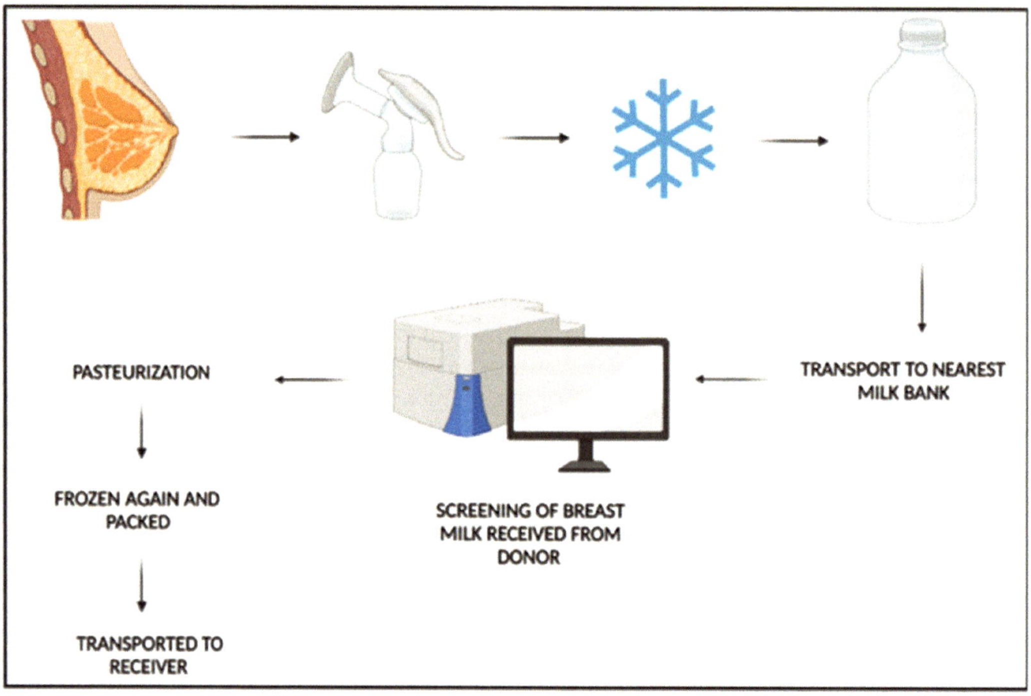

FIGURE 11.4 Process for donation of breast milk.

(Created using BioRender website.)

ABBREVIATIONS

AAP	American Association of Paediatrics
Acid EPA	Eicosapentaenoic acid
BCL2 gene	B-cell lymphoma 2 gene
BIM BCL2	interacting mediator of cell death
BMI	Body Mass Index
CAT	Catalase
DHA	Docosahexaenoic
FAA	Free amino acid
FAO	Food and Agriculture Organization of the United Nations
FDA	Food and Drug Administration
GPX	Glutathione peroxidase
Hep B	Hepatitis B virus
HIV	Human immunodeficiency virus
HM	Human milk
HMBANA	Human Milk Banking Association of North America
HMOs	Human milk oligosaccharides
IgA	Immunoglobulin A
IGF-1	Insulin-like growth factor-1
LCPUFAs	Long-chain polyunsaturated fatty acids
LDL	Low-density lipoprotein
LF	Lactoferrin
LPO	Lipid peroxidation

miRNA	microRNA
NEC	Necrotizing enterocolitis
ROS	Reactive oxygen species
SIDS	Sudden infant death syndrome
SNP	Single nucleotide protein
SOD	Superoxide dismutase
TAC	Total antioxidant capacity
TNFa	Tumour necrosis factor-a
WHO	World Health Organization

REFERENCES

Alsaweed, Mohammed, Ching Tat Lai, *et al.* "Human milk cells and lipids conserve numerous known and novel miRNAs, some of which are differentially expressed during lactation." PLoS One 11, no. 4 (2016): e0152610.

Cacho, Nicole Theresa, and Robert M. Lawrence. "Innate immunity and breast milk." Frontiers in Immunology 8 (2017): 584.

Chrustek, A., A. Dombrowska-Pali, and D. Olszewska-Słonina. "Analysis of the composition and antioxidant status of breast milk in women giving birth prematurely and on time." PLos One 16, Jul 23, no. 7 (2021): e0255252.

Engwa, Godwill Azeh. "Free radicals and the role of plant phytochemicals as antioxidants against oxidative stress-related diseases." Phytochemicals – Source of Antioxidants and Role in Disease Prevention 7 (2018): 49–74.

Eriksen, Kamilla G., Sophie H. Christensen, *et al.* "Human milk composition and infant growth." Current Opinion in Clinical Nutrition and Metabolic Care 21, no. 3 (2018): 200–206.

G., Nandini, Gopenath T. S., Nagalambika Prasad, Murugesan Karthikeyan, Ashok Gnanasekaran, Ranjith M. S., Pradeep Palanisamy, and Kanthesh M. Basalingappa. "Phytochemical analysis and antioxidant properties of leaf extracts of *Carica papaya* phytochemical analysis and antioxidant properties of leaf extracts of *Carica.*" Asian Journal of Pharmaceutical and Clinical Research 13, no. 11 (2020): 58–62. https://doi.org/10.22159/ajpcr.2020.v13i11.38956

Gates, Amy, Terri Marin, Gianluca De Leo, and Brian K. Stansfield. "Review of preterm human-milk nutrient composition." Nutrition in Clinical Practice 36, no. 6 (2021): 1163–1172.

Gila-Diaz, Andrea, Silvia M. Arribas, Alba Algara, María A. Martín-Cabrejas, Ángel Luis López de Pablo, Miguel Saenz de Pipaon, and David Ramiro-Cortijo. "A review of bioactive factors in human breast-milk: A focus on prematurity." Nutrients 11, no. 6 (2019): 1307.

Gruse, Jeannine, Ellen Kanitz, Joachim M. Weitzel, Armin Tuchscherer, Tadeusz Stefaniak, Paulina Jawor, Siegfried Wolffram, and Harald M. Hammon. "Quercetin feeding in newborn dairy calves cannot compensate colostrum deprivation: Study on metabolic, antioxidative and inflammatory traits." PLoS One 11, no. 1 (2016): 1–20.

Haiden, Nadja, and Ekhard E. Ziegler. "Human milk banking." Annals of Nutrition and Metabolism 69, no. Suppl. 2 (2017): 7–15.

Iguacel, Isabel, Laura Monje, María Jesús Cabero Pérez, Luis A. Moreno Aznar, María Pilar Samper, María Rodríguez-Palmero, Montserrat Rivero, Jesús Lino Álvarez Granda, and Gerardo Rodríguez. "Feeding patterns and growth trajectories in breast-fed and formula-fed infants during the introduction of complementary food." (2019). ISSN: 1699-5198 - ISSN : 0212-1611 - CODEN NUHOEQ S.V.R. 318.

Kim, S. Y., and D. Y. Yi. "Components of human breast milk: From macronutrient to microbiome and microRNA." Clinical and Experimental Pediatrics 63, no. 8 (2020): 301.

Li, Y., X. Deng, X. Zeng, and X. Peng. "The role of Mir-148a in cancer." Journal of Cancer 7, no. 10 (2016): 1233.

Logeshwaran, A., P. Selvasekaran, and R. Chidambaram. "Infant milk formulas." Food Science, Technology and Nutrition for Babies and Children (2020): 3–4. https://doi.org/10.1007/978-3-030-35997-3

Lorenzetti, Stefano, Torsten Plösch, and Inga C. Teller. "Antioxidative molecules in human milk and environmental contaminants." Antioxidants 10, no. 4 (2021): 1–37.

Ma, Xiao, Long-Gang Zhao, Jiang-Wei Sun, Yang Polach, Jia-Li Zheng, Jing Gao, and Yong-Bing Xiang. "Association between breastfeeding and risk of endometrial cancer: A meta-analysis of epidemiological studies." European Journal of Cancer Prevention 27, no. 2 (2018): 144–151.

Meneses, Tatiana Mota Xavier de, Maria Inês Couto de Oliveira, and Cristiano Siqueira Boccolini. "Prevalence and factors associated with breast milk donation in banks that receive human milk in primary health care units." Jornal de Pediatria 93 (2017): 382–388.

Moberg, Kerstin Uvnäs, Linda Handlin, Kathleen Kendall-Tackett, and Maria Petersson. "Oxytocin is a principal hormone that exerts part of its effects by active fragments." Medical Hypotheses 133 (2019): 109394.

Mosca, Fabio, and Maria Lorella Gianni. 2017. "Human milk: Composition and health benefits." La Pediatria Medica e Chirurgica 39 (2). https://doi.org/10.4081/pmc.2017.155

Nguyen, Binh, Kai Jin, and Ding SLY. "Breastfeeding and maternal cardiovascular risk factors and outcomes: A systematic review." PLoS One 12, no. 11 (2017): e0187923.

Nguyen, T. "Unravelling the mysteries of microRNA in breast milk." Nature 582, Jun 18, no. 7812 (2020): S12.

Palmeira, P., and M. Carneiro-Sampaio. "Immunology of breast milk." Revista da Associação Médica Brasileira 62, no. 6 (2016):) 584–593. https://doi.org/10.1590/1806-9282.62.06.584

Planat, M., M. M. Amaral, and K. Irwin. "Algebraic morphology of DNA–RNA transcription and regulation." Symmetry 15, no. 3 (2023): 770.

Poniedziałek, Barbara, Piotr Rzymski, Małgorzata Pięt, Monika Gąsecka, Anna Stroińska, Przemysław Niedzielski, Mirosław Mleczek, Paweł Rzymski, and Maciej Wilczak. "Relation between polyphenols, malondialdehyde, antioxidant capacity, lactate dehydrogenase and toxic elements in human colostrum milk." Chemosphere 191 (2018): 548–554.

Ríos, Jimena, Viviana Valero-Jara, and Samanta Thomas-Valdés. "Phytochemicals in breast milk and their benefits for infants." Critical Reviews in Food Science and Nutrition 62, no. 25 (2022): 6821–6836.

Roy, Madhumita, Amitava Datta, Madhumita Roy, and Amitava Datta. "Fundamentals of phytochemicals." Cancer Genetics and Therapeutics: Focus on Phytochemicals (2019): 49–81. https://doi.org/10.1007/978-981-13-9471-3

Sánchez-Hernández, Silvia, Adelaida Esteban-Muñoz, Cristina Samaniego-Sánchez, Rafael Giménez-Martínez, Beatriz Miralles, and Manuel Olalla-Herrera. "Study of the phenolic compound profile and antioxidant activity of human milk from Spanish women at different stages of lactation: A comparison with infant formulas." Food Research International 141 (2021): 110149.

Santhiravel S, Bekhit AEA, Mendis E, et al. The Impact of Plant Phytochemicals on the Gut Microbiota of Humans for a Balanced Life. Int J Mol Sci. 2022;23(15):8124. Published 2022 Jul 23. doi:10.3390/ijms23158124.

Schaefer, Ella, Hans Demmelmair, Jeannie Horak, Lesca Holdt, Veit Grote, Karoline Maar, Christoph Neuhofer, *et al.* "Multiple micronutrients, lutein, and docosahexaenoic acid supplementation during lactation: A randomized controlled trial." Nutrients 12, no. 12 (2020): 3849.

Sibeko, Lindiwe, Timothy Johns, and Lorraine S. Cordeiro. "Traditional plant use during lactation and post-partum recovery: Infant development and maternal health roles." Journal of Ethnopharmacology 279 (2021) (October 2020): 114377.

Siraneh, Yibeltal, Mirkuzie Woldie, and Zewdie Birhanu. "Effectiveness of positive deviance approach to promote exclusive breastfeeding practice: A cluster randomized controlled trial." Risk Management and Healthcare Policy (2021): 3483–3503.

S., Rex Jeya Rajkumar, Muthukumar Nadar M. S. A., and Paulraj Mosae Selvakumar. "Phytochemicals as a potential source for anti-microbial, anti-oxidant and wound healing – a review." MOJ Bioorganic & Organic Chemistry 2, no. 2 (2019): 61–70.

Tan, Bee Ling, Mohd Esa Norhaizan, Winnie-Pui-Pui Liew, and Heshu Sulaiman Rahman. "Antioxidant and oxidative stress: A mutual interplay in age-related diseases." Frontiers in Pharmacology 9 (2018): 1162.

Tsopmo, Apollinaire. "Phytochemicals in human milk and their potential antioxidative protection." Antioxidants 7, no. 2 (2018).

Vass, Réka A., Gabriella Kiss, Edward F. Bell, Robert D. Roghair, Attila Miseta, József Bódis, Simone Funke, and Tibor Ertl. "Breast milk for term and preterm infants—own mother's milk or donor milk?" Nutrients 13, no. 2 (2021): 424.

Vélez-Ixta, J. M., T. Benítez-Guerrero, A. Aguilera-Hernández, H. Martínez-Corona, K. Corona-Cervantes, C. J. Juárez-Castelán, M. N. Rangel-Calvillo, and J. García-Mena. "Detection and quantification of immunoregulatory miRNAs in human milk and infant milk formula." BioTech 11, Apr 20, no. 2 (2022): 11.

Victora, Cesar G., Rajiv Bahl, Aluísio J. D. Barros, Giovanny V. A. França, Susan Horton, Julia Krasevec, Simon Murch, Mari Jeeva Sankar, Neff Walker, and Nigel C. Rollins. "Breastfeeding in the 21st century: Epidemiology, mechanisms, and lifelong effect." The Lancet 387, no. 10017 (2016): 475–490.

Yu, S., L. Chen, K. Song, T. Shu, Z. Fang, L. Ding, J. Liu, L. Jiang, G. Zhang, B. Zhang, and Z. Qin. "Irreversible electroporation mediates glioma apoptosis via upregulation of *AP-1* and *Bim*: Transcriptome evidence." Brain Sciences 12, Oct 29, no. 11 (2022): 1465.

Yuksel, Sevda, Ayse Arzu Yigit, Miyase Cinar, Nurgul Atmaca, and Yuksel Onaran. "Oxidant and antioxidant status of human breast milk during lactation period." Dairy Science and Technology 95, no. 3 (2015): 295–302.

Zachou, Georgia, Eleni Armeni, and Irene Lambrinoudaki. "Lactation and maternal cardiovascular disease risk in later life." Maturitas 122 (2019): 73–79.

Zielinska, M. A., J. Hamulka, I. Grabowicz-Chądrzyńska, J. Bryś, and A. Wesolowska. "Association between breastmilk LC PUFA, carotenoids and psychomotor development of exclusively breastfed infants." International Journal of Environmental Research and Public Health 16, no. 7 (2019): 1144.

12 Potential Health Benefits of Phytochemical Compounds on Endocrine Disruption and Female Reproductive Health

Barathi Seetharaman[1], *Akshaya Chandrakanth*[1], *Winkins Santosh*[2], *and Ramasamy Vasantharekha*[1]
[1]Endocrine Disruption and Reproductive Toxicology Laboratory, Department of Biotechnology, SRM Institute of Science & Technology Kattankulathur, India
[2]PG & Research Department of Advanced Zoology & Biotechnology, Government Arts College for Men Nandanam, Chennai, India

12.1 INTRODUCTION

Phytochemicals encompass a diverse range of primary and secondary metabolites synthesized by plants, serving functions beyond basic nutrition. These non-essential compounds contribute significantly to plant resilience and defense, primarily acting as protective mechanisms against herbivory and environmental stressors. Most commonly consumed foods, such as whole grains, beans, fruits, vegetables, and herbs, are rich sources of diverse phytochemicals. These bioactive molecules, acting alone or in synergistic combinations, have demonstrated promising therapeutic efficacy in addressing numerous health conditions [Prakash et al. 2012]. On the other hand, the term "nutraceutical," coined by Dr. Stephen de Felice, stems from the combination of "nutrition" and "pharmaceutical," encompassing food or food products with demonstrably beneficial health and potentially medical applications, including disease prevention and treatment [Biesalski, 2001]. These products encompass a diverse spectrum, ranging from isolated nutrients, dietary supplements, and specific dietary regimens to genetically modified foods, herbal products, and even processed foods like cereals, soups, and beverages. Nutraceuticals have been scientifically confirmed for their physiological benefits or protection against chronic diseases. These benefits are often attributed to specific bioactive ingredients, which offer health-promoting properties and bridge the gap between the food and pharmaceutical industries [Kalra, 2003, Prakash, & Sharma, 2014]. Adequate intake of phytochemicals and phytonutrients is crucial for a balanced lifespan and immune function. However, deficits or excessive ingestion of these phytochemicals might have a deleterious impact on health. Nutrigenomics, fueled by advances in genomic information, has emerged as a cutting-edge field in nutraceutical therapy, holding promise in the fight against disease and aging and promoting healthy aging. This field investigates the link between the dietary environment and cellular/genetic processes, giving a molecular-genetic knowledge of how bioactive plant chemicals impact health. Nutrigenomics delves into how these compounds alter gene expression and/or structure, potentially affecting the initiation, development, and progression of specific diseases. This opens the door to the fascinating concept of "personalized" nutrition and medicine, which tailors dietary and therapeutic approaches to everyone's genetic composition.

DOI: 10.1201/9781003486237-12

This method has tremendous potential for preventing illness onset and maintaining optimal human health [Dijsselbloem et al. 2004, Kaput & Rodriguez, 2004].

Compelling evidence has emerged in recent years suggesting a concerning association between ubiquitous environmental chemicals and declining human reproductive health. This trend necessitates robust scientific interrogation, potentially prompting a reevaluation of current environmental safety standards and a paradigm shift in our approach to reproductive health [Piazza, & Urbanetz, 2019]. But as we begin to talk about reproductive health, it is important to understand that precise definitions are crucial in reproductive health research. Fertility encompasses the ability of a couple to achieve a live birth [Tingen et al. 2004], while fecundity reflects a woman's biological potential to conceive, quantified as the monthly probability of conception. Notably, clinical infertility is defined as the failure to achieve pregnancy after 12 months of unprotected intercourse [Gnoth et al. 2005]. Industrialization and its subsequent societal shifts, lifestyle, and junk food habits have introduced a wider range of potential risk factors impacting female reproductive health. Well-recognized lifestyle factors like environmental and occupational exposures, reproductive tract infections, cigarette smoking, obesity, psychological stress, and advanced maternal age influence fecundity and fertility [Kelly-Weeder & Cox, 2007]. Similarly, occupational exposures such as heavy physical work, anesthetic gases, anti-neoplastic drugs, heavy metals, and solvents have established detrimental effects [Caserta et al. 2011]. In developed nations, approximately 15% of the population experiences infertility, highlighting the substantial burden of this issue [Evers 2002]. Despite evidence from toxicological studies implicating environmental and dietary chemicals in female reproductive impairment, this realm receives considerably less attention [Foster et al. 2008]. An endocrine disruptor (ED) is an exogenous entity that alters endocrine system function, leading to adverse health effects across generations [Caserta et al. 2008]. Endocrine-disrupting chemicals (EDCs) target sex hormones and thyroid balance, putting reproductive health at risk. This vulnerability pertains to gamete formation, fertilization, and offspring development [Caserta et al. 2011].

However, research has repeatedly proved that several phytochemicals have protective effects against endocrine disturbance and can improve female reproductive health [Ogunlakin et al. 2023]. It is interesting to note that recent research has discovered the mechanisms of actions and routes of action of various herbs and plants whose phytochemicals have a good impact on female reproductive health, this information has been the foundation of traditional medicine in many nations [Kumar & Khanum 2012]. Ayurveda, with historical roots in ancient India, evolved into a comprehensive medical system incorporating herbal remedies, lifestyle practices, and philosophical principles [Chopra & Doiphode, 2002]. This alternative stream of medicine boasts, in its long texts, several phytochemicals with the ability to protect female reproductive health [Pandey et al. 2018]. The increasing concern regarding ubiquitous EDCs and their potentially detrimental effects on female reproductive health, a detailed investigation into the interplay between endocrine disruption, the female reproductive system, and the potential ameliorative effects of phytochemicals is required.

12.2 ENDOCRINE DISRUPTION AND FEMALE REPRODUCTIVE HEALTH

12.2.1 Female Reproductive Health

12.2.1.1 Hormonal Regulation of Female Reproductive System

The female reproductive system operates through a coordinated network of organs and hormones. This network, spanning from the hypothalamus-pituitary gland, in the brain, ovaries, and reproductive tract, ensures ovulation, hormone production, and successful pregnancy. It coordinates all the hormones by releasing gonadotropin-releasing hormone (GnRH). GnRH travels to the pituitary gland, where it stimulates the production of follicle-stimulating hormone (FSH) and luteinizing hormone (LH). Within the ovaries, FSH plays a vital role in nurturing follicles, tiny structures sheltering maturing egg cells. LH triggers ovulation, by releasing the mature egg for potential fertilization. Both FSH and LH also collaborate to regulate the ovary's production of sex hormones like estrogen and progesterone. Estrogen exhibits a characteristic pulsatile pattern, with a mid-cycle peak that sends a feedback signal back to the brain. This surge prompts a temporary increase in GnRH, LH, and FSH, setting the stage for ovulation. Beyond stimulating ovulation, LH also fuels the transformation of the ruptured follicle into the corpus luteum. This temporary structure dedicates itself to producing progesterone, another crucial hormone for preparing the uterine lining for a potential pregnancy. If fertilization occurs, the developing placenta takes on hormonal production duties, mimicking both pituitary and ovarian functions. These placental hormones support fetal development and cater to the physiological changes leading to childbirth. Finally, during lactation, the mammary glands, under the influence of pituitary hormones and ovarian steroids, provide nourishing breast milk for the newborn. This process, regulated by prolactin and oxytocin, ensures maternal sustenance for the baby's critical early stages [Terranova, 2003].

In general, the female reproductive system is cyclically regulated by hormones from puberty to menopause and is called the menstrual cycle. While the classic 28-day cycle with 14-day follicular and luteal phases holds true for young, fertile women, accurate cycle stage or ovulation prediction in individual women remains challenging due to inherent variability [Mihm et al. 2011]. The follicular phase, encompassing the first day of menstruation to ovulation, initiates a meticulously choreographed interplay of hormones within the female reproductive system. This phase revolves around the selection and maturation of a dominant follicle within each ovary, laying the groundwork for potential pregnancy. As menstruation subsides, rising FSH levels recruit a cohort of ovarian follicles, each harboring an immature oocyte. FSH promotes aromatase activity within granulosa cells, facilitating the conversion of androgens to estradiol. This rising estradiol exerts negative feedback, suppressing further FSH secretion and fostering the emergence of a single dominant follicle. Inhibin B, secreted by the dominant follicle, further dampens FSH production and enriches the local microenvironment with androgens, potentially hindering the development of rival follicles. Estradiol plays a pivotal role here. Its initial rise induces the formation of LH receptors on granulosa cells within the dominant follicle. This paves the way for the secretion of low-level progesterone and 17-hydroxyprogesterone, acting as a positive feedback loop on the estrogen-primed pituitary gland. This culminates in a dramatic surge of LH, peaking approximately 10–12 hours before ovulation. The LH surge triggers the release of the mature oocyte from the dominant follicle, marking the culmination of the follicular phase [Reed & Carr, 2015]. Following ovulation, the ruptured follicle undergoes a transformation, adopting the mantle of the corpus luteum. This transient endocrine gland flourishes under the influence of LH, dedicating itself to the production of progesterone. Progesterone reigns supreme in the second half of the cycle, diligently preparing the endometrial lining for potential embryo implantation. Vascularization within the corpus luteum peaks around the anticipated window of implantation, roughly eight to nine days after ovulation. While progesterone dominates the luteal phase, estradiol doesn't relinquish the stage entirely. It experiences two distinct peaks: one mirroring the rise of progesterone and another shortly before the onset of the next menstrual cycle. Studies of ovarian vein blood flow confirm the corpus luteum as the primary source of steroid production during this phase. Its ultimate fate hinges on the success or failure of fertilization [Niswender 1994].

12.2.2 Endocrine Disruptors

EDCs, encompassing both synthetic and naturally occurring molecules, pose a pressing concern due to their potential for disrupting hormonal homeostasis in humans and wildlife alike. There are many mechanisms by which EDCs wield their effects on the endocrine system (Table 12.1). These mechanisms are listed as follows [Combarnous & Nguyen, 2019]:

1. Direct receptor agonism: EDCs bind to hormone receptors, mimicking the natural ligand and activating the downstream signaling pathways, potentially causing unwanted effects.
2. Direct receptor antagonism: EDCs bind to hormone receptors and block the binding of the natural ligand, inhibiting downstream signaling pathways and potentially disrupting normal hormonal responses.

3. Post-receptor modulation: EDCs interact with components of the signaling pathway downstream of the receptor, altering the signal transmission and potentially disrupting hormonal responses.
4. Hormone biosynthesis modulation: EDCs can either stimulate or inhibit the biosynthesis of endogenous hormones, leading to imbalances in circulating hormone levels.
5. Hormone-binding protein interaction: EDCs may bind to circulating hormone-binding proteins, affecting the availability and transport of hormones and potentially altering their hormonal actions.
6. Regulation of hormone-binding protein metabolism: EDCs can stimulate or inhibit the synthesis or degradation of hormone-binding proteins, further affecting hormone bioavailability and action.
7. Modulation of receptor expression: EDCs may influence the expression levels of hormone receptors, altering the sensitivity of the target tissues to hormonal signals.
8. Epigenetic modifications: EDCs can interact with and modify epigenetic marks on DNA and histone proteins, potentially leading to long-term changes in gene expression and hormonal responses.
9. Non-receptor-mediated effects: Some EDCs may exert effects through non-receptor-mediate mechanisms, such as direct interactions with enzymes or membrane channels, potentially disrupting cellular functions and causing hormonal imbalances (Figure 12.1).

TABLE 12.1
EDCs and Their Effects on Female Reproductive Health

S. No.	EDCs	Effects on Female Reproductive Health	References
1.	PCBs	Fewer pregnancies	Neblett et al. 2020
		Longer, irregular cycles	Cooper et al. 2005
		Lower FSH, LH, and DHEA levels	Gerhard et al. 1999
2.	Dioxins	Endometriosis	Chandrakanth et al. 2023
		PCOS	Yang et al. 2015
3.	PFAS	Preeclampsia	Wang et al. 2016
		Pregnancy-induced hypertension	
4.	BPA	Endometriosis	Caserta et al. 2014
		Oligomenorrhea	Rashidi et al. 2017
5.	Phthalates	Decrease in the number of antral follicles	Hlisníková et al. 2020
		Precocious puberty	
		Inhibited placental angiogenesis	
		Induces miscarriage	
7.	PBDE	Longer time to conception	Harley et al. 2010
		Decrease in FSH	Gao et al. 2016
		Increased risk of abortion	
		Increased risk of premature birth	
8.	PBB	Low FSH level during follicular phase	Howards et al. 2019
9.	Organochlorine pesticides	Spontaneous abortions	Toft 2014
		Delayed pregnancy	
		Reduced birth weight	

PCB: Polychlorinated biphenyl
PFAS: Per- and polyfluoroalkyl substances
BPA: Bisphenol A
PBDE: Polybrominated diphenyl ether
DHEA: Dehydroepiandrosterone
PBB: Polybrominated biphenyl

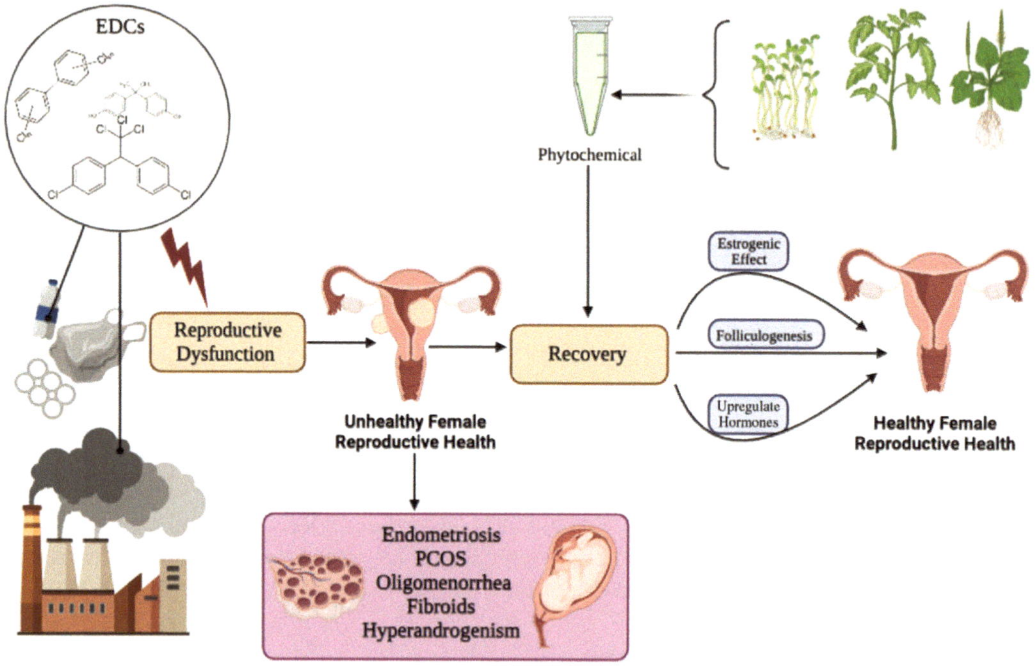

FIGURE 12.1 Diagrammatic representation of effects of phytochemicals on female reproductive health.

12.2.3 Female Reproductive Issues Associated with EDCs

According to World Health Organization (WHO) reports as of 2023, one in six people today experience problems with infertility. This is consistent with increasing industries and production of synthetic chemicals that act as EDs. Hence it is clear that EDCs may be one of the leading causes of infertility in adults. Supporting this, mounting evidence from clinical and experimental studies suggests that EDCs can exert prenatal and prepubertal effects on both male and female children, manifested primarily as altered pubertal timing. Specifically, EDCs with estrogenic properties, such as alkylphenols, bisphenol A (BPA), and phthalates, have been linked to precocious puberty [Ahn & Jeung, 2023]:

- Decreased serum estradiol and thyroxine concentrations
- Significant delay in vaginal opening
- Disruptions in estrous cycle length
- Reduced number of corpora lutea

Prenatal exposure to EDCs, notably diethylstilbestrol (DES) and BPA, has been linked to adverse reproductive outcomes in human. These include congenital malformations of the reproductive tract, as well as the development of cysts, adenomas, and carcinomas in reproductive tissues. Furthermore, both clinical and experimental evidence suggests that EDCs disrupt ovarian function and fertility by interfering with diverse reproductive processes, including folliculogenesis (follicle development), steroidogenesis (hormone production), ovulation, fertilization, and gestation. Compelling evidence suggests that exposure to EDCs may increase the incidence of uterine fibroids through diverse mechanisms. Additionally, EDC exposure has been linked to endometriosis, premature ovarian failure (POF), and polycystic ovary syndrome (PCOS), each of which is associated with an increased risk of fibroids due to altered hormonal milieu [Chandrakanth et al. 2023, Gao et al. 2010, Hirke et al. 2023,

Kawa et al. 2021]. Furthermore, prenatal exposure to EDCs may disrupt intrauterine implantation and uterine receptivity, potentially leading to implantation failure [Tran et al. 2018a]. Moreover, emerging EDCs have been demonstrated, through both in vitro and in vivo investigations, to exert modulatory effects on the immune system. These effects encompass diverse aspects, including immune cell development, humoral and cellular responses, cytokine production, and even lifespan, thereby the immune system often neglects its presence and the deleterious effects that it can have on the system. Of particular interest are xenoestrogens, a subset of EDCs exhibiting estrogenic activity, that have garnered particular interest due to their interaction with classical estrogen receptors (ERα and ERβ) and the membrane-associated G protein-coupled estrogen receptor (GPER). While the mechanisms underlying xenoestrogen action via ERα and ERβ have been extensively explored, the specific pathways activated by GPER in mammalian immune cells remain largely uncharacterized. Nonetheless, the ubiquitous presence of GPER on various immune cell types necessitates further investigation into its potential role in mediating xenoestrogen-induced immune modulation [Nowak et al. 2019].

12.3 CLASSIFICATION OF PHYTOCHEMICALS

12.3.1 Terpenes

The word "terpene" has its roots in the Latin "terebinthina," which identified the resinous exudate of turpentine trees (*Balsamum terebinthinae*), later giving rise to the scientific term for these diverse hydrocarbon compounds. Traditionally, terpenes encompass natural compounds built from C5 isoprene units, predominantly originating from plants [Breitmaier 2006]. Their biosynthetic ubiquity underscores their essential roles in various physiological functions. This versatility, coupled with the fact that isoprene units can be coupled in many ways, has yielded a vast range of terpene structures [Gershenzon & Dudareva 2007]. Terpenes, also known as terpenoids or isoprenoids, are also the most populous class of natural products, boasting over 36,000 documented structures [Wink, M. 1999]. However, despite their abundance, only a fraction of these fascinating molecules have been thoroughly investigated for their functional properties. Unveiling the chemical blueprints of terpenes relies on a classification system based on the number of isoprene units incorporated into their structures. The major categories showcase this diversity, ranging from monoterpenes (two units) to tetraterpenes (eight units), with sesquiterpenes (three units), diterpenes (four units), sesterterpenes (five units), and triterpenes (six units) completing the numerical sequence [Ashour et al. 2010]. Several estrogen-modulating terpenes and terpenoids, encompassing a diverse array of chemical classes including hemi-, mono-, sesqui-, di-, tri-, tetra-, and polyterpenes, along with their derivatives and meroterpenes exist whose signaling pathways and cellular mechanisms underpin their estrogen-mimetic activity [Kiyama, 2017].

12.3.2 Polyphenols

Polyphenols, a vast class of secondary plant metabolites, constitute the most abundant antioxidants present in the human diet. These intriguing molecules are characterized by an aromatic ring bearing one or more hydroxyl groups. Their structural diversity gives rise to a multitude of subgroups, classified based on the number of phenol rings and the connecting elements. Traditionally, polyphenols are categorized into two major groups: flavonoids and non-flavonoids. Flavonoids possess a distinctive C6-C3-C6 carbon skeleton and encompass several subclasses, including flavanones, flavones, isoflavones, and proanthocyanidins. Non-flavonoids, categorized by their total carbon count, are further subdivided into simple phenols, benzoic acids, hydrolyzable tannins, acetophenones and phenylacetic acids, cinnamic acids, coumarins, lignans, and secoiridoids [Andrés-Lacueva et al. 2009]. Polyphenols exhibit estrogenic activity, potentially modulating diverse responses within the female reproductive system [Aguiar, Jr et al. 2014]. A common source of polyphenols today, is from green tea which is widely consumed worldwide, especially in many Asian countries. These polyphenols exhibit the following qualities among many others [Rahman et al. 2018]:

1. Enhanced egg quality and reduced reproductive organ damage: Polyphenols may improve egg viability and decrease cellular damage within reproductive organs, potentially supporting fertility.
2. Protection against reproductive disorders: Animal studies indicate that green tea polyphenols (GrTPs) might safeguard fertility and offer some degree of protection from reproductive disorders.
3. Fetal protection from harmful substances: Green tea extract has been shown to significantly protect developing fetuses from the toxic effects of high-dose Indomethacin in animal studies. Additionally, it prevented parental mortality associated with the drug and exhibited ameliorative effects on various reproductive performance and oxidative stress parameters.

Additionally, mounting evidence from diverse model systems, including in vitro experiments, animal models, and human studies, indicates that polyphenols exert regulatory effects on key molecular targets implicated in oxidative stress, inflammation, and advanced glycation end products (AGEs). While the bulk of these investigations have focused on chronic diseases like cancer and diabetes, it is noteworthy that several of the affected pathways also play crucial roles in obstetrical complications. These include preeclampsia, intrauterine growth restriction (IUGR), and preterm birth, suggesting potential therapeutic or preventative avenues for polyphenols. Furthermore, emerging data highlight the influence of polyphenols on fertility, sexual development, fetal health, and nutrient bioavailability, further expanding their potential impact on reproductive health [Ly et al. 2015].

12.3.3 Carotenoids

Carotenoids, a class of over 600 naturally occurring fat-soluble pigments, grace the plant and algal world with vibrant hues and potent antioxidant properties [Krinsky & Yeum, 2003]. These molecules fall into two main categories: xanthophylls, adorned with oxygen atoms, and carotenes, devoid of oxygen and composed solely of hydrocarbons. Their primary ecological roles encompass two crucial functions [Armstrong & Hearst, 1996]:

- Light harvesting: Carotenoids efficiently absorb blue light within the photosynthetic spectrum, channeling this captured energy to chlorophyll molecules, fueling the light-dependent reactions of photosynthesis.
- Photoprotection: Beyond their light-harvesting role, carotenoids act as vigilant guardians, shielding chlorophyll and other key biomolecules from photooxidative damage caused by excessive light exposure.

In females, carotenoids and their derivatives exert multifaceted regulatory roles in reproduction, encompassing both folliculogenesis and oogenesis, as well as steroidogenesis. Furthermore, they contribute to enhanced fertility by mitigating embryonic mortality risk [Pasquariello et al. 2022]. Moreover, in a particular study, overweight women (BMI $\geq$ 25 kg/m^2) exhibited a significantly shorter time to pregnancy with increasing β-carotene intake, as evidenced by a hazard ratio (HR) of 1.29 [Schaefer & Deborah 2019].

12.3.4 Glucosinolates

Glucosinolates (GLS) constitute a prominent class of sulfur- and nitrogen-containing organic compounds found in many plants, particularly among vegetables [Larsen, 1981]. Over 100 distinct GLS have been identified since the initial isolation of sinalbin, the first crystalline glucosinolate, from white mustard seeds in 1831. GLS primarily occurs within the plant order Capparales, with the highest concentrations found in the Cruciferae, Resedaceae, and Capparidaceae families. Nevertheless, their presence has also been reported in other plant families [Larsen, 1981]. This widespread distribution, coupled with their structural diversity, highlights the important role that GLS plays in plant metabolism and defense mechanisms.

12.3.5 Dietary Fibers

Dietary fiber, a key component of healthy diets, encompasses edible or analogous carbohydrates that resist digestion and absorption in the small intestine. These undigested components undergo complete or partial fermentation in the large intestine, contributing to a range of beneficial physiological effects. In general, dietary fiber is a broad category encompassing polysaccharides, oligosaccharides, lignin, and associated plant substances, with the potential to promote lactation, attenuate blood cholesterol levels, and/or modulate blood glucose [Tiwari et al. 2015]. Chemically, dietary fibers are classified as polymers of monosaccharides linked together through glycosidic bonds. Their intricate structure is described by several key features [Choct, M. 1997]:

a. Monosaccharide composition: Identifying the types of monosaccharides present in the fiber molecule.
b. Monosaccharide ring forms: Distinguishing between pyranose (six-membered) and furanose (five-membered) ring forms of the monosaccharide units.
c. Glycosidic linkage positions: Specifying the positions on the monosaccharide rings where glycosidic bonds connect them.
d. Glycosidic linkage configurations: Differentiating between alpha (α) and beta (β) configurations of the glycosidic bonds.
e. Monosaccharide sequence: Determining the order in which monosaccharide residues are linked within the fiber chain.
f. Presence of non-carbohydrate substituents: Noting the presence or absence of non-sugar components attached to the fiber molecule.

A particular study observed a negative association between dietary fiber intake and female infertility, with an odds ratio (OR) of 0.643 (95% CI 0.480–0.861, $P = 0.004$) after adjusting for potential confounders. This statistically significant association translates to a 35.7% reduction in the odds of infertility for individuals with higher dietary fiber intake compared to those with lower intake. Further subgroup analysis revealed interesting interactions with weight status. For women with normal or underweight, dietary fiber intake did not significantly impact infertility risk (OR = 0.620, 95% CI 0.332–1.157, $P = 0.130$). Similarly, overweight women did not demonstrate a significant association between fiber intake and infertility (OR = 0.764, 95% CI 0.523–1.117, $P = 0.160$). However, a strikingly inverse association emerged for obese women, with a significant reduction in infertility risk with higher fiber intake (OR = 0.610, 95% CI 0.443–0.841, $P = 0.003$). This suggests that the protective effect of dietary fiber against infertility might be potentiated in the context of obesity [Cai et al. 2023].

12.3.6 Lectins

Lectins, aptly named after their selective binding of carbohydrates, are omnipresent proteins found in viruses, bacteria, plants, and animals. These versatile molecules influence myriad biological processes, including [Ambrosi et al. 2005]:

- Glycoprotein clearance: Streamlining waste management in the circulatory system.
- Pathogen adhesion: Facilitating viral and bacterial entry into host cells.

- Leukocyte recruitment: Directing immune cells to inflammation sites.
- Immune system interactions: Orchestrating complex cell-cell communication.
- Development and cancer: Potentially influencing cell growth and migration.

12.3.7 Other Phytochemicals

There are a variety of other phytochemicals ranging from alkaloids to chlorophyll, each with its own structural identity and chemical properties [Tiwari et al. 2015]. Another phytochemical popularly used in many Asian cultures is virgin coconut oil (VCO). Traditionally, dry-processed coconut oil held dietary and medicinal value in Southeast Asia, likely due to its medium-chain fatty acid content. VCO, produced via wet processing, boasts retention of bioactive components like tocopherols, tocotrienols, and polyphenols, potentially contributing to its reported antioxidant, anti-inflammatory, antipyretic, analgesic, and hypolipidemic properties. Additionally, monolaurin, formed from glycerol and lauric acid, exhibits immunomodulatory and antimicrobial effects, potentially contributing to VCO's benefits. Compared to traditional coconut oil, VCO displays enhanced antioxidant activity attributed to its higher content of bioactive polyphenols, tocopherols, sterols, and squalene. However, despite documented claims of VCO's medicinal properties, research has yet to comprehensively elucidate the mechanisms by which it interacts with the endocrine and immune systems through intracellular signaling pathways to mediate these purported health benefits [Hima et al. 2020; Karrunanithi et al. 2020].

12.4 BENEFITS OF PHYTOCHEMICALS IN FEMALE REPRODUCTIVE HEALTH

12.4.1 Folliculogenesis

Studies have shown that administration of *Justicia insularis* aqueous extract in albino rats resulted in a statistically significant elevation of pre-and post-implantation loss rates. This observation implies a possible adverse influence of the extract on ovarian follicle development [Telefo et al. 2012]. Moreover, supplementation with flaxseed, *Linum usitatissimum*, oil, or its major component, α-linolenic acid, significantly increases the proportion of oocytes reaching the blastocyst stage in dairy cows, suggesting an enhanced folliculogenesis process [Moallem et al. 2013]. Studies on female albino rats demonstrated that both *Citrus hystrix* and the hydroethanol extract from *Foeniculum vulgare* when administered to pregnant and lactating mice, enhanced folliculogenesis in their offspring. This suggests the potential of these extracts to promote healthy follicle development within the ovaries [Ogunlakin et al. 2023]. Research using PCOS-afflicted rats revealed that spearmint extract exhibited intriguing therapeutic effects. It significantly reduced body weight and testosterone levels, while simultaneously increasing the size of the corpus luteum (essential for progesterone production) and decreasing the number of ovarian cysts and atretic follicles [Alaee et al. 2020]. These findings raise hope for the potential of spearmint in managing PCOS-related issues like hyperandrogenism and cystic ovaries. One promising avenue in treating an endocrine disorder, as a result of EDC exposure, lies in *Phyllanthus muellerianus*, a plant extract shown in studies to reduce the number of cystic follicles in albino rats with letrozole-induced PCOS [Ndeingang et al. 2019]. This observation suggests a potential role in disrupting the development of cystic follicles, a hallmark of PCOS and a factor impacting fertility. Another interesting candidate is *Calendula officinalis*, the commonly known marigold flower. In studies utilizing albino rats with PCOS and non-ovulation, its hydroalcoholic extract demonstrated the ability to stimulate the process of folliculogenesis (follicle development) and even increase ovulation rates [Gharanjik et al. 2022]. Studies in mice demonstrate the ability of fennel, *F. vulgare*, to not only induce folliculogenesis, the process of follicle development, but also significantly increase the number of growing follicles. Diosgenin, an estrogenic component within the fennel, is believed to be a key driver of this effect. Additionally, the lesser-known plant, *Schumanniophyton magnificum*, exhibits remarkable potential in regulating hormonal imbalances and supporting ovarian function. In Wistar rats, it has been shown to have a significant positive impact on puberty onset, gonadotropin synthesis, and folliculogenesis. Studies in mice and rabbits have revealed quercetin's ability to significantly enhance this crucial process, paving the way for potential applications in optimizing reproductive health. The leaf methanol extract from *Tetracera potatoria* has been shown to reverse impaired folliculogenesis in albino rats with PCOS. Apigenin, a component of the extract, is believed to play a crucial role in this process by safeguarding the structural integrity of ovarian follicles, ultimately facilitating fertility recovery [Choct, 1997].

12.4.2 Upregulation of Sex Hormones

Studies in Chinchilla rabbits have shown that methanol extract from the leaves of *Mangifera indica* can increase serum estradiol concentrations, a key female sex hormone while reducing progesterone levels [Ngokere et al. 2014]. Similarly, extracts of *Alchornea cordifolia*, a plant found in tropical regions, have demonstrated hormonal effects in albino rats. Both aqueous and ethyl acetate extracts were found to significantly elevate testosterone, FSH, LH, and progesterone levels [Ebenyi et al. 2016]. The plant extract of *Eucommia ulmoides* has demonstrated the ability to boost sex hormone release in an albino rat model of PCOS-insulin resistance [Peng et al. 2021]. Studies in albino mice indicate that the root extract of the plant *Mimosa pudica* affects the release of gonadotropins (hormones stimulating follicle development) and estradiol [Ganguly et al. 2007]. The plant *Myrianthus arboreus*, traditionally employed for promoting female fertility, has shown promising results in preclinical studies. In albino rats, it induced earlier puberty onset, increased progesterone and gonadotropin levels, and promoted ovarian follicle maturation [Awounfack et al. 2018]. These findings align with its traditional use and suggest a potential role in supporting ovarian function and reproductive health in PCOS. Studies in cynomolgus monkeys suggest that the plant *Pueraria mirifica*, when administered daily, may influence the hormonal landscape of menopause. Notably, it appears to decrease gonadotropin and LH levels while increasing FSH levels. *Plukenetia conophora*'s leaves, in the form of an aqueous ethanolic extract, have been shown to significantly elevate testosterone levels in female albino rats. Additionally, ellagic acid, a compound found in the plant, exhibits a similar effect. In female Wistar rats, the plant extract of *M. arboreus* has demonstrated the ability to increase serum estradiol levels [Tiwari et al. 2015].

12.4.3 Restoration of the Estrous Cycle

In this section, there are a few studies on non-human mammals that aim to achieve a regular and healthy estrous cycle by the administration of phytochemicals. This stands as a testament that such treatment may also be used to regulate the human female menstrual cycle. The *M. arboreus* plant emerges as a potential regulator of menstrual regularity in PCOS. In studies, it has been shown to revive the estrous cycle in Wistar rats, offering hope for restoring hormonal balance and menstrual rhythm in women with PCOS. Additionally, it promotes uterine epithelial cell hypertrophy, suggesting a positive impact on uterine health and potentially improving receptivity for embryo implantation [Mvondo et al. 2020; Tran et al. 2018b]. Additionally, the extract of *P. muellerianus* shines a light on restoring disrupted menstrual cycles in a letrozole-induced PCOS model in albino rats. Studies have shown its ability to reestablish normal estrous cyclicity within just 14 days of treatment [Ndeingang et al. 2019]. Studies in Wistar rats have shown that an aqueous extract of the plant *J. insularis* can regulate the estrous cycle, reversing hormonal imbalances and even restoring fertility [Meguem et al. 2021]. Moreover, methanol extract from the leaves of *T. potatoria* holds promise for alleviating menstrual inconsistencies. When administered to PCOS albino rats, it demonstrably improved the regularity of their menstrual cycles, offering hope for managing this hormonal disorder and its associated reproductive challenges [Ogunlakin et al. 2023].

12.4.4 Estrogenic Effect

Studies in female albino rats suggest that parsley, *Petroselinum sativum*, may possess some estrogenic activity. However, the specific mechanisms and potential implications for human health are yet to be fully understood. In addition, research indicates that ginseng, *Panax ginseng*, may exert an estrogenic effect on the reproductive tissues of immature mice. This effect appears to involve stimulating estrogen biosynthesis and upregulating estrogen receptors. However, the precise molecular mechanisms remain unclear. Consuming yam, *Dioscorea villosa*, as a replacement for staple foods has been shown to influence sex hormone levels, lipids, and antioxidant status in a 30-day study. While this suggests potential hormonal modulation, further research is needed to elucidate the specific effects and long-term implications. The isoflavone genistein in soybean exhibits estrogenic activity on both ERα and ERβ receptors and even induces an estrogen-responsive gene in MCF-7 cells. Bamboo extract (*Bambusa bambos*) has been shown to impact MCF-7 cell proliferation rates similarly to standard estrogen therapy. While this suggests a potential alternative for hormone replacement, rigorous research is crucial to establish safety and assess potential side effects compared to established therapies [Tiwari et al. 2015]. Irregularity in endogenous estrogen levels or exposure to environmental estrogens has been linked to an increased risk of multiple chronic health conditions, including type 2 diabetes, hypertension, and dyslipidemia. *Pueraria thunbergiana*, a plant containing the estrogenic precursors puerarin and daidzin, may undergo activation by intestinal bacterial metabolism. This activation could potentially contribute to the observed health risks associated with Pueraria thunbergiana consumption. [Alonso-Magdalena et al. 2006, Park et al. 2006].

12.5 FUTURE DIRECTIONS AND CONCLUSION

Phytochemicals have been widely utilized in various forms of alternative medicine globally, demonstrating significant benefits. Despite the repeated success of alternative medical approaches, the incorporation of many highly valuable phytochemicals into mainstream therapy has been limited. Often, phytochemicals are underestimated as contemporary medicine increasingly focuses on personalized approaches. It is crucial to recognize that, unlike numerous mainstream medicines and personalized therapies, phytochemicals do not pose significant risks or side effects upon administration. Nevertheless, the integration of most, if not all, valuable phytochemicals into mainstream medicine necessitates widespread awareness, dispelling misconceptions, and conducting additional clinical trials. Many phytochemicals are primarily tested on model animals rather than humans, rendering their effects unavailable to the public. Consequently, we are still in the process of unlocking the potential of phytochemicals and comprehending how they could revolutionize global medical care [Tiwari et al. 2023].

REFERENCES

Aguiar, Jr, Odair, Vanessa Cardoso Pires, and Daniel Araki Ribeiro. "Polyphenols and reproductive health." In *Polyphenols in human health and disease*, pp. 707–714. Academic Press, 2014.

Ahn, Changhwan, and Eui-Bae Jeung. "Endocrine-disrupting chemicals and disease endpoints." *International Journal of Molecular Sciences* 24, no. 6 (2023): 5342.

Alaee, Sanaz, Mohammad Jafar Bagheri, Mahmood Sadeghi Ataabadi, and Farhad Koohpeyma. "Capacity of *Mentha spicata* (spearmint) extract in alleviating hormonal and folliculogenesis disturbances in polycystic ovarian syndrome rat model." *World* 10, no. 3 (2020): 451–456.

Alonso-Magdalena, Paloma, Sumiko Morimoto, Cristina Ripoll, Esther Fuentes, and Angel Nadal. "The estrogenic effect of bisphenol A disrupts pancreatic β-cell function in vivo and induces insulin resistance." *Environmental Health Perspectives* 114, no. 1 (2006): 106–112.

Ambrosi, Moira, Neil R. Cameron, and Benjamin G. Davis. "Lectins: Tools for the molecular understanding of the glycocode." *Organic & Biomolecular Chemistry* 3, no. 9 (2005): 1593–1608.

Andrés-Lacueva, Cristina, Alex Medina-Remon, Rafael Llorach, Mireia Urpi-Sarda, Nasiruddin Khan, Gemma Chiva-Blanch, Raul Zamora-Ros, Maria Rotches-Ribalta, and Rosa M. Lamuela-Raventós. "Phenolic compounds: Chemistry and occurrence in fruits and vegetables." *Fruit and Vegetable Phytochemicals: Chemistry, Nutritional Value, and Stability* (2009): 53–88.

Armstrong, Gregory A., and John E. Hearst. "Genetics and molecular biology of carotenoid pigment biosynthesis." *The FASEB Journal* 10, no. 2 (1996): 228–237.

Ashour, Mohamed, Michael Wink, and Jonathan Gershenzon. "Biochemistry of terpenoids: Monoterpenes, sesquiterpenes and diterpenes." *Annual Plant Reviews Volume 40: Biochemistry of Plant Secondary Metabolism* (2010): 258–303.

Awounfack, Charline Florence, Marie Alfrede Mvondo, Stéphane Zingue, Sylvin Benjamin Ateba, Sefirin Djiogue, Rosette Megnekou, Derek Tantoh Ndinteh, and Dieudonné Njamen. "*Myrianthus arboreus* P. Beauv (Cecropiaceae) extracts accelerates sexual maturation, and increases fertility index and gestational rate in female Wistar rats." *Medicines* 5, no. 3 (2018): 73.

Biesalski, Hans K. "Nutraceuticals: The link." *Nutraceuticals in Health and Disease Prevention* 6 (2001): 1.

Breitmaier, Eberhard. *Terpenes: flavors, fragrances, pharmaca, pheromones*. John Wiley & Sons, 2006.

Cai, Qian, and Tianquan Chen. "Association between dietary fiber and female fertility: A NHANES-based study." *Reproductive Sciences* 30, no. 5 (2023): 1555–1564.

Caserta, D., A. Mantovani, R. Marci, A. Fazi, F. Ciardo, C. La Rocca, F. Maranghi, and M. Moscarini. "Environment and women's reproductive health." *Human Reproduction Update* 17, no. 3 (2011): 418–433.

Caserta, Donatella, Luca Maranghi, Alberto Mantovani, R. Marci, F. Maranghi, and Massimo Moscarini. "Impact of endocrine disruptor chemicals in gynaecology." *Human Reproduction Update* 14, no. 1 (2008): 59–72.

Chandrakanth, Akshaya, Sana Firdous, Ramasamy Vasantharekha, Winkins Santosh, and Barathi Seetharaman. "Exploring the effects of endocrine-disrupting chemicals and miRNA expression in the pathogenesis of endometriosis by unveiling the pathways: A systematic review." *Reproductive Sciences* 31 (2023): 1–10.

Choct, Mingan. "Feed non-starch polysaccharides: Chemical structures and nutritional significance." *Feed Milling International* 191, no. 1 (1997): 13–26.

Chopra, Arvind, and Vijay V. Doiphode. "Ayurvedic medicine: core concept, therapeutic principles, and current relevance." *Medical Clinics* 86, no. 1 (2002): 75–89.

Combarnous, Yves, and Thi Mong Diep Nguyen. "Comparative overview of the mechanisms of action of hormones and endocrine disruptor compounds." *Toxics* 7, no. 1 (2019): 5.

Dijsselbloem, Nathalie, Wim Vanden Berghe, An De Naeyer, and Guy Haegeman. "Soy isoflavone phyto-pharmaceuticals in interleukin-6 affections: Multi-purpose nutraceuticals at the crossroad of hormone replacement, anti-cancer and anti-inflammatory therapy." *Biochemical Pharmacology* 68, no. 6 (2004): 1171–1185.

Ebenyi, L. N., E. I. Akubugwo, M. E. Ogbanshi, K. N. Agbafor, and O. R. Inya-Agha. "Fertility enhancing potentials of *Alchornea cordifolia* on albino rats." *Middle-East Journal of Scientific Research* 24, no. 5 (2016): 1802–1808.

Evers, Johannes LH. "Female subfertility." *The Lancet* 360, no. 9327 (2002): 151–159.

Foster, Warren G., Michael S. Neal, Myoung-Soek Han, and Miguel M. Dominguez. "Environmental contaminants and human infertility: Hypothesis or cause for concern?" *Journal of Toxicology and Environmental Health, Part B* 11, no. 3–4 (2008): 162–176.

Ganguly, Mausumi, Nirada Devi, Rita Mahanta, and Mridul K. Borthakur. "Effect of *Mimosa pudica* root extract on vaginal estrous and serum hormones for screening of antifertility activity in albino mice." *Contraception* 76, no. 6 (2007): 482–485.

Gao, Xiaohua, Linda Yu, Lysandra Castro, Alicia B. Moore, Tonia Hermon, Carl Bortner, Maria Sifre, and Darlene Dixon. "An endocrine-disrupting chemical, fenvalerate, induces cell cycle progression and collagen type I expression in human uterine leiomyoma and myometrial cells." *Toxicology Letters* 196, no. 3 (2010): 133–141.

Gershenzon, Jonathan, and Natalia Dudareva. "The function of terpene natural products in the natural world." *Nature Chemical Biology* 3, no. 7 (2007): 408–414.

Gharanjik, Fatemeh, Manzar Banoo Shojaeifard, Narges Karbalaei, and Marzieh Nemati. "The effect of hydroalcoholic *Calendula officinalis* extract on androgen-induced polycystic ovary syndrome model in female rat." *BioMed Research International* 2022 (2022).

Gnoth, Christian, Erhard Godehardt, Petra Frank-Herrmann, K. Friol, Jürgen Tigges, and G. Freundl. "Definition and prevalence of subfertility and infertility." *Human Reproduction* 20, no. 5 (2005): 1144–1147.

Hima, Lalgi, Uday P. Pratap, Sunil Karrunanithi, Kishore A. Ravichandran, Ramasamy Vasantharekha, and Srinivasan ThyagaRajan. "Virgin coconut oil supplementation in diet modulates immunity mediated through survival signaling pathways in rats." *Journal of Complementary and Integrative Medicine* 17, no. 1 (2020).

Hirke, Amol, Bincy Varghese, Shruti Varade, and Ramu Adela. "Exposure to endocrine-disrupting chemicals and risk of gestational hypertension and preeclampsia: A systematic review and meta-analysis." *Environmental Pollution* 317 (2023): 120828.

Kalra, Ekta K. "Nutraceutical-definition and introduction." *AAPS PharmSci* 5, no. 3 (2003): 25.

Kaput, Jim, and Raymond L. Rodriguez. "Nutritional genomics: The next frontier in the postgenomic era." *Physiological Genomics* 16, no. 2 (2004): 166–177.

Karrunanithi, Sunil, Kishore A. Ravichandran, Lalgi Hima, Uday P. Pratap, Ramasamy Vasantharekha, and Srinivasan ThyagaRajan. "Virgin coconut oil enhances neuroprotective and anti-inflammatory factors in the thymus and mesenteric lymph nodes of rats." *Clinical and Experimental Neuroimmunology* 11, no. 1 (2020): 65–72.

Kawa, Iram Ashaq, Qudsia Fatima, Shahnaz Ahmad Mir, Humira Jeelani, Saika Manzoor, and Fouzia Rashid. "Endocrine disrupting chemical bisphenol A and its potential effects on female health." *Diabetes & Metabolic Syndrome: Clinical Research & Reviews* 15, no. 3 (2021): 803–811.

Kelly-Weeder, Susan, and Cheryl Lorane Cox. "The impact of lifestyle risk factors on female infertility." *Women & Health* 44, no. 4 (2007): 1–23.

Kiyama, Ryoiti. "Estrogenic terpenes and terpenoids: Pathways, functions and applications." *European Journal of Pharmacology* 815 (2017): 405–415.

Krinsky, Norman I., and Kyung-Jin Yeum. "Carotenoid–radical interactions." *Biochemical and Biophysical Research Communications* 305, no. 3 (2003): 754–760.

Kumar, G. Phani, and Farhath Khanum. "Neuroprotective potential of phytochemicals." *Pharmacognosy Reviews* 6, no. 12 (2012): 81.

Larsen, P. O. (1981). Glucosinolates.

Ly, Christina, Julien Yockell-Lelievre, Zachary M. Ferraro, John T. Arnason, Jonathan Ferrier, and Andree Gruslin. "The effects of dietary polyphenols on reproductive health and early development." *Human Reproduction Update* 21, no. 2 (2015): 228–248.

Meguem, Stéphanie Ouabo, Landry Lienou Lienou, Marie Stéphanie Chekem Goka, Richard Simo Tagne, Didiane Mefokou Yemele, and Phélix Bruno Telefo. "Effects of the aqueous extract of *Dicliptera verticillata* on fertility and different stages of gestation in female rats." *Zygote* 29, no. 4 (2021): 307–313.

Mihm, M., S. Gangooly, and S. Muttukrishna. "The normal menstrual cycle in women." *Animal Reproduction Science* 124, no. 3–4 (2011): 229–236.

Moallem, U., A. Shafran, M. Zachut, I. Dekel, Y. Portnick, and A. Arieli. "Dietary α-linolenic acid from flaxseed oil improved folliculogenesis and IVF performance in dairy cows, similar to eicosapentaenoic and docosahexaenoic acids from fish oil." *Reproduction* 146, no. 6 (2013): 603–614.

Mvondo, Marie Alfrede, Flavie Ingrid Mzemdem Tsoplfack, Charline Florence Awounfack, and Dieudonné Njamen. "The leaf aqueous extract of *Myrianthus arboreus* P. Beauv. (Cecropiaceae) improved letrozole-induced polycystic ovarian syndrome associated conditions and infertility in female Wistar rats." *BMC Complementary Medicine and Therapies* 20 (2020): 1–13.

Ndeingang, Eveline Christiane, Patrick Brice Defo Deeh, Pierre Watcho, and Albert Kamanyi. "*Phyllanthus muellerianus* (Euphorbiaceae) restores ovarian functions in letrozole-induced polycystic ovarian syndrome in rats." *Evidence-Based Complementary and Alternative Medicine* 2019 (2019).

Ngokere, A. A., C. P. Ezeofor, J. O. Okoye, S. N. Ibekailo, T. Ude, C. J. Awalu, and M. S. Amadi. "Antiprogesteronic and estrogenic effect of *Mangifera indica* in female rabbits." *Journal of Pharmacology and Toxicology* 9, no. 2 (2014): 82–89.

Niswender, G. D. "Corpus luteum and its control in infraprimate species." *The Physiology of Reproduction* (1994. https://doi.org/10.1152/physrev.2000.80.1.1

Nowak, Karolina, Ewa Jabłońska, and Wioletta Ratajczak-Wrona. "Immunomodulatory effects of synthetic endocrine disrupting chemicals on the development and functions of human immune cells." *Environment International* 125 (2019): 350–364.

Ogunlakin, Akingbolabo D., Mubo A. Sonibare, and Oluwafemi A. Ojo. "Review on effect of medicinal plants on female reproductive system." *Tropical Journal of Natural Product Research* 7, no. 3 (2023).

Pandey, Ajai K., Anumegha Gupta, Meenakshi Tiwari, Shilpa Prasad, Ashutosh N. Pandey, Pramod K. Yadav, Alka Sharma, et al. "Impact of stress on female reproductive health disorders: Possible beneficial effects of shatavari (*Asparagus racemosus*)." *Biomedicine & Pharmacotherapy* 103 (2018): 46–49.

Park, Eun-Kyung, Jieun Shin, Eun-Ah Bae, Young-Chul Lee, and Dong-Hyun Kim. "Intestinal bacteria activate estrogenic effect of main constituents puerarin and daidzin of *Pueraria thunbergiana*." *Biological and Pharmaceutical Bulletin* 29, no. 12 (2006): 2432–2435.

Pasquariello, Rolando, Polina Anipchenko, Georgia Pennarossa, Martina Crociati, Massimo Zerani, Tiziana AL Brevini, Fulvio Gandolfi, and Margherita Maranesi. "Carotenoids in female and male reproduction." *Phytochemistry* (2022): 113459.

Peng, Meng-fan, Shuo Tian, Ya-gang Song, Cai-xia Li, Zhen Ren, and Ming Li. "Effects of total flavonoids from *Eucommia ulmoides* Oliv. leaves on polycystic ovary syndrome with insulin resistance model rats induced by letrozole combined with a high-fat diet." *Journal of Ethnopharmacology* 273 (2021): 113947.

Piazza, Mauri José, and Almir Antônio Urbanetz. "Environmental toxins and the impact of other endocrine disrupting chemicals in women's reproductive health." *JBRA Assisted Reproduction* 23, no. 2 (2019): 154.

Prakash, Dhan, Charu Gupta, and Girish Sharma. "Importance of phytochemicals in nutraceuticals." *Journal of Chinese Medicine Research and Development* 1, no. 3 (2012): 70–78.

Prakash, Dhan, and Girish Sharma, eds. *Phytochemicals of nutraceutical importance*. CABI, 2014.

Rahman, Sajid Ur, Yingying Huang, Lei Zhu, Shibin Feng, Ibrar Muhammad Khan, Jinjie Wu, Yu Li, and Xichun Wang. "Therapeutic role of green tea polyphenols in improving fertility: A review." *Nutrients* 10, no. 7 (2018): 834.

Reed, Beverly G., and Bruce R. Carr. "The normal menstrual cycle and the control of ovulation." (2015).

Schaefer, Ella, and Deborah Nock. "The impact of preconceptional multiple-micronutrient supplementation on female fertility." *Clinical Medicine Insights: Women's Health* 12 (2019): 1179562X19843868.

Telefo, Phelix Bruno, Simon Richard Tagne, Olga Elodie Sandrine Koona, Didiane M. Yemele, and Felicite Mbiapo Tchouanguep. "Effect of the aqueous extract of *Justicia insularis* T. Anders (Acanthaceae) on ovarian folliculogenesis and fertility of female rats." *African Journal of Traditional, Complementary and Alternative Medicines* 9, no. 2 (2012): 197–203.

Terranova, Paul F. "The female reproductive system." *Medical physiology*, 2nd edition, pp. 667–683. Lippincott William & Wilkins, 2003.

Tingen, Candace, Joseph B. Stanford, and David B. Dunson. "Methodologic and statistical approaches to studying human fertility and environmental exposure." *Environmental Health Perspectives* 112, no. 1 (2004): 87–93.

Tiwari, Brijesh K., Nigel Brunton, and Charles S. Brennan. *Handbook of plant food phytochemicals*. Wiley-Blackwell, 2015.

Tran, Dinh Nam, Eui-Man Jung, Changhwan Ahn, Jae-Hwan Lee, Yeong-Min Yoo, and Eui-Bae Jeung. "Effects of bisphenol A and 4-tert-octylphenol on embryo implantation failure in mouse." *International Journal of Environmental Research and Public Health* 15, no. 8 (2018): 1614.

Tran, Dinh Nam, Eui-Man Jung, Yeong-Min Yoo, Changhwan Ahn, Hee Young Kang, Kyung-Chul Choi, Sang-Hwan Hyun, Vu Hoang Dang, Thi Ngoc Pham, and Eui-Bae Jeung. "Depletion of follicles accelerated by combined exposure to phthalates and 4-vinylcyclohexene diepoxide, leading to premature ovarian failure in rats." *Reproductive Toxicology* 80 (2018): 60–67.

Wink, Michael, ed. *Biochemistry of plant secondary metabolism*. Vol. 2. CRC Press, 1999.

13 Natural Compounds and Biogenic Nanoparticles in Treatment of Polycystic Ovarian Syndrome*

Sayani Mondal, Akansha Chakraborty, and Kousik Pramanick
Integrative Biology Research Unit
Department of Life Sciences, Presidency University
Kolkata, India

13.1 POLYCYSTIC OVARIAN SYNDROME (PCOS): AN OVERVIEW

Polycystic ovarian syndrome or PCOS is a widespread reproductive disorder in females, beginning from puberty and continuing throughout the reproductive years. It is an inflammatory, autoimmune disease characterised by hyperinsulinemia, hyperandrogenaemia, chronic oligo anovulation along with depression and mood disorder (De Leo et al., 2016; Liu et al., 2016). PCOS was first described by Stein and Leventhal as oligo amenorrhea and polycystic ovaries along with hirsutism, acne and obesity (Azziz & Adashi, 2016; Rosenfield & Ehrmann, 2016; Stein & Leventhal, 1935). According to the report of US National Institutes of Health (NIH), the rate of incidence of PCOS worldwide is 6–10%, most of the patients having obesity and insulin resistance (Hartmann & McEwen, 2019; Moran et al., 2011; Pasquali et al., 2011). Biochemical tests and ultrasound scanning are the main diagnostic methods for the detection of PCOS. Detection methods include hormone profiling, checking estrous cyclicity and ovarian morphology (Andhalkar et al., 2021). Therapeutic approach for the disease includes oral contraceptive pills, hormone therapy, metformin; depending on the comorbidities (Mihanfar et al., 2021). According to the study of Kahsar-Miller et al. (2001) the incidence of PCOS is increased by fivefold to sixfold in the first-degree female relatives of the affected patients compared to the general population. Along with endocrine abnormalities, metabolic dysfunction is also associated with PCOS. Among them, the presence of insulin resistance, hyper-insulinemia, and obesity are most common. As the evidence suggests, Insulin resistance is generally associated with reduced action of androgen in glucose transport, with a compensatory hyperinsulinemia (Ciaraldi et al., 1992, 1997; Dunaif et al., 1995; Marsden et al., 1994). Type II Diabetes mellitus is one of the long-term risks associated with poly cystic ovarian syndrome. Several studies show that patients with PCOS have reduced beta cell function increasing the chances of developing type II diabetes mellitus fivefold to tenfold considering the age, obesity and family history (Ovalle & Azziz, 2002). These types of metabolic dysfunctions and obesity are reported to directly or indirectly affect fertility in women by altering pituitary-hypothalamic function or ovarian function. Type I Diabetes mellitus is often related to fewer pregnancies and live births (Cena et al., 2020; Jonasson et al., 2007). Teenagers associated with exercise and healthy lifestyle are proven to have less occurrence of PCOS compared to teenagers with sedentary lifestyle which is also positively linked to the prevalence of obesity (Eleftheriadou et al., 2012; Samdal et al., 2007). PCOS affected women are often detected with higher levels of intercellular adhesion molecule (ICAM)-1, tumour necrosis factor (TNF)-α,

* These authors contributed equally to this work.

DOI: 10.1201/9781003486237-13

interleukin-6 (IL-6) and monocyte chemoattractant protein (MCP)-1 indicating an inflammatory condition in the body. The patients are also reported to have increased mean platelet volume (MPV) and neutrophil to lymphocyte ratio as compared to healthy women (Patel, 2018). Apart from physical effects PCOS can cause psychological effects including depression, bipolar disorder, binge eating disorder as comorbidities (Blay et al., 2016; Patel, 2018).

13.2 PATHOPHYSIOLOGY

Pathophysiology of PCOS is dependent mainly on the nature of changes in ovarian function along with external factors, defects in the hypothalamus-pituitary-gonadal axis as well as insulin secretion. Specific hormones like LH, androgen, insulin or insulin like growth factors also give light to this field (Balen, 2004; STRAUSS, 2003).

13.2.1 Ovarian Function

The presence of unilateral or bilateral Poly cystic ovary structure in patients has excess androgen (hyperandrogenism) and adiposity. Theca cells produce androgens, which due to dysregulation in PCOS patients cause hyperandrogenism. In PCOS theca cells there is an increased production of progesterone, 17α-hydroxyprogesterone and testosterone thereby the correlation between androgen synthesis and follicular development gets hampered resulting in poor follicular maturation and increased follicular atresia (GILLING-SMITH et al., 1994). There is a higher amount of estradiol throughout follicular maturation because of androgen substrate and extreme response of follicular development and estradiol secretion to FSH (Erickson et al., 1992). Hirsutism as well as the presence of acne are the clinical markers of hyperandrogenism (Slayden et al., 2001).

13.2.2 Hypothalamus–Gonadal–Pituitary Axis

Follicle-stimulating hormone (FSH) and luteinising hormone (LH) secretion is directly regulated by Gonadotropin Releasing Hormone (GnRH) secreted from pituitary and regulated by combined feedback mechanism. Pulsatile secretion of GnRH by controlling the GnRH secretion maintains the ratio of FSH and LH throughout the cycle. The slow pulsatility results in increased FSH secretion whereas rapid for LH secretion (Dalkin et al., 1989). Pituitary GnRH sensitivity in menstrual cycle changes with the changes of oestradiol concentrations (Wang et al., 1976). b-endorphin and opiate peptides, angiotensin II, serotonin, neuropeptide Y, neurotensin, somatostatin, corticotropin releasing factor, dopamine, melatonin, norepinephrine (noradrenaline), oxytocin and substance P are some other factors which possess an impact on GnRH secretion. Steroids are one of the reasons for the changes in GnRH level in PCOS patients (Danforth et al., 1990). PCOS patients have a higher sensitivity to external dopamine and dopaminergic inhibition of GnRH secretion is hampered (Quigley et al., 1981). Thereby hypothalamic defect can be cause of increased secretion of LH by reducing opioid and dopaminergic control of GnRH (Cumming et al., 1984; Balen, 2004).

13.3 CONVENTIONAL THERAPEUTIC APPROACH

Through the years, management of PCOS has involved symptomatic treatment for various related conditions such as ovarian dysfunction, disturbed menstrual cyclicity, androgen related issues, metabolic disorders, infertility, obesity and psychological disorders. Weight loss has been proposed as an initial treatment strategy only in patients who are overweight (BMI 25–27 kg/m2) as obesity is associated with several complications like anovulation, miscarriage or late pregnancy (Badawy & Elnashar, 2011). Losing as little as 5% of the initial weight can improve menstrual cyclicity and ovulation rates (Patel & Nestler, 2006). Lifestyle intervention for weight loss includes implying a low-calorie diet with reduced carbohydrate intake, proper exercise and sometimes bariatric surgery

(Badawy & Elnashar, 2011; Hoeger et al., 2021). The symptoms of PCOS can also be reversed by administering medications such as clomiphene citrate (CC), tamoxifen, aromatase inhibitors, metformin, glucocorticoids, or gonadotropins or surgically by laparoscopic ovarian drilling (Badawy & Elnashar, 2011). Treatment with metformin can have a significant effect on menstrual cyclicity, ovulation rate and hyperandrogenism. It shows the clinical effects mainly by preventing hepatic glucose uptake and reducing insulin levels (Grundy, 2002; Sam & Dunaif, 2003). Clomiphene citrate acts as an oestrogen receptor antagonist that can increase levels of FSH followed by an LH surge and ovulation, ultimately achieving high rates of live birth in infertile women affected with PCOS (Badawy & Elnashar, 2011; Legro et al., 2007). The PCOS affected women who do not wish to conceive can be treated with oral contraceptive pills (OCP) that can decrease serum androgen levels by direct negative feedback on LH secretion (Badawy & Elnashar, 2011). In the PCOS patients with clomiphene resistance, laparoscopic ovarian drilling can be an effective alternative treatment. This surgical method is reported to decrease serum dehydroepiandrosterone (DHEA) as well as androgen levels by the termination of androgen-producing ovarian stroma and the follicles having high androgen levels (Felemban et al., 2000). The ultimate treatment option for achieving a healthy pregnancy in PCOS women stands as In Vitro Fertilisation (IVF) techniques, when treatment with CC, metformin or gonadotropins have failed (Badawy & Elnashar, 2011).

Although these therapeutic approaches have been used in treating PCOS since long, they are often reported to have potential side effects. The primary side effect related to metformin administration is gastrointestinal symptoms like stomach pain, nausea, vomiting, (Hoeger et al., 2021) whereas it can also lead to lactic acidosis, tiredness, severe drowsiness, muscle pain and irregular heartbeat like symptoms (Nasri & Rafieian-Kopaei, 2014). Hormone therapy for long duration can cause psychiatric distress, obesity and even cancer (Patel, 2018). Constant use of oral contraceptives can have adverse effects on insulin resistance, glucose tolerance, vascular reactivity, and coagulability (Badawy & Elnashar, 2011). One of the major concerns with laparoscopic ovarian surgery is that it can lead to intra-abdominal adhesion formation that can cause stomach pain and bowel obstruction (Al-Took et al., 1999; Felemban et al., 2000). Due to the presence of the adverse effects of this conventional treatment, the researchers are concerned to figure out an advanced therapeutic approach with potential benefits and insignificant side effects.

13.4 NATURAL COMPOUNDS: AN EMERGING FIELD OF TREATMENT IN PCOS

In the course of time, various therapeutic approaches have come into the picture to overcome the pathological features of PCOS. Nowadays, supplementation with natural compounds in treatment of this heterogeneous endocrine disorder has reached a turning point due to its benefits with lesser side effects over conventional therapy (Goswami et al., 2012). Natural molecules, representing different chemical compounds like vitamins, inositol, herbal extracts (mainly plant secondary metabolites) are either being used by themselves or in combination with the traditional medications against various pathological aspects of PCOS (Iervolino et al., 2021). Many studies have suggested, natural compounds like vitamin C (Olaniyan et al., 2019), omega-3 fatty acids (Khani et al., 2017), curcumin (Wang et al., 2017), catechin (Dou, Q.P., 2019), inositol (Nestler et al., 1999), cinnamon (Wang et al., 2007), liquorice (Yang et al., 2018) can be considered as an effective alternative treatment for PCOS.

13.4.1 Inositol

Inositols chemically defined as hexahydroxy cyclohexane have recently come into the spotlight as precursors of inositol 3-phosphate that is involved as a second messenger in various intracellular pathways (Bevilacqua & Bizzarri, 2018; Iervolino et al., 2021; Kamenov & Gateva, 2020). It comprises a family of nine stereoisomers, with myo-inositol (myo-ins) being the most abundant one (Bevilacqua & Bizzarri, 2018; Halsted, C.H., 1987). Myo-inositol is associated with cellular uptake of glucose by means of glucose transporters, on the other hand glucose remarkably regulates the uptake of inositol. Myo-inositol can be converted to another stereoisomer D-chiro-inositol (d-chiro-ins) by insulin-dependent epimerase (Cammarata et al., 1992; Iervolino et al., 2021). D-chiro-inositol also has a role in glucose metabolism being involved in synthesis and storage of glycogen (Facchinetti et al., 2020). Insulin resistance is one of the most crucial pathological aspects of PCOS due to which an altered ratio of myo-inositol and d-chiro-inositol can be seen in patients suffering from the disease (Laganà et al., 2018). D-chiro-ins supplementation in PCOS affected women reported to increase insulin action and decrease serum androgen (free testosterone) and triglyceride concentrations as well as blood pressure (Nestler et al., 1999). Treatment with myo-ins is reported to significantly enhance the rate of ovulation and maintain proper frequency of menstrual cycle in women affected by PCOS (Pundir et al., 2018). Many studies have also reported that therapy with inositols (with myo-ins and d-chiro-ins in an effective ratio of 40:1) contributes to maintenance of FSH ovarian signalling (Benelli et al., 2016), regulation in the levels of serum low density lipoprotein (LDL), high density lipoprotein (HDL) and triglycerides along with reduction in fasting and circulating insulin levels (Tabrizi et al., 2018). The available data suggests that a combination of myo-inositol and d-chiro-inositol is beneficial in regulating several pathological aspects of PCOS.

13.4.2 Vitamins

13.4.2.1 Vitamin C

Vitamin C or ascorbic acid is a micronutrient required in very small quantities due to its physiological role in healthy growth of cells and tissues (Iervolino et al., 2021). It acts as an effective antioxidant by scavenging aqueous peroxyl radicals as well as re-establishing the antioxidant properties of fat-soluble vitamin E. Vitamin C has a crucial role in synchronisation of menstrual cycle and ovarian functionality. Supplementation with vitamin C in pre-ovulatory phase is most likely to facilitate proper ovulation (Olaniyan et al., 2019). Ascorbic acid also contributes to collagen synthesis that is required for follicle growth as well as for repairing ovarian tissue in post-ovulation phase (Chen et al., 2005). Supplementation of vitamin C in combination with DHEA (Dehydroepiandrosterone) is reported to significantly decrease the levels of malondialdehyde, cytokines, and oestrogen and increase antioxidant and metabolic enzymes as compared to DHEA treated group (Olaniyan et al., 2019).

13.4.2.2 Vitamin D

Vitamin D is a steroid hormone well known for its role in maintaining the levels of calcium phosphate and bone mineralisation (Holick, 2007). Studies suggest that supplementation with vitamin D can reduce insulin resistance and hyperandrogenism and improve lipid metabolic parameters (such as LDL, HDL, triglycerides) in PCOS affected patients (Miao et al., 2020). According to the study of Asadi et al. (2014), vitamin D was found to significantly induce endometrial proliferation in PCOS women indicating its role in success of intrauterine insemination (IUI) (Asadi et al., 2014; Lerchbaum & Rabe, 2014). Supplementation of vitamin D in combination with calcium is reported to significantly decrease total testosterone and androstenedione levels (Pal et al., 2012).

13.4.2.3 Vitamin E

Vitamin E, also termed as tocopherol, is well known for its non-enzymatic antioxidant properties. It is widely used in the field of reproductive medicine due to its role in altering the adverse effects created by oxidative stress. Short-term supplementation with vitamin E can reverse antioxidant imbalance by preventing the action of phospholipase A and lipoxygenase and maintain normal physiological function of the reproductive system (Chen et al., 2020). Vitamin E in combination with coenzyme Q10 is reported to significantly decrease total testosterone level in serum if administered for 8 weeks (Izadi et al., 2019). Vitamin E may have a role in improving endometrial thickness in women having idiopathic infertility due to its anticoagulant and antioxidant properties (Cicek et al., 2012).

Although a growing number of clinical studies have tried to establish a connection between supplementation vitamins and PCOS, more studies are needed to find out the definite benefits of vitamin administration in PCOS management.

13.4.3 Herbal Supplements

Traditional herbal therapy has reached a turning point as a modern remedy for PCOS. These are naturally occurring substances with little or no chemical or industrial modification. Herbal supplements include several plant extracts or polyphenols (secondary metabolites produced by plants mainly for defence) such as flavonoids, phenolic acids, stilbenes etc with potentiating effect in treatment of PCOS (Goswami et al., 2012; Mihanfar et al., 2021).

13.4.3.1 Curcumin

Curcumin is a yellow-orange polyphenolic substance extracted from rhizomes of *Curcuma longa*, widely used as an Indian spice, turmeric. It has a wide range of biological effects like anti-inflammatory, antioxidant, hypoglycaemic, and antihyperlipidemic properties, thus having a beneficial role in treatments of diabetes, neurological disorders and cancer (Mihanfar et al., 2021). According to studies, curcumin has shown potential effects in treatment of PCOS symptomatically. It is reported to significantly decrease the levels of inflammatory mediators like IL-6, TNF-α and CRP (Mohammadi et al., 2017), upregulate the levels of antioxidant enzymes and scavenge ROS from ovarian tissue (Wang et al., 2017). Curcumin administration also showed to improve the hormone profile, lipid profile (Kafali et al., 2004) and the thickness of thecal layer in PCOS induced rats (Nabiuni et al., 2015).

13.4.3.2 Cinnamon

Cinnamon is a widely used spice and flavouring agent obtained from the inner bark of *Cinnamomum verum* also having antioxidant, anti-inflammatory, and anticancer properties (Dugoua et al., 2007; Sadeghi et al., 2019). Cinnamon extracts have potentiating effect in increasing insulin sensitivity by escalating phosphatidylinositol 3-kinase activity in the insulin signalling pathway (Wang et al., 2007) as well as significantly decreasing hormone profile (LH, FSH, progesterone, 17β-estradiol, testosterone, dihydrotestosterone) (Heibashy et al., 2013). Cinnamon supplementation has also been beneficial in improving menstrual cyclicity and ovarian morphology in PCOS affected patients. (Dou et al., 2018; Wang et al., 2007).

13.4.3.3 Quercetin

Quercetin is a flavonoid present in a wide range of fruits and vegetables such as tomato, onion, broccoli, lettuce, grapes, apple, and blueberries, having biological effects like antioxidant, anti-inflammatory, anti-apoptotic, anti-cancer properties (Darband et al., 2018). It has gained special attention in efficiently treating PCOS because of its beneficial effects on ovarian function and architecture. Quercetin can reverse insulin resistance by suppressing the toll-like receptor/NF-kB signalling pathway (Wang et al., 2017). It is also reported to have potentiating antioxidant effects and role in improving lipid profile showing parallel therapeutic effects with metformin (Jahan et al., 2018).

13.4.3.4 Resveratrol

Resveratrol is one of the most abundant polyphenolic compounds extracted from red wine, grapes, peanuts, and a number of medicinal herbs having antioxidant, anti-apoptotic, and anti-inflammatory properties (Pezzuto, J.M., 2019). According to the study of Ghowsi et al. (2018) with rats induced PCOS by testosterone enanthate, resveratrol can be used as a potential therapeutic to decrease ROS production and improve insulin sensitivity in ovarian tissue (Ghowsi et al., 2018). It is also reported to alter serum testosterone, LH, LH/FSH, TNF-α and tissue AMH levels and decrease the number of graafian follicles in rats with dehydroepiandrosterone-induced PCOS (Furat Rencber et al., 2018).

13.4.3.5 Rutin

Rutin (quercetin-3-rhamnosyl-glucoside) is another secondary metabolite (flavonol) derived from tea, onions, apples, and some other plants having a wide range of biological effects like antibacterial, antitumour, antioxidant, immunomodulatory properties (Ananth et al., 2015; Ganeshpurkar & Saluja, 2017). According to a study in rats with letrozole-induced PCOS, rutin administration was reported to increase antioxidant activities and maintain lipid profile. It also caused significant decrease in CRP levels and number of cystic follicles in ovarian tissue of treated rats (Jahan et al., 2016). Treatment with rutin was also found to significantly improve fertility and birth defects (Hu et al., 2017).

Studies suggest that there are a number of other medicinal plant extracts or plant secondary metabolites that can be used as potential therapeutics in treatment of PCOS. Table 13.1 summarises the natural compounds that can improve ovarian function, androgen excess, obesity, insulin resistance, blood lipids and inflammation in PCOS affected patients.

TABLE 13.1
Diverse Roles of Natural Compounds in Treatment of PCOS

Herbal Extracts	Source	Applications in Treatment of PCOS	Reference
Aloe Vera Gel	*Aloe barbadensis*	Reduce ovarian cyst, Restore ovarian steroid balance	Maharjan et al. (2010)
Welsh onion	*Allium fistulosum*	Normalise follicular growth, Alter hormonal imbalance and ovarian function	Lee et al. (2018)
Gymnema	*Gymnema sylvestre*	Modulate insulin sensitivity, Lower triglyceride levels, Maintain menstrual cyclicity	Stracquadanio et al. (2018); Sudhakar et al. (2018)
Flaxseed	*Linum usittassimum*	Lower serum testosterone levels, Reduce hirsutism, Improve menstrual cyclicity.	Jelodar et al. (2018); Nowak et al. (2007)
Peppermint	*Mentha piperita*	Necrosis in stromal mesenchymal cells, Hyperplasia of luminal epithelial cells.	Amoura et al. (2015)
Liquorice	*Glycyrrhiza glabra*	Antiandrogenic activity, Reduce hirsutism.	Yang et al. (2018)
Pomegranate	*Punica granatum*	Improve hormonal imbalance	Hossein et al. (2015)
Chaste berry	*Vitex agnus*	Normalise menstrual cycle	Westphal et al. (2006)
	Vitex negundo	Improve glucose tolerance, estrous cyclicity, LH/FSH ratio, Ovarian functionality	Kakadia et al. (2019)
Spearmint tea	*Mentha spicata*	Reduce serum testosterone levels, Increase LH and FSH, Reduce hirsutism.	Grant (2010)
Ginseng	*Panax ginseng* (Roots)	Improve ovulation.	Choi et al. (2020)
Milk thistle	*Silybum marianum*	Improve ovulation rate and insulin sensitivity.	Taher et al. (2012)
Chamomile	*Matricaria chamomilla*	Improve ovarian functionality, LH secretion.	Zangeneh et al (2010)
Puncture vine	*Tribulus terrestris*	Improve ovulation, Normalise estrous cyclicity and steroid hormone levels.	Saiyed et al. (2016)
Kacip Fatimah	*Labisia pumila*	Improve insulin sensitivity and lipid profile.	Mannerås et al. (2010)
Astragalus polysaccharide	*Astragalus* sp.	Increase insulin sensitivity, Improve lipid profile and serum androgen hormone status.	Goswami et al. (2012)
Gallic acid	Bearberry, Pomegranate, Gallnuts, Sumac	Decrease TNF-α, IL-1β, and IL-6 levels.	Mazloom et al. (2017)
Naringenin	Grapefruit	Decrease serum testosterone, estradiol levels, Improve ovarian architecture.	Hong et al. (2019)
Catechin	*Camellia sinensis*	Reduce TNF-α, IL-1β, and IL-6 levels.	Raheleh and Damoon (2022)
Lodh tree	*Symplocos racemosa*	Improve fertility and hormonal balance, Prevent ovarian cell dysfunction.	Jadhav et al. (2013)

13.5 BIOGENIC NANOPARTICLES IN TREATMENT OF POLYCYSTIC OVARIAN SYNDROME

Nanotechnology has effectively knocked on the door and is currently being experienced by us due to intense and comprehensive study by the scientific community (Chandrasekaran et al., 2019; Patil & Chandrasekaran, 2020). The characteristics of materials have altered and shown amazing optical, magnetic, and electrical qualities when they are controlled at the nanoscale level. Computers, batteries, fuel additives, solar cells, catalysts, electrochemical industries, defence, cosmetics, pharmaceuticals, food additives and packaging, biosensors, diagnostic imaging, vaccines, antimicrobial and chemotherapy, and drug delivery are just a few of the industries that use these kinds of special nano properties (Khan et al., 2019; Patil & Chandrasekaran, 2020). An emerging area called "green nanotechnology" uses a green chemistry method to build new nanoparticles. A novel avenue for nanoparticle synthesis employing natural reducing and stabilising agents is offered by biological processes. It uses no harmful chemicals or energy and is a cost-effective and green substitute for chemical and physical methods (Patil & Chandrasekaran, 2020).

Biogenic nanoparticles are synthesised either by a top-down approach, bulk materials with the help of photolithographic techniques, grinding, sputtering, and milling are made fine-grained (Meyers et al., 2006) or the bottom-up approach, by assembling atom-atom, molecule-molecule and cluster by techniques like self-assembly of monomer/polymer molecules, chemical or electrochemical nano structural precipitation, sol-gel processing, laser pyrolysis, chemical vapour deposition (CVD), plasma or flame spraying synthesis, and bio-assisted synthesis to generate nanoparticles (Decarolis et al., 2019; Dhand et al., 2015). There are physical (deposition, sputtering, ball milling, and plasma-based techniques) (Dhand et al., 2015), chemical (chemical reduction, pyrolysis, sol-gel method, microemulsion, polyol synthesis, hydrothermal synthesis, chemical vapour deposition) and biological methods of synthesising nanoparticles from the natural resources. Biogenic nanoparticles are widely used in therapeutics for their antimicrobial, anti-inflammatory, anti-cancer, anti-viral, and antioxidant properties (Patil & Chandrasekaran, 2020).

Nowadays, Biogenic Nanoparticles is an emerging field of study for the treatment of PCOS and a potential therapeutic approach with fewer side effects than allopathy and other conventional medications.

13.5.1 Biogenic Metal Nanoparticles

Using plant systems for the biosynthesis of nanoparticles is environmentally beneficial and has been regarded as a safe and dependable approach. Having a huge biodiversity of plants means that their potential for producing noble metal nanoparticles has not yet been completely investigated. A deep comprehension of the molecular pathways behind plant-mediated nanoparticle production will be needed to make this strategy economically viable compared to traditional techniques (Akhtar et al., 2013; Bhattacharya & Gupta, 2005; Mandal et al., 2006). For example, gold (Au) nanoparticles are extracted from leaves of aloe vera (Chandran et al., 2006), *Pelargonium graveolens* (Shankar et al., 2003), *Mangifera indica* (Phillip, 2010), *Azadirachta indica* (Thirumurugan et al., 2010) etc. and from the seeds of *Cuminum cyminum* (Krishnamurthy et al., 2011), Palladium (Pd) nanoparticles are extracted from the leaf extract of *Gardenia jasminoides* (Jia et al., 2009) etc. Several research studies show that there is a positive correlation between the increase of inflammatory cytokines such as Tumour Necrosis Factor-α (TNF-α), Interleukin-6 (IL-6), and Interleukin-8 (IL-8) and the occurrence of PCOS (Thathapudi et al., 2014). *C. zeylanicum* bark has been shown to have anti-inflammatory properties (Tung et al., 2008). Plant-derived silver nanoparticles (AgNPs) having immunomodulatory and antibacterial activities work against antibiotic-resistant bacteria and chronic inflammation. According to the study of Alwan and Al-Saeed (2023), AgNP shows effectiveness against inflammatory cytokines and decreases the release of IL-6, IL-8 and TNF-α in 30 days (Alwan and Al-Saeed, 2023). Biogenic metal nanoparticles are an emerging field of potential therapeutic development against PCOS.

13.5.2 Natural Compound-Based Nanoparticles

Several plant metabolites are antioxidant, anti-carcinogenic, and possess therapeutic potential. Nanocarriers are effective for the bioavailability and drug delivery of these metabolites (Maity et al., 2022). Flavonoid dietary polyphenols have pharmacological properties like antioxidants, anti-cancer, anti-inflammatory, etc. Nanoengineering of such components helps in drug delivery against the targeted disease (Amawi et al., 2017; Bunkar et al., 2019; Khan et al., 2021). Polysaccharide nanoparticles are produced from polysaccharides with the help of mechanisms like ionic or covalent crosslinking, aggregation and coacervation, have several beneficial roles as drug delivery, cosmetics as well as food (Plucinski et al., 2021).

13.5.2.1 Naringenin Nanoparticles

Several metabolic, cardio-metabolic and endocrine disorders influence the pathogenesis of PCOS. Naringenin is a flavonoid compound isolated from different plant species aiming to produce Poly lactic co glycolic acid (PLGA) nanoparticle drug delivery systems. Naringenin nanoparticles are not only found to alleviate ovarian dysfunction, but they also act as a therapeutic agent against PCOS as they regulate the sex hormone, enhancing and decreasing the intestinal barrier and disorders, respectively (Dashputre et al., 2023).

13.5.2.2 Chitosan Nanoparticles

Chitosan is a polysaccharide whose polymer has several biological and physicochemical properties, making it one of the unique drug delivery systems with low toxicity. Chitosan in the presence of sodium triphosphate nanoparticles and fennel seed extract have been proven to help in PCOS treatment and ovulation. Estradiol valerate treatment for 42 days increases the estradiol level and decreases the progesterone level. Chitosan, along with propolis, can reverse the negative effects of estradiol valerate in PCOS (Anbu & Venkatachalam, 2016; Bayrami et al., 2020; Hosseini et al., 2023).

13.5.3 Extracellular Vesicles

Extracellular vesicles (EVs) are nanoscale membrane-bound phospholipid vesicles secreted by the cells by outward blebbing. They are further classified into several categories based on structure, size and properties (Colombo et al., 2014; Raposo & Stoorvogel, 2013; Shao et al., 2018; Théry et al., 2002, 2009).

13.5.3.1 Exosomes

Exosomes are naturally occurring mediums of endocrine intracellular communication that help deliver target-specific cargos. With this phenomenon, exosomes can be designed for target-specific drug delivery (Braicu et al., 2015; Gupta & Pulliam, 2014). According to the study of Hang-Soo Park et al. (2023), the steroidogenesis gene is downregulated in H295R by MSC-derived exosomes, showing similar effectiveness with the whole secretome to decrease the androgen synthesis pathway (Figure 13.1). Intravenous and intra-ovarian injection methods showed effectiveness in bringing PCOS-treated mice back to fertile condition along with the modulation of PCOS-related obesity. Whereas intravenous injection is effective in controlling blood glucose, intraovarian injection helps in the re-establishment of ovarian function (Park et al., 2023). ADSC-EVs have been shown to increase the proliferation of cumulus cells by inhibiting apoptosis in PCOS patients by elevating the expression of miR-323-3p in exosomes (Liao et al., 2021; Zhao et al., 2019) (Figure 13.1). Follicle fluid-derived exosomes containing S100A9 lessen the steroid production and stimulate the inflammatory reactions by activating kappa-Beta of steroid human granulosa tumour cells (Li et al., 2020) (Figure 13.1). Exosomes containing CircLDLR induce the expression of CYP19A1 by hindering the expression of miR-1294 in KGN cells (Yuan et al., 2021). miRNAs in exosomes such as miR-424-5p and miR-323-3p can act as therapeutic targets

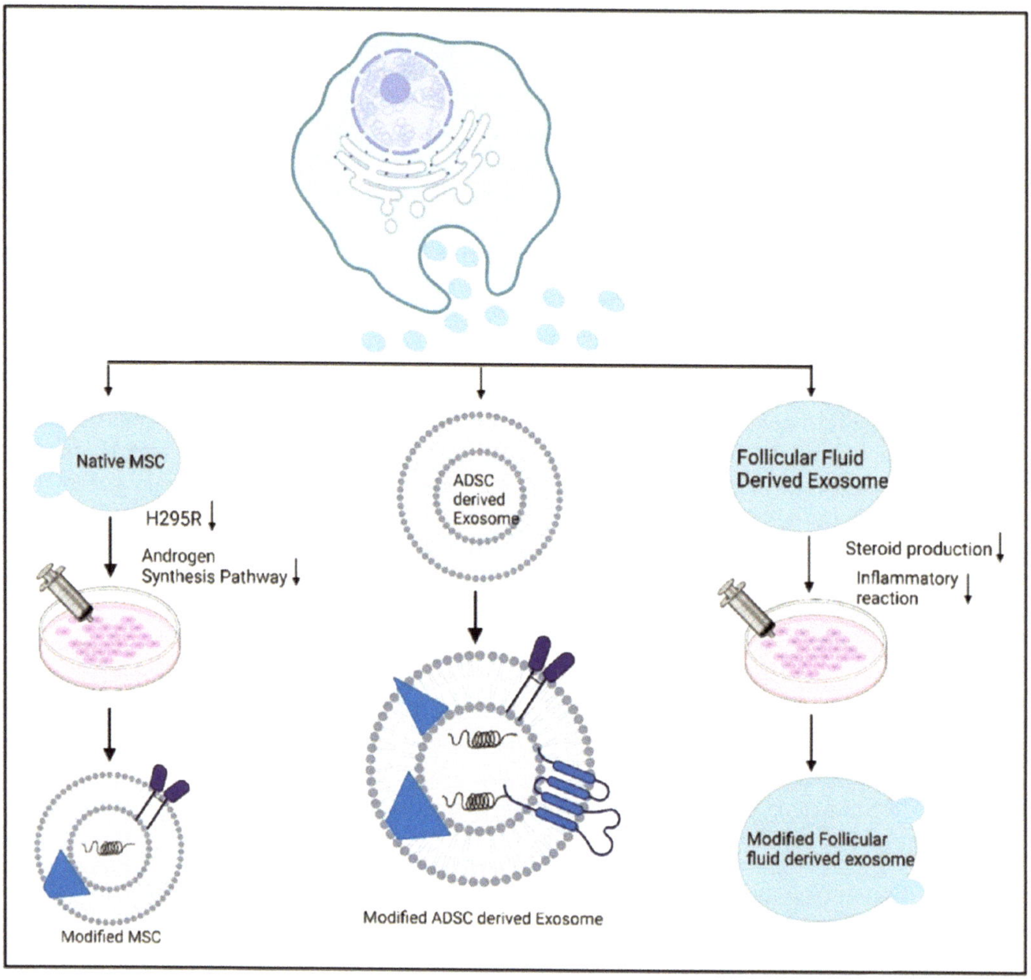

FIGURE 13.1 Modification of exosomes for the treatment of polycystic ovarian syndrome.

of PCOS (Zhang et al., 2024; Zhao et al., 2019). miRNA profiling between the groups of women with PCOS and the control group showed significant changes, and thereby, miRNA is used as a potential therapeutic approach in the treatment of PCOS. miRNAs are involved in different aspects of this heterogeneous disorder showing differential expressions, such as the proliferation of granulosa cells with the help of the expression of miR-376(upregulated in PCOS condition) by upregulation of PCNA (Zhang et al., 2014). miR-103/miR-155 express in blood or granulosa cells play a dual role as it increases the release of progesterone as well as reduces the release of estradiol, and is thought to be correlated with the rise of testosterone with its increase in PCOS (Abdalla et al., 2020; Murri et al., 2018).

13.5.3.2 Microvesicles (MVs)

As this area of research remains unclear as it is currently under review, little evidence shows microvesicles to have therapeutic potential against PCOS (Zhang et al., 2024). According to Faruk et al. (2018), M-MSCs-derived MVs have the ability to regain ovarian function in rat models.

13.5.3.3 Apoptotic Bodies

This area of research is still at its infancy and is thought to emerge as one of the potential approaches along with other nanoparticles in this field of PCOS therapeutics.

TABLE 13.2
Diverse Roles of Biogenic Nanoparticles in Treatment of PCOS

Biogenic Nanoparticles	Source	Therapeutic Potential in Treatment of PCOS	Reference
AgNP	*C. zeylanicum*	target and decrease inflammatory cytokines (IL6, IL8, and TNF-α) release in 30 days	Alwan and Al-Saeed (2023)
Naringenin	Plant flavonoid	Regulates the sex hormone, enhancing and decreasing the intestinal barrier	Dashputre et al. (2023)
Chitosan	Plant polysaccharide	Chitosan, along with propolis, can revert back the negative effects of estradiol valerate in PCOS	Anbu and Venkatachalam (2016)
MSC-derived Exosomes		Effectiveness with the whole secretome to decrease the androgen synthesis pathway	Park et al. (2023)
ADSC-secreted Extracellular Vesicles		Increase the proliferation of cumulus cells, inhibit the apoptosis in PCOS, and elevate miR-323-3 expression	Zhao et al. (2019)
Follicle fluid-derived exosomes		Lessen the steroid production and stimulate the inflammatory reactions	Li et al. (2020)
circLDLR		Induce the expression of CYP19A1	Yuan et al. (2021)
miR-103/miR-155		Increases the release of progesterone as well as reduces the release of estradiol	Murri et al. (2018)
miR-376		Granulosa cells proliferation	Zhang et al. (2014)
miR-518f-3p		Positive correlation with total free androgen in serum	Sørensen et al. (2016)
M-MSCs-derived MVs		Regain ovarian function in PCOS	Faruk et al. (2018)

Here, Table 13.2 summarises the overall potential role of biogenic nanoparticles in treatment of PCOS.

13.6 CONCLUSION

Since PCOS is one of the most prevalent heterogeneous lifestyle disorders affecting reproductive physiology of women, finding beneficial cures with lesser potential side effects is very essential. Emerging fields of curative approaches include natural compounds, biogenic nanoparticles, homoeopathy, acupuncture, reflexology etc. Changes in sedentary lifestyle can also be a regulatory factor in management of PCOS. This book chapter focuses on the two upcoming methods, natural compounds and biogenic nanoparticles, in targeting various pathological aspects of PCOS with minimal side effects.

Though natural compounds and biogenic nanoparticles have been showing promising effects as therapeutics of PCOS, the research on this area is still at its infancy. Therefore, for making these products commercially available, a wide range of research, clinical trials, meta-analysis is required in future.

REFERENCES

Abdalla, M., Deshmukh, H., Atkin, S.L. and Sathyapalan, T., 2020. miRNAs as A novel clinical biomarker and therapeutic targets in polycystic ovary syndrome (PCOS): A review. *Life sciences*, *259*, p.118174.

Akhtar, M.S., Panwar, J. and Yun, Y.S., 2013. Biogenic synthesis of metallic nanoparticles by plant extracts. *ACS sustainable chemistry & engineering*, *1*(6), pp.591–602.

Al-Took, S., Platt, R. and Tulandi, T., 1999. Adhesion-related small-bowel obstruction after gynecologic operations. *American Journal of obstetrics and gynecology*, *180*(2), pp.313–315.

Alwan, S.H. and Al-Saeed, M.H., 2023. Silver nanoparticles biofabricated from Cinnamomum zeylanicum reduce IL-6, IL-18, and TNF-α in female rats with polycystic ovarian syndrome. *International journal of fertility & sterility*, *17*(1), p.80.

Amawi, H., Ashby, C.R. and Tiwari, A.K., 2017. Cancer chemoprevention through dietary flavonoids: what's limiting?. *Chinese journal of cancer*, *36*, pp.1–13.

Amoura, M., Lotfy, Z.H., Neveen, E. and Khloud, A., 2015. Potential effects of Mentha piperita (peppermint) on letrozole-induced polycystic ovarian syndrome in female albino rat. *International journal*, 3(10), pp.211–226.

Ananth, D.A., Rameshkumar, A., Jeyadevi, R., Jagaadeeswari, S., Nagarajan, N., Renganathan, R. and Sivasudha, T., 2015. Antibacterial potential of rutin conjugated with thioglycolic acid capped cadmium telluride quantum dots (TGA-CdTe QDs). *Spectrochimica acta part a: Molecular and biomolecular spectroscopy*, *138*, pp.684–692.

Anbu, A.S. and Venkatachalam, P., 2016. Biological macromolecule cross linked TPP–chitosan complex: A novel nanohybrid for improved ovulatory activity against PCOS treatment in female rats. *RSC advances*, 6(97), pp.94301–13.

Andhalkar, S., Chaware, V. and Redasani, V., 2021. A review on medicinal plants of natural origin for treatment of polycystic ovarian syndrome (PCOS). *Asian journal of pharmaceutical research and development*, *9*(3), pp.76–81.

Asadi, M., Matin, N., Frootan, M., Mohamadpour, J., Qorbani, M. and Tanha, F.D., 2014. Vitamin D improves endometrial thickness in PCOS women who need intrauterine insemination: A randomized double-blind placebo-controlled trial. *Archives of gynecology and obstetrics*, *289*, pp.865–870.

Azziz, R. and Adashi, E.Y., 2016. Stein and leventhal: 80 years on. *American journal of obstetrics and gynecology*, *214*(2), pp.247–e1.

Badawy, A. and Elnashar, A., 2011. Treatment options for polycystic ovary syndrome. *International journal of women's health*, pp.25–35.

Balen, A., 2004. The pathophysiology of polycystic ovary syndrome: Trying to understand PCOS and its endocrinology. *Best practice & research clinical obstetrics & gynaecology*, *18*(5), pp.685–706.

Bayrami, A., Shirdel, A., Rahim Pouran, S., Mahmoudi, F., Habibi-Yangjeh, A. and Singh, R., et al., 2020. Co-regulative effects of chitosan-fennel seed extract system on the hormonal and biochemical factors involved in the polycystic ovarian syndrome. *Materials science & engineering c, materials for biological applications*, *117*, p.111351.

Benelli E, Del Ghianda S, Di Cosmo C, Tonacchera M. A Combined Therapy with Myo-Inositol and D-Chiro-Inositol Improves Endocrine Parameters and Insulin Resistance in PCOS Young Overweight Women. Int J Endocrinol. 2016;2016:3204083. doi:10.1155/2016/3204083.

Bevilacqua, A. and Bizzarri, M., 2018. Inositols in insulin signaling and glucose metabolism. *International journal of endocrinology*, *2018*.

Bhattacharya, D. and Gupta, R.K., 2005. Nanotechnology and potential of microorganisms. *Critical reviews in biotechnology*, *25*(4), pp.199–204.

Blay, S.L., Aguiar, J.V.A. and Passos, I.C., 2016. Polycystic ovary syndrome and mental disorders: A systematic review and exploratory meta-analysis. *Neuropsychiatric disease and treatment*, pp.2895–2903.

Braicu, C., Tomuleasa, C., Monroig, P., Cucuianu, A., Berindan-Neagoe, I. and Calin, G.A., 2015 Exosomes as divine messengers:Are they the hermes of modern molecular oncology?. *Cell death & differentiation*, *22*(1), pp.34–45.

Bunkar, N., Shandilya, R., Bhargava, A., Samarth, R.M., Tiwari, R., Mishra, D.K., Srivastava, R.K., Sharma, R.S., Lohiya, N.K. and Mishra, P.K., 2019. Nano-engineered flavonoids for cancer protection. *Frontiers in bioscience-Landmark*, *24*, pp.1097–1157.

Cammarata, P.R., Chen, H.Q., Yang, J. and Yorio, T., 1992. Modulation of myo-[3H] inositol uptake by glucose and sorbitol in cultured bovine lens epithelial cells. II. Characterization of high-and low-affinity myo-inositol transport sites. *Investigative ophthalmology & visual science*, *33*(13), pp.3572–3580.

Cena, H., Chiovato, L. and Nappi, R.E., 2020. Obesity, polycystic ovary syndrome, and infertility: A new avenue for GLP-1 receptor agonists. *The journal of clinical endocrinology & metabolism*, *105*(8), pp.e2695–e2709.

Chandran, S.P., Chaudhary, M., Pasricha, R., Ahmad, A. and Sastry, M., 2006. Synthesis of gold nanotriangles and silver nanoparticles using aloevera plant extract. *Biotechnology progress*, *22*(2), pp.577–583.

Chandrasekaran, R., Yadav, S.A. and Sivaperumal, S., 2019. Phytosynthesis and characterization of copper oxide nanoparticles using the aqueous extract of Beta vulgaris l and evaluation of their antibacterial and anticancer activities. *Journal of cluster science*, *6*(2019). https://doi.org/10.1007/s10876-019-01640-6

Chen, J., Guo, Q., Pei, Y.H., Ren, Q.L., Chi, L., Hu, R.K. and Tan, Y., 2020. Effect of A short-term vitamin e supplementation on oxidative stress in infertile PCOS women under ovulation induction: A retrospective cohort study. *BMC women's health*, *20*, pp.1–9.

Chen, Q., Espey, M.G., Krishna, M.C., Mitchell, J.B., Corpe, C.P., Buettner, G.R., Shacter, E. and Levine, M., 2005. Pharmacologic ascorbic acid concentrations selectively kill cancer cells: Action as a pro-drug to deliver hydrogen peroxide to tissues. *Proceedings of the national academy of sciences*, *102*(38), pp.13604–13609.

Choi, J.H., Jang, M., Kim, E.J., Lee, M.J., Park, K.S., Kim, S.H., In, J.G., Kwak, Y.S., Park, D.H., Cho, S.S. and Nah, S.Y., 2020. Korean red Ginseng alleviates dehydroepiandrosterone-induced polycystic ovarian syndrome in rats via its antiinflammatory and antioxidant activities. *Journal of ginseng research, 44*(6), pp.790–798.

Ciaraldi, T.M., El-Roeiy, A., Madar, Z., Reichart, D. and Olefsky, J.M., 1992. Yen SCC. Cellular mechanisms of insulin resistance in polycystic ovarian syndrome. *The journal of clinical endocrinology and metabolism, 75*, pp.577–583.

Ciaraldi, T.P., Morales, A.J., Hickman, M.G., Odom-Ford, R., Olefsky, J.M. and Yen, S.S., 1997. Cellular insulin resistance in adipocytes from obese polycystic ovary syndrome subjects involves adenosine modulation of insulin sensitivity. *The journal of clinical endocrinology and metabolism, 82*, pp.1421–1425.

Cicek, N., Eryilmaz, O.G., Sarikaya, E., Gulerman, C. and Genc, Y., 2012. Vitamin e effect on controlled ovarian stimulation of unexplained infertile women. *Journal of assisted reproduction and genetics, 29*, pp.325–328.

Colombo, M., Raposo, G. and Théry, C., 2014. Biogenesis, secretion, and intercellular interactions of exosomes and other extracellular vesicles. *The annual review of cell and developmental biology, 30*, pp.255–289.

Cumming, D.C., Reid, R.L., Quigley, M.E., Rebar, R.W. and Yen, S.S., 1984. Evidence for decreased endogenous dopamine and opioid inhibitory influences on LH secretion in polycystic ovary syndrome. *Clinical endocrinology, 20*, pp.643–648.

Dalkin, A.C., Haisenleder, D.J., Ortolano, G.A., Ellis, T.R. and Marshall, J.C., 1989. The frequency of GnRH stimulation differentially regulates gonadotropin subunit mRNA expression. *Endocrinology, 125*, pp.917–924.

Danforth, D.R., Elkind-Hirsch, K. and Hodgen, G., 1990. In-vivo and in-vitro modulation of GnRH metabolism by estradiol and progesterone. *Endocrinology, 127*, pp.319–324.

Darband, S.G., Kaviani, M., Yousefi, B., Sadighparvar, S., Pakdel, F.G., Attari, J.A., Mohebbi, I., Naderi, S. and Majidinia, M., 2018. Quercetin: A functional dietary flavonoid with potential chemo-preventive properties in colorectal cancer. *Journal of cellular physiology, 233*(9), pp.6544–6560.

Darroudi M., Sabouri Z., Oskuee R.K., Zak A.K., Kargar, H. and Abd Hamid, M.H., 2014. Green chemistry approach for the synthesis of ZnO nanopowders and their cytotoxic effects. Ceramics International. *40*(3), pp.4827–4831.

Dashputre, N.L., Laddha, U.D., Patil, S.B., Kadam, J.D. and Kshirsagar, S.J., 2023. An insight to development and in-vitro, ex-vivo, in-vivo study of naringenin nanoparticles against letrozole induced polycystic ovarian syndrome in female Wistar rats. *Journal of drug delivery science and technology, 90*, p.105129.

De Leo, V., Musacchio, M.C., Cappelli, V., Massaro, M.G., Morgante, G. and Petraglia, F.J.R.B., 2016. Genetic, hormonal And metabolic aspects of PCOS: An update. *Reproductive biology and endocrinology, 14*(1), pp.1–17.

Decarolis, D., Odarchenko, Y., Herbert, J.J., Qiu, C., Longo, A., and Beale, A.M., 2019. Identification of the key steps in the self-assembly of homogeneous gold metal nanoparticles produced using inverse micelles. *Physical chemistry chemical physics*, https://doi.org/10.1039/c9cp03473k.

Dhand, C., Dwivedi, N., Loh, X.J., Jie Ying, A.N., Verma, N.K., Beuerman, R.W., Lakshminarayanan, R. and Ramakrishna, S., 2015. Methods and strategies for the synthesis of diverse nanoparticles and their applications: A comprehensive overview. *RSC advances 5* (2015):105003–105037. https://doi.org/10.1039/c5ra19388e.

Dou, L., Zheng, Y., Li, L., Gui, X., Chen, Y., Yu, M. and Guo, Y., 2018. The effect of cinnamon on polycystic ovary syndrome in a mouse model. *Reproductive biology and endocrinology, 16*, pp.1–10.

Dou, Q.P., 2019. Tea in health and disease. *Nutrients, 11*(4), p.929.

Dugoua, J.J., Seely, D., Perri, D., Cooley, K., Forelli, T., Mills, E. and Koren, G., 2007. From type 2 diabetes to antioxidant activity: A systematic review of the safety and efficacy of common and cassia cinnamon bark. *Canadian journal of physiology and pharmacology, 85*(9), pp.837–847.

Dunaif, A., Xia, J., Book, C.-B., Schenker, E. and Tang, Z., 1995. Excessive insulin receptor phosphorylation in cultured fibroblasts and in skeletal muscle: A potential mechanism for insulin resistance in the polycystic ovary syndrome. *Journal of clinical investigation, 96*, pp.801–810.

Eleftheriadou, M., Michala, L., Stefanidis, K., Iliadis, I., Lykeridou, A. and Antsaklis, A., 2012. Exercise and sedentary habits among adolescents with PCOS. *Journal of pediatric and adolescent gynecology, 25*(3), pp.172–174.

Erickson, G.F., Magoffin, D.A. and Garza, V.G., et al., 1992. Granulosa cells of polycystic ovaries: Are they normal or abnormal? *Human reproductionl, 7*, pp. 293–299

Facchinetti, F., Appetecchia, M., Aragona, C., Bevilacqua, A., Bezerra Espinola, M.S., Bizzarri, M., D'Anna, R., Dewailly, D., Diamanti-Kandarakis, E., Hernández Marín, I. and Kamenov, Z.A., 2020. Experts' opinion on inositols in treating polycystic ovary syndrome and non-insulin dependent diabetes mellitus: A further help for human reproduction and beyond. *Expert opinion on drug metabolism & toxicology, 16*(3), pp.255–274.

Faruk, E.M., El desoky, R.E., El-Shazly, A.M. and Taha, N.M., 2018. Does exosomes derived bone marrow mesenchymal stem cells restore ovarian function by promoting stem cell survival on experimentally induced polycystic ovary in adult female albino rats? (Histological and immunohistochemical study). *Journal of stem cell research and therapy*, 8, p.1000442.

Felemban, A., Tan, S.L. and Tulandi, T., 2000. Laparoscopic treatment of polycystic ovaries with insulated needle cautery: A reappraisal. *Fertility and sterility*, *73*(2), pp.266–269.

Furat Rencber, S., Kurnaz Ozbek, S., Eraldemır, C., Sezer, Z., Kum, T., Ceylan, S. and Guzel, E., 2018. Effect of resveratrol And metformin on ovarian reserve And ultrastructure in PCOS: An experimental study. *Journal of ovarian research*, *11*, pp.1–16.

Ganeshpurkar, A. and Saluja, A.K., 2017. The pharmacological potential of rutin. *Saudi pharmaceutical journal*, *25*(2), pp.149–164.

Ghowsi, M., Khazali, H. and Sisakhtnezhad, S., 2018. The effect of resveratrol on oxidative stress in the liver and serum of a rat model of polycystic ovary syndrome: An experimental study. *International journal of reproductive biomedicine*, *16*(3), p.149.

Gilling-smith, C., Willis, D.S. and Beard, R.W., et al. 1994. Hypersecretion of androstenedione by isolated thecal cells from polycystic ovaries. *The journal of clinical endocrinology and metabolism*, *79*: 1158–1165.

Goswami, P.K., Khale, A. and Ogale, S., 2012. Natural remedies for polycystic ovarian syndrome (PCOS): A review. *International journal of pharmaceutical and phytopharmacological research*, *1*(6), pp.396–402.

Grant, P., 2010. Spearmint herbal tea has significant anti-androgen effects in polycystic ovarian syndrome. A randomized controlled trial. *Phytotherapy research: An international journal devoted to pharmacological and toxicological evaluation of natural product derivatives*, *24*(2), pp.186–188.

Grundy, S.M., 2002. Obesity, metabolic syndrome, and coronary atherosclerosis. *Circulation*, *105*(23), pp.2696–2698.

Gupta, A. and Pulliam, L., 2014. Exosomes as mediators of neuro-inflammation. *Journal of neuroinflammation*, *11*, p.68.

Halsted, C.H., 1987. Handbook of vitamins: Nutritional, biochemical, and clinical aspects: LJ machlin, ed. New York: Marcel dekker, 1984, 614. *The journal of nutrition*, *117*(2), pp.397–397.

Hartmann, G. and McEwen, B., 2019. Insulin resistance and polycystic ovary syndrome (PCOS): Part 2. Diet and nutritional medicine. *Journal of Australian traditional-medicine society*, *25*(1), p.18.

Heibashy, M., Mazen, G. and Shahin, M., 2013. Metabolic changes and hormonal disturbances in polycystic ovarian syndrome rats and the amelioration effects of metformin and/or cinnamon extraction. *American Journal of Science*, *9*(12), pp.54–62.

Hoeger, K.M., Dokras, A. and Piltonen, T., 2021. Update on PCOS: Consequences, challenges, and guiding treatment. *The journal of clinical endocrinology & metabolism*, *106*(3), pp.e1071–e1083.

Holick, M.F., 2007. Vitamin d deficiency. *New England journal of medicine*, *357*(3), pp.266–281.

Hong, Y., Yin, Y., Tan, Y., Hong, K. and Zhou, H., 2019. The flavanone, naringenin, modifies antioxidant and steroidogenic enzyme activity in a rat model of letrozole-induced polycystic ovary syndrome. *Medical science monitor: International medical journal of experimental and clinical research*, *25*, p.395.

Hossein, K.J., Leila, K.J., koukhdan Ebrahim, T., Nazanin, S.J., Farzad, P. and Elham, R., 2015. The effect of pomegranate juice extract on hormonal changes of female Wistar rats caused by polycystic ovarian syndrome. *Biomedical and pharmacology journal*, *8*(2), pp.971–977.

Hosseini, S.F., Khodaei, F., Hasansagha, Z., Khosravizadeh, H., Abdollahi, M. and Azaryan, E., 2023. Ameliorative Effect of Chitosan-Propolis Nanoparticles on the Estradiol Valerate-Induced PCOS in Rat.

Hu, T., Yuan, X., Ye, R., Zhou, H., Lin, J., Zhang, C., Zhang, H., Wei, G., Dong, M., Huang, Y. and Lim, W., 2017. Brown adipose tissue activation by rutin ameliorates polycystic ovary syndrome in rat. *The journal of nutritional biochemistry*, *47*, pp.21–28.

Iervolino, M., Lepore, E., Forte, G., Laganà, A.S., Buzzaccarini, G. and Unfer, V., 2021. Natural molecules in the management of polycystic ovary syndrome (PCOS): An Analytical review. *Nutrients*, *13*(5), p.1677.

Izadi, A., Ebrahimi, S., Shirazi, S., Taghizadeh, S., Parizad, M., Farzadi, L. and Gargari, B.P., 2019. Hormonal and metabolic effects of coenzyme Q10 and/or vitamin e in patients with polycystic ovary syndrome. *The journal of clinical endocrinology & metabolism*, *104*(2), pp.319–327.

Jadhav, M., Menon, S. and Shailajan, S., 2013. Anti-androgenic effect of Symplocos racemosa roxb. Against letrozole induced polycystic ovary using rat model. *Journal of coastal life medicine*, *1*(4), pp.309–314.

Jahan, S., Abid, A., Khalid, S., Afsar, T., Shaheen, G., Almajwal, A. and Razak, S., 2018. Therapeutic potentials of quercetin in management of polycystic ovarian syndrome using letrozole induced rat model: A histological and A biochemical study. *Journal of ovarian research*, *11*(1), pp.1–10.

Jahan, S., Munir, F., Razak, S., Mehboob, A., Ain, Q.U., Ullah, H., Afsar, T., Shaheen, G. and Almajwal, A., 2016. Ameliorative effects of rutin against metabolic, biochemical and hormonal disturbances in polycystic ovary syndrome in rats. *Journal of ovarian research*, *9*, pp.1–9.

Jelodar, G., Masoomi, S. and Rahmanifar, F., 2018. Hydroalcoholic extract of flaxseed improves polycystic ovary syndrome in a rat model. *Iranian Journal of basic medical sciences*, *21*(6), p.645.

Jia, L., Zhang, Q., Li, Q. and Song, H., 2009. The biosynthesis of palladium nanoparticles by antioxidants in Gardenia jasminoides ellis: Long lifetime nanocatalysts forp-nitrotoluene hydrogenation. *Nanotechnology*, *20*, p.385601.

Jonasson, J.M., Brismar, K., Sparen, P., Lambe, M., Nyren, O., Ostenson, C.G. and Ye, W., 2007. Fertility in women with type 1 diabetes: A population-based cohort study in Sweden. *Diabetes care*, *30*(9), pp.2271–2276.

Kafali, H., Iriadam, M., Ozardalı, I. and Demir, N., 2004. Letrozole-induced polycystic ovaries in the rat: A new model for cystic ovarian disease. *Archives of medical research*, *35*(2), pp.103–108.

Kahsar-Miller, M.D., Nixon, C., Boots, L.R., Go, R.C. and Azziz, R., 2001. Prevalence of polycystic ovary syndrome (PCOS) in first-degree relatives of patients with PCOS. *Fertility and sterility*, *75*(1), pp.53–58.

Kakadia, N., Patel, P., Deshpande, S. and Shah, G., 2019. Effect of Vitex negundo l. Seeds in letrozole induced polycystic ovarian syndrome. *Journal of traditional and complementary medicine*, *9*(4), pp.336–345.

Kamenov, Z. and Gateva, A., 2020. Inositols in PCOS. *Molecules*, *25*(23), p.5566.

Khan, H., Ullah, H., Martorell, M., Valdes, S.E., Belwal, T., Tejada, S., Sureda, A. and Kamal, M.A., 2021, February. Flavonoids nanoparticles in cancer: Treatment, prevention and clinical prospects. In *Seminars in cancer biology* (Vol. 69, pp. 200-211). Academic Press.

Khan, I., Saeed, K. and Khan, I., 2019. Nanoparticles: Properties, applications and toxicities. *The Arabian journal of chemistry*, *12*, pp. 908–931. https://doi.org/10.1016/j.arabjc.2017.05.011.

Khani, B., Mardanian, F. and Fesharaki, S.J., 2017. Omega-3 supplementation effects on polycystic ovary syndrome symptoms and metabolic syndrome. *Journal of research in medical sciences: The official journal of Isfahan university of medical sciences*, *22*.

Krishnamurthy, S., Sathishkumar, M., Lee, S.Y., Bae, M.A. and Yun, Y.S., 2011. Biosynthesis of au nanoparticles using cumin seed powder extract. *Journal of nanoscience and nanotechnology*, *11*(2), pp.1811–1814.

Laganà, A.S., Garzon, S., Casarin, J., Franchi, M. and Ghezzi, F., 2018. Inositol in polycystic ovary syndrome: Restoring fertility through a pathophysiology-based approach. *Trends in endocrinology & metabolism*, *29*(11), pp.768–780.

Lee, Y.H., Yang, H., Lee, S.R., Kwon, S.W., Hong, E.J. and Lee, H.W., 2018. Welsh Onion root (Allium fistulosum) restores ovarian functions from letrozole induced-polycystic ovary syndrome. *Nutrients*, *10*(10), p.1430.

Legro, R.S., Barnhart, H.X., Schlaff, W.D., Carr, B.R., Diamond, M.P., Carson, S.A., Steinkampf, M.P., Coutifaris, C., McGovern, P.G., Cataldo, N.A. and Gosman, G.G., 2007. Clomiphene, metformin, or both for infertility in the polycystic ovary syndrome. *New England journal of medicine*, *356*(6), pp.551–566.

Lerchbaum, E. and Rabe, T., 2014. Vitamin D and female fertility. *Current opinion in obstetrics and gynecology*, *26*(3), pp.145–150.

Li, H., Huang, X., Chang, X., Yao, J., He, Q., Shen, Z., Ji, Y. and Wang, K., 2020. S100-A9 protein in exosomes derived from follicular fluid promotes inflammation via activation of NF-κB pathway in polycystic ovary syndrome. *Journal of cellular and molecular medicine*, *24*, pp.114–125.

Liao, Z., Liu, C., Wang, L., Sui, C. and Zhang, H., 2021. Therapeutic role of mesenchymal stem cell-derived extracellular vesicles in female reproductive diseases. *Frontiers in endocrinology*, *12*, p.665645.

Liu, A.L., Liao, H.Q., Li, Z.L., Liu, J., Zhou, C.L. and Guo, Z.F., et al., 2016. New insights into mTOR signal pathways in ovarian-related diseases: Polycystic ovary syndrome and ovarian cancer. *The Asian Pacific journal of cancer prevention (APJCP)*, *17*(12), p.5087.

Maharjan, R., Nagar, P.S. and Nampoothiri, L., 2010. Effect of Aloe barbadensis mill. Formulation on letrozole induced polycystic ovarian syndrome rat model. *Journal of ayurveda and integrative medicine*, *1*(4), p.273.

Maity, S., Acharyya, A. and Chakraborti, A.S., 2022. Flavonoid-based polymeric nanoparticles: A promising approach for cancer and diabetes treatment. *European Polymer journal*, p.111455.

Mandal, D., Bolander, M.E., Mukhopadhyaya, D., Sarkar, G. and Mukherjee, P., 2006. The use of microorganism for the formation of metal nanoparticles and their applications. *Applied microbiology and biotechnology*, *69*(5), pp.485–492.

Mannerås, L., Fazliana, M., Nazaimoon, W.W., Lönn, M., Gu, H.F., Östenson, C.G. and Stener-Victorin, E., 2010. Beneficial metabolic effects of the Malaysian herb labisia pumila var. alata in a rat model of polycystic ovary syndrome. *Journal of ethnopharmacology*, *127*(2), pp.346–351.

Marsden, P.J., Murdoch, A. and Taylor, R., 1994. Severe impairment of insulin action in adipocytes from amenorrheic subjects with polycystic ovary syndrome. *Metabolism clinical and experimental, 43*, pp.1536–1542.
Mazloom, B.F., Edalatmanesh, M.A. and Hosseini, E., 2017. The effect of gallic acid on pituitary-ovary axis and oxidative stress in rat model of polycystic ovary syndrome. *Report of health care, 3*(1), pp.41–47.
Meyers, M.A., Mishra, A. and Benson, D.J., 2006. Mechanical properties of nanocrystalline materials. *Progress in materials science, 51*:427–556. https://doi.org/10.1016/j.pmatsci.2005.08.003
Miao, C.Y., Fang, X.J., Chen, Y. and Zhang, Q., 2020. Effect of vitamin D supplementation on polycystic ovary syndrome: A meta-analysis. *Experimental and therapeutic medicine, 19*(4), pp.2641–2649.
Mihanfar, A., Nouri, M., Roshangar, L. and Khadem-Ansari, M.H., 2021. Polyphenols: Natural compounds with promising potential in treating polycystic ovary syndrome. *Reproductive biology, 21*(2), p.100500.
Mohammadi, S., Kayedpoor, P., Karimzadeh-Bardei, L. and Nabiuni, M., 2017. The effect of curcumin on TNF-α, IL-6 and CRP expression in a model of polycystic ovary syndrome as an inflammation state. *Journal of reproduction & infertility, 18*(4), p.352.
Moran, L.J., Hutchison, S.K., Norman, R.J. and Teede, H.J., 2011. Lifestyle changes in women with polycystic ovary syndrome. *Cochrane database system review, 7*), pp.1–61.
Murri, M., Insenser, M., Fernandez-Duran, E., San-Millan, J.L., Luque-Ramirez, M. and Escobar-Morreale, H.F., 2018. Non-targeted profiling of circulating microRNAs in women with polycystic ovary syndrome (PCOS): Effects of obesity and sex hormones. *Metabolism, 86*, pp.49–60.
Nabiuni, M., Mohammadi, S., Kayedpoor, P. and Karimzadeh, L., 2015. The effect of curcumin on the estradiol valerate-induced polycystic ovary in rats. *Feyz journal of Kashan university of medical sciences, 18*(6).
Nasri, H. and Rafieian-Kopaei, M., 2014. Metformin: Current knowledge. *Journal of research in medical sciences: The official journal of Isfahan university of medical sciences, 19*(7), p.658.
Nestler, J.E., Jakubowicz, D.J., Reamer, P., Gunn, R.D. and Allan, G., 1999. Ovulatory and metabolic effects of d-chiro-inositol in the polycystic ovary syndrome. *New England journal of medicine, 340*(17), pp.1314–1320.
Nowak, D.A., Snyder, D.C., Brown, A.J. and Demark-Wahnefried, W., 2007. The effect of flaxseed supplementation on hormonal levels associated with polycystic ovarian syndrome: A case study. *Current topics in nutraceutical research, 5(4), p.*177.
Olaniyan, O.T., Femi, A., Iliya, G., Ayobami, D., Godam, E., Olugbenga, E., Bamidele, O. and Mali, P.C., 2019. Vitamin C suppresses ovarian pathophysiology in experimental polycystic ovarian syndrome. *Pathophysiology, 26*(3–4), pp.331–341.
Ovalle, F. and Azziz, R., 2002. Insulin resistance, polycystic ovary syndrome, and type 2 diabetes mellitus. *Fertility and sterility, 77*(6), pp.1095–1105.
Pal, L., Berry, A., Coraluzzi, L., Kustan, E., Danton, C., Shaw, J. and Taylor, H., 2012. Therapeutic implications of vitamin d and calcium in overweight women with polycystic ovary syndrome. *Gynecological endocrinology, 28*(12), pp.965–968.
Park, H.S., Cetin, E., Siblini, H., Seok, J., Alkelani, H., Alkhrait, S., Liakath Ali, F., Mousaei Ghasroldasht, M., Beckman, A. and Al-Hendy, A., 2023. Therapeutic potential of mesenchymal stem cell-derived extracellular vesicles to treat PCOS. *International journal of molecular sciences, 24*(13), p.11151.
Pasquali, R., Stener-Victorin, E., Yildiz, B.O., Duleba, A.J., Hoeger, K., Mason, H., Homburg, R., Hickey, T., Franks, S., Tapanainen, J.S. and Balen, A., 2011. PCOS forum: Research in polycystic ovary syndrome today and tomorrow. *Clinical endocrinology, 74*(4), pp.424–433.
Patel, S., 2018. Polycystic ovary syndrome (PCOS), an inflammatory, systemic, lifestyle endocrinopathy. *The journal of steroid biochemistry and molecular biology, 182*, pp.27–36.
Patel, S.M. and Nestler, J.E., 2006. Fertility in polycystic ovary syndrome. *Endocrinology and metabolism clinics, 35*(1), pp.137–155.
Patil, S. and Chandrasekaran, R., 2020. Biogenic nanoparticles: A comprehensive perspective in synthesis, characterization, application and its challenges. *Journal of genetic engineering and biotechnology, 18*, pp.1–23.
Pezzuto, J.M., 2019. Resveratrol: Twenty years of growth, development and controversy. *Biomolecules & therapeutics, 27*(1), p.1.
Phillip, D., 2010. Rapid green synthesis of spherical gold nanoparticles using Mangifera indica leaf. *Spectrochim acta, part a, 77*(4), pp.807–810.
Plucinski, A., Lyu, Z. and Schmidt, B.V., 2021. Polysaccharide nanoparticles: From fabrication to applications. *Journal of materials chemistry b, 9*(35), pp.7030–7062.
Pundir, J., Psaroudakis, D., Savnur, P., Bhide, P., Sabatini, L., Teede, H., Coomarasamy, A. and Thangaratinam, S., 2018. Inositol treatment of anovulation in women with polycystic ovary syndrome: A meta-analysis of randomised trials. *BJOG: An international journal of obstetrics & gynaecology, 125*(3), pp.299–308.

Quigley, M., Rakoff, J. and Yen, S.S.C., 1981. Increased luteinising hormone sensitivity to dopamine inhibition in the polycystic ovary syndrome. *The journal of clinical endocrinology and metabolism*, *52*, p.231.

Raheleh, R. and Damoon, S., 2022. Effect of catechin on serum levels of inflammatory cytokines, antioxidant enzymes activity and DNA oxidative damage of ovarian tissue in polycystic ovarian syndrome rat model. *Pars journal of medical sciences*, *15*(1), pp.23–35.

Raposo, G. and Stoorvogel, W., 2013. Extracellular vesicles: Exosomes,Microvesicles, and friends. *Journal of cell biology*, *200*, pp.373–383.

Rosenfield, R.L. and Ehrmann, D.A., 2016. The pathogenesis of polycystic ovary syndrome (PCOS): The hypothesis of PCOS as functional ovarian hyperandrogenism revisited. *Endocrine reviews*, *37*(5), pp.467–520.

Sadeghi, S., Davoodvandi, A., Pourhanifeh, M.H., Sharifi, N., ArefNezhad, R., Sahebnasagh, R., Moghadam, S.A., Sahebkar, A. and Mirzaei, H., 2019. Anti-cancer effects of cinnamon: Insights into its apoptosis effects. *European journal of medicinal chemistry*, *178*, pp.131–140.

Saiyed, A., Jahan, N., Makbul, S.A.A., Ansari, M., Bano, H. and Habib, S.H., 2016. Effect of combination of *Withania somnifera Dunal and Tribulus* terrestris linn on letrozole induced polycystic ovarian syndrome in rats. *Integrative medicine research*, *5*(4), pp.293–300.

Sam, S. and Dunaif, A., 2003. Polycystic ovary Syndrome: Syndrome XX? *Trends in endocrinology & metabolism*, *14*(8), pp.365–370.

Samdal, O., Tynjala, J., Roberts, C., Sallis, J. F., Villberg, J., & Wold, B. (2007). Trends in vigorous physical activity and TV watching of adolescents from 1986 to 2002 in seven European Countries. *European journal of public health*, 17(3), 242–248. https://doi.org/10.1093/eurpub/ckl245

Shankar, S.S., Ahmad, A., Pasricha, R. and Sastry, M., 2003. Bioreduction of chloroaurate ions by geranium leaves and its endophytic fungus yields gold nanoparticles of different shapes. *Journal of materials chemistry*, *13*(7), pp.1822–1826.

Shao, H., Im, H., Castro, C.M., Breakefield, X., Weissleder, R. and Lee, H., 2018. New technologies for analysis of extracellular vesicles. *Chemical reviews*, *118*(4), pp.1917–1950.

Slayden, S.M., Moran, C., Sams, W.M. Jr, Boots, L.R. and Azziz, R., 2001. Hyperandrogenemia in patients presenting with acne. *Fertility and sterility*, *75*, pp.889–892.

Sørensen, A.E., Wissing, M.L., Englund, A.L.M. and Dalgaard, L.T., 2016. MicroRNA species in follicular fluid associating with polycystic ovary syndrome and related intermediary phenotypes. *The journal of clinical endocrinology & metabolism*, *101*(4), pp.1579–1589.

Stein, I.F. and Leventhal, M.L., 1935. Amenorrhea associated with bilateral polycystic ovaries. *American journal of obstetrics & gynecology*, *29*, pp.181–191.

Stracquadanio, M., Ciotta, L. and Palumbo, M.A., 2018. Effects of myo-inositol, gymnemic acid, and l-methylfolate in polycystic ovary syndrome patients. *Gynecological endocrinology*, *34*(6), pp.495–501.

Strauss, J.F. III, 2003. Some new thoughts on the pathophysiology and genetics of polycystic ovary syndrome. *Annals of the new York academy of sciences*, *997*, pp.42–48. https://doi.org/10.1196/annals.1290.005

Sudhakar, P., Suganeswari, M., Pushkalai, P.S. and Haripriya, S., 2018. Regulation of estrous cycle using combination of Gymnema sylvestre and *Pergularia* daemia in estradiol valerate induced PCOS rats. *Asian journal of research in pharmaceutical science*, *8*(1), pp.4–8.

Tabrizi, R., Ostadmohammadi, V., Lankarani, K.B., Peymani, P., Akbari, M., Kolahdooz, F. and Asemi, Z., 2018. The effects of inositol supplementation on lipid profiles among patients with metabolic diseases: A systematic review and meta-analysis of randomized controlled trials. *Lipids in health and disease*, *17*, pp.1–11.

Taher, M.D.A., Atia, Y.A. and Amin, M.K., 2012. Improving an ovulation rate in women with polycystic ovary syndrome by using silymarin. *Global journal of medical research*, *12*(6), pp.15–21.

Thathapudi, S., Kodati, V., Erukkambattu, J., Katragadda, A., Addepally, U. and Hasan, Q., 2014. Tumor necrosis factor-alpha and polycystic ovarian syndrome: A clinical, biochemical, and molecular genetic study. *Genetic testing and molecular biomarkers*, *18*(9), pp.605–609.

Théry, C., Ostrowski, M. and Segura, E., 2009. Membrane vesicles as conveyors of immune responses. *Nature reviews immunology*, *9*, pp.581–593.

Théry, C., Zitvogel, L. and Amigorena, S., 2002. Exosomes: Composition, biogenesis and function. *Nature reviews immunology*, 2, pp.569–579.

Thirumurugan, A., Jiflin, G.J., Rajagomathi, G., Tomy, N.A., Ramachandran, S. and Jaiganesh, R., 2010. Biotechnological synthesis of gold nanoparticles of Azadirachta indica leaf extract. *International journal of biological technology*, *1*(1), pp.75–77.

Tung, Y.T., Chua, M.T., Wang, S.Y. and Chang, S.T., 2008. Anti-inflammation activities of essential oil and its constituents from indigenous cinnamon (Cinnamomum osmophloeum) twigs. *Bioresource technology*, *99*(9), pp.3908–3913.

Wang, C.F., Lasley, B.L., Lein, A. and Yen, S.S.C., 1976. The functional changes of the pituitary gonadotrophs during the menstrual cycle. *The journal of clinical endocrinology and metabolism*, 42, pp.718–728.

Wang, J.G., Anderson, R.A., Graham, G.M. III, Chu, M.C., Sauer, M.V., Guarnaccia, M.M. and Lobo, R.A., 2007. The effect of cinnamon extract on insulin resistance parameters in polycystic ovary syndrome: A pilot study. *Fertility and sterility*, *88*(1), pp.240–243.

Wang, X.N., Zhang, C.J., Diao, H.L. and Zhang, Y., 2017. Protective effects of curcumin against sodium arsenite-induced ovarian oxidative injury in a mouse model. *Chinese medical journal*, *130*(9), pp.1026–1032.

Wang, Z., Zhai, D., Zhang, D., Bai, L., Yao, R., Yu, J., Cheng, W. and Yu, C., 2017. Quercetin decreases insulin resistance in a polycystic ovary syndrome rat model by improving inflammatory microenvironment. *Reproductive sciences*, *24*(5), pp.682–690.

Westphal, L.M., Polan, M.L. and Trant, A.S., 2006. Double-blind, placebo-controlled study of fertilityblend: A nutritional supplement for improving fertility in women. *Clinical and experimental obstetrics & gynecology*, *33*(4), pp.205–208.

Yang, H., Kim, H.J., Pyun, B.J. and Lee, H.W., 2018. Licorice ethanol extract improves symptoms of polycytic ovary syndrome in letrozole-induced female rats. *Integrative medicine research*, *7*(3), pp.264–270.

Yuan, D., Luo, J., Sun, Y., Hao, L., Zheng, J. and Yang, Z., 2021. PCOS follicular fluid derived exosomal miR-424-5p induces granulosa cells senescence by targeting CDCA4 expression. *Cell signal*, *85*, p.110030.

Zangeneh, F.Z., Minaee, B., Amirzargar, A., Ahangarpour, A. and Mousavizadeh, K., 2010. Effects of chamomile extract on biochemical and clinical parameters in a rat model of polycystic ovary syndrome. *Journal of reproduction & infertility*, *11*(3), p.169.

Zhang, H., Jiang, X., Zhang, Y., Xu, B., Hua, J. and Ma, T., et al., 2014. microRNA 376a regulates follicle assembly by targeting Pcna in fetal and neonatal mouse ovaries, *Reproduction*, *148* (1), pp.43–54.

Zhang, Y., Zhao, J., Han, L., Zhang, Z., Wang, C., Long, W., Meng, K. and Wang, X., 2024. Research progress of extracellular vesicles in the treatment of ovarian diseases. *Experimental and therapeutic medicine*, *27*(1), pp.1–14.

Zhao, Y., Tao, M., Wei, M., Du, S., Wang, H. and Wang, X., 2019. Mesenchymal stem cells derived exosomal miR-323-3p promotes proliferation and inhibits apoptosis of cumulus cells in polycystic ovary syndrome (PCOS). *Artificial cells, nanomedicine, and biotechnology*, *47*, pp.3804–3813.

14 The Hallmarks of Amyloidosis
Focus on Treatment Strategies Using Natural Products

Saranya Udayakumar, Agnishwar Girigoswami, Shangavy P, J Jashwanth Raj, and Koyeli Girigoswami
Medical Bionanotechnology, Faculty of Allied Health Sciences, Chettinad Hospital and Research Institute, Chettinad Academy of Research and Education
Chettinad Health City, Rajiv Gandhi Salai
Kelambakkam, Chengalpattu District, Chennai, India

14.1 INTRODUCTION

Amyloids are insoluble fibrous protein aggregates sharing specific structural traits. Abnormal accumulation of amyloid in organs may lead to amyloidosis and may play a role in various neurodegenerative diseases. The amyloids have common characteristics, like (1) all stain with Congo red and demonstrate red/green color under polarized light, (2) all contain predominant beta-pleated sheet secondary structure, and (3) all give enhanced fluorescence when stained with thioflavin T dye (Metkar et al. 2017b; Metkar et al. 2017a). The amyloid diseases include Alzheimer's disease, Down's syndrome, and Hereditary cerebral hemorrhage; where specific amyloid is referred to as β-amyloid protein or Aβ (Murphy and LeVine 2010). It is also associated with chronic inflammation and with the prion diseases including Creutzfeldt-Jakob disease, Gerstmann-Straussler syndrome, kuru, and animal scrapie wherein specific amyloid is referred to as PrP amyloid (Prusiner 1998). In type II diabetes also insulin amyloids are deposited at the site of repeated subcutaneous injection with insulin. Thus, any agent that can dissociate these amyloids can act as a possible therapeutic agent (Metkar et al. 2023; Sipe et al. 2016).

Various cellular and extracellular components are thought to interact with one another to form amyloid fibrils. However, the exact mechanism is not fully understood. Amyloidosis manifests differently depending on which organs and tissues are affected. Heart failure, arrhythmias, and cardiomyopathy are the most common symptoms of systemic amyloidosis. A nephrotic syndrome or renal failure can also result from involvement of the renal system. Proteinuria is also common. Weight loss, diarrhea, and malabsorption can be caused by gastrointestinal involvement. It is possible to suffer from autonomic dysfunction, peripheral neuropathy, and cognitive impairment if you have neurological involvement (Picken 2020). Amyloidosis is treated according to its underlying cause and extent of organ damage. Imaging techniques are available for the detection of amyloid in Alzheimer's disease (Chitradevi et al. 2021; Subramanian et al. 2020). There are several methods of managing cardiac and renal symptoms, including ACE inhibitors, diuretics, and beta-blockers, which are used to treat them. Amyloidosis that is caused by multiple myeloma as well as amyloidosis related to primary amyloidosis can be treated with chemotherapy. Hereditary amyloidosis associated with mutations in transthyretin is treated with liver transplantation in the case of the disease. In addition to these treatments, small molecule inhibitors, monoclonal antibodies, and gene therapy are also being explored as potential therapies.

DOI: 10.1201/9781003486237-14

Symptoms of amyloidosis are managed, disease progression is slowed, and underlying causes are addressed even though there is no cure. A brief overview of the signs and symptoms of amyloidosis is provided here, as well as a list of some natural treatments that may be effective.

14.2 HALLMARKS OF AMYLOIDOSIS

14.2.1 Amyloid Fibril Formation

As a result of protein misfolding, normally soluble proteins become insoluble fibrillar aggregates called amyloid fibrils. Although many amyloidogenic proteins have common characteristics, the exact mechanisms underlying their misfolding can vary based on the protein involved. In summary, here are the mechanisms involved in protein misfolding in amyloidosis (Han et al. 2017; Jiménez et al. 2002; Makin and Serpell 2005; Toyama and Weissman 2011): (i) Structure of native proteins: A protein's normal function depends on its specific three-dimensional structure. As a result of the misfolding process, hydrophobic regions normally buried in the native structure are exposed. (ii) Modifications in conformation: Mutations, post-translational modifications, or environmental stress can alter the native state of proteins, causing conformational changes. Typically buried within the protein's core, hydrophobic amino acid residues may be exposed during conformational changes. (iii) Creating oligomers: Amyloid fibril formation begins with the formation of small, soluble oligomers from misfolded proteins. It is known that oligomers are toxic to cells and that they are intermediates in the process of forming fibrils (Picken 2020). (iv) Structural nucleation: An important nucleus or seed structure is formed during the nucleation phase of misfolded proteins. In order to add more misfolded proteins, this nucleus serves as a template. It is often thought that nucleation is the rate-limiting step in the formation of amyloid fibrils. (v) Elongation of the fibrils: Upon the formation of the nucleus, misfolded proteins quickly add themselves to it, causing the fibril to elongate as a result. Fibrils tend to have a cross β-sheet structure in which the strands of the fibril line up perpendicularly to the axis of the fibril itself. (vi) The process of aggregation and maturation: Through lateral association, fibrils grow into larger aggregates. Amyloid fibrils are composed of long, unbranched fibrils when the aggregates mature. (vii) Deposition and damage to the tissues: Cellular function and tissue architecture can be disrupted by the formation of mature amyloid fibrils in tissues and organs. Amyloidosis is characterized by amyloid fibril deposition, which can cause organ dysfunction as mentioned earlier. (viii) Toxicity to cells: It is possible for cells to become toxic as a result of misfolded oligomers and intermediate structures. An interference with the function of mitochondria, membrane integrity, and signal transduction can lead to cellular dysfunction and death. (ix) Chaperones and clearing mechanisms: To prevent misfolded proteins from aggregating, molecular chaperones assist in protein folding. In order to remove misfolded proteins from the cell, a number of mechanisms are present, including ubiquitin-proteasomes and autophagy. There are many different types of amyloidosis, all of which can result in protein misfolding due to different triggers and pathways. The misfolding process can be triggered by a number of factors, such as environmental factors, genetic mutations (Husain et al. 2019), and the aging process. For the development of therapeutic strategies to prevent or treat diseases as devastating as amyloidosis, understanding the mechanisms of protein misfolding is crucial.

14.2.2 Disease Heterogenicity

Amyloidosis is frequently classified into two types: AL (immunoglobulin light chain) and ATTR (transthyretin), and each has its own clinical features and underlying cause (Ciccone et al. 2020; Dungu et al. 2012; Hasib Sidiqi and Gertz 2021). As a result of AL amyloidosis, abnormal immunoglobulin light chains (AL) are depositions of amyloid fibrils that are deposited on the surface. Typically, it is associated with a plasma cell dyscrasia, such as multiple myeloma or monoclonal gammopathy with undetermined significance (MGUS). On the other hand, the cause of ATTR amyloidosis is the deposition of transthyretin (TTR) protein, either the wild-type protein (ATTRwt) or the mutant protein (ATTRm). In addition to affecting the peripheral nervous system (ATTR polyneuropathy), ATTR can affect the heart (ATTR cardiomyopathy).

14.2.3 Organ Involvement

Several organs, such as kidneys, liver, heart, and nervous system, become dysfunctional because of the deposition of amyloid fibrils (Merlini, Seldin, and Gertz 2011; Figure 14.1). (i) Cardiac amyloidosis (amyloidosis in the heart): There are two types of amyloidosis that affect the heart: light chain (AL) and transthyretin (ATTR). Cardiomyopathy, arrhythmias, and heart failure are symptoms of cardiac involvement. In order to diagnose this condition, echocardiography, cardiac magnetic resonance imaging, and biopsy may be used (Dungu et al. 2012). (ii) Renal amyloidosis (kidney amyloidosis): Renal involvement is most commonly caused by AL amyloidosis, but ATTR and other forms can also occur. Renal failure, proteinuria, and nephrotic syndrome are the most common symptoms. Laboratory tests to determine kidney function and a renal biopsy are used for diagnosis (Dember 2006; Scarpioni et al. 2016). (iii) Neuropathic amyloidosis (amyloidosis of the nervous system): Peripheral and autonomic neuropathy are often associated with ATTR amyloidosis. It is characterized by numbness, tingling, autonomic dysfunction, and sometimes motor neuropathy as well. An electromyography, nerve conduction study, and genetic test are used for diagnosis (Pinney and Lachmann 2012). (iv) Gastrointestinal amyloidosis (amyloidosis of the gastrointestinal tract): The gastrointestinal tract (GI) can be affected by AL and ATTR amyloidosis. Abdominal pain, malabsorption, diarrhea, and weight loss can be symptoms of GI involvement. Imaging studies, endoscopy, and biopsy are used for diagnosis. (v) Hepatic amyloidosis (amyloidosis of the liver): Amyloidosis of the liver can be of two types, AL and ATTR. An enlarged liver, increased liver enzyme levels, and hepatomegaly are symptoms of this condition. Imaging studies and liver biopsy are used in the diagnosis (Merlini et al. 2011). (vi) Spleen amyloidosis: Amyloid deposition in the spleen can occur with AL and

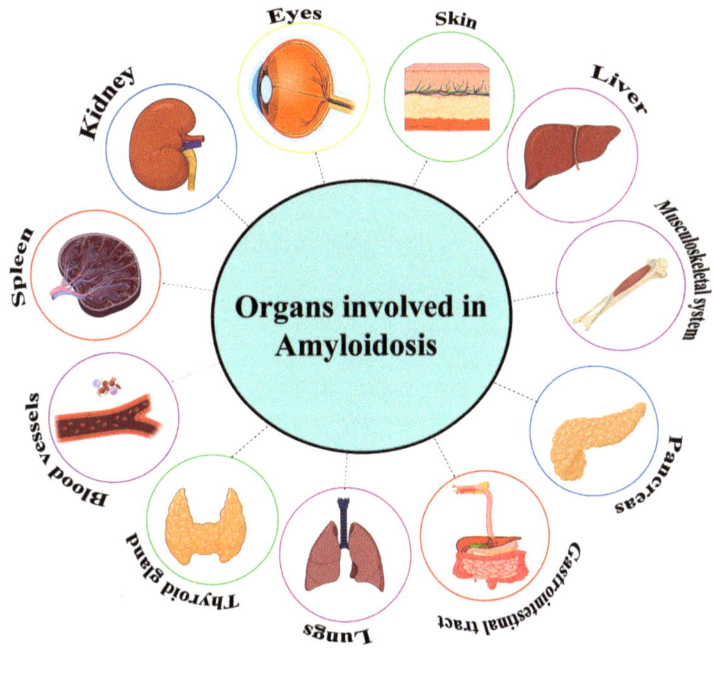

FIGURE 14.1 The different types of organs involved in amyloidosis.

ATTR amyloidosis. Splenomegaly (enlarged spleen) is one of the symptoms. Imaging studies and biopsy are used to diagnose the disease. (vii) Amyloid arthropathy (amyloidosis of musculoskeletal system): Joints, tendons, and ligaments can be affected by AL and ATTR amyloidosis. Pain in the joints, stiffness of the joints and muscles, and limited range of motion are some of the symptoms. A joint aspiration and imaging studies are used to diagnose the condition (Accardi et al. 2018). (viii) Cutaneous amyloidosis (skin amyloidosis): Amyloidosis of the AL and ATTR types can involve the skin. It is characterized by lesions on the skin, purpura, and easy bruising. Biopsy of the skin confirms the diagnosis. (ix) Pulmonary amyloidosis (lung amyloidosis): Lung amyloidosis can be caused by AL or ATTR amyloidosis. A pleural effusion, shortness of breath, and cough are common symptoms. A biopsy as well as imaging studies and pulmonary function tests can be used in the diagnosis. (x) Thyroid amyloidosis: Amyloidosis of the thyroid gland is classified as AL and ATTR. The thyroid is enlarged and dysfunctional. Imaging studies and biopsy are used to diagnose the disease. (xi) Vascular amyloidosis (amyloidosis of blood vessels): A blood vessel can be affected by AL or ATTR amyloidosis. It may cause bleeding disorders and vascular fragility. An imaging study and a biopsy are used to diagnose the disease. (xii) Ocular amyloidosis (amyloidosis of the eyes): Amyloid deposits can develop in the eyes due to AL and ATTR amyloidosis. As far as the ocular manifestations are concerned, depositions on the cornea and other parts of the eye are the most common. It is diagnosed by ophthalmic examination and biopsies. (xiii) Endocrine amyloidosis (amyloidosis of the endocrine glands): In addition to AL amyloidosis, ATTR amyloidosis can also cause organ damage to the endocrine system. An imbalance in hormones and an imbalance in the endocrine system are symptoms. An imaging study and a biopsy are used for diagnosis (Ihne et al. 2020). Table 14.1 summarizes the symptoms and diagnosis of different types of amyloidosis occurring in the various organs.

TABLE 14.1
The Organs Involved, Symptoms, and Diagnosis of Various Types of Amyloidosis

S. No.	Organs Involved in Amyloidosis	Symptoms	Diagnosis
1.	Cardiac amyloidosis	Cardiomyopathy, arrhythmias, and heart failure	Echocardiography, cardiac magnetic resonance imaging, and biopsy
2.	Renal amyloidosis	Renal failure, proteinuria, and nephrotic syndrome	Renal biopsy
3.	Neuropathic amyloidosis	numbness, tingling, autonomic dysfunction	Electromyography, nerve conduction study, and genetic test
4.	Gastrointestinal amyloidosis	Abdominal pain, malabsorption, diarrhea, and weight loss	Endoscopy, and biopsy
5.	Hepatic amyloidosis	Enlarged liver, increased liver enzyme levels, and hepatomegaly	Liver biopsy and Imaging studies
6.	Spleen amyloidosis	Splenomegaly	Imaging studies and biopsy
7.	Amyloid arthropathy	Pain in the joints, stiffness of the joints and muscles, and limited range of motion	Joint aspiration and imaging studies
8.	Cutaneous amyloidosis	Lesions on the skin, purpura, and easily bruising	Biopsy
9.	Pulmonary amyloidosis	A pleural effusion, shortness of breath, and cough	Imaging studies and pulmonary function tests
10.	Thyroid amyloidosis	Enlarged thyroid	Biopsy and Imaging studies
11.	Vascular amyloidosis	Bleeding disorders and vascular fragility	Imaging studies and biopsy
12.	Ocular amyloidosis	Lid malpositions, globe displacement, increased intraocular pressure	Ophthalmic examination and biopsies
13.	Endocrine amyloidosis	Hormonal and endocrine system imbalance	Imaging study and a biopsy

As a result of this condition, the management can vary depending on the type of amyloidosis, and it may include chemotherapy, treatment specific to a particular organ, as well as supportive care. In many cases, comprehensive management requires multidisciplinary care involving cardiologists, nephrologists, neurologists, gastroenterologists, and other specialists. Amyloidosis can be better managed if it is diagnosed and treated early.

14.3 CURRENT THERAPEUTIC APPROACH TO AMYLOIDOSIS

Identifying the specific variant of the amyloid fibrils in a very precise manner, typically achieved using methods like immunohistochemistry and genetic examination, in addition to evaluating the dispersion of amyloid accumulations across the body, plays a pivotal role in selecting appropriate courses of treatment (Hirschfield and Hawkins 2003). Contemporary endeavors aimed at diminishing amyloid β-peptide production in Alzheimer's disease predominantly focus on experimental strategies, which center on pinpointing and suppressing the β- and γ-secretase enzyme systems responsible for cleaving Aβ from its larger precursor protein. Interrupting fibril formation presents a prospective therapeutic avenue for addressing amyloidosis (Ghosh and De 2020; Sacchettini and Kelly 2002). Improved understanding of the molecular mechanisms that lead to the formation of amyloid fibrils, coupled with the realization that the relative instability of precursor molecules plays a pivotal role in amyloidogenesis, has prompted researchers to seek out small molecules capable of binding and stabilizing fibril precursors. Significant sugar analogs resembling N-acetylglucosamine play a pivotal role in disrupting the binding interactions occurring between heparan sulfate or the heparan sulfate proteoglycan perlecan and amyloid precursor proteins (Annaval et al. 2020; Kisilevsky et al. 2007). This disruption effectively hampers the facilitation or preservation of specific structural changes associated with the process of amyloidogenesis.

Developing synthetic peptides known as aggregation inhibitor peptides presents a treatment avenue for preventing the formation of highly organized aggregates featuring a predominant β-pleated sheet structure, thereby targeting and impeding the abnormal conformational alterations occurring during fibrillogenesis (Kundu et al. 2020). In compound libraries, screening assays employing dyes such as thioflavin T are utilized to seek out aggregation inhibitors, with a primary focus on Aβ in Alzheimer's disease, resulting in the identification of various active compounds, including derivatives of amyloid-binding dyes and small peptides homologous to Aβ (Maniam and Maniam 2024). An immunotherapy-based investigative approach explores the potential of antibodies targeting amyloid deposits, aiming to enhance their clearance despite the lack of significant host immune responses elicited by these highly abnormal structures due to their common structural motifs among all amyloid fibrils (Plotkin and Cashman 2020). Antibodies have been generated to target common conformational epitopes present on amyloid fibrils in general, demonstrating potential interest as therapeutic tools (Ma, Zhao, and Nussinov 2016). Advanced immunotherapy strategies for Alzheimer's disease, including active or passive vaccination against Aβ, have shown promising results in reducing amyloid plaque burden in transgenic mice expressing mutations associated with the disease, alongside successful passive vaccine therapy employing anti-amyloid antibodies (Bittar, Sengupta, and Kayed 2018). Phase I human trials showed tolerability of Aβ immunizations, but phase II trials were halted due to cases of aseptic meningoencephalitis, although confirming patients' ability to generate high titers of selective anti-amyloid antibodies against aggregated Aβ (Pinheiro and Faustino 2019). The non-fibrillar plasma glycoprotein SAP is consistently found in amyloid deposits due to its specific binding to all types of amyloid fibrils, remaining unmodified and resistant to degradation; studies in SAP knock-out mice suggest its potential role in amyloidosis pathogenesis and persistence. The distinctive approach involves inhibiting SAP in vivo with the aim of impeding new amyloid deposition and/or destabilizing existing amyloid deposits to promote their regression. A high throughput screening assay identified compounds that hinder the binding of SAP to amyloid fibrils, resulting in the development of the drug CPHPC, or R-1-[6-[R-2-carboxy-pyrrolidin-1-yl]-6-oxo-hexanoyl] pyrrolidine-2-carboxylic acid (Sahota et al. 2015). Several of these novel therapeutic strategies are currently undergoing clinical testing in patients in addition to being used to treat systemic amyloidosis and related disorders; however, they are hopeful that they could also be effective in treating a broader range of disorders.

14.4 NATURAL PRODUCTS AS THERAPEUTIC CANDIDATES FOR AMYLOIDOSIS

Polyphenols are natural compounds obtained from various fruits and vegetables and have shown multifaceted effects in ameliorating diseased conditions (Agraharam, Girigoswami, and Girigoswami 2022; De et al. 2021; Subramanian et al. 2019). The degradation or dissociation of amyloid fibrils *in vitro* using polyphenols has been shown in some studies (Henríquez et al. 2020). The effects of polyphenols are primarily exerted through their antioxidant properties and their influence on stress response through Nrf-2 signaling, which are directed towards the induction of antioxidant potential as well as an enhancement of cellular stress response, which can prevent cellular damage by increasing its antioxidant defense (Simoni et al. 2017). Polyphenols can directly or indirectly strengthen the degradation of misfolded and damaged proteins. It inhibits the formation of insoluble Aβ fibrils and/or destabilize β-sheet rich Aβ fibrils to form non-cytotoxic aggregates (Hajieva 2017; Kobayashi et al. 2020). There are some limitations in using the polyphenols directly for the dissociation of amyloids because they cannot cross the blood–brain barrier (BBB). The BBB cutoff is 400 Da and any amyloid degrading agent has to be lower than this size to enable its action by crossing the BBB. Various tissues and organs accumulate abnormally large amounts of amyloid proteins in amyloidosis. Infections, autoimmune disorders, and genetic mutations can cause the body to produce this protein. It is possible for these proteins to misfold and aggregate, resulting in a buildup of insoluble fibrils within the organs that disrupt the normal functioning of these organs. There are several organs that may be affected by amyloidosis, including the liver, heart, nervous system, kidneys, and gastrointestinal tract, which could lead to progressive organ damage and failure over time. Various organs and tissues are affected by amyloidosis due to the deposition of insoluble protein aggregates. In the body, the proteolytic enzymes are unable to degrade the beta-sheet fibrils of amyloid. In order to form amyloid fibrils, precursor proteins must be misfolded, aggregated into oligomers, and then converted into fibrils. Figure 14.2 shows the different natural products used for the management of amyloidosis.

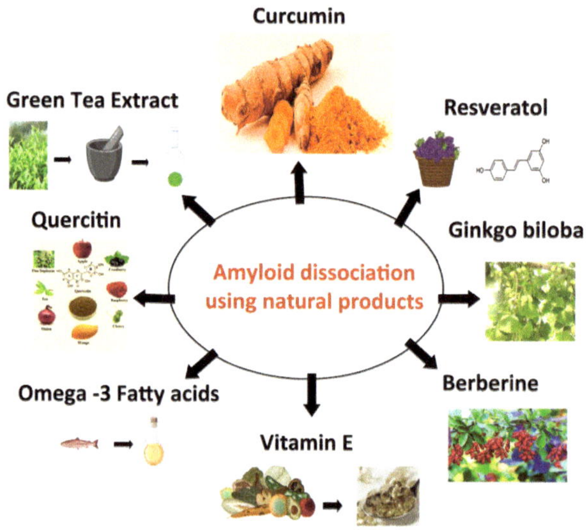

FIGURE 14.2 The range of natural products used so far for the dissociation of amyloid aggregates.

The following section discusses a few natural compounds and their derivatives which have been evidenced to protect against amyloidosis.

14.4.1 Curcumin

Curcumin (CUR), a natural compound, shows promise in protecting nerves and treating AD. Its fluorescence properties also make it valuable for AD diagnosis. However, CUR faces challenges such as low water solubility, fluid instability, rapid metabolism, low bioavailability, and difficulty crossing biological barriers. Nanocarriers offer a potential solution to enhance CUR's biocompatibility and ability to penetrate biological barriers. Utilizing nanocarriers for CUR delivery presents an effective approach to improve its overall efficacy in addressing AD as it satisfies the limitations of the compound (Lv et al. 2023). Curcumin's anti-inflammatory, antioxidant properties, and ability to influence cellular pathways through epigenetic mechanisms make it a promising candidate for AD management (Abdul-Rahman et al. 2024). Curcumin offers insights into membrane-mediated aggregation by embedding into membranes and modifying their structure, influencing the aggregation of these proteins. Researchers exploited how curcumin affects the interaction between Aβ fibrillar oligomers (FO), and anionic membranes. They found that curcumin inhibits FO interactions with the membrane, by embedding itself to the 1,2-dimyristoyl-*sn*-glycero-3-phosphorylglycerol (DMPG) membrane causing thinning and disorder of the membrane, thereby reducing FO accumulation nearly by 66 percent by Liquid surface X-ray scattering experiments and molecular dynamics simulation. These simulations conclude that curcumin clusters near phenylalanine residues on the membrane-embedded FO structure (Sallaberry et al. 2023). Curcumin's impact on AD and depression was assessed by monitoring its therapeutic potential in addressing shared pathogenic features, including reduced BDNF levels, oxidative stress, inflammation, beta-amyloid aggregation, Tau protein accumulation, and aluminium neurotoxicity. Curcumin supplementation restores BDNF levels in AD and depression, modulates monoamines, reduces oxidative stress, beta-amyloid aggregation, Tau protein accumulation, and aluminium neurotoxicity, and improves symptoms of these conditions, according to the data analyzed (Oliveira and Pieniz 2024).

14.4.2 Omega-3 Fatty Acids

Alzheimer's disease has been linked beneficially to polyunsaturated fatty acids (PUFAs). There is evidence that some PUFAs reduce the toxicity of amyloid-beta (Aβ) by encouraging its dissociation and clearance. Studies investigating direct relations between PUFAs and Aβ peptides are limited and inconsistent. Previous researchers studied the interactions of Aβ peptides with different fatty acids, such as eicosatetraenoic acid (ETA), docosahexaenoic acid (DHA), α-linolenic acid (ALA), linoleic acid (LNA) arachidonic acid (ARA), and oleic acid (OA) using transmission electron microscopy, fluorescence-based aggregation kinetic experiments, and molecular docking studies. As a result of their research, they were able to demonstrate that all the fatty acids tested had anti-aggregation properties, therefore they prevented both Aβ40 and Aβ42 fibrillogenesis (~16–84% inhibition). The use of OA and DHA as fibrillogen inhibitors was found to be excellent (84% and 81%) at 25 μM (El Shatshat, Pham, and Rao 2019). Docosahexaenoic acid (DHA) and eicosapentaenoic acid (EPA) are two omega-3 polyunsaturated fatty acids (PUFAs) found in fish and other marine organisms. Human health is guaranteed by EPA and DHA, which can be purchased as dietary supplements. PUFAs have been reported to play an important role in the prevention of neurodegenerative conditions associated with amyloidogenic proteins, such as beta-amyloid in AD, synuclein in Parkinson's disease, and transthyretin (TTR) in TTR amyloidosis. Scientists examined whether EPA and DHA could directly interact with TTR, binding in thyroxin-binding pockets (T_4BPs) that stabilize tetramers. According to the results obtained, EPA and DHA interact with T_4BP to stabilize the TTR tetramer (Ciccone et al. 2020; Ciccone et al. 2023).

The blood-brain barrier (BBB) transport mechanism plays a vital role in the clearance of Aβ from brain to blood. In an early disease stage of AD, researchers explored dynamic alterations in the function of the BBB transport system in APPswe/PS1dE9 C57BL/6J (APP/PS1) transgenic mice. A key efflux transporter of BBB, lipoprotein receptor-related protein 1 (LRP-1), started to degrade at the age of 4 months. However, fish oil supplementation which is filled with omega-3 polyunsaturated fatty acids (PUFAs) caused significant enhancement in the LRP-1 expression level. There was a noticeable clearance of Aβ from the brain to circulation, evidenced by a reduction in soluble/insoluble Aβ levels as well as senile plaque parenchyma of the brain and a simultaneous increase in the plasma levels of Aβ. A step forward, the supplementation with fish oil inhibited the activation of NF-κB significantly, caused a reduction in interleukin-1β expression and TNF-α, and resulted in glial activation suppression in APP/PS1 mice. This study suggests that BBB transport function might be impaired from an early disease stage, contributing to Aβ pathological accumulation in AD. Increasing the blood–brain ratio through omega-3 PUFAs could be a promising intervention to prevent AD progression (Yan et al. 2020). Patients with Alzheimer's disease and mild cognitive impairment (MCI) have defective T macrophages (Mφs) that lack the ability to phagocytose amyloid-1-42 and are more susceptible to apoptosis by Aβ. MCI patients show increased phagocytosis of Aβ by the Mϕs when there is in vitro and in vivo administration of resolvin D1, a ω-3 mediator, and Omega-3 fatty acids (ω-3s). There was a longitudinal study in which researchers studied the unfolded protein response by Mϕs to endoplasmic reticulum (ER) stress in patients with MCI who consumed fish-derived, omega-3-supplemented formulas that contained these fatty acids. The results indicated that ω-3s derived from fish could enhance the cytoprotective gene expression and decrease the proapoptotic genes, improving the Aβ immune clearance. The fish-derived ω3s were also found to be related to improved Mini-Mental State Examination (MMSE) rate of change in ApoEe3/e3 vs. ApoEe3/e4 patients (Olivera-Perez et al. 2017).

14.4.3 Quercetin

There is significant clinical potential for quercetin as a therapeutic flavonoid. Fruits and vegetables are the primary sources of this compound in daily diets. Several *in vitro* studies have demonstrated quercetin's neuroprotective properties. In addition to reducing lipid peroxidation, it provides protection from oxidative damage to neurons. Apart from its antioxidant properties, it also inhibits the formation of fibrils from amyloid protein peptides, thereby counteracting the destruction of cells and the production of inflammatory cascades (Khan et al. 2019). Using bovine insulin as a model protein, quercetin was investigated in vitro for its anti-amyloidogenic effects. Quercetin inhibited insulin's amyloid formation dose-dependently (Wang, Wang, and Zeng 2011). Researchers have also observed quercetin's protective effects on endothelial cells causing wounds in brain microvascular endothelial cells exposed to fibrillar Aβ1-40 (f Aβ1-40). A quercetin-treated human brain microvascular endothelial cell (hBMEC) model was shown to alleviate f Aβ1-40-induced cytotoxicity. Cell viability is increased by quercetin, lactate dehydrogenase release is reduced, and nuclear condensation is relieved by quercetin. Moreover, quercetin increases the activity of superoxide dismutase and reduces intracellular reactive oxygen species (ROS) generation. Additionally, it strengthens the integrity of the barrier by preserving the transendothelial electrical resistance value, ameliorating the deteriorated permeability, and increasing the levels of characteristic enzymes that increase after being exposed to f Aβ1–40. The findings indicate quercetin protects hBMECs from toxicity caused by f Aβ1–40-induced stress (Li et al. 2015). By treating PC12 neuronal cells with Aβ, quercetin greatly reduced the death rate. After receiving 100 mg/kg quercetin daily for 30 days, mice showed significant improvement in learning and memory loss caused by Aβ. In mice, 575 mg/kg body weight of quercetin was determined to be the acute oral dose. There is, therefore, evidence that quercetin can help treat neurological disorders, including AD, but more research is needed (Li et al. 2017).

In addition to being a major cholesterol carrier, apoE is also an important part of lipid homeostasis by mediating the transfer of lipids from one cell type to another. ApoE, which belongs to the low-density lipoprotein receptor family, is primarily produced by astrocytes in the central nervous system (CNS) and transports cholesterol to neurons via apoE receptors. Alzheimer's disease (AD)-associated APOEε4 gene products are strong genetic risk factors, likely because they affect the accumulation of amyloid peptides. APOEε4 carriers and patients with AD have reduced ApoE protein levels in plasma and cerebrospinal fluid (CSF). Alterations in cholesterol levels are also associated with AD risk. The pathogenesis of AD is characterized by the accumulation, oligomerization, and deposition of Aβ in the brain. A growing body of evidence indicates that apoE and apoE receptors play an important role in these processes. The apoE contained in astrocytes is crucial for the metabolism of cholesterol in the brain and the clearance of Aβ. A therapeutic strategy for AD could therefore involve increasing brain apoE levels. The study by researchers shows that quercetin inhibits apoE degradation in immortalized astrocytes, thereby increasing apoE levels. A major finding is that oral quercetin administration led to significant increases in brain apoE levels and reduced insoluble Aβ levels in the cortex of 5xFAD amyloid model mice. As a novel class of AD therapy, quercetin can be explored as a potential agent that increases apoE levels through a novel mechanism (Zhang et al. 2016).

14.4.4 Resveratrol

The relationship between resveratrol (Res), a naturally occurring polyphenol found mainly in grapes and red wine, and Alzheimer's disease has been established in a number of in vivo and in vitro models. It enhances the clearance of amyloid beta-peptides, inhibits the cleavage of the amyloid precursor protein (APP), and reduces aggregation of Aβ. Additionally, Res has antioxidant properties that protect neuronal functions (Jia, Wang, and Liu 2017). Res is an appropriate treatment candidate for neurological disorders (NDs) due to its excellent therapeutic effects and ability to penetrate the blood–brain barrier. The relationship between Res administration and inhibition of amyloidogenic pathways and the stimulation of autophagy has been elaborately described by researchers. Their studies demonstrate that Res supplementation promotes dephosphorylation and suppresses the aggregation of tau proteins (Ashrafizadeh et al. 2020).

Alzheimer's disease (AD) is believed to be proximately caused by the progressive accumulation and deposition of amyloid in the central nervous system (CNS) with aging. A treatment approach that targets molecular mechanisms of aging may be a viable option. In animal models, caloric restriction prevents diseases of aging (Agraharam, Girigoswami, and Girigoswami 2023), including Alzheimer's disease, possibly by activating sirtuins. Energy balance (NAD+/NADH) and gene transcription are regulated by sirtuins (such as mammalian SIRT1). By activating SIRT1, resveratrol may prevent diseases of aging by mimicking caloric restriction. Resveratrol was tested in a phase II randomized, double-blind, placebo-controlled trial on individuals with mild-to-moderate Alzheimer's disease. As well as being detectable in CSF (at low nanomolar levels), resveratrol has a few other benefits, including preventing AD biomarker trajectory changes, preserving blood–brain barrier integrity, and modulating CNS immunity. It is necessary to conduct further research on resveratrol in order to determine its safety and efficacy in the treatment and prevention of AD and other aging-related diseases (Sawda, Moussa, and Turner 2017). In a study, researchers used surface plasmon resonance (SPR) and proton nuclear magnetic resonance (1H NMR) to identify direct RES binding to Aβ. SPR was used to analyze the ability of RES to bind to fibrils as well as monomers of Aβ (1–40 and 1–42). The RES binding response towards fibrillar Aβ(1–42) was higher compared to the monomeric Aβ (1–42), whereas in the case of Aβ (1–40), the RES binding response was lower for fibrillar form than the monomeric form. The data indicated that the binding capacity of RES to Aβ present in different forms is a direct binding and RES exerts its protective properties to alleviate the symptoms of AD (Ge et al. 2012).

A transgenic strain of *Caenorhabditis elegans*, CL2006, that can express Aβ1–42 when controlled by the muscle-specific promoter and can respond to Aβ-toxicity by paralysis manifestation, was taken as a model organism for the study by Regitz et al. In this study, specific RNA interference knockdowns of proteostasis genes were used to identify target genes. ProteoStat® Detection

Reagent was used to identify the effects of resveratrol on protein aggregation, and a fluorogenic peptide substrate was used to prove the activation of proteasomal degradation by resveratrol. Paralysis was reduced by 40% when 100 μM Resveratrol was administered. UBL-5, which is involved in the mitochondrial unfolded protein response (UPR) was necessary for the Aβ-toxicity prevention by resveratrol. It has also been determined that XBP-1, an endoplasmic reticulum-resident factor involved in UPR, is also needed for resveratrol's effects. Accordingly, resveratrol was unable to reduce paralysis when the macroautophagy and chaperone-mediated autophagy pathways were inhibited, along with proteasomal degradation. In addition, resveratrol reduced the number of lysosomes, indicating that autophagy pathways were activated and proteasomal degradation was activated by resveratrol. Resveratrol could target the proteostasis-involved proteins and reduce the Aβ aggregation amount, thereby reducing the Aβ-induced toxicity in the *C. elegans* model of AD (Regitz et al. 2016). Not only resveratrol, the derivatives of resveratrol have also shown some promising effects in the inhibition of amyloids. According to researchers, dietary supplements such as pterostilbene (a derivative of resveratrol) have the same inhibitory properties as tafamidis for inhibiting the fibrillization of V30M-TTR. According to X-ray crystallography and thermodynamic analysis, the binding of the resveratrol derivative to TTR is enthalpy-driven, with hydrogen bonding to S117 providing the binding enthalpy (Yokoyama et al. 2023).

14.4.5 Berberine

The unraveling mechanism was done by introducing the berberine in Amyloid beta (Aβ), which delayed the equilibrium of Aβ16-21 oligomerization, as revealed using computational approaches. β-sheet content gradually increases from 12.89% to 22.19% in the 10 ns of simulations to 20 ns in the initial stage of fibrillation. β-sheets formation was reduced to 1.36% after the introduction of berberine. Hence, berberine had the potency against Aβ16-21 oligomerization which was an important step for the β-sheet formation. The berberine was added by replacing the water molecules which reduced the total number of hydrogen bonds. Due to the formation of stable structures, it led to a slight reduction in Aβ16-21 flexibility. This computational approach was very crucial for drug development in neurological disorders as it provided all the insights at their mechanical level (Qais et al. 2024). Multifaceted toxicity like mitochondrial dysfunction, activation of premature apoptosis, generation of reactive oxygen species (ROS), and interfering with synaptic signaling in neuronal cells due to Aβ aggregation, are the major reasons for AD. They can be ameliorated by the berberine-derived (Ber-D) natural product which inhibits the multifaceted toxicity. The rescue of biomacromolecules from oxidative stress and ROS generation prevention by arresting the redox cycle was done by efficient Cu chelation of the Ber-D(polyphenolic structures). *In silico* studies proved that the metal-dependent and metal-independent Aβ aggregation was inhibited by Ber-D. These crucial multifunctionalities of Ber-D made the amelioration of multifaceted Aβ toxicity in AD (Rajasekhar et al. 2020).

Berberine (BBR) extracted from a Chinese medicinal herb *Coptis chinensis*, is given to the 2 months old TgCRND8 mice (transgenic mice which are well-known model for Alzheimer's disease) for 6 months by the process of oral gavaging at different doses (low (25 mg/kg) and high (100 mg/kg)) per day. This intervention remarkably ameliorated the defects in learning, maintenance of long-term spatial memory, as well as plaque load as compared with vehicle control treatment. The BBR-treated mice showed a reduction of detergent-insoluble and soluble levels in brain homogenates evidenced by ELISA. In addition, it inhibited the tau phosphorylation and APP in GSK3, decreased the level of C-terminal fragments, hyperphosphorylation of APP, and tau in N2a mouse neuroblastoma cells which express the APP695 (N2a-SweAPP) Human Swedish mutant through the signaling pathway of GSK3. The APP processing was thus regulated by BBR treatment which provided neuroprotective effects in TgCRND8 mice and was shown by these results (Durairajan et al. 2012). Researchers aimed to determine BBR's neuroprotective benefits on AD by associating its ability to decrease Aβ pathogenesis. The Morris water maze (MWM) test was used to evaluate mice's cognitive abilities.

ELISA was used to measure the amounts of Aβ; Western blotting was used to identify the expression of ADAM10, BACE1, sAPPβ, APP, and ADAM17 and immunohistochemistry was done. Western blotting was used to assess the activity of the γ-secretase complex Aph-1α, NCT, PS1, and Pen-2. The APP/PS1 mice's learning and memory deficiencies were alleviated by BBR. BBR reduced the amounts of Aβ in the APP/PS1 mice's hippocampal regions. The hippocampal regions of AD mice treated with BBR exhibited a considerable decrease in BACE1 and sAPP-β levels. BBR showed little impact on NCT, but it significantly reduced the expression of Aph-1α, Pen-2, and PS1. In the hippocampal regions of BBR-treated mice ADAM17, sAPPα, and ADAM10 levels were much higher than those of the control group ($P < 0.05$; Cai et al. 2018).

14.4.6 Ginkgo biloba

Many neurological disorders, including Alzheimer's disease (AD), have been treated with extracts of Ginkgo biloba leaves (Nowak et al. 2021). The cellular and molecular mechanisms of Ginkgo extracts are undetermined; however, there is substantial experimental evidence of their neuroprotective effects in humans and animals. Clinical trials have shown EGb761, an extract of the leaves of the Ginkgo biloba tree, to be beneficial for the cognitive functions of the brain, especially in the treatment of dementias caused by aging and Alzheimer's disease. A number of experiments have shown that EGb761 protects against neuronal damage from a variety of insults; however, the mechanism underlying this protection remains unknown. Scientists report that EGb761 inhibits amyloid-β (Aβ) fibril formation in neuroblastoma cells with a double mutation associated with AD. EGb761 decreased Aβ fibrillogenesis in both conditioned medium and solution in vitro of this Aβ-secreting cell line. Cells treated with EGb761 showed significant reductions in apoptosis initiated by mitochondria and in caspase-3 activity, a key enzyme involved in signaling. This study suggests that EGb761 is antioxidative, anti-amyloidogenic, and antiapoptotic in nature (Luo et al. 2002). *Caenorhabditis elegans* was used as a model organism to study EGb 761 and its constituents. In a transgenic *C. elegans*, EGb 761 and ginkgolide A (a component of EGb 761) reduce pathological behaviors induced by Aβ, such as paralysis and chemotaxis. Furthermore, they found that EGb 761 prevented Aβ oligomers from forming and that Aβ deposits in worms were inhibited by EGb 761. It is also important to note that EGb 761 and ginkgolide A are not suppressing Aβ-induced paralysis by reducing oxidative stress because the antioxidant L-ascorbic acid did the same thing as EGb 761 in reducing intracellular levels of hydrogen peroxide. According to these findings, (1) EGb 761 inhibits Aβ-related pathological behaviors, (2) Aβ toxicity is protected by EGb761 primarily through modulating Aβ oligomeric species, and (3) ginkgolide A has therapeutic potential for preventing and treating Alzheimer's disease (Wu et al. 2006).

Several in vivo and in vitro studies have shown that Ginkgo biloba leaf extract (EGb761) is effective in treating symptoms of impaired cerebral function in primary degenerative dementia syndromes. There are two main components in EGb761: flavonoid glycosides and terpene trilactones (TTLs), each accounting for 24 and 6% of the total content, respectively. A previous study isolated and characterized by spectroscopic analysis seven flavonoid glycosides from the leaves of Ginkgo biloba. For the simultaneous quantification of these seven flavonoids, an ultra-high-performance liquid chromatography method was developed. A thioflavin T fluorescence assay was used to evaluate the inhibitory activity of these flavonoids, along with four TTLs, namely ginkgolides A, B, and C, and bilobalide (compounds 8–11). Three flavonoids (flavonoids 1, 3, and 4) showed moderate inhibitory activity, while the other four flavonoids, 2, 5, 6, and 7, and four terpene trilactones, had poor inhibitory properties. Using Ginkgo leaves as a model system, these flavonoids 1–7 were investigated in terms of their structure-activity relationship for the very first time (Xie et al. 2014). As part of the Ginkgo Evaluation of Memory Study, researchers examined whether plasma levels of amyloid-β1–42, 1–40, and the ratio of amyloid-β1–42/1–40 among participants in Pittsburgh were associated with the extent of brain fibrillar amyloid plaques measured using Pittsburgh compound-B PET imaging in 2009, hippocampal volume, cortical thickness in the temporal lobe, and white matter lesions. Cognitively normal individuals with low plasma amyloid-β1–42 levels had inverse associations with brain amyloid deposition. At the same time, the plasma level alterations in amyloid-β1–42 over 8 years time were very little and there was no relation between these levels and Pittsburgh compound-B amount (Lopez et al. 2020).

14.4.7 Green Tea

Green tea is a well-known nutritious beverage with significant antioxidant properties. The infusion is made from dry leaves of Kuntze, Camellia sinensis plants. These green tea extracts have abundant polyphenols in them, including flavonols, epigallocatechin gallate (EGCG), and catechins. Moreover, catechins reduce the toxicity of amyloid-β-derived peptides caused in rat hippocampus cell cultures. Gallic acid and EGCG inhibit the production of Aβ-derived neurotoxic ligands and the aggregation of amyloid-β. Gancar et al. (2020) demonstrated the combined effect of green tea polyphenols mixture on the inhibition of insulin amyloid fibril aggregation. Out of all the molecules that were evaluated separately, only EGCG prevented the process of fibrillization. Epicatechin, gallic acid, and caffeine were ineffective when they were treated individually against insulin amyloid fibrils. Therefore, the action of EGCG was potentiated by individually as well as their equimolar combination with epicatechin, gallic acid, and caffeine (Gancar et al. 2020). A low risk of neurodegenerative diseases is associated with the consumption of green tea. The protective benefits of green tea against amyloid β (Aβ) accumulation and cognitive impairment are used to study the amyloid precursor protein cleavage, glial cell activation, neuroinflammation, and synaptic changes. In a recent study conducted by Kan et al, green tea extract (GTE) was administrated to 5XFAD mice for 16 weeks. In that therapy, green tea extract significantly improved memory and spatial learning reduced the IL-1β levels, and upregulated the immune response. GTE minimizes cognitive impairment in 5XFAD mice by suppressing Aβ levels. These advantageous effects include a decrease in synapse loss, regulation of APP cleavage pathways, and suppression of pro-inflammatory and glial cell activation (Kan et al. 2021).

One of the numerous natural substances that have been tested for stopping or delaying the amyloidogenic process is epigallocatechin-3-gallate (EGCG), a powerful and prevalent polyphenolic molecule found in green tea that has a wide range of biological effects. Research supports that EGCG is capable of preventing amyloid-β, huntingtin, and α-synuclein proteins. EGCG is an antioxidant, but when it becomes oxidized, it can help fibrils remodel by creating Schiff bases and crosslinking the fibrils. In addition, oxidized EGCG was used to create microparticles for medicinal administration, which were then loaded with a second anti-amyloidogenic chemical to produce a synergistic therapeutic effect (Fernandes et al. 2021). Zhan et al. demonstrated the protofibril-disrupting effect of EGCG/EGC. EGCG exhibited a greater protofibril disruptive effect than EGC. EGCG caused the L-shaped Aβ42 protofibril to be deformed by breaking the hydrogen bond between H6 and E11 through hydrogen-bonding interaction with E11 and π- π interactions with residues H14/Y10 on the surface of Aβ42 protofibril. Altering the L-shaped portion, however, is a result of EGC's insertion into the hydrophobic core created by A2, F4, L34, and V36. EGC interacts preferentially with aromatic residues H6 and Y10. The K28-A42 salt bridge is disrupted by EGCG's hydrogen bonding with Aβ42 and its cation-π interaction with K28 through its RingGA, whereas EGC interacts with hydrophobic residues V39 and I41, as well as I32, M35, and V40 in the C-terminal hydrophobic core, disrupted the salt bridges. While cation-π connections are not found between EGC and any positively charged residues, RingGA of EGCG creates a cation-π connection with Aβ42 protofibril. This suggests that the gallic acid ester group of EGCG plays a crucial role in disrupting Aβ42 protofibril (Zhan et al. 2020).

The most prevalent precursor protein for hereditary amyloidosis is transthyretin (TTR). It also causes other nongenetic disorders with the deposition of wild-type TTR amyloid in the heart region. To overcome this cardiac wild-type TTR amyloidosis green tea has been utilized in many studies. Epigallocatechin-3-gallate (EGCG) is a prominent catechin present in green tea and prevents various amyloidogenic proteins. The study conducted by Siepen et al. evaluated a decrease in left ventricular mass in patients with wild-type TTR by taking green tea (Siepen et al. 2015).

14.4.8 Vitamin E

Numerous investigations have assessed the connection between vitamin E deficiency or levels in bodily fluids and neurodegenerative illnesses. According to certain research, vitamin E may be able to prevent neurodegeneration in conditions like amyotrophic cerebral leukemia, stroke, Lateral Sclerosis (ALS), Parkinson's disease, and Alzheimer's disease. Supplementing with vitamin E has also been linked to risk factors for some neurological illnesses. Being one of the most potent dietary antioxidants, it scavenges oxygen radicals to prevent oxidation of proteins, lipids, and nucleic acids. Research has demonstrated that incorporating vitamin E into cell membranes shields the cells from harmful oxidation. Vitamin E is produced by plants and comes in eight different forms: α, β, γ, δ tocopherol, and α, β, γ, δ tocotrienol. All of these forms are fat-soluble. Long back, α-tocopherol the most prevalent form of vitamin E in human tissues has been thought to have cytoprotective properties, possibly reducing oxidation, inflammation, and degenerative processes. Vitamin E was eventually found to be an antioxidant of polyunsaturated lipids after it was initially recognized in 1922 by Evans and Bishop as a dietary component necessary to prevent fetal reabsorption in rats (İçer, Arslan, and Karadağ 2021).

As an effective antioxidant, vitamin E may help to reduce oxidative stress and the formation of ROS associated with AD. Both the treatment and prevention of AD may benefit from this vitamin E. Jahanshahi et al. used Wistar rats in which 3 different types of drugs (scopolamine, Sham scopolamine+sesame oil, scopolamine+vitamin E) were administrated for 14 days. The number of neurofibrillary tangles and congophilic amyloid plaques in the hippocampal region significantly increased after scopolamine therapy. A high dose of vitamin E was more effective in reducing congophilic amyloid plaque and neurofibrillary tangle density. Hence, the result suggested that vitamin E can be an effective therapeutic agent for congophilic amyloid plaque and neurofibrillary tangle density reduction (Jahanshahi, Nikmahzar, and Sayyahi 2020). Studying vitamin E's effects on the risk of AD and progression of AD through five cohorts and three randomized controlled trials (RCTs; total $n = 14{,}262$), Wang et al demonstrated 1313 cases related to vitamin E's effect on the risk of AD, and 244 cases related to its effect on the progression of AD (Wang et al. 2021).

AD development is well known to be influenced by traumatic brain injury. Most commonly, Tg2576 mice with repetitive concussive brain injury (RCBI) aggravate cognitive impairments, and brain lipid peroxidation and also support amyloid aggregation. The modified controlled cortical impact model of closed injury was used by the author to assess the effects of vitamin E over 4 weeks. Animals treated with vitamin E showed large increases in brain lipid peroxidation levels and significant decreases in vitamin E levels in the brain. When compared to the control group, the mice receiving vitamin E showed no increase in Aβ peptides and a reduction in learning impairments. Therefore, the study conducted by Conte et al. proved that vitamin E can reduce lipid peroxidase levels in the brain (Conte et al. 2004). Shahidi et al. administrated β 25-35 to Wistar rats to cause Alzheimer's disease. After ingesting 200 mg/kg of vitamin E and ginkgo biloba orally for ten days, memory impairments in rats were massively reduced (Shahidi et al. 2021). Similarly, a Wistar rat with bilateral common carotid artery brain damage leading to chronic cerebral hypoperfusion was tested by Annaházi et al. to determine the impact of α-Toc treatment. The learning process was enhanced by the injection of α-Toc (100 mg/kg) five days before the injury and five days following the damage (Annaházi et al. 2007). Wang and colleagues observed the oral administration of α-Toc (100 mg/kg for four weeks) to transgenic APPswe/PS1dE9 mice, which had previously been induced to develop AD. They reported a protective effect that reduced lipid peroxidation and the release of inflammatory mediators (IL-1 and IL-6). This led to an improvement in cognitive dysfunction associated with spatial learning (Wang et al. 2016). Garcia-Alloza et al. have used APPswe/PS1d9 mice that had been induced with AD. Neurite structural damage, oxidative stress, and the development of

AD were all reduced by the oral administration of vitamin E (210 mg/kg) one day before and fifteen days following the surgery (da Cunha Germano et al. 2023; Garcia-Alloza et al. 2006).

Certain dietary habits and minerals have been related to the presence of Aβ. The reduction of Aβ levels may be facilitated by eating a healthy diet (such as a Mediterranean diet or consuming less sodium, saturated fat, high-glycemic foods, and junk/fast food), HDL cholesterol, polyunsaturated fatty acids, and alcohol. Conversely, increased consumption of total cholesterol, triglycerides, LDL cholesterol, saturated fatty acids, sugar, and fructose are some reasons for worsening Aβ profiles. Numerous minerals, such as folic acid, cobalamin, vitamin E, vitamin D, zinc, copper, aluminium, and magnesium, are still not well understood in this field. To address potential nutritional methods to avoid Aβ disruption, cognitive decline, and ultimately AD, more studies on nutrition and Aβ are required (Díaz et al. 2022).

14.5 LIMITATIONS OF NATURAL PRODUCTS AND FUTURE DIRECTIONS

Natural products have been used in medicine for many years and their derivatives called nutraceuticals have gained a lot of popularity in the past few years (De et al. 2021). The major limitation of natural products is their poor solubility, bioavailability, and stability. To improve these drawbacks nano formulations have come into rescue and have yielded promising results. Nano formulation of nutraceuticals like lycopene (Deepika, Agnishwar, and Koyeli 2023a) (Balasubramanian, Girigoswami, and Girigoswami 2023a), resveratrol (Balasubramanian, Girigoswami, and Girigoswami 2023b), lemongrass oil (Balasubramanian, Girigoswami, and Girigoswami 2022), carrageenan (Udayakumar, Girigoswami, and Girigoswami 2023b), and myricetin (Agraharam et al. 2021, 2022) has shown beneficial effects in several types of human diseases. Nanotechnology is the field of science that has revolutionized the biomedical field with its variety of applications in drug delivery (Kannan, Geeva, and Narayan 2022; Thirumalai, Girigoswami, et al. 2023b; Thirumalai, Elboughdiri, et al. 2023; Thirumalai et al. 2023a), biosensing (Harini et al. 2023; Mercy, Kiran et al. 2023b), theranostics (Chatterjee et al. 2023; Mercy et al. 2023a,b; Harini et al. 2023a; Pallavi et al. 2022; Sakthi Devi et al. 2022), cancer treatment (Janani, Girigoswami, and Girigoswami 2023a,b, Biswas et al. 2023; Ps et al. 2023), and also in management of amyloidosis (Agraharam et al. 2022; Harini et al. 2022).

Common medications for AD are symptomatic and aim at the disrupted neurotransmission between the degenerated neurons. Examples of such medications are acetylcholinesterase inhibitors, including Tacrine, Donepezil, Rivastigmine, and Galantamine (Britt and Day 2016). This therapeutic is for targeting different stages of the AD pathogenic process and to cease the progress of the disease. The focus of nanotechnology therapeutic approaches for every disease has been on drug discovery and monitoring and targeting drug delivery. Nanotechnology has shown promising applications in targeted drug delivery for AD, and several nanocarrier systems have been studied to increase the bioavailability and efficacy of different AD therapeutic agents (Ma et al. 2018). The physicochemical characteristics of drugs in nanocarrier systems have biodegradability, bioavailability, ionization, high molecular weight, and other beneficial effects. For example, the interaction of Cu^{2+} and Fe^{3+} with Aβ leads to the production of oxidative stress, and Aβ oligomerizes in the lipid bilayer of the cell plasma membrane leading to the form calcium (Ca) channels. These channels cause an imbalance in Ca homeostasis which ends in oxidative stress. Aβ interacts through an amino acid (Methionine) with membrane lipid molecules and lipid peroxidation produces 4HNE(4-hydroxy-2-nonenal). This interaction ends in membrane disruption, and finally oxidative stress in the brain tissue will be involved in the production of Reactive Oxygen Species (ROS; Hureau and Faller 2009). Enzymes from natural sources and ZnO nanoflowers have also been found to be effective in the degradation of insulin amyloids, amyloid β1-42, and prion peptides in vitro as well as in vivo (only insulin amyloids) (Girigoswami et al. 2019; Metkar et al. 2017a; Metkar et al. 2017b; Metkar et al. 2019; Metkar et al. 2020; Metkar et al. 2022). Shuttle peptides that can cross the BBB, have also been proposed as a therapeutic agent for the dissociation of amyloids (Prades et al. 2015; Sánchez-Navarro, Giralt, and Teixidó 2017). Small stress molecules or extrapolates have also been exploited to dissociate insulin amyloids, prion amyloid, and amyloid β 1-42, *in vitro* (Arora, Ha, and Park 2004; Kanapathipillai et al. 2008; Ryu et al. 2008; Ryu, Park, and Lentzen 2007). These studies could indicate a future direction for the effective use of natural products by entrapping them in proper nanostructures that can increase their stability in biological fluids, improve their bioavailability by higher absorption in the body, and also increase their shelf life.

ACKNOWLEDGMENT

Authors acknowledge Chettinad Hospital and Research Institute, Chettinad Academy of Research and Education for providing the research infrastructure for carrying out the work. S.U. acknowledges Chettinad Academy of Research and Education for providing her with PhD fellowship.

AUTHOR'S CONTRIBUTION

Conceptualization, data curation, writing review and editing, writing original draft, supervision: K.G., A.G.; investigation, writing original draft: K.G., A.G., S.U, J.J.R., S.P. All authors have read and agreed to the published version of the manuscript.

CONFLICT OF INTEREST

Authors declare no conflict of interest(s) among the co-authors.

REFERENCES

Abdul-Rahman T, Awuah WA, Mikhailova T, et al. Antioxidant, anti-inflammatory and epigenetic potential of curcumin in Alzheimer's disease. Biofactors. 2024;50(4):693–708. doi:10.1002/biof.2039.

Accardi F, Papa V, Capozzi AR, et al. A Rare Case of Systemic AL Amyloidosis with Muscle Involvement: A Misleading Diagnosis. Case Rep Hematol. 2018;2018:9840405. Published 2018 Jan 31. doi:10.1155/2018/9840405.

Agraharam, Gopikrishna, Agnishwar Girigoswami, and Koyeli Girigoswami. 2021. "Nanoencapsulated myricetin to improve antioxidant activity and bioavailability: a study on zebrafish embryos." *Chemistry* 4 (1):1–17. https://doi.org/10.3390/chemistry4010001

Agraharam, Gopikrishna, Agnishwar Girigoswami, and Koyeli Girigoswami. 2022. "Myricetin: a multifunctional flavonol in biomedicine." *Current Pharmacology Reports*: 1–14. https://doi.org/10.1007/s40495-021-00269-2

Agraharam, Gopikrishna, Agnishwar Girigoswami, and Koyeli Girigoswami. 2023. "Calorie restriction and extracellular ATP on health and longevity-a perspective." *Current Nutrition & Food Science* 19 (1):4–8.

Agraharam, Gopikrishna, Nishakavya Saravanan, Agnishwar Girigoswami, and Koyeli Girigoswami. 2022. "Future of Alzheimer's disease: nanotechnology-based diagnostics and therapeutic approach." *BioNanoScience* 12 (3):1002–1017.

Annaházi, Anita, Éva Mracskó, Zoltán Süle, Eszter Karg, Botond Penke, Ferenc Bari, and Eszter Farkas. 2007. "Pre-treatment and post-treatment with α-tocopherol attenuates hippocampal neuronal damage in experimental cerebral hypoperfusion." *European Journal of Pharmacology* 571 (2–3):120–128.

Annaval, Thibault, Rebekka Wild, Yoann Crétinon, Rabia Sadir, Romain R. Vivès, and Hugues Lortat-Jacob. 2020. Heparan sulfate proteoglycans biosynthesis and post synthesis mechanisms combine few enzymes and few core proteins to generate extensive structural and functional diversity. *Molecules* 25 (18). https://doi.org/10.3390/molecules25184215

Arora, Anubhav, Chanki Ha, and Chan Beum Park. 2004. "Inhibition of insulin amyloid formation by small stress molecules." *FEBS Letters* 564 (1–2):121–125.

Ashrafizadeh, Milad, Ali Zarrabi, Masoud Najafi, Saeed Samarghandian, Reza Mohammadinejad, and Kwang Seok Ahn. 2020. "Resveratrol targeting tau proteins, amyloid-beta aggregations, and their adverse effects: an updated review." *Phytotherapy Research* 34 (11):2867–2888.

Balasubramanian, Deepika, Agnishwar Girigoswami, and Koyeli Girigoswami. 2022. "Antimicrobial, pesticidal and food preservative applications of lemongrass oil nanoemulsion: A mini-review." *Recent Advances in Food Nutrition & Agriculture* 13 (1):51–58. https://doi.org/10.2174/2212798412666220527154707

Balasubramanian, Deepika, Agnishwar Girigoswami, and Koyeli Girigoswami. 2023a. "Antioxidant and anticancer activity of nano lycopene." *Research Journal of Biotechnology* 18:6. 10.25303/1806rjbt980104

Balasubramanian, Deepika, Agnishwar Girigoswami, and Koyeli Girigoswami. 2023b. "Nano resveratrol and its anticancer activity." *Current Applied Science and Technology* 23 (3). https://doi.org/10.55003/cast.2022.03.23.010

Biswas, Kunal, Gopalarethinam Janani, Saranya Udayakumar, Balasubramanian Deepika, and Koyeli Girigoswami. 2023. "Rough edges of reduced graphene oxide (rGO) sheets elicit anticancerous activities: an in vitro study." *Results in Chemistry* 6:101207.

Bittar, Alice, Urmi Sengupta, and Rakez Kayed. 2018. "Prospects for strain-specific immunotherapy in Alzheimer's disease and tauopathies." *Npj Vaccines* 3 (1):9. https://doi.org/10.1038/s41541-018-0046-8

Britt, Daniel MI, and Gregory S Day. 2016. "Over-prescribed medications, under-appreciated risks: a review of the cognitive effects of anticholinergic medications in older adults." *Missouri Medicine* 113 (3):207.

Cai, Zhiyou, Chuanling Wang, Wenbo He, and Yi Chen. 2018. "Berberine alleviates amyloid-beta pathology in the brain of APP/PS1 transgenic mice via inhibiting β/γ-secretases activity and enhancing α-secretases." *Current Alzheimer Research* 15 (11):1045–1052.

Chatterjee, Sampreeti, Karthick Harini, Agnishwar Girigoswami, Moupriya Nag, Dibyajit Lahiri, and Koyeli Girigoswami. 2023. "Nanodecoys: a quintessential candidate to augment theranostic applications for a plethora of diseases." *Pharmaceutics* 15 (1):73. https://doi.org/10.3390/pharmaceutics15010073

Chitradevi, D, S Prabha, and Alex Daniel Prabhu. 2021. "Diagnosis of Alzheimer disease in MR brain images using optimization techniques." *Neural Computing and Applications* 33:223–237.

Ciccone, Lidia, Nicoló Tonali, Susanna Nencetti, and Elisabetta Orlandini. 2020. "Natural compounds as inhibitors of transthyretin amyloidosis and neuroprotective agents: analysis of structural data for future drug design." *Journal of Enzyme Inhibition and Medicinal Chemistry* 35 (1):1145–1162.

Ciccone, Lidia, Susanna Nencetti, Armando Rossello, Lucia Barlettani, Nicolò Tonali, Paola Nieri, and Elisabetta Orlandini. 2023. "Omega-3 PUFAs as a dietary supplement in senile systemic amyloidosis." *Nutrients* 15 (3):749.

Conte, Valeria, Kunihiro Uryu, Scott Fujimoto, Yuemang Yao, Joshua Rokach, Luca Longhi, John Q Trojanowski, Virginia M-Y Lee, Tracy K McIntosh, and Domenico Praticò. 2004. "Vitamin E reduces amyloidosis and improves cognitive function in Tg2576 mice following repetitive concussive brain injury." *Journal of Neurochemistry* 90 (3):758–764.

da Cunha Germano, Bianca Caroline, Lara Cristina Carlos de Morais, Francisca Idalina Neta, Amélia Carolina Lopes Fernandes, Francisco Irochima Pinheiro, Amália Cinthia Meneses do Rego, Irami Araújo Filho, Eduardo Pereira de Azevedo, José Rodolfo Lopes de Paiva Cavalcanti, and Fausto Pierdona Guzen. 2023. "Vitamin E and its molecular effects in experimental models of neurodegenerative diseases." *International Journal of Molecular Sciences* 24 (13):11191.

De Shaoli, Agraharam Gopikrishna, Vedhantham Keerthana, Agnishwar Girigoswami, and Koyeli Girigoswami. 2021. "An overview of nanoformulated nutraceuticals and their therapeutic approaches." *Current Nutrition & Food Science* 17 (4):392–407. https://doi.org/10.2174/1573401316999200901120458

Dember, Laura M. 2006. "Amyloidosis-associated kidney disease." *Journal of the American Society of Nephrology* 17 (12):3458–3471.

Díaz, Gustavo, Laetitia Lengele, Sandrine Sourdet, Gaëlle Soriano, and Philipe de Souto Barreto. 2022. "Nutrients and amyloid β status in the brain: a narrative review." *Ageing Research Reviews*: 101728.

Dungu, Jason N, Lisa J Anderson, Carol J Whelan, and Philip N Hawkins. 2012. "Cardiac transthyretin amyloidosis." *Heart* 98 (21):1546–1554.

Durairajan, Siva Sundara Kumar, Liang-Feng Liu, Jia-Hong Lu, Lei-Lei Chen, Qiuju Yuan, Sookja K Chung, Ling Huang, Xing-Shu Li, Jian-Dong Huang, and Min Li. 2012. "Berberine ameliorates β-amyloid pathology, gliosis, and cognitive impairment in an Alzheimer's disease transgenic mouse model." *Neurobiology of Aging* 33 (12):2903–2919.

El Shatshat, Amna, Amy Trinh Pham, and Praveen PN Rao. 2019. "Interactions of polyunsaturated fatty acids with amyloid peptides Aβ40 and Aβ42." *Archives of Biochemistry and Biophysics* 663:34–43.

Fernandes, Luiza, Thyago R Cardim-Pires, Debora Foguel, and Fernando L Palhano. 2021. "Green tea polyphenol epigallocatechin-gallate in amyloid aggregation and neurodegenerative diseases." *Frontiers in Neuroscience* 15:718188.

Gancar, Miroslav, Elena Kurin, Zuzana Bednarikova, Jozef Marek, Pavel Mucaji, Milan Nagy, and Zuzana Gazova. 2020. "Amyloid aggregation of insulin: an interaction study of green tea constituents." *Scientific Reports* 10 (1):9115.

Garcia-Alloza, Monica, Sarah A Dodwell, Melanie Meyer-Luehmann, Bradley T Hyman, and Brian J Bacskai. 2006. "Plaque-derived oxidative stress mediates distorted neurite trajectories in the Alzheimer mouse model." *Journal of Neuropathology & Experimental Neurology* 65 (11):1082–1089.

Ge, Jin-Fang, Jin-Ping Qiao, Cong-Cong Qi, Chen-Wei Wang, and Jiang-Ning Zhou. 2012. "The binding of resveratrol to monomer and fibril amyloid beta." *Neurochemistry International* 61 (7):1192–1201.

Ghosh, Pooja, and Priyadarsi De. 2020. "Modulation of amyloid protein fibrillation by synthetic polymers: recent advances in the context of neurodegenerative diseases." *ACS Applied Bio Materials* 3 (10):6598–6625. https://doi.org/10.1021/acsabm.0c01021

Girigoswami, Agnishwar, M Ramalakshmi, Najim Akhtar, Sanjay Kisan Metkar, and Koyeli Girigoswami. 2019. "ZnO nanoflower petals mediated amyloid degradation-an in vitro electrokinetic potential approach." *Materials Science and Engineering: C* 101:169–178. https://doi.org/10.1016/j.msec.2019.03.086

Hajieva, Parvana. 2017. "The effect of polyphenols on protein degradation pathways: implications for neuroprotection." *Molecules* 22 (1):159.

Han, Shen, Marius Kollmer, Daniel Markx, Stephanie Claus, Paul Walther, and Marcus Fändrich. 2017. "Amyloid plaque structure and cell surface interactions of β-amyloid fibrils revealed by electron tomography." *Scientific Reports* 7 (1):43577.

Harini, Karthick, Koyeli Girigoswami, A Vijaya Anand, Pragya Pallavi, Pemula Gowtham, Noureddine Elboughdiri, and Agnishwar Girigoswami. 2022. "Nano-mediated strategies for metal ion–induced neurodegenerative disorders: focus on Alzheimer's and Parkinson's diseases." *Current Pharmacology Reports* 8 (6):450–463.

Harini, Karthick, Koyeli Girigoswami, Pragya Pallavi, Pemula Gowtham, Anbazhagan Thirumalai, Kamalakkannan Charulekha, and Agnishwar Girigoswami. 2023. "MoS2 nanocomposites for biomolecular sensing, disease monitoring, and therapeutic applications." *Nano Futures*. https://doi.org/10.1088/2399-1984/ace178

Hasib Sidiqi, M, and Morie A Gertz. 2021. "Immunoglobulin light chain amyloidosis diagnosis and treatment algorithm 2021." *Blood Cancer Journal* 11 (5):90.

Henríquez, Gabriela, Alejandra Gomez, Erick Guerrero, and Mahesh Narayan. 2020. "Potential role of natural polyphenols against protein aggregation toxicity: in vitro, in vivo, and clinical studies." *ACS Chemical Neuroscience* 11 (19):2915–2934.

Hirschfield, G. M., & Hawkins, P. N. (2003). Amyloidosis: new strategies for treatment. The international journal of biochemistry & cell biology, 35(12), 1608–1613. https://doi.org/10.1016/s1357-2725(03)00169-9

Hureau, Christelle, and Peter Faller. 2009. "Aβ-mediated ROS production by Cu ions: structural insights, mechanisms and relevance to Alzheimer's disease." *Biochimie* 91 (10):1212–1217.

Husain, RS Akram, K Subramaniyan, Shiek SSJ Ahmed, and V Ramakrishnan. 2019. "Association of PSEN1 rs165932 polymorphism with Alzheimer's disease susceptibility: An extensive meta-analysis." *Meta Gene* 19:123–133.

İçer, MEHMET, Neslihan Arslan, and Gezmen Karadağ. 2021. "Effects of vitamin E on neurodegenerative diseases: an update." *Acta Neurobiologiae Experimentalis* 81.

Ihne, Sandra, Caroline Morbach, Claudia Sommer, Andreas Geier, Stefan Knop, and Stefan Störk. 2020. "Amyloidosis—the diagnosis and treatment of an underdiagnosed disease." *Deutsches Ärzteblatt International* 117 (10):159.

Jahanshahi, Mehrdad, Emsehgol Nikmahzar, and Ali Sayyahi. 2020. "Vitamin E therapy prevents the accumulation of congophilic amyloid plaques and neurofibrillary tangles in the hippocampus in a rat model of Alzheimer's disease." *Iranian Journal of Basic Medical Sciences* 23 (1):86.

Janani, Gopalarethinam, Agnishwar Girigoswami, and Koyeli Girigoswami. 2023a. "Advantages of nanomedicine over the conventional treatment in acute myeloid leukemia." *Journal of Biomaterials Science, Polymer Edition*: 1–27.

Janani, Gopalarethinam, Agnishwar Girigoswami, and Koyeli Girigoswami. 2023b. "Supremacy of nanoparticles in the therapy of chronic myelogenous leukemia." *ADMET and DMPK* 11 (4):499–511. 10.5599/admet.2013

Jia, Yongming, Na Wang, and Xuewei Liu. 2017. "Resveratrol and amyloid-beta: mechanistic insights." *Nutrients* 9 (10):1122.

Jiménez, José L, Ewan J Nettleton, Mario Bouchard, Carol V Robinson, Christopher M Dobson, and Helen R Saibil. 2002. "The protofilament structure of insulin amyloid fibrils." *Proceedings of the National Academy of Sciences* 99 (14):9196–9201.

Kan, Zhipeng, Yijun Wang, Qian Chen, Xiaoyu Tang, Henry J Thompson, Jinbao Huang, Jinsong Zhang, Feng Gao, Yong Shen, and Xiaochun Wan. 2021. "Green tea suppresses amyloid β levels and alleviates cognitive impairment by inhibiting APP cleavage and preventing neurotoxicity in 5XFAD mice." *Molecular Nutrition & Food Research* 65 (19):2100626.

Kanapathipillai, Mathumai, Sook Hee Ku, Koyeli Girigoswami, and Chan Beum Park. 2008. "Small stress molecules inhibit aggregation and neurotoxicity of prion peptide 106–126." *Biochemical and Biophysical Research Communications* 365 (4):808–813.

Kannan, Rathi Usha, Suresh Sethurajan Geeva, and Shoba Narayan. 2022. "Design and Fabrication of pH-Sensitive Curcumin Coated Gold Nanorods as Anti-Angiogenic Targeting Nanostructures." *Biointerface Research in Applied Chemistry* 13(3):286. https://doi.org/10.33263/BRIAC133.286

Khan, Haroon, Hammad Ullah, Michael Aschner, Wai San Cheang, and Esra Küpeli Akkol. 2019. "Neuroprotective effects of quercetin in Alzheimer's disease." *Biomolecules* 10 (1):59.

Kisilevsky, Robert, John B. Ancsin, Walter A. Szarek, and Suzana Petanceska. 2007. "Heparan sulfate as a therapeutic target in amyloidogenesis: prospects and possible complications." *Amyloid* 14 (1):21–32. https://doi.org/10.1080/13506120601116419

Kobayashi, Hatasu, Mariko Murata, Shosuke Kawanishi, and Shinji Oikawa. 2020. "Polyphenols with anti-amyloid β aggregation show potential risk of toxicity via pro-oxidant properties." *International Journal of Molecular Sciences* 21 (10):3561.

Kundu, Debanjan, Kumari Prerna, Rahul Chaurasia, Manoj Kumar Bharty, and Vikash Kumar Dubey. 2020. "Advances in protein misfolding, amyloidosis and its correlation with human diseases." *3 Biotech* 10 (5):193. https://doi.org/10.1007/s13205-020-2166-x

Li, Yan-Ling, Hua Guo, Yong-Qiang Zhao, Ai-Fan Li, Yue-Qin Ren, and Jie-Wen Zhang. 2017. "Quercetin protects neuronal cells from oxidative stress and cognitive degradation induced by amyloid β-peptide treatment." *Molecular Medicine Reports* 16 (2):1573–1577.

Li, Yongjie, Sibai Zhou, Jinze Li, Yuhua Sun, Hamlati Hasimu, Rui Liu, and Tiantai Zhang. 2015. "Quercetin protects human brain microvascular endothelial cells from fibrillar β-amyloid1–40-induced toxicity." *Acta Pharmaceutica Sinica B* 5 (1):47–54.

Lopez, Oscar L, William E Klunk, Chester A Mathis, Beth E Snitz, Yuefang Chang, Russell P Tracy, and Lewis H Kuller. 2020. "Relationship of amyloid-β1–42 in blood and brain amyloid: Ginkgo evaluation of memory study." *Brain Communications* 2 (1):fcz038.

Luo, Yuan, Julie V Smith, Vijaykumar Paramasivam, Adam Burdick, Kenneth J Curry, Justin P Buford, Ikhlas Khan, William J Netzer, Huaxi Xu, and Peter Butko. 2002. "Inhibition of amyloid-β aggregation and caspase-3 activation by the Ginkgo biloba extract EGb761." *Proceedings of the National Academy of Sciences* 99 (19):12197–12202.

Lv, Hongqian, Yan Wang, Xiaotong Yang, Guixia Ling, and Peng Zhang. 2023. "Application of curcumin nano-formulations in Alzheimer's disease: prevention, diagnosis and treatment." *Nutritional Neuroscience* 26 (8):727–742.

Ma, Buyong, Jun Zhao, and Ruth Nussinov. 2016. "Conformational selection in amyloid-based immunotherapy: survey of crystal structures of antibody-amyloid complexes." *Biochimica et Biophysica Acta (BBA) - General Subjects* 1860 (11, Part B):2672–2681. https://doi.org/10.1016/j.bbagen.2016.05.040

Ma, Tian-Jun, Jie Gao, Yan Liu, Jian-Hua Zhuang, Chuan Yin, Peng Li, Lei Mao, Jin Xu, Yi-Xin Xu, and Yan-Peng Li. 2018. "Nanomedicine strategies for sustained, controlled and targeted treatment of Alzheimer's disease." *Mini Reviews in Medicinal Chemistry* 18 (12):1035–1046.

Makin, O Sumner, and Louise C Serpell. 2005. "Structures for amyloid fibrils." *The FEBS Journal* 272 (23): 5950–5961.

Maniam, Sandra, and Subashani Maniam. 2024. "Screening techniques for drug discovery in Alzheimer's disease." *ACS Omega* 9 (6):6059–6073. https://doi.org/10.1021/acsomega.3c07046

Mercy, Devadass Jessy, Karthick Harini, Swaminathan Madhumitha, Chinnadurai Anitha, Jayakumar Iswariya, Koyeli Girigoswami, and Agnishwar Girigoswami. 2023a. "pH-responsive polymeric nanostructures for cancer theranostics." *Journal of Metals, Materials and Minerals* 33 (2):1–15. https://doi.org/10.55713/jmmm.v33i2.1609

Mercy, Devadass Jessy, Venkatakrishnan Kiran, Anbazhagan Thirumalai, Karthick Harini, Koyeli Girigoswami, and Agnishwar Girigoswami. 2023b. "Rice husk assisted carbon quantum dots synthesis for amoxicillin sensing." *Results in Chemistry* 6:101219. https://doi.org/10.1016/j.rechem.2023.101219

Merlini, Giampaolo, David C Seldin, and Morie A Gertz. 2011. "Amyloidosis: pathogenesis and new therapeutic options." *Journal of Clinical Oncology* 29 (14):1924.

Metkar, Sanjay Kisan, Agnishwar Girigoswami, Devanand D Bondage, Umakant G Shinde, and Koyeli Girigoswami. 2022. "The potential of lumbrokinase and serratiopeptidase for the degradation of Aβ 1–42 peptide–an in vitro and in silico approach." *International Journal of Neuroscience*: 1–12. https://doi.org/10.1080/00207454.2022.2089137

Metkar, Sanjay Kisan, Agnishwar Girigoswami, R Vijayashree, and Koyeli Girigoswami. 2020. "Attenuation of subcutaneous insulin induced amyloid mass in vivo using lumbrokinase and serratiopeptidase." *International Journal of Biological Macromolecules* 163:128–134.

Metkar, Sanjay Kisan, Agnishwar Girigoswami, Ramachandran Murugesan, and Koyeli Girigoswami. 2017a. "Lumbrokinase for degradation and reduction of amyloid fibrils associated with amyloidosis." *Journal of Applied Biomedicine* 15 (2):96–104.

Metkar, Sanjay Kisan, Agnishwar Girigoswami, Ramachandran Murugesan, and Koyeli Girigoswami. 2017b. "In vitro and in vivo insulin amyloid degradation mediated by serratiopeptidase." *Materials Science and Engineering: C* 70:728–735.

Metkar SK, S Ghosh, A Girigoswami, K. Girigoswami. The Potential of Serratiopetidase and Lumbrokinase for the Degradation of Prion Peptide 106–126 - an In Vitro and In Silico Perspective. CNS Neurol Disord Drug Targets. 2019;18(9):723–731. doi:10.2174/1871527318666191021150002.

Metkar, Sanjay Kisan, Saranya Udayakumar, Agnishwar Girigoswami, and Koyeli Girigoswami. 2023. "Amyloidosis-history and development, emphasis on insulin and prion amyloids." *Brain Disorders*: 100106.

Murphy, M Paul, and Harry LeVine III. 2010. "Alzheimer's disease and the amyloid-β peptide." *Journal of Alzheimer's Disease* 19 (1):311–323. 10.3233/JAD-2010-1221

Nowak, Anna, Klaudyna Kojder, Joanna Zielonka-Brzezicka, Jacek Wróbel, Mateusz Bosiacki, Marta Fabiańska, Mariola Wróbel, Joanna Sołek-Pastuszka, and Adam Klimowicz. 2021. "The use of Ginkgo biloba L. as a neuroprotective agent in the Alzheimer's disease." *Frontiers in Pharmacology* 12:775034.

Oliveira, Jordan Teixeira, and Simone Pieniz. 2024. "Curcumin in Alzheimer's disease and depression: therapeutic potential and mechanisms of action." *Brazilian Archives of Biology and Technology* 67:e24220004.

Olivera-Perez, Henry M, Larry Lam, Johnny Dang, Weilan Jiang, Fabian Rodriguez, Elizabeth Rigali, Sarah Weitzman, Verna Porter, Liudmilla Rubbi, and Marco Morselli. 2017. "Omega-3 fatty acids increase the unfolded protein response and improve amyloid-β phagocytosis by macrophages of patients with mild cognitive impairment." *The FASEB Journal* 31 (10):4359.

Pallavi, Pragya, Karthick Harini, Sultan Alshehri, Mohammed M Ghoneim, Areej Alshlowi, Pemula Gowtham, Koyeli Girigoswami, Faiyaz Shakeel, and Agnishwar Girigoswami. 2022. "From synthetic route of silica nanoparticles to theranostic applications." *Processes* 10 (12):2595.

Picken, Maria M. 2020. "The pathology of amyloidosis in classification: a review." *Acta Haematologica* 143 (4):322–334.

Pinheiro, Lídia, and Célia Faustino. 2019. "Therapeutic strategies targeting amyloid-β in Alzheimer's disease." *Current Alzheimer Research* 16 (5):418–452. https://doi.org/10.2174/1567205016666190321163438

Pinney JH, Lachmann HJ. Systemic AA amyloidosis. Protein aggregation and fibrillogenesis in cerebral and systemic amyloid disease. 2012:541–64.

Plotkin, Steven S., and Neil R. Cashman. 2020. "Passive immunotherapies targeting Aβ and tau in Alzheimer's disease." *Neurobiology of Disease* 144:105010. https://doi.org/10.1016/j.nbd.2020.105010

Prades, Roger, Benjamí Oller-Salvia, Susanne M Schwarzmaier, Javier Selva, María Moros, Matilde Balbi, Valeria Grazú, Jesus M de La Fuente, Gustavo Egea, and Nikolaus Plesnila. 2015. "Applying the retro-enantio approach to obtain a peptide capable of overcoming the blood–brain barrier." *Angewandte Chemie International Edition* 54 (13):3967–3972.

Prusiner, Stanley B. 1998. "Prions." *Proceedings of the National Academy of Sciences* 95 (23):13363–13383.

Ps, Sharon Sofini, Arina Guha, Balasubramanian Deepika, Saranya Udayakumar, Moupriya Nag, Dibyajit Lahiri, Agnishwar Girigoswami, and Koyeli Girigoswami. 2023. "Nanocargos designed with synthetic and natural polymers for ovarian cancer management." *Naunyn-Schmiedeberg's Archives of Pharmacology*:19. https://doi.org/10.1007/s00210-023-02608-0

Qais, Faizan Abul, Nagma Parveen, Mohd Afzal, Mohammad Furkan, and Rizwan Hasan Khan. 2024. "Preventing amyloid-β oligomerization and aggregation with berberine: Investigating the mechanism of action through computational methods." *International Journal of Biological Macromolecules* 258:128900.

Rajasekhar, Kolla, Sourav Samanta, Vardhaman Bagoband, N Arul Murugan, and Thimmaiah Govindaraju. 2020. "Antioxidant berberine-derivative inhibits multifaceted amyloid toxicity." *Iscience* 23 (4).

Regitz, Charlotte, Elena Fitzenberger, Friederike Luise Mahn, Lisa Marie Dußling, and Uwe Wenzel. 2016. "Resveratrol reduces amyloid-beta (Aβ 1–42)-induced paralysis through targeting proteostasis in an Alzheimer model of caenorhabditis elegans." *European Journal of Nutrition* 55:741–747.

Ryu, Jungki, Chan Beum Park, and Georg Lentzen. 2007. "Extremolytes give us a hint for designing synthetic compounds as potential inhibitors of amyloid formation." 한국생물공학회 학술대회: 18–18.

Ryu, Jungki, Mathumai Kanapathipillai, Georg Lentzen, and Chan Beum Park. 2008. "Inhibition of β-amyloid peptide aggregation and neurotoxicity by α-d-mannosylglycerate, a natural extremolyte." *Peptides* 29 (4):578–584.

Sacchettini, James C., and Jeffery W. Kelly. 2002. "Therapeutic strategies for human amyloid diseases." *Nature Reviews Drug Discovery* 1 (4):267–275. https://doi.org/10.1038/nrd769

Sahota, T., A. Berges, S. Barton, L. Cookson, S. Zamuner, and D. Richards. 2015. "Target mediated drug disposition model of CPHPC in patients with systemic amyloidosis." *CPT: Pharmacometrics & Systems Pharmacology* 4 (2):116–126. https://doi.org/10.1002/psp4.15

Sakthi Devi, R, Agnishwar Girigoswami, M Siddharth, and Koyeli Girigoswami. 2022. "Applications of gold and silver nanoparticles in theranostics." *Applied Biochemistry and Biotechnology* 194 (9):4187–4219. https://doi.org/10.1007/s12010-022-03963-z

Sallaberry, Chad A, Barbie J Voss, William B Stone, Fabiola Estrada, Advita Bhatia, J Daniel Soto, Charles W Griffin, and Crystal M Vander Zanden. 2023. "Curcumin reduces amyloid beta oligomer interactions with anionic membranes." *ACS Chemical Neuroscience* 14 (22):4026–4038.

Sánchez-Navarro, Macarena, Ernest Giralt, and Meritxell Teixidó. 2017. "Blood–brain barrier peptide shuttles." *Current Opinion in Chemical Biology* 38:134–140.

Sawda, Christine, Charbel Moussa, and R Scott Turner. 2017. "Resveratrol for Alzheimer's disease." *Annals of the New York Academy of Sciences* 1403 (1):142–149.

Scarpioni, Roberto, Marco Ricardi, Vittorio Albertazzi, Sara De Amicis, Francesco Rastelli, and Lara Zerbini. 2016. "Dialysis-related amyloidosis: challenges and solutions." *International Journal of Nephrology and Renovascular Disease*: 319–328.

Shahidi, Siamak, Fatemeh Ghahremanitamadon, Sara Soleimani Asl, Alireza Komaki, Simin Afshar, and Nasrin Hashemi-Firouzi. 2021. "Electrophysiological, behavioral and molecular study of vitamin E and Ginkgo biloba in a rat model of Alzheimer's disease." Research Journal of Pharmacognosy 8 (1):39–51. 10.22127/RJP.2020.250269.1630

Siepen, Fabian aus dem, Ralf Bauer, Matthias Aurich, Sebastian J Buss, Henning Steen, Klaus Altland, Hugo A Katus, and Arnt V Kristen. 2015. "Green tea extract as a treatment for patients with wild-type transthyretin amyloidosis: an observational study." *Drug Design, Development and Therapy*: 6319–6325.

Simoni, Elena, Melania M Serafini, Roberta Caporaso, Chiara Marchetti, Marco Racchi, Anna Minarini, Manuela Bartolini, Cristina Lanni, and Michela Rosini. 2017. "Targeting the Nrf2/amyloid-beta liaison in Alzheimer's disease: a rational approach." *ACS Chemical Neuroscience* 8 (7):1618–1627.

Sipe, Jean D, Merrill D Benson, Joel N Buxbaum, Shu-ichi Ikeda, Giampaolo Merlini, Maria JM Saraiva, and Per Westermark. 2016. "Amyloid fibril proteins and amyloidosis: chemical identification and clinical classification International Society of Amyloidosis 2016 Nomenclature Guidelines." *Amyloid* 23 (4):209–213.

Subramanian, Manickam, Balaji Thotakura, Swathi Priyadarshini Chandra Sekaran, and Indumathi Sundaramurthi. 2019. "Naringin (4′, 5, 7-trihydroxyflavanone 7-rhamnoglucoside) attenuates β-cell dysfunction in diabetic rats through upregulation of PDX-1." *Cells Tissues Organs* 206 (3):133–143.

Subramanian, Saraswathi, Karunanithi Rajamanickam, Joy Sebastian Prakash, Murugesan Ramachandran, and for Alzheimer's Disease Neuroimaging Initiative. 2020. "Study on structural atrophy changes and functional connectivity measures in Alzheimer's disease." *Journal of Medical Imaging* 7 (1): 016002–016002.

Thirumalai, Anbazhagan, Karthick Harini, Pragya Pallavi, Pemula Gowtham, Koyeli Girigoswami, and Agnishwar Girigoswami. 2024. "Bile salt-mediated surface-engineered bilosome-nanocarriers for delivering therapeutics." *Nanomedicine Journal* 11 (1):1–12.

Thirumalai, Anbazhagan, Koyeli Girigoswami, Pragya Pallavi, Karthick Harini, Pemula Gowtham, and Agnishwar Girigoswami. 2023b. "Cancer therapy with iRGD as a tumor-penetrating peptide." *Bulletin du Cancer* 110(12):1288–1300. https://doi.org/10.1016/j.bulcan.2023.08.009.

Thirumalai, Anbazhagan, Noureddine Elboughdiri, Karthick Harini, Koyeli Girigoswami, and Agnishwar Girigoswami. 2023a. "Phosphorus-carrying cascade molecules: inner architecture to biomedical applications." *Turkish Journal of Chemistry* 47 (4):667–688.

Toyama, Brandon H, and Jonathan S Weissman. 2011. "Amyloid structure: conformational diversity and consequences." *Annual Review of Biochemistry* 80:557–585.

Udayakumar, Saranya, Agnishwar Girigoswami, and Koyeli Girigoswami. 2024. "Biological activities of carrageenan from red algae: a mini review." *Current Pharmacology Reports* 10 (1):12–26.

Wang, Jian-Bo, Yi-Min Wang, and Cheng-Ming Zeng. 2011. "Quercetin inhibits amyloid fibrillation of bovine insulin and destabilizes preformed fibrils." *Biochemical and Biophysical Research Communications* 415 (4):675–679.

Wang, Shao-wei, Shi-gao Yang, Wen Liu, Yang-xin Zhang, Peng-xin Xu, Teng Wang, Tie-jun Ling, and Rui-tian Liu. 2016. "Alpha-tocopherol quinine ameliorates spatial memory deficits by reducing beta-amyloid oligomers, neuroinflammation and oxidative stress in transgenic mice with Alzheimer's disease." *Behavioural Brain Research* 296:109–117.

Wang, Wanyu, Jiao Li, Huizhen Zhang, Xiaokai Wang, and Xiaofeng Zhang. 2021. "Effects of vitamin E supplementation on the risk and progression of AD: a systematic review and meta-analysis." *Nutritional Neuroscience* 24 (1):13–22.

Wu, Yanjue, Zhixin Wu, Peter Butko, Yves Christen, Mary P Lambert, William L Klein, Christopher D Link, and Yuan Luo. 2006. "Amyloid-β-induced pathological behaviors are suppressed by Ginkgo biloba extract EGb 761 and ginkgolides in transgenic *Caenorhabditis elegans*." *Journal of Neuroscience* 26 (50):13102–13113.

Xie, Haiyan, Jing-Rong Wang, Lee-Fong Yau, Yong Liu, Liang Liu, Quan-Bin Han, Zhongzhen Zhao, and Zhi-Hong Jiang. 2014. "Quantitative analysis of the flavonoid glycosides and terpene trilactones in the extract of Ginkgo biloba and evaluation of their inhibitory activity towards fibril formation of β-amyloid peptide." *Molecules* 19 (4):4466–4478.

Yan, Lingli, Youna Xie, Senthil Kumaran Satyanarayanan, Haitao Zeng, Qiang Liu, Miaodan Huang, Yuemeng Ma, Jian-Bo Wan, Xiaoli Yao, and Kuan-Pin Su. 2020. "Omega-3 polyunsaturated fatty acids promote brain-to-blood clearance of β-amyloid in a mouse model with Alzheimer's disease." *Brain, Behavior, and Immunity* 85:35–45.

Yokoyama, Takeshi, Katsuhiro Kusaka, Mineyuki Mizuguchi, Yuko Nabeshima, and Satoru Fujiwara. 2023. "resveratrol derivatives inhibit transthyretin fibrillization: structural insights into the interactions between resveratrol derivatives and transthyretin." *Journal of Medicinal Chemistry* 66 (22):15511–15523.

Zhan, Chendi, Yujie Chen, Yiming Tang, and Guanghong Wei. 2020. "Green tea extracts EGCG and EGC display distinct mechanisms in disrupting Aβ42 protofibril." *ACS Chemical Neuroscience* 11 (12):1841–1851.

Zhang, Xilin, Jin Hu, Li Zhong, Na Wang, Longyu Yang, Chia-Chen Liu, Huifang Li, Xin Wang, Ying Zhou, and Yunwu Zhang. 2016. "Quercetin stabilizes apolipoprotein E and reduces brain Aβ levels in amyloid model mice." *Neuropharmacology* 108:179–192.

15 Regulation of Cell Signaling Pathways by Phytochemicals in Aging Brain

Anushka Agrawal[1], *Ankita Panigrahi*[1], *Dhanu AS*[1], *Gobianand Kuppannan*[2], *Malarvizhi Arthanari*[3], and *Kanthesh M Basalingappa*[1]

[1]Division of Molecular Biology, School of Life Sciences, JSS Academy of Higher Education & Research Mysuru, India

[2]Department of Microbiology, Noorul Islam College of Dental Sciences Aralummoodu, Thiruvananthapuram, India

[3]Vivekandha College of Arts and Sciences for Women (Autonomous) Elayampalayam, Tiruchengode, India

15.1 INTRODUCTION

The brain records and stores each person's individual experiences are one of the world's greatest scientific mysteries. This organ provides our ability to understand other vast mysteries in the world around us, which also gives us our unique identity. The accumulation of damage in molecules, cells, and tissues over a lifetime is what is known as aging, which affects physiological functions. This accumulation of damage frequently reduces an organism's ability to maintain homeostasis under stress conditions and increases the risk of many diseases such as cancer, cardiovascular, and neurodegenerative disorders as well as early mortality [1]. Recent years have seen an unparalleled advancement in aging research, especially with the realization that genetic pathways and metabolic processes that have been preserved through evolution, at least in part, regulate the rate of aging [2]. Changes in dynamic biological, physiological, environmental, psychological, behavioral, and social systems are linked to aging. The loss of tissue homeostasis that comes with aging is unavoidable, and the brain is more susceptible to the grave consequences of aging than other organs. The high demand that neurons have for oxidative metabolism in the synthesis of energy, even though this vulnerability is not universal in the brain, is a significant factor that significantly contributes to brain aging. Hence, inherent, and inevitable mild oxidative stress-increasing processes connected to senescence and aging are associated with normal brain metabolism. More than one process that causes redox state changes can have an impact on the central nervous system [3]. The various signaling pathways that contribute to brain aging are illustrated in Figure 15.1.

Ca^{2+} homeostasis in neurons is hampered by the accumulation of stress and the normal aging-related antioxidant defense mechanisms, making neurons more susceptible to neurodegenerative disorders. Age-related alterations in the expression of intracellular receptors, such as TRPM2, IP3R, RyR, and others, disrupt calcium homeostasis and impair normal brain function [4]. Another reason for aging could be an impaired waste disposal system where PET of prodromal AD reveals that beta-amyloid builds abnormally between the ages of 40 and 50, although symptoms do not manifest until a few molecular changes first arise [5]. Age-related dysfunctions like atherosclerosis, cataract genesis, AD, PD, etc. are caused by the accumulation of altered proteins both intracellularly and extracellularly, which occurs as people age and is either accompanied by an increase in their population

DOI: 10.1201/9781003486237-15

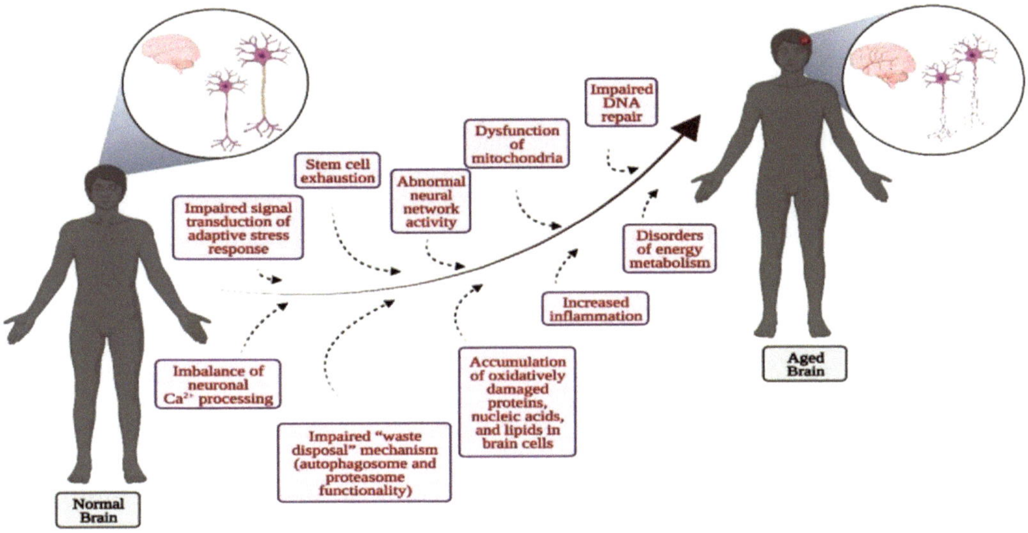

FIGURE 15.1 Signaling pathways that contribute to brain aging process.

or a decrease in the cell's capacity to remove them [6]. Pro-inflammatory and anti-inflammatory cytokines are in balance in the normal brain, but as we grow, the balance of inflammatory mediators shifts from being homeostatic to being pro-inflammatory. Although microglia are normally in the resting stage, foreign antigen-like stress can activate them, causing the release of cytokines and neuroinflammation as a result. Moreover, the amounts of anti-inflammatory mediators are reduced, which also harms the brain as we age [7]. Age-dependent declines in brain cell energy metabolism (glucose utilization) in the prefrontal cortex and hippocampus of rats are linked to poorer performance on learning and memory tests. According to studies, decreased levels of GLUT 1 in mice are linked to age-dependent declines in cerebral capillary density, decreased glucose absorption, and decreased cerebral blood flow [8].

The moles of cholesterol: phospholipid fall by 30% as we age due to membrane thinning, which has a direct impact on the coupling and uncoupling characteristics of G proteins, as well as their receptors and effectors, and is what causes impairments in signal transmission during aging [9]. In comparison to younger animals, older animals exhibit significantly decreased levels of neural stem cells (NSCs), stem cell proliferation, neuronal differentiation, and neonatal neuron survival. This drop is accompanied by an epigenetic loss of DNA methylation, which suggests that NSCs may begin to age molecularly before they physically do. Yet, it is still unknown what cellular factors caused early neurogenesis reduction [10]. Normal brain function depends on the structural integrity and effective integration of the synaptic activity of these several neurotransmitter systems. The consistency of neural network activity within and between brain areas is hampered by brain aging. Normal aging-related G protein-coupled serotonergic, noradrenergic, dopaminergic, and cholinergic neuron degeneration and malfunction can put people at risk for developing one or more age-related neurodegenerative illnesses [11]. Due to malfunctioning mitochondria, the brain's ability to make ATP through oxidative phosphorylation declines with age and causes respiration to diminish. Low mtNOS activity in old neurons slows mitochondrial synthesis, increases oxidative damage, and impairs mitochondrial function [12]. Oxidative injury is particularly widespread in the brain. Brain cells have very efficient BER mechanisms to deal with the heightened oxidation stress and oxidative DNA damage, but it lowers as we age. Endogenous metabolic activity damages DNA in the brain, which results in genomic instability and cellular malfunction [13].

Recent research suggests that certain food-derived tiny molecules, commonly known as phytochemicals, may be a useful strategy to slow down aging and age-related neurodegeneration.

Phytochemicals are regarded as guardians of our health. Most phytochemicals, in particular, are secondary plant metabolites, and these bioactive compounds can be found in a wide range of foods. There are many different types of phytochemicals that have received the most research. Although many different mechanisms for the health advantages of phytochemicals have been put forth, their capacity to boost the production of antioxidant proteins and activate transcription factors that combat inflammation has drawn significant attention due to their part in brain aging. Fruits, vegetables, whole grains, nuts, legumes, and herbal seasonings all contain a variety of beneficial phytochemicals, making frequent consumption of these foods crucial for maintaining a healthy lifestyle [3, 14]. This chapter focuses on the cell signaling pathways that are regulated by phytochemicals in order to delay or inhibit the brain aging process.

15.2 AGING THEORIES

The intricate process of aging is made up of numerous cellular processes that participate in the process of senescence in an integrated way. The large degree of longevity variation among individuals of the same species demonstrates that mechanisms that hurt stem cell function, harm repair mechanisms, and cause mistakes to accumulate over time have a considerable impact on the aging process. To understand how genetic and epigenetic mechanisms, which are influenced by genes, the environment, and stochastic events, may cause these changes, future studies will need to determine the relative contributions of each of these factors. There have been many different theories put up regarding the causes of aging [1]. Significant ones are listed in Figure 15.2 and summarized in Table 15.1.

There can be four types of aging. The biological age of a person is an estimation of where that person stands right now in relation to his or her possible lifespan. Psychological age relates to a person's capacity for adaptation, or how effectively they can adjust to shifting environmental demands as compared to the average. Psychological age and functional age are tightly related. The level of a person's capacities for functioning in a certain human civilization relative to others his age is known

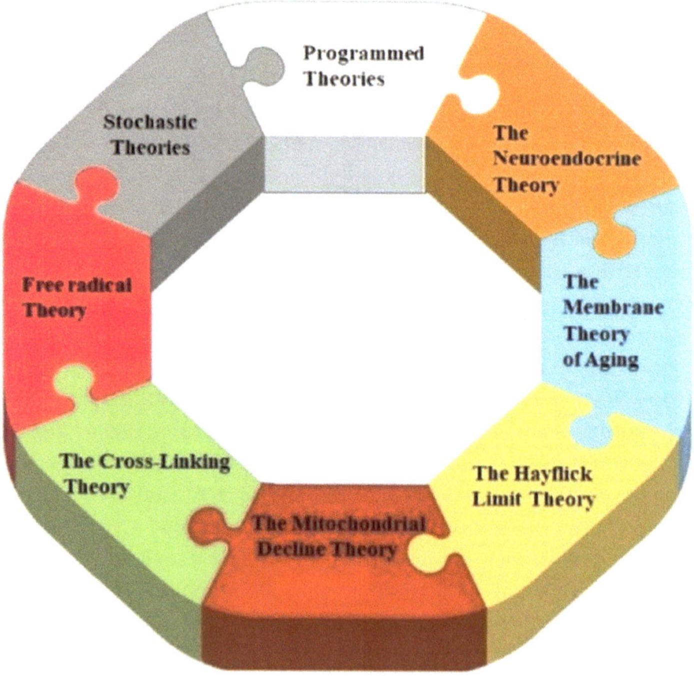

FIGURE 15.2 Theories of aging.

TABLE 15.1
Theories of Aging

SL. No.	Theories of Aging	Characteristics of the Proposed Theory
1.	Programmed Theories	Some programmable theories of aging contend that every living thing has a biological clock that controls the aging process. Extremely recent research indicates that it might include, at least in part, DNA caps called telomeres that cover the ends of our chromosomes. The telomeres shorten with each cell division; when this shortening reaches a certain threshold, the cell is no longer able to divide, which may speed up the aging process [16].
2.	The Neuroendocrine Theory	As most neuroendocrine activities start to decline as people become older, the neuroendocrine system, in conjunction with the brain, plays a special role in moderating the impacts of the environment on the body [17].
3.	The Membrane Theory of Aging	The impairment comes from the aging-related changes in the cells' capacity to transport chemicals, heat, and electrical activities. The lipid content of the cell membrane decreases with age This makes it less effective at carrying out routine tasks, and there is a toxic build-up in particular [18].
4.	The Hay flick Limit Theory	According to the Hay Flick Limit Theory of Aging, a human cell can only divide so many times. Cell waste build-up, as suggested by the membrane aging theory, may have an impact on a portion of this idea [19].
5.	The Mitochondrial Decline Theory	Mitochondria lack most of the defenses found in other parts of the body, so as we age, mitochondria become less efficient and fewer in number. As a result, ATP production falls. Because organs cannot borrow energy from one another, the efficiency of each organ's mitochondria is critical to that organ's repair processes and functions. If the mitochondria in a specific organ fail, the organ fails as well [20, 21, 22].
6.	The Cross-Linking Theory	It states that aging is caused by the progressive linking together of large vital molecules. When many of these molecules in a cell become entangled and form massive tangles, intracellular transport is hampered. This slows down life processes and reduces the safety margins of vital functions [23].
7.	Free Radical Theory	According to the free radical theory, oxidative damage to the body's cells and tissues, which predominantly results from aerobic metabolism, leads to aging over time [24, 25, 26].
8.	Stochastic Theories	The body's incapacity to repair its systems and prevent the development of aging, according to stochastic theories of aging, is the result of small alterations that develop gradually over time. Organ function is impacted by the injury to cells and tissues that results from the accumulation of damage over time [27].

as functional age. The term "social age" describes a person's roles and social behaviors in relation to other people in a culture [15].

15.3 PHYTOCHEMICALS ARE STEWARDS OF OUR HEALTH

Plants produce two kinds of metabolites. The primary metabolites are proteins, fats and carbohydrates, and dietary fibers and they are in charge of energy metabolism and maintaining the structure of plant cells. On the other hand, plants frequently create modest amounts of secondary metabolites, also known as phytochemicals, which are non-nutritive food components with low molecular weight in reaction to bacteria, insects, growth regulators, and colors as a kind of defense [28]. Studies demonstrated that the Mediterranean diet (MedDiet) is rich in olive oil, berries, and honey and is responsible for the prevention and progression of cancer and cardiovascular diseases by inhibiting angiogenesis and carcinogens and activating apoptosis and cell cycle arrest [29]. Another study with almost 2000 participants discovered that those who consumed chocolate, wine, and tea had a better mean test score than the ones who didn't consume [30].

As previously mentioned, brain aging is linked to neuronal dysfunction that can result from a variety of pathways, as seen in Figure 15.1. This is because neurons have high energy and oxygen requirements, and even little stress can result in the creation of ROS and RNS, which endanger neurons. Since high ROS and RNS increase the synthesis of pro-inflammatory cytokines (IL-1, IL-6, and TNF), which have the potential to activate microglia and astrocytes and cause neuroinflammation, communication between the brain and immune system is essential for maintaining CNS homeostasis, according to an extensive literature review. It has been demonstrated that phytochemicals can activate a number of signaling pathways that result in the development of cytoprotective and repair proteins, preventing neurodegeneration [31].

Oxidative stress and excessive ROS generation can be detrimental to cells and result in cellular damage and dysfunction. The body has defense mechanisms to eliminate ROS, but when these defenses are overworked, ROS can interact with cellular components and cause harm that can be a factor in several illnesses and health issues. Therefore, ROS have the potential to interact with and harm crucial cellular components like DNA, lipids, and proteins. DNA harm brought on by ROS can result in mutations that may aid in the growth of numerous diseases as well as damage to the cell. Lipid peroxides, which are toxic and can harm cell membranes, can be produced when ROS -induced injury to lipids occurs. Protein aggregation and misfolding brought on by ROS degradation can compromise the functionality of these molecules. All these effects have the potential to negatively impact cellular function and initiate the process of aging. India has a long history of employing different herbs and herbal extracts to treat a variety of illnesses. The usefulness of many Indian plants as sources of antioxidants or as antioxidants themselves has been studied [32]. Herbal remedies provide a variety of brain targets, nutritional advantages, safe dose, long-term usage, and effective anti-aging treatments. The phytochemicals used in conventional Chinese medicine have shown to be safe for intake by humans. Dietary herbs and calorie restriction can prevent brain aging, which is connected to the buildup of beta-amyloid proteins, cardiovascular disorders, hypertension, and diabetes [33].

15.4 PHYTOCHEMICALS REGULATE CELL SIGNALING PATHWAYS IN THE AGING BRAIN

Numerous studies have been carried out over the past few decades to investigate the effects of natural compounds that are plant secondary metabolites or phytochemicals that affect brain aging pathways and prevent people from developing age-related diseases. Changes in neural circuit activities and the cellular structure of the brain are regulated by complicated intercellular molecules and intercellular signal transduction pathways, and during brain aging several signal transduction systems are altered which is shown in Figure 15.1 [34]. Significantly these changes play a significant role in the development of these systemic deficits and are antecedents of the age-related deterioration in all other physiological systems [35]. Based on the molecular properties of the aging brain, we summarize the function of herbal/natural substances in brain aging in this review article in the hopes that it may improve research into age-related diseases. This review, on the other hand, focuses on numerous dietary components for which substantial experimental data has been published. Here we are focusing on the effects of herbal compounds that regulate the following brain aging process; (1) Nrf-2 regulation by phytochemicals against oxidative stress, (2) NF-kB regulation by phytochemicals against neuroinflammation, (3) Autophagy induction by phytochemicals, (4) Telomerase activation by phytochemicals, (5) Anti-apoptotic property of carotenoids, (6) Phytochemicals promote pro-survival of neurotropic factors, (7) MMP prevention by phytochemicals. Regulation of cell signaling pathways by phytochemicals in the aging brain is illustrated in Figure 15.3.

15.4.1 Nrf-2 Regulation by Phytochemicals against Oxidative Stress

To function properly our body cells, need oxygen and some cells like neurons and astrocytes demand high oxygen to function efficiently. Such a high demand for oxygen generally produces reactive oxygen species, hydroxyl radicals, and hydrogen peroxide that causes cellular stress often known as oxidative stress. These reactive species can directly interfere with the cellular machinery

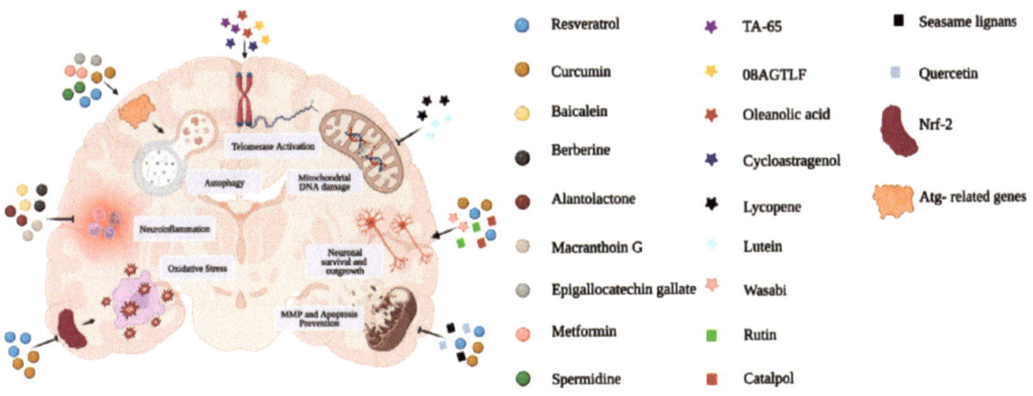

FIGURE 15.3 Regulation of cell signaling pathways by phytochemicals in aging brain.

and often lead to damage to cellular molecules and organelles. As a defense mechanism, our cells have a certain antioxidant system that attenuates oxidative stress-mediated injury. One such mechanism involves proteins like Nrf-2. Under normal conditions, Nrf-2 binds with Keap-1 and stays in the cytoplasm in an inactive form. As soon as the cell experiences oxidative stress Nrf-2 dissociates from Keap-1 and translocate to nucleus where it combines with MAF and interacts with ARE, leading to the expression of antioxidant enzymes that release the overload of stress on the cell by neutralizing reactive oxygen species. But as we age the ability of the cells to do so decreases, so phytochemicals can directly activate Nrf-2 and hence lead to the activation of antioxidant enzymes. Various studies have identified such potential phytochemicals such as Resveratrol (5-[(*E*)-2-(4-hydroxyphenyl) ethenyl] benzene-1,3-diol) which is present in berries, and red wine is a polyphenolic stilbenoid which acts as agonists and leads to the expression of Nrf-2 in the nucleus that activates its downstream targets and activates antioxidant enzymes, thus leading to the protection of neuron cells against oxidative stress. Another phytochemical is Curcumin ((1*E*,6*E*)-1,7-bis(4-hydroxy-3-methoxyphenyl) hepta-1,6-diene-3,5-dione) obtained from the rhizome of turmeric has the property to cross the blood–brain barrier (BBB) and causes the upregulation of Nrf-2 thereby acting as neuroprotectants. In BTBR mice models, neonatal curcumin treatment has been shown to enhance neurogenesis and neural progenitor cell proliferation. Nrf-2 upregulation by using phytochemicals can delay the normal aging process [36–38].

15.4.2 NF-κB Regulation by Phytochemicals against Neuroinflammation

Activation of the NF-κB signaling pathway is associated with the degeneration of neurons and ultimately results in aging and is also one of the prime signaling in glioblastoma. NF-κB is associated with IκB and remains in the cytoplasm in its inactive form. As soon as the cell receives the signal of stress IκB is ubiquitin and is degraded by the proteasome and NF-kB translocate to the nucleus and activates its downstream targets which are the majority pro-inflammatory cytokines and cause neuroinflammation in the brain. Certain phytochemicals have been studied for their anti-inflammatory property, these include Quercetin (2-(3,4-dihydroxyphenyl)-3,5,7-trihydroxychromen-4-one) is a phenolic compound and studies have shown that low doses of quercetin glioblastoma cell invasion and angiogenesis by regulating PI3K pathway and it was also shown to suppress the expression of PD-L1 by inhibiting NF-kB signaling pathway. Studies have shown that free Keap-1 prevents the degradation of IκB which ultimately leads to the inactivation of the NF-kB signaling pathway. Baicalein (5,6,7-trihydroxy-2-phenylchromen-4-one) is a flavone that is known for its anti-cancer property, and it has been shown to accelerate cellular apoptosis by blocking the NF-κB pathway followed by a reduction in the levels of Bcl-2 and increases the level of Bax. Another phytochemical berberine (16,17-dimethoxy-5,7-dioxa-13-azoniapentacyclo [11, 8.002, 10.04, 8.015, 20] henicosa-1(13),2,4(8),9,14,16,18,20-octaene) isolated from *Hydrastis canadensis* decreases the phosphorylation of p65 and also inhibited phosphorylation of IκB, leading to inactivation of NF-kb pathway inhibit the inflammatory response in microglial cells. A sesquiterpene lactone Alantolactone ((3*aR*,5*S*,8*aR*,9*aR*)-5,8*a*-dimethyl-3-methylidene-5,6,7,8,9,9*a*-hexahydro-3*aH*-benzo[f] [1] benzofuran-2-one) that is extracted from *Inula helenium L* and it has been shown to downregulate the expression of COX-2 that is the downstream target of NF-kB thereby inhibiting neuroinflammation during initial stages of glioblastoma. Also, it has been able to trigger apoptosis by activating the caspase/cytochrome c signaling pathway. Macranthoin G (methyl 3,5-bis[3-(3,4-dihydroxyphenyl) prop-2-enoyloxy]-1,4-dihydroxycyclohexane-1-carboxylate) that is isolated from *Eucommia ulmoides* activate the phosphorylation of ERK, p38, and IκBα and inhibit NFkB pathway and thereby minimize the cell damage [39–41].

15.4.3 Autophagy Induction by Phytochemicals

Eukaryotic cells have conserved mechanisms to eliminate dysfunctional organelles, long-lived proteins, and aggregates. All these waste products are sequestered into a double membraned structure known as phagophore which matures to form an autophagosome which then binds with a lysosome and causes the degradation of the above-mentioned waste products. Autophagy can be divided into 3 major classes- macro-autophagy, micro-autophagy, and chaperon-mediated autophagy. All of them follow a similar mechanism for the degradation process. The pre-autophagosome is formed with the help of the proteins Fip200, Atg 13, Afg 101, and ULK-1; ULK-1 phosphorylates Beclin 1, which results in PIP3 formation from PIP2, which recruits p62; LC3 is activated by Atg 4 and leads to the activation of LC3-1. Atg-7 causes the formation of mature autophagosomes by activating LC3-II. Mature autophagosome fuses with lysosomes with the help of SNARE proteins and Rab 7 and thus causes the removal of altered biomolecules or organelles. Epigallocatechin gallate (EGCG) ([(2*R*,3*R*)-5,7-dihydroxy-2-(3,4,5-trihydroxyphenyl)-3,4-dihydro-2*H*-chromen-3-yl] 3,4-dihydroxy-5-methoxybenzoate) from *Camellia sinensis* has been shown to activate autophagy by activating LC3-I, which causes the synthesis of LC3-II and thus provide neuroprotection to primary neuron cells. Curcumin ((1*E*,6*E*)-1,7-bis(4-hydroxy-3-methoxyphenyl) hepta-1,6-diene-3,5-dione) from *Curcuma longa* upregulates autophagy-related genes like Atg-5 and Atg-8 (LC3) and inhibits mTOR and induces autophagy thus protect from oxidative stress caused by missense proteins. AMPK activity has been shown to increase the life span and its inhibition shortens the life span. It phosphorylates PGC-1α, a key regulator of FOXO transcription factor, and directly activates ULK-1 that inhibits mTOR and autophagy. Metformin (3-(diaminomethylidene)-1,1-dimethylguanidine) isolated from *Galega officinalis* induces autophagy and is administered to type-2 diabetes patients as they have a high risk of developing neurodegenerative diseases. Resveratrol (5-[(*E*)-2-(4-hydroxyphenyl)ethenyl]benzene-1,3-diol) which is present in berries, and red wine is a polyphenolic stilbenoid and Spermidine (*N*′-(3-aminopropyl)butane-1,4-diamine) which is a naturally occurring polyamine found in wheatgerm activates Sirtuin 1 (SIRT-1), a nicotine amide NAD+ -dependent histone deacetylase that has been shown to activate Atg-related genes and FOXO transcription factors that induces autophagy thus preventing neural cells to undergo aging [42–45].

15.4.4 Telomerase Activation by Phytochemicals

A telomere is a nucleotide sequence that repeats frequently at the end of a chromosome. It protects against chromosome fusion and retains chromosome disintegration. Telomere length is approximately 15,000 base pairs at birth and decreases to 4000 base pairs as people age. Because a portion of the chromosome's end does not replicate with each replication, the length of the chromosomes decreases. Telomere length reduction has a negative impact on health, causes genetic instability, and jeopardizes the ability of the cell cycle to function. As a result, the cells enter the aging and apoptotic phases. To avoid this ribonucleotide enzyme telomerase is present which consists of two components: telomerase reverse transcriptase protein (TERT) and telomerase RNA component (TERC) that contains repetitive telomeric sequence. With age, the telomeres of microglia in the cerebellum and cortex become shorter and stress induced by free radicals which can lead to activation of DNA damage response and cell cycle arrest and may result in aging-related diseases like AD, PD, etc. Thus, the balance between telomeres and telomerase is necessary for proper brain function. As mentioned earlier telomerase can repair short and dysfunctional telomeres. Phytochemicals act as inducers or activators of telomerase that can enhance their function and prevent the shortening of telomeres and thus delay the aging process. The small molecule TA-65 isolated from *Astragalus membranaceus* is a potent telomerase activator that increases the length of telomeres. Another molecule is 08AGTLF, which is extracted from *Centella asiatica and* has been shown to have a strong effect on telomerase activity. Another study showed that oleanolic acid (OA) ((4*aS*,6*aR*,6*aS*,6*bR*,8*aR*,10*S*,12*aR*,14*bS*)-10-hydroxy-2,2,6*a*,6*b*,9,9,12*a*-heptamethyl-1,3,4,5,6,6*a*,7,8,8*a*,10,11,12,13,14*b*-tetradecahydropicene-4*a*-carboxylic acid), a pentacyclic triterpenoid extracted from *Olea europaea* L. decreases the expression of MMP-1 and MMP-2 and increases the expression of COL1A1 and promotes collagen synthesis, thereby inhibiting aging process and is also a potent activator of telomerase enzyme. The primary component of astragalus, astragaloside IV, which is converted into Cycloastragenol CAG ((1*S*,3*R*,6*S*,8*R*,9*S*,11*S*,12*S*,14*S*,15*R*,16*R*)-15-[(2*R*,5*S*)-5-(2-hydroxypropan-2-yl)-2-methyloxolan-2-yl]-7,7,12,16-tetramethylpentacyclo[9.7.0.01, 3.03, 8.012, 16]octadecane-6,9,14-triol), boosted telomerase activity and counteracted the effects of aging [46–48].

15.4.5 Anti-Apoptotic Property of Carotenoids

Apoptosis is generally referred to as programmed cell death and is often involved in the maturation and development of the normal brain. When the balance between apoptotic proteins and anti-apoptotic proteins is disrupted, it can lead to the death of the neurons which is one of the main reasons for aging. When stress is induced the brain cells undergo either receptor-mediated apoptosis (extrinsic) or mitochondria-mediated apoptosis (intrinsic). In receptor-mediated apoptosis death receptor TNFR1 recruits its adapter proteins like FADD which activates caspases that induce apoptosis whereas in the case of intrinsic pathway death signal comes from mitochondria whereas, in the pro-apoptotic proteins like Bax, Bak gets activated and makes the mitochondrial membrane permeable that causes the release of cytochrome c which causes the activation of caspases and leads to apoptosis. Carotenoids are the family of yellow, red, and orange pigments found in plants and are often involved in the process of photosynthesis. They are known to be important for the human body. Lycopene (6*E*,8*E*,10*E*,12*E*,14*E*,16*E*,18*E*,20*E*,22*E*,24*E*,26*E*)-2,6,10,14,19,23,27,3 octamethyldotriaconta -2,6,8,10,12,14,16,18,20,22,24,26,30-tridecaene) which is found in tomatoes is known to protect mitochondria from damage. As we know mitochondria demand high oxygen consumption, therefore, more ROS is produced by them which could lead to apoptosis and DNA damage. Lycopene prevents the activation of pro-apoptotic proteins and thus protects mitochondrial DNA as well. During neurotoxicity, these molecules enhance the production of BDNF and NGF which causes neurons to survive. Green leafy foods include the carotenoid lutein ((1*R*)-4-[(1*E*,3*E*,5*E*,7*E*,9*E*,11*E*,13*E*,15*E*,17*E*)-18-[(1*R*,4*R*)-4-hydroxy-2,6,6-trimethylcyclohex-2-en-1-yl]-3,7,12,16-tetramethyloctadeca-1,3,5,7,9,11,13,15,17-nonaenyl]-3,5,5-trimethylcyclohex-3-en-1-ol), which is derived from plants. When administered orally its presence was detected in the striatum, cerebellum, cortex, and hippocampus of rhesus monkeys thus showing it can cross BBB. It has been associated to decrease mitochondria ROS and thus protects the cells from death. During neuro degradation, excess glutamate-activated NMDA receptors cause excess calcium influx and activate calpain and phospholipases that cause the death of the neurons. In retinal ganglion cells, lutein has been shown to prevent the activation of NMDA receptors [49, 50].

15.4.6 Phytochemicals Promote Pro Survival of Neurotropic Factors

A large amount of the nutrients that a person consumes are used by the brain. Consequently, some diets may enhance brain function. It has been demonstrated that consuming dietary macro and micronutrients derived from various conventionally used medicinal herbs can improve cognitive performance and can partially cross the BBB. Neurotrophins are a group of closely related proteins that were initially discovered to be essential for the survival of sympathetic and sensory neurons. Since then, it has been discovered that they also regulate a variety of aspects of neuronal survival, development, and function throughout the central and peripheral nervous systems. These include NGF, BDNF, NT-3, NT-4, and NT-5 which bind to tropomyosin-related kinase (Trk) receptor, and GDNF which binds to its specific receptor GDNF family receptor α (GFRα). These receptors, in turn, activate the majority of three pathways that are: MAPK, PI3K/Akt, and PLC-γ. MAPK and PI3K/Akt cause neurite outgrowth and neuronal survival and PLC-γ regulates intracellular calcium levels. The study conducted in vitro showed that flavonoids behave as NTF and activate TrkB which in turn activates PI3K/Akt causes the survival of cultured motor neurons, hippocampal neuronal cells, and spinal ganglion neurons. Wasabi (*Wasabia japonica*), a Japanese horseradish, contains 6-Methylsufinylhexyl isothiocyanate, which binds to the TrkA receptor and promotes neurite outgrowth. Certain phytochemicals can initiate neurotrophic signaling cascades. Extracellular signal-regulated kinase (ERK)-1, CREB, and BDNF gene expression in the rat hippocampus was dramatically elevated by the flavanol rutin (3,3,4,5,7-pentahydroxyflavone-3-rhamnoglucoside), which is present in buckwheat, passionflower, apple, and tea. Increased level of BDNF through Curcumin protects the rat cerebral cortical neurons from glutamate excitotoxicity by activating TrkB-MAPK and PI3K pathways. GDNF is primarily increased by curcumin, resveratrol, catalpol, and other non-flavonoid phytochemicals, BDNF is both in vivo and in vitro induced by flavonoids [51–53].

15.4.7 MMP Prevention by Phytochemicals

Mitochondria is the cell's powerhouse; during aging, mitochondrial dysfunction is most common, leading to damage to neurons. Neurons of the brain follow a definite death-signal pathway, where apoptosis in mitochondria is induced through MMP which leads to the opening of the mitochondrial permeability transition pore (mPTP) through which caspases are released from mitochondria to the cytoplasm and leads to entering of various molecules and metals inside the mitochondria. Studies showed that in SH-SY5Y cells cellular apoptosis was inhibited by phytochemicals by preventing the opening of the mitochondrial permeability transition pore. In the study, they tried to induce apoptosis by using PK11195 ligand which was inhibited by sesame lignans showing the anti-apoptotic property of phytochemicals [54]. In another study, scientists found out that curcumin interacted with VDAC that is present on the OMM and changed its confirmation by interacting with amino acids that were present at the N terminal and inhibited mPTP opening. Similarly, Resveratrol increased the anti-apoptotic proteins like bcl-2 and inhibited cytochrome release by deacetylating VADC1 through Sirt-1. It has been shown that quercetin (2-(3,4-dihydroxyphenyl)-3,5,7-trihydroxychromen-4-one) which is widely found in red wine, onion, broccoli, apples, and tea can protect the neuron cells to undergo cell death by upregulating Parvalbumin protein levels in the cytoplasm. Parvalbumin protein binds to calcium ions and neutralizes them thereby reducing the load of calcium in neurotransmission and also reducing the cytochrome c concentrations thereby protecting the cells against cell death [55]. Mitochondrial fusion is the process wherein two mitochondria fuse to form elongated mitochondria to maintain their number. Resveratrol has been shown to increase mitochondrial fusion and biogenesis and balance the mitochondrial DNA thereby protecting the neuronal cells. It has been shown that Resveratrol also prevented mitophagy in the neuronal cells by activating mitochondrial Sirt-3, Sirt1, and Foxo3/PINK1 (PTEN-induced kinase 1)/Parkin pathway [56].

15.5 ROLE OF PHYTOCHEMICALS IN TREATING NEURODEGENERATIVE DISEASES

A significant aspect of the tradition and culture of most people on Earth is the use of plants in traditional remedies. Tradition herbal medicines are less toxic than synthetic drugs and they often have a wide variety of therapeutic applications like anti-aging, anti-inflammatory, anti-apoptotic, and antioxidants. Many incurable neurodegenerative diseases can be managed by using these traditional medicines. Here we describe the role of phytochemicals in managing some age-related neurodegenerative diseases.

15.5.1 Alzheimer's Disease

Alzheimer's disease is one of the most complex neurodegenerative diseases worldwide and currently it does not have any cure. The only way that works is the management of early symptoms. Deficiency of Acetylcholine is the major cause of Alzheimer's disease. An enzyme Acetylcholinesterase (AChE) cleaves the neurotransmitter Acetylcholine into an acetate group and choline group which leads to the deficiency of acetylcholine. An in-silco work carried by *John J Georrge et al* showed 3 phytochemicals Voacangine, hydroxyindolenine, and 1-Acetoxylycorine which can be used instead of synthetic drugs to manage AD [52]. Another group of scientists tried to explore the phytochemicals that can act as AChE inhibitors. *Pulok K. Mukherjee et al* reported that physostigmine that is obtained from the *Physostigma venenosum* plant can act as a potent anti-acetylcholinesterase [57]. *Sidra Rehman et al* showed that phytochemicals, glycine betaine, and betanin extracted from *Beta vulgaris* possessed a very strong therapeutic value in inhibiting AChE [58]. 6-shogaol extracted from *Zingiber officinale* has been shown to possess neuroinflammation and neuroprotection by activating neurotrophic factors like GDNF, BDNF, and NGF. Ginsenoside Rg3, panaxynol isolated from *Panax ginseng* also possesses neuroprotection and neuroinflammation. Quercetin, gallic acid isolated from *Morus alba (L)* also possesses neuroprotection and antiaging effects [59].

15.5.2 Parkinson's Disease

One of the second most prevalent neurodegenerative diseases is Parkinson's Disease after Alzheimer's disease. One of the most prevalent pathophysiology associated with this disease is the reduction in the dopaminergic neurons in the substantia nigra of the brain. This loss gives rise to symptoms like slow movement, and accumulation of misfolded α-synuclein that often form insoluble Lewy bodies. Till now no cure has been found, only the symptoms of this disease are controlled by drugs [60]. Scientists carried out an experiment on *Drosophila melanogaster* to validate their hypothesis that phytochemicals can regulate PD conditions. He used the dUCH knockdown model of Drosophila melanogaster and *Polyscias fruticose* (L.) leaf extract and found out that this extract had the ability to protect the dopaminergic neurons from degradation in larval as well as the adult stage [61]. In *Azadirachta indica* approximately 300 phytochemicals have been extracted and the major ones are triterpenes, flavonoids, phenols, and gallic acid. Most of them have an anti-Parkinson effect by targeting the TrkB signaling pathway. Neuroprotective benefits are produced via stigmasterol's actions on the NMDAR, AMPK, and JNK pathways, which guard against inflammation, and oxidative damage, and modulate apoptosis [62]. Another study conducted showed that apples are rich in phloridzin which has a remarkable property of reducing oxidative stress and ROS and also has been shown to have anti-aging effects [63]

15.5.3 Ischemic Brain Injury

Oxidative stress and ROS are the leading causes of neurodegenerative diseases. Ischemic stroke is one of them caused by the same and due to blockage in blood flow. As brain tissues are sensitive, they are damaged more as compared with other body tissues [64]. Studies used the phytochemical Leonurine extracted from *Herba Leonuri* and showed that it increases anti-oxidant enzymes and also inhibited mitochondrial ROS and also possesses antiplatelet aggregation activities [65]. Microglia activation causes the release of cytokines and elicits an inflammatory action. Another study showed that Ginsenoside Rb1 (GRb1) that is extracted from *Panax ginseng* was able to inhibit the activation of microglial cells thereby providing neuroprotection. Here they used a rat stroke model where in the phytochemical GRb1 was given intranasally one week before the stroke onset [66]. Puerarin, isolated from *pueraria* which is a Chinese herb has been shown to provide neuroprotective activity after cerebral ischemia. Study conducted on rats and found out that when Puerarin was administered intraperitoneally it was able to attenuate the process of apoptosis and elevate the expression of erythropoietin at dose concentrations of 200 mg/kg and 400 mg/kg [67]. In the study, curcumin was shown to decrease the microglia/macrophage activation and changed the phenotype of distal MCAO mice from M1 pro-inflammatory state to M2 anti-inflammatory state. It was also shown that they did not cause any toxicity to the microglial cell survival [68].

15.6 NEUROTOXIC EFFECTS OF ACONITINE

Aconitine, which is an alkaloid obtained from *Aconitum* plants has been shown to possess various medicinal properties, but it has also been to cause aconitine poisoning. Aconitine could be detected in the brain tissues suggesting that it could pass through BBB. Studies showed that aconitine induces Endoplasmic reticulum stress in the PC12 cells. The cell morphology clearly indicated that the brain of rat deteriorated as aconitine concentration was increased [69]. Another study showed that aconitine is able to excite the amino acids that induce excitotoxicity in the HT22 cells and this causes an increase in the process of glycolysis and accumulation of lactic acid causing cell damage. This process is also accompanied by the overload of intracellular calcium that damages the mitochondria and ultimately causes cell death [70]. It is well known that the roots of these plants possess the highest toxicity so in modern practices the roots are often processed under high pressure by autoclaving the roots in order to decrease the toxicity. During this process, toxic alkaloids are converted to less toxic alkaloids [71].

15.7 CONCLUSION AND FUTURE DIRECTIONS

We truly have very little hope today of predicting, preventing, much less treating brain illnesses like Parkinson's, Alzheimer's, or senile dementia compared to other branches of medicine like oncology and cardiovascular medicine. In fact, neurodegeneration is still somewhat of a scientific mystery. We still don't fully understand the genetics of these disorders, the effects of environmental factors, or even why neurons tend to degrade and stick out so much more than other tissues in our bodies. Several experimental approaches have provided information on the health-promoting potential of isolated phytochemicals by using cell and tissue cultures, animal experiments, and human studies. It is evident that the prevalence of many of the main brain disorders, including Alzheimer's disease, vascular dementia, and other illnesses, rises with age and that lifestyle factors like obesity, decreased insulin sensitivity, and metabolic syndrome may play a role in their causation. Unfortunately, there is a clear lack of understanding of these problems among clinicians and doctors, where dietary advice might be combined with conventional neurogerontology methods.

As we already know in the past few decades, more research has been done on Chinese traditional herbs and Ayurvedic medicine. The dietary lifestyle affects most of the physiological processes in our body and during initial disease symptoms most plant-based medicines are given, and they tend to improve the condition. We have already discussed the role of phytochemicals in managing most diseases including diseases that have not got any cure. Phytochemicals are plants' secondary metabolites that are often produced in response to defense mechanisms. Brain aging which is a complex as well as natural process could be delayed by these active compounds that may prevent the onset of several neurological diseases. However most of the studies that are carried out describing the neuroprotective role of phytochemicals are in-silico which should be further validated in in-vitro and in-vivo non-primate models. Here we have discussed the role of active phytochemicals that could delay the brain aging processes which can lead to neurological disorders. Many of these phytochemicals are consumed by all individuals irrespective of their geographical location and these compounds regulate the majority of the signaling cascade that delays brain aging and eliminates ROS and other oxidative stress that protects the neuronal cell's DNA as well as the mitochondria.

Traditional medicinal plants are in high demand because of their availability, affordability, and accessibility. Whatever drugs are now formulated contains one or more phytochemicals as their major constituent as phytochemicals provide less toxicity as compared to synthetic chemical molecules and have provided long-term benefits. Most of the phytochemicals can cross BBB and are able to regulate the majority of the signaling transduction pathways. However most of the research is targeted at the crude extract of the plant rather than focusing on the active ingredient of the plant which should be considered in future phytochemical research. Also, the active compound should not produce toxicity in other body tissues.

ABBREVIATIONS

BBB	Blood–brain barrier
TRPM2	Transient receptor potential melastatin 2
IP3R	Inositol 1,4,5-trisphosphate receptor
RyR	Ryanodine Receptor
AD	Alzheimer's disease
PET	Positron emission tomography
PD	Parkinson's disease
GLUT 1	Glucose transporter 1
NSCs	Neural stem cells
DNA	Deoxyribonucleic Acid
ATP	Adenosine triphosphate
mtNOS	Mitochondrial calcium-dependent NO synthase
BER	Base excision repair
MedDiet	Mediterranean diet
ROS	Reactive oxygen species
RNS	Reactive nitrogen species
IL	Interleukin
TNF	Tumor necrosis factor
CNS	Central nervous system
Nrf-2	Nuclear factor erythroid 2-related factor 2
NF-kB	Nuclear factor kappa-light-chain-enhancer
MMP	Mitochondrial Membrane Permeabilization
Keap-1	Kelch-like ECH-associated protein 1
MAF	MAF BZIP Transcription Factor
ARE	Antioxidant response element
IκB	Ikappa B kinase
PD-L1	Programmed death-ligand 1
COX-2	Cyclooxygenase-2
ERK	Extracellular-signal regulated kinases
Atg	Autophagy-related gene
PIP2	Phosphatidylinositol 4,5-bisphosphate
PIP3	Phosphatidylinositol (3,4,5)-trisphosphate
EGCG	Epigallocatechin gallate

LC3	Microtubule-associated protein 1A/1B-light chain 3
mTOR	Mammalian target of rapamycin
AMPK	AMP-activated protein kinase
PGC-1α	Peroxisome proliferator-activated receptor-gamma coactivator
NAD	Nicotinamide Adenine Dinucleotide
TERT	Telomerase reverse transcriptase protein
TERC	Telomerase RNA component
TNFR1	Tumor necrosis factor receptor 1
BDNF	Brain-derived neurotrophic factor
NGF	Nerve Growth Factor
NMDA	N-methyl-D-aspartate receptor
NT	Neurotrophin
Trk	Tropomyosin-related kinase
GFRα	GDNF family receptor α
PI3K	Phosphatidylinositol 3-kinase
PLC-γ	Phospholipase C gamma
NTF	Neurotrophic factors
CREB	cAMP response element-binding protein
mPTP	mitochondrial permeability transition pore
VDAC	Voltage-dependent anion channels
Sirt-1	Sirtuin 1
AChE	Acetylcholinesterase
GDNF	Glial cell line-derived neurotrophic factor
JNK	Jun N-terminal Kinase
GRb1	Ginsenoside Rb1
MCAO	Middle cerebral artery occlusion

REFERENCES

1. Rodríguez-Rodero S, Fernández-Morera JL, Menéndez-Torre E, Calvanese V, Fernández AF, Fraga MF. Aging genetics and aging. *Aging Disease.* 2011;2(3):186.
2. Partridge LO. L Serrano M Kroemer G The hallmarks of aging. *Cell.* 2013;153(6):1194.
3. Davinelli S, Scapagnini G, Koverech G, Luca M, Calandra C, Calabrese V. Neuroprotective mechanisms of dietary phytochemicals: *implications for successful brain aging*: *In Molecular basis of nutrition and aging* 2016 (pp. 251–261). Academic Press.
4. Chandran R, Kumar M, Kesavan L, Jacob RS, Gunasekaran S, Lakshmi S, Sadasivan C, Omkumar RV. Cellular calcium signaling in the aging brain. *J Chem Neuroanat.* 2019;95:95–114.
5. Valenza M, Facchinetti R, Steardo L, Scuderi C. Altered waste disposal system in aging and Alzheimer's disease: Focus on astrocytic aquaporin 4. *Frontiers Pharmacol.* 2020; Jan 29; 10:1656.
6. Hipkiss AR. Accumulation of altered proteins and aging causes and effects. *Exp Gerontol.* 2006;41(5): 464–473.
7. Sparkman NL, Johnson RW. Neuroinflammation associated with aging sensitizes the brain to the effects of infection or stress. *Neuroimmunomodulation.* 2008;15(4-6):323–330.
8. Camandola S, Mattson MP. Brain metabolism in health, aging, and neurodegeneration. *EMBO J.* 2017; 36(11):1474–1492.
9. Roth GS, Joseph JA, Mason RP. Membrane alterations as causes of impaired signal transduction in Alzheimer's disease and aging. *Trends Neurosci.* 1995;18(5):203–206.
10. Ibrayeva A, Bay M, Pu E, Jörg DJ, Peng L, Jun H, Zhang N, Aaron D, Lin C, Resler G, Hidalgo A. Early stem cells aging in the mature brain. *Cell Stem Cell.* 2021;28(5):955–966.
11. Mattson MP, Arumugam TV. Hallmarks of brain aging: Adaptive and pathological modification by metabolic States. *Cell Metabolism.* 2018;27(6):1176–1199.
12. Boveris A, Navarro A. Brain mitochondrial dysfunction in aging. *IUBMB Life.* 2008;60(5):308–314.
13. Canugovi C, Misiak M, Ferrarelli LK, Croteau DL, Bohr VA. The role of DNA repair in brain related disease pathology. *DNA Repair.* 2013;12(8):578–587.

14. Lee SH, Min KJ. Phytochemicals. *Encyclopedia Biomed Gerontol.* 2019;35–47.
15. Muravchick S. Theories of aging. *Geriatric Anesthesiol.* 2008;29–37.
16. Mitteldorf J. Programmed and non-programmed theories of aging. *Russian J Gen Chem.* 2010;80: 1465–1475.
17. Macieira-Coelho A. The implications of the 'Hayflick limit 'for aging of the organism have been misunderstood by many gerontologists. *Gerontology.* 1995;41(2):94–97.
18. Zs-Nagy I. Aging of cell membranes: Facts and theories. *Interdiscip Topics Gerontol.* 2014; 39:62–85.
19. Shay JW, Wright WE. Hayflick, his limit, and cellular aging. *Nat Rev Mol Cell Biol.* 2000;1(1):72–76.
20. de Souza-Pinto NC, Bohr VA. The mitochondrial theory of aging: Involvement of mitochondrial DNA damage and repair. *Int Rev Neurobiol.* 2002;53:519–534.
21. Sun N, Youle RJ, Finkel T. The mitochondrial basis of aging. *Mol Cell.* 2016;61(5):654–666.
22. Gemma C, Vila J, Bachstetter A, Bickford PC. Oxidative stress and the aging brain: From theory to prevention.
23. Bjorksten J, Tenhu H. The crosslinking theory of aging—Added evidence. *Exp Gerontol.* 1990;25(2): 91–95.
24. Poon HF, Calabrese V, Scapagnini G, Butterfield DA. Free radicals, and brain aging. *Clinics Geriatric Med.* 2004;20(2):329–359.
25. Andrew PW. Aging and the free radical theory. *Respiration Physiol.* 2001;128(3):379–391.
26. Beckman KB, Ames BN. The free radical theory of aging matures. *Physiol Rev.* 1998;78(2):547–581.
27. Yepuri G, Velagapudi S, Xiong Y, Rajapakse AG, Montani JP, Ming XF, Yang Z. Positive crosstalk between arginase-II and S6K1 in vascular endothelial inflammation and aging. *Aging Cells.* 2012;11(6): 1005–1016.
28. Leitzmann C. Characteristics and health benefits of phytochemicals. *Forsch Komplementarmed.* 2016; 23(2):69–74.
29. Battino M, Forbes-Hernández TY, Gasparrini M, Afrin S, Cianciosi D, Zhang J, Manna PP, Reboredo-Rodríguez P, Varela Lopez A, Quiles JL, Mezzetti B, Bompadre S, Xiao J, Giampieri F. Relevance of functional foods in The mediterranean diet: The role of olive oil, berries, and honey in The prevention of cancer and cardiovascular diseases. *Crit Rev Food Sci Nutr.* 2019;59(6):893–920.
30. Nurk E, Refsum H, Drevon CA, Tell GS, Nygaard HA, Engedal K, et al. Cognitive performance among the elderly in relation to the intake of plant foods. The hordaland health study. *Br J Nutr.* 2010; 104(8):1190–1201.
31. Davinelli S, Maes M, Corbi G, Zarrelli A, Willcox DC, Scapagnini G. Dietary phytochemicals and neuro-inflammaging: From mechanistic insights to translational challenges. *Immun Aging.* 2016;13(1): 1–17.
32. Ali SS, Kasoju N, Luthra A, Singh A, Sharanabasava H, Sahu A, et al. Indian Medicinal herbs as sources of antioxidants. *Food Res Int.* 2008;41(1):1–15.
33. Hügel HM. Brain food for Alzheimer-free aging: Focus on herbal medicines. *Adv Exp Med Biol.* 2015; 863:95–116.
34. Mattson MP, Chan SL, Duan W. Modification of brain aging and neurodegenerative disorders by genes, diet, and behavior. *Physiol Rev.* 2002;82(3):637–672.
35. Qiao J, Wang C, Chen Y, Yu S, Liu Y, Yu S, Jiang L, Jin C, Wang X, Zhang P, Zhao D. Herbal/Natural compounds resist hallmarks of brain aging: From molecular mechanisms to therapeutic strategies. *Antioxidants.* 2023;12(4):920.
36. Hannan MA, Dash R, Sohag AA, Haque MN, Moon IS. Neuroprotection against oxidative stress: Phytochemicals targeting TrkB signaling and the Nrf2-ARE antioxidant system. *Frontiers Mol Neurosci.* 2020;13:116.
37. Yang J, Fu X, Liao X, Li Y. Nrf2 activators as dietary phytochemicals against oxidative stress, inflammation, and mitochondrial dysfunction in autism spectrum disorders: A systematic review. *Frontiers Psychiatry.* 2020; 11:561998.
38. Ren L, Zhan P, Wang Q, Wang C, Liu Y, Yu Z, et al. Curcumin upregulates the Nrf2 system by repressing inflammatory signaling-mediated Keapl expression in insulin-resistant conditions. *Biochem Biophys Res Commun.* 2019;514(3):691–698.
39. Uddin MS, Kabir MT, Mamun AA, Sarwar MS, Nasrin F, Emran TB, et al. Natural small molecules targeting NF-κB signaling in glioblastoma. *Front Pharmacol.* 2021;12(August):1–14.
40. Salminen A, Kauppinen A, Kaarniranta K. Phytochemicals suppress nuclear factor-κB signaling: Impact on health span and the aging process. *Curr Opin Clin Nutr Metab Care.* 2012;15(1):23–28.
41. Seo EJ, Fischer N, Efferth T. Phytochemicals as inhibitors of NF-κB for treatment of Alzheimer's disease. *Pharmacol Res.* 2018;129:262–273.

42. Limanaqi F, Biagioni F, Busceti CL, Ryskalin L, Polzella M, Frati A, Fornai F. Phytochemicals bridging autophagy induction and alpha-synuclein degradation in parkinsonism. *Int J Mol Sci.* 2019;20(13):3274.
43. Ntsapi C, Toit D, Loos A B. Dietary impact on neuronal autophagy control and brain health.
44. Nourbakhsh F, Read MI, Barreto GE, Sahebkar A. Boosting the autophagy-lysosomal pathway by phytochemicals: A potential therapeutic strategy against Alzheimer's disease. *IUBMB Life.* 2020;72(11): 2360–2281.
45. Sarubbo F, Esteban S, Miralles A, Moranta D. Effects of resveratrol, and other polyphenols on Sirt1: Relevance to brain function during aging. *Curr Neuropharmacol.* 2017;16(2):126–136.
46. Tsoukalas D, Fragkiadaki P, Docea AO, Alegakis AK, Sarandi E, Thanasoula M, Spandidos DA, Tsatsakis A, Razgonova MP, Calina D. Discovery of potent telomerase activators: Unfolding new therapeutic and anti-aging perspectives. *Mol Med Rep.* 2019;20(4):3701–3708.
47. Hong M, Yu J, Wang X, Liu Y, Zhan S, Wu Z, Zhang X. Tea polyphenols as prospective natural attenuators of brain aging. *Nutrients.* 2022;14(15):3012.
48. Maleki M, Khelghati N, Alemi F, Bazdar M, Asemi Z, Majidinia M, Sadeghpoor A, Mahmoodpoor A, Jadidi-Niaragh F, Targhazeh N, Yousefi B. Stabilization of telomere by the antioxidant property of polyphenols: Anti-aging potential. *Life Sci.* 2020; 259:118341.
49. Park HA, Hayden MM, Bannerman S, Jansen J, Crowe-White KM. Anti-apoptotic effects of carotenoids in neurodegeneration. *Molecules.* 2020;25(15):1–19.
50. Cho KS, Shin M, Kim S, Lee SB. Recent advances in studies on the therapeutic potential of dietary carotenoids in neurodegenerative diseases. *Oxidative Med Cellular Longevity.* 2018;2018.
51. Lee J, Jo DG, Park D, Chung HY, Mattson MP. Adaptive cellular stress pathways as therapeutic targets of dietary phytochemicals: Focus on the nervous system. *Pharmacol Rev.* 2014;66(3):815–868.
52. Gupta VK, Sharma B. Role of phytochemicals in neurotrophins mediated regulation of Alzheimer's disease. *Int. J. Complement. Alt Med.* 2017;7(4):00231.
53. Naoi M, Inaba-Hasegawa K, Shamoto-Nagai M, Maruyama W. Neurotrophic function of phytochemicals for neuroprotection in aging and neurodegenerative disorders: Modulation of intracellular signaling and Gene expression. *J Neural Transmission.* 2017;124:1515–1527.
54. Wu Y, Shamoto-Nagai M, Maruyama W, Osawa T, Naoi M. Phytochemicals prevent mitochondrial membrane permeabilization and protect SH-SY5Y cells against apoptosis induced by PK11195, a ligand for outer membrane translocator protein. *J Neural Transmission.* 2017;124:89–98.
55. Riche K, Lenard NR. Quercetin's effects on glutamate cytotoxicity. *Molecules.* 2022;27(21):7620.
56. Naoi M, Wu Y, Shamoto-Nagai M, Maruyama W. Mitochondria in neuroprotection by phytochemicals: Bioactive polyphenols modulate mitochondrial apoptosis system, function and structure. *I J Mol Sci.* 2019;20(10):2451.
57. Mukherjee PK, Kumar V, Mal M, Houghton PJ. Acetylcholinesterase inhibitors from plants. *Phytomedicine.* 2007;14(4):289–300.
58. Rehman S, Ali Ashfaq U, Sufyan M, Shahid I, Ijaz B, Hussain M. The insight of in silico and in vitro evaluation of Beta vulgaris phytochemicals against Alzheimer's disease targeting acetylcholinesterase. *PLoS One.* 2022;17(3):e0264074.
59. Georrge JJ. Docking studies, ADMET prediction of phytochemical inhibitors for Alzheimer's disease. *In Proceedings of 8th National Level Science Symposium on Recent Trends in Science and Technology* 2015 Feb 22 (Vol. 2, pp. 115–120).
60. Surguchov A, Bernal L, Surguchev AA. Phytochemicals as regulators of genes involved in synucleinopathies. *Biomolecules.* 2021;11(5):1–15.
61. Ly HT, Nguyen TT, Le VM, Lam BT, Mai TT, Dang TP. Therapeutic potential of *Polyscias fruticosa* (L.) Harms leaf extract for Parkinson's disease treatment by Drosophila Melanogaster model. *Oxidative Med Cellular Longevity.* 2022;2022.
62. Sandhir R, Khurana M, Singhal NK. Potential benefits of phytochemicals from Azadirachta indica against neurological disorders. *Neurochem Int.* 2021;146:105023.
63. Prabhakar P, Ahmed AB, Chidambaram SB. The role of phloridzin and its possible potential therapeutic effect on Parkinson's disease. *Int J Nutrition Pharmacol Neurolog Diseases.* 2020;10(2):69–74.
64. Subedi L, Gaire BP. Phytochemicals as regulators of microglia/macrophages activation in cerebral ischemia. *Pharmacol Res.* 2021;165:105419.
65. Loh KP, Qi J, Tan BKH, Liu XH, Wei BG, Zhu YZ. Leonurine protects the middle cerebral artery occluded rats through antioxidant effect and regulation of mitochondrial function. *Stroke.* 2010;41(11):2661–2668.
66. Zhu J, Jiang Y, Wu L, Lu T, Xu G, Liu X. Suppression of local inflammation contributes to the neuroprotective effect of ginsenoside Rb1 in rats with cerebral ischemia. *Neuroscience.* 2012;202:342–351.

67. Gao L, Ji X, Song J, Liu P, Yan F, Gong W, et al. Puerarin protects against ischemic brain injury in a rat model of transient focal ischemia. *Neurol Res.* 2009;31(4):402–406.
68. Liu Z, Ran Y, Huang S, Wen S, Zhang W, Liu X, et al. Curcumin protects against ischemic stroke by titrating microglia/macrophage polarization. *Front Aging Neurosci.* 2017;9(JUL):1–10.
69. Zhang X, Jiang X, Yu A, Duan H. Aconitine induces brain tissue damage by increasing the permeability of the cerebral blood-brain barrier and over-activating endoplasmic reticulum stress. *Am J Transl Res.* 2022;14(5):3216–3224.
70. Zhang Y, Chen S, Fan F, Xu N, Meng XL, Zhang Y, Lin JM. Neurotoxicity mechanism of aconitine in HT22 cells studied by microfluidic chip-mass spectrometry. *J Pharm Anal.* 2023;13(1):88–98.
71. Rodríguez-Rodero S, Fernández-Morera JL, Menéndez-Torre E, Calvanese V, Fernández AF, Fraga MF. Aging genetics and aging. *Aging Disease.* 2011;2(3):186.

16 Recent Advancement in Ethnopharmacology on Age-Related Brain Disorders

Ankita Panigrahi, Anushka Agrawal, and Kanthesh M Basalingappa
JSS AHER
Mysuru, Karnataka

16.1 INTRODUCTION

All living things eventually age because it is an inevitable part of their existence. It's linked to deterioration in health and a higher probability of getting sick and dying. Degeneration associated with ageing progresses at varying rates depending on species, individual, and tissue type. The brain's structure and function both undergo changes as we become older. A major loss of cellular function occurs in brain cells as a result of ageing processes such cellular senescence and stem cell exhaustion. This dysfunctional decline speeds up the deterioration of the brain as a whole. The financial burden of healthcare issues associated with ageing is growing in tandem with the elderly share of the population. Therefore, there is an urgent need to either mitigate or forestall the onset of age-related health issues. Cognitive impairments due to age-related brain problems are widespread among the elderly. This affects not only how long individuals live in good health, but also how well they live throughout that time. Neurodegeneration is a complex brain disorder that has not yet been fully elucidated. The majority of neurodegenerative diseases have age as their key risk factor. These diseases include Alzheimer's disease (AD), Parkinson's disease (PD), Huntington's disease (HD), and amyotrophic lateral sclerosis (ALS). In older populations, neurological disorders like Alzheimer's and Parkinson's are increasingly common. According to world health organisation each year there are 10 million new cases of dementia and 60% of them are from people living in low economic conditions.

Before the advent of modern medicine, humans relied on a vast assortment of naturally occurring cures for the treatment of conditions that affected the central nervous system (CNS). In western societies, there has been a growing interest in herbal medicines, which are frequently viewed as more 'natural' and' soft' treatments when contrasted to synthetic pharmaceuticals. This has led to an increase in the use of herbal medicines (1). And also there are not many medicines that have been found to be successful in treating these epidemic neurological crises (2, 3). Let's see some examples that are time-tested in case of neurological issues. One such example that is effective in treating vascular as well as Alzheimer's type dementia includes extracts from *Gingko biloba* marketed as EGb 761, shown very much effectiveness in clinical trials (4). Another excellent example of a plant secondary chemical that has been utilised well for the treatment of mild to severe Alzheimer's disease is galantamine, which is an alkaloid that can be produced from the bulbs and blooms of the *Caucasian snowdrop, Galanthus woronowii* (Amaryllidaceae). Atropine, pilocarpine, cardiac glycosides, curare, and reserpine are only a few examples of the important discoveries that have been made in the field of pharmacological therapy thanks to the application of ethnomedical expertise. Discovering novel medications for the treatment of Alzheimer's disease and other cognitive diseases could benefit from the use of ethnomedical material as a source of inspiration, and this could be accomplished by consulting ethnomedical sources.

 DOI: 10.1201/9781003486237-16

What exactly is ethnopharmacology? "Ethnopharmacology," which is the study of traditional medical practises from a variety of cultures as well as the utilisation of plants and other naturally occurring compounds for therapeutic purposes, is a field of research. The knowledge of anthropology, pharmacology, botany, and other related subjects are combined in the field of ethnopharmacology, which is an example of an interdisciplinary field (5). Understanding the cultural and traditional applications of medicinal plants, in addition to the chemical and pharmacological qualities of these plants, is the primary emphasis of this field. The goal of researchers working in the field of ethnopharmacology is to discover novel natural medicine sources, verify the efficacy of traditional remedies, and examine the biological processes behind the therapeutic effects of natural substances that are employed in a wide variety of traditional healing practises. This field is also concerned with addressing concerns pertaining to cultural preservation, intellectual property rights, and sustainable practises (6). Researchers in India are researching the chemical and pharmacological qualities of a variety of traditional remedies in this modern period, which has led to a revived interest in the study of ethnopharmacology. It has been shown that many medicinal herbs from India contain significant biological activities, such as anti-inflammatory, analgesic, anti-cancer, and anti-diabetic benefits. In addition, India is home to a number of institutes that are dedicated to the research of ethnopharmacology, such as the National Institute of Ethnopharmacology in Chhindwara and the Society for Ethnopharmacology in Lucknow. These institutions do study on the traditional medicinal practises that are used in India and strive towards maintaining the knowledge and practises that are unique to indigenous cultures (7, 8).

In this chapter, we are going to discuss the recent advancements in ethnopharmacology which holds promise to prevent brain-related disorders. For example, as we know Age-related cognitive decline is a common problem, but traditional medical practises like Ayurveda and Traditional Chinese Medicine (TCM) have been successfully treating and preventing this condition for centuries. Recent studies have aimed to scientifically verify the efficacy of these time-honoured treatments. For instance, one recent study indicated that the traditional Chinese medicine (TCM) formula TianWangBuXin Dan (TWBXD) enhanced cognitive performance and decreased the risk of dementia in the elderly (9, 10). There are many such examples lets discuss what are the examples, their safety profiles, research and challenges faced in today's world.

16.2 EXAMPLES OF AGE RELATED BRAIN DISORDERS AND THEIR TREATMENT USING HERBAL MEDICINE

The probability of a person experiencing a condition associated with the brain rises with age. A person's daily functioning and capacity to enjoy life may be severely impaired by one of these conditions. Here are some examples:

16.2.1 Alzheimer's Disease

Alzheimer's disease, one of the most prevalent age-related brain illnesses, is marked by the slow and steady decline in cognitive function that occurs over the course of the disease. People over 65 have a higher risk of developing Alzheimer's disease. Alzheimer's disease is believed to be in part caused by the buildup of beta-amyloid plaques and tau tangles, both of which are proteins found in the brain. Alzheimer's disease is characterised by changes in cognition and behaviour that may begin with a disruption in normal brain cell communication caused by these proteins. It is believed that both genetic and environmental variables have a part in the development of Alzheimer's disease, while the exact roles remain unknown. There is a growing interest in using traditional medicines to treat Alzheimer's disease (11). Many plant-derived compounds have been shown to have neuroprotective effects, raising the possibility that they could be effective in the treatment of Alzheimer's disease. The anti-inflammatory and antioxidant benefits of curcumin, found in the spice turmeric, have been associated to a lower risk of acquiring Alzheimer's disease. Traditional treatments such as Ginkgo biloba and ginsengs have also been studied for their potential to alleviate the symptoms of Alzheimer's disease. Ginkgo biloba has been demonstrated to improve cognitive abilities and memory, especially in patients with Alzheimer's disease. This may be because the plant improves cerebral perfusion, allowing more oxygen and nutrients to reach the brain's cells. Physostigma venenosum Balf, sometimes known as the Calabar bean, is a prototypical cholinesterase (ChE) inhibitor. The discovery and 150 years of use of this drug in the treatment of Alzheimer's disease (AD) is a classic narrative of ethnopharmacology. The discovery of a new beneficial function in a traditional medicine can be utilised to search for compounds in other plants that have the same effect, even if those plants have never been used medicinally previously. Strong ChE inhibitors (ChEIs) have been found in a wide range of plants that are not typically used as medicine (12, 13), providing evidence for the effectiveness of ChE inhibition in AD. Let's have a look Table 16.1 to learn about the phytochemicals and their derivatives that are useful in Alzheimer's treatment.

In addition to conventional treatments, natural substances including omega-3 fatty acids, vitamin E, and resveratrol have also been investigated for the role they could play in the treatment of Alzheimer's disease. Fish oil contains omega-3 fatty acids, which have been discovered to have anti-inflammatory characteristics and may help protect against the development of Alzheimer's disease. These fatty acids can be found in plenty. It has been shown that the antioxidant qualities of vitamin E, which may be found in foods such as nuts, seeds, and vegetables with leafy greens, may help to protect against the development of Alzheimer's disease. It has been discovered that the compound known as resveratrol, which may be found in red wine and grapes, possesses anti-inflammatory and antioxidant effects. These features may help guard against the development of Alzheimer's disease (25) (Figure 16.1).

16.2.2 Parkinson's Disease

Parkinson's disease is a degenerative neurological ailment that mostly manifests in movement symptoms. Dopamine-producing neurons in the substantia nigra, a brain area essential in motor control, gradually die off, causing the disease. Due to a lack of dopamine due to the death of dopamine-producing neurons, the motor symptoms of Parkinson's disease manifest themselves. Several factors have been linked in the degeneration of dopamine-producing neurons in Parkinson's disease, although their precise roles remain unclear. The accumulation of the protein alpha-synuclein in the brain is commonly believed to play a significant role in the development of the disease. The protein alpha-synuclein plays a role in controlling how synapses in the brain communicate with one another. Pathological protein deposits called Lewy bodies are formed when alpha-synuclein clumps together in the substantia nigra and other areas of the brain in people with Parkinson's disease. It is believed that the death of dopamine-producing neurons is caused by an accumulation of Lewy bodies, which interfere with normal cellular function. Oxidative stress, mitochondrial dysfunction, inflammation, and altered protein degradation pathways are some of the other factors that have been linked to the development of Parkinson's disease. The motor symptoms of Parkinson's disease, such as tremors, rigidity, bradykinesia (slowness of movement), and postural instability, are caused by the death of dopamine-producing neurons and the consequent dopamine shortage in the brain (26).

16.2.2.1 Treatment of PD Using Ethnopharmacology

Ashwagandha, or *Withania somnifera* (L.) Dunal, is a medicinal herb with a long history of usage in Ayurvedic medicine. It has been studied for its ability to treat a variety of medical diseases, including neurological illnesses like Parkinson's disease. In one study conducted rats induced with Parkinsonism neurotoxin called 6-hydroxydopamine (6-OHDA) which destroys dopaminergic neurons in the substantia nigra, is a key feature of Parkinson's disease pathology and found that treatment with an extract of Ashwagandha root improved their motor function and reversed Parkinsonian behaviour. Additionally, the researchers found that Ashwagandha extract increased Hsp70 expression in the brain, which may have a protective effect on dopaminergic neurons (27). The heat shock protein Hsp70 has been linked to preventing cellular damage in the face of a variety of stresses. Ashwagandha extract therapy upregulated Hsp70 expression in rat brains, as discovered by the researchers. Because dopaminergic neurons are known to be impaired in Parkinson's

TABLE 16.1

Examples of All the Phytochemicals and Their Derivatives Which Are Effective in Treatment of Alzheimer's

Phytochemical	Plant	Traditional Use of Plant	Effect on Alzheimer's	Clinical Observation	Conclusion	References
Vincamine	Apocynaceae	Traditional European medicine to treat problems with the brain	Improves memory via increasing blood flow to the brain blocking voltage-gated Na+ channels	Improve cognitive function in dementia patients.	Boost dementia patients' cognitive abilities, while further RCTs are needed to corroborate the overall scant data and usefulness for AD Especially AChEIs, demonstrate more favourable and consistent benefits on cognition in AD.	14, 15
Huperzine A	Club moss	Involves the preparation of the herbal remedy "Qian ceng ta," which is taken for memory loss, to increase blood flow throughout the body.	Blocking AChE. Neuroprotective and neurotrophic effects against oxidative stress	Increased MMSE** scores and activities of daily life in Alzheimer's patients	Show efficacy in AD with few adverse effects observed, which is essential because existing AChEIs for AD produce unpleasant symptoms, particularly GI.	16, 17
Cannabidiol	Cannabis sativa L.	Classified as a psychoactive	Protects neurons from damage, has anti-psychotic and anxiolytic properties.	Helpful for anxiety sensations in individuals, have not been shown to be effective in blind RCTs for treating psychosis associated with dementia	While preliminary evidence suggests that cannabidiol (CBD) and other cannabinoids (including synthetic variants) may be effective in the treatment of BPSD	18, 19
Magnolol	Magnoliaceac	Treats anxiety and neuron issues	Prevent memory loss, protect against cholinergic dysfunction, have anti-inflammatory properties, improves sleep habits, perhaps through GABAergic pathways.	The absence of H and 4-MH) or standardised extracts Premenopausal women who experience mild, temporary anxiety may benefit from a combination of M. officinalis and Phellodendron amurense Rupr	Relevant to AD pathology and cognition	20, 21
Physostigmine	Calabar bean	Used in Africa	Inhibition of acetylcholinesterase	Physostigmine helps with Alzheimer's disease	Synthetic equivalent rivastigmine has shown clinical efficacy in AD, despite the fact that physostigmine and several synthetic variants have been investigated for therapeutic potential in AD	22, 23
Curcumin	Turmeric	Used in Ayurveda	Inhibition of BACE1 overexpression, prevents cognitive decline in animal models, inflammation suppression, protection of neural cells	Better cognitive function and a decreased Alzheimer's disease	Studies show curcumin may be more useful for preventing dementia than treating it.	24

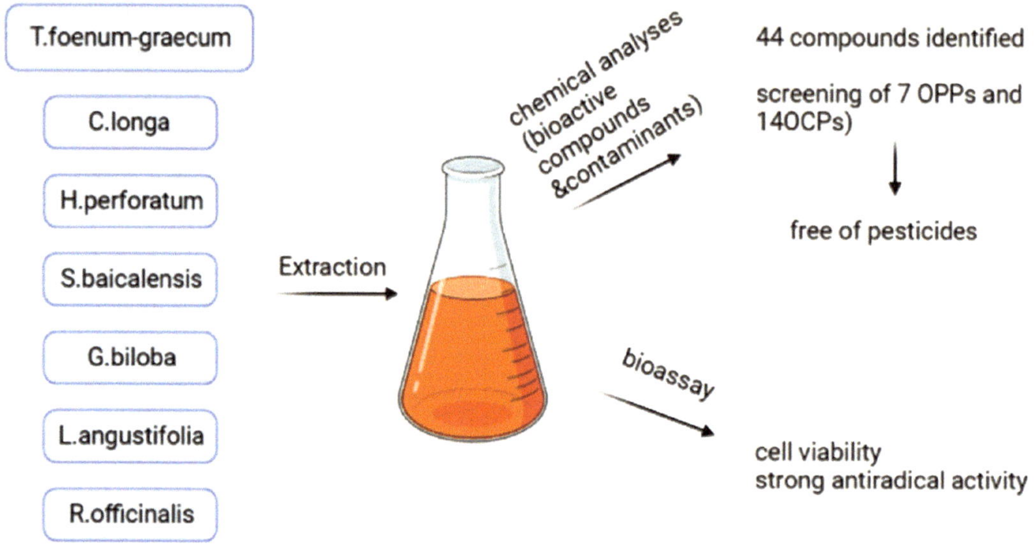

FIGURE 16.1 The seven different herbs shown above play a role in targeting different molecular mechanisms that cause Alzheimer's.

disease, this is relevant because it shows that Ashwagandha may have a neuroprotective impact on these neurons. One way by which Ashwagandha may counteract Parkinson's symptoms in rats is through an uptick in Hsp70 expression, according to the study's authors.

Banisteriopsis caapi, also known as ayahuasca, is a plant commonly used in traditional South American medicine for spiritual and medicinal purposes. While there is limited research on the potential effects of *B. caapi* on Parkinson's disease, there have been some preliminary studies suggesting that its active ingredients may have neuroprotective and anti-inflammatory effects. A study published in the Journal of Ethnopharmacology in 2013 investigated the effect of harmine on a mouse model of Parkinson's disease. Harmine is a beta-carboline alkaloid found in several plant species, including *B. caapi*. It is considered the main psychoactive component of *B. caapi* and is believed to contribute to its medicinal properties. Harmine has been shown to have neuroprotective effects in several studies, including those related to Parkinson's disease. It has been found to increase the activity of key enzymes involved in dopamine metabolism, as well as to reduce oxidative stress and inflammation in the brain, both of which are thought to contribute to the development of Parkinson's disease. Harmine also has been shown to increase the expression of proteins involved in neuronal survival and to promote the growth and differentiation of neurons. These effects suggest that harmine and *B. caapi* may have potential therapeutic applications in the treatment of Parkinson's disease. For instance, The study found that harmine improved motor symptoms and reduced neuroinflammation in the mouse model (28). Another study published in the Journal of Neural Transmission in 2018 examined the effect of *B. caapi* tea on motor symptoms and cognitive function in patients with Parkinson's disease. The study found that *B. caapi* tea improved motor symptoms and cognitive function in patients, although the exact mechanism of action is still unclear (29).

Uncaria rhynchophylla is a plant that has been traditionally used in Chinese medicine for its neuroprotective effects. Possible therapeutic value in the management of Parkinson's disease has been proposed by certain research. Uncaria rhynchophylla's effects on a mouse model of Parkinson's disease were studied in a 2012 article published in the Journal of Ethnopharmacology. The results demonstrated that the motor function was enhanced and the loss of dopamine neurons in the substantia nigra (a brain region affected in Parkinson's disease) was decreased after treatment with Uncaria rhynchophylla extract (30). A cell culture model of Parkinson's disease was used in another study published in 2017 in the Journal of Medicinal Food, and it was shown that Uncaria rhynchophylla extract showed neuroprotective properties. The extract prevented cell death and inflammation in dopaminergic neurons, which were both caused by oxidative stress. The results of these experiments are encouraging, but further work is required to fully understand the mechanism of action and therapeutic potential of Uncaria rhynchophylla for the treatment of Parkinson's disease (31). There are a number of different plants and herbs that have been traditionally utilised for their therapeutic properties and that have shown potential effects in treating Parkinson's disease. *Mucuna pruriens*, *Bacopa monnieri*, *W. somnifera*, *Uncaria rhynchophylla*, *B. caapi*, and *Acanthopanax* are only few of the plants that fall under this category. These plants contain bioactive chemicals that are capable of interacting with a variety of molecular targets in the brain, which results in neuroprotective as well as anti-inflammatory actions. In addition to this, they are capable of elevating levels of neurotransmitters such as dopamine, which is essential for maintaining motor control but suffers from a deficiency in patients with Parkinson's disease (32).

16.2.3 Dementia

Memory loss, difficulty thinking and speaking, and impaired judgement are only few of the cognitive signs of dementia. It's a degenerative brain disease that usually affects the elderly but can occasionally manifest in younger persons as well. Several diseases, such as Alzheimer's, Parkinson's, Huntington's, vascular dementia, Lewy body dementia, and frontotemporal dementia, have been linked to the onset of dementia in patients. Frontotemporal dementia and Lewy body disease also play a role in the general dementia rate. Dementia is distinguished by a reduction in cognitive abilities such as memory, ability to carry out everyday tasks, clarity of thought, personality, and the ability to express oneself verbally. Therapy, lifestyle changes, and medication may all have a role in treating dementia by getting to the root of the issue (33).

The traditional Chinese medicine herb ginkgo biloba has been studied extensively for its possible use in the treatment of dementia and other cognitive disorders. Ginkgo biloba's potential as a treatment for dementia has been the subject of multiple studies. Its extract improved cognitive function and memory in aged female rats, most likely as a result of its antioxidant and anti-inflammatory effects. Ginkgo biloba extract has been demonstrated to increase expression of brain-derived neurotrophic factor (BDNF), a protein crucial to the survival, development, and maintenance of neurons. Overall, the study's findings (34) suggest that Ginkgo biloba extract has therapeutic potential for treating cognitive decline and dementia in the elderly. Journal of Molecular Neuroscience reported the results of a study on the neuroprotective effects of ginkgo biloba extract (GBE) in rats with aluminium chloride (AlC13)-induced dementia. GBE therapy was observed to improve cognitive and memory impairment caused by AlC13, and it also reduced oxidative stress and brain inflammation. Scientists attribute GBE's therapeutic efficacy against dementia to the compound's antioxidant and anti-inflammatory properties. Further research is needed to verify the findings of this study in order to determine the optimal GBE therapy settings (35) (Figure 16.2).

B. monnieri, also known as Brahmi, is a herb commonly used in Ayurvedic medicine for its cognitive-enhancing properties. Several studies have investigated the effects of *B. monnieri* in dementia patients. The exact mechanism through which it exerts its effects is not completely understood. On the other hand, it is thought to exert its effects via a number of distinct mechanisms, among which include the promotion of cholinergic neurotransmission, the mitigation of oxidative stress, and the reduction of beta-amyloid build-up in the brain. In addition to improving cognitive function, it has been shown to have antioxidant and anti-inflammatory properties, which may help protect the brain from oxidative stress and neuroinflammation, both of which have been implicated in the development and progression of dementia (36).

One randomised, double-blind, placebo-controlled trial conducted in Australia evaluated the effects on cognitive function in elderly individuals with memory problems. The study found that Brahmi supplementation for 12 weeks resulted in significant improvements in cognitive function, including working memory, attention, and verbal learning and memory, compared to the placebo

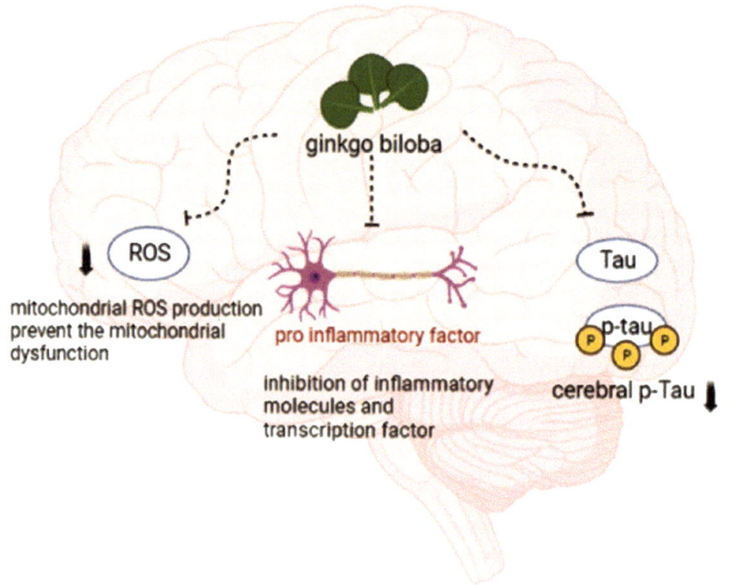

FIGURE 16.2 This figure shows different pathways that are responsible for causing dementia and how consumption of ginkgo biloba heals it.

group (37). Another study conducted in India investigated the effects on cognitive function in patients with Alzheimer's disease. The study found that if supplemented for six months there is significantly improved cognitive function, including attention, verbal memory, and logical memory, compared to the placebo group (38).

Turmeric, a spice commonly used in Indian cuisine, has also been investigated for its potential in treating dementia. Its bioactive compound curcumin has been found to have antioxidant and anti-inflammatory properties, and its extract has been shown to improve cognitive function and reduce plaque buildup in the brain, which is associated with dementia. Other natural products that have shown potential in the treatment of dementia include Huperzine A, a compound found in the plant Huperzia serrata, and Resveratrol, a compound found in red wine and grapes (39).

16.2.4 Mild Cognitive Impairment (MCI)

Mild cognitive impairment (MCI) is a kind of dementia that typically affects seniors. Individuals with this disorder experience impairments in memory, language, reasoning, and judgement that go beyond those associated with ageing but are nevertheless manageable in everyday life. Although MCI is often a precursor to dementia, it does not always lead to dementia. Ginkgo biloba, for instance, has been the subject of much research, and some findings show it may help people with mild cognitive impairment. *B. monnieri*, meanwhile, has been shown to enhance cognitive performance in healthy aged people. The potential benefits of these and other traditional medicinal plants for treating moderate cognitive impairment, however, require further study. Some traditional medicines may combine with other medications or have potential negative effects, thus it's crucial to always use ethnopharmacological treatments in collaboration with a skilled healthcare provider (40).

16.2.5 Huntington's Disease

Huntington's disease is a degenerative brain ailment that typically affects older people. Cognitive, behavioural, and motor symptoms emerge in midlife (about age 30 to 50) and deteriorate over time due to this degenerative neurological illness. A genetic abnormality passed down through generations disrupts normal brain cell activity, eventually causing cell death. There is currently no treatment for the condition because its exact causes remain unknown. Curcumin, resveratrol, and ginsenoside Rb1 are just a few of the natural compounds that have shown promise as potential therapeutics for Huntington's disease. In animal models of Huntington's disease, for instance, the polyphenolic component curcumin present in turmeric was discovered to have neuroprotective effects by lowering oxidative stress and inflammation. Animal models of Huntington's disease have also revealed that the natural polyphenol resveratrol, which is present in grapes and red wine, can have neuroprotective benefits by altering cellular signalling pathways and lowering inflammation (41). Animal models of Huntington's disease have revealed anti-inflammatory and anti-apoptotic properties of ginsenoside Rb1, a significant active component of ginseng. These tests, however, have only been performed on animals; their efficacy in people has not yet been determined. To know if these agents are safe and effective for treating Huntington's disease, more study is required. The safe and effective use of any prospective therapy in addition to routine care requires close collaboration with medical experts (42, 43).

16.2.6 Creutzfeldt–Jakob Disease (CJD)

Creutzfeldt–Jakob disease, also known as CJD, is a fatal brain ailment that is extremely uncommon. It is a degenerative disease that mainly strikes people who are over the age of 50. It is a kind of the sickness known as prion disease, which indicates that the aberrant folding of prion proteins in the brain is the root cause of the condition. CJD can result in a wide range of symptoms, including as shifts in personality and behaviour, memory loss, decreased coordination and mobility, as well as difficulties with speech and vision. It is a quickly advancing disease that, in most cases, results in death within one year after the disease's initial manifestation (44). Studies on the efficacy of ethnopharmacological agents in the management of CJD are scant. Curcumin and resveratrol are two examples of natural substances that have shown promise in inhibiting the aberrant folding of prion proteins and halting their accumulation in the brain (45). Some research suggests that TCM herbs like Radix Polygalae and Radix Bupleuri may have neuroprotective benefits and traditional Chinese medicine (TCM) have also been investigated as a potential therapeutic option for CJD. These results are encouraging, but they are still preliminary; more study is needed to confirm the compounds' safety and effectiveness in treating CJD. The standard treatment for this fatal illness is still palliative care (46).

16.3 RECENT ADVANCEMENTS IN ETHNOPHARMACOLOGY

The use of these plant-based remedies dates back centuries, not just in India, but all throughout the world. Terpenes, steroids, glycosides, flavonoids, alkaloids, amino acids, fatty acids, aryl esters, and carbohydrates are only some of the phytochemicals found in plants. The pharmacological data gathered thus far indicates that all plant components are neuroprotective, anxiolytic, antidepressant, antioxidant, immunomodulatory, sedative, antiproliferative, and anticonvulsant in nature. In recent years, there have been several advancements in the field of ethnopharmacology, lets discuss about them one by one (47).

16.3.1 Integration of Traditional Medicine with Modern Medicine

There is a growing movement to combine conventional and alternative approaches to the treatment of neurological diseases. Herbal medicine and acupuncture are two examples of traditional medicine that have been used to treat neurological and psychiatric diseases for ages (48). By combining the best of both worlds, traditional and Western medicine may offer patients more comprehensive care that takes into account the whole person. It's a chance to experiment with alternative treatments and create innovative cures that incorporate the best of both conventional and alternative medicine (49). Using Ayurvedic medicine in the treatment of Alzheimer's disease is an example of combining conventional and alternative approaches to the disease. Ayurveda is an ancient Indian medical practise that incorporates dietary and lifestyle changes in addition to herbal remedies to treat a wide range of conditions (50). Research into the effects of several Ayurvedic medicines, including *B. monnieri* and *W. somnifera*, on Alzheimer's disease has yielded promising results. Interest in combining Ayurvedic and conventional approaches to Alzheimer's disease treatment has been on the rise in recent years (51). For instance, patients with Alzheimer's disease who received a combination of Ayurvedic medicine and conventional pharmacological treatment showed significant improvements in cognitive function and reductions in behavioural symptoms, according to a randomised controlled experiment published in the Journal of Alzheimer's Disease in 2018. According to the results of this research, combining traditional and Western medicine to treat neurological conditions could be effective (52).

16.3.2 Identification of New Compounds

Alzheimer's disease, Parkinson's disease, and depression are just some of the brain illnesses that may benefit from the therapeutic use of newly discovered natural compounds derived from plants and other sources. For example, Epigallocatechin-3-gallate (EGCG) is a component of green tea that has shown promise in animal studies for treating neurological illnesses including Alzheimer's disease. It's anti-inflammatory and good at fighting free radicals (53). Others include the traditional Chinese herb salvia miltiorrhiza that has been the subject of research into its possible use in the treatment of neurological illnesses like Alzheimer's and stroke. Tanshinones and salvianolic acids are among its active ingredients (54), Cannabidiol (CBD) which is a non-psychoactive cannabinoid present in cannabis, has shown promise in the treatment of a variety of neurological conditions such as epilepsy, anxiety, and neurodegenerative disease (55).

16.3.3 Use of Advanced Technologies

The study of ethnopharmacology and neurological illnesses has advanced greatly because to the use of cutting-edge technology. These fields make extensive use of cutting-edge technologies, such as:

- Researchers can rapidly and effectively examine a large number of substances using high-throughput screening (HTS). New chemicals with therapeutic promise for brain illnesses have been discovered with the use of this method (56).
- Genomics, transcriptomics, proteomics, and metabolomics are all examples of Omics technologies that enable researchers to analyse massive volumes of data at the molecular level. Biomarkers for brain illnesses have been discovered using these technologies, and their underlying mechanisms have become better understood (57).
- Magnetic resonance imaging (MRI), positron emission tomography (PET), and computed tomography (CT) are just a few examples of imaging technologies that let scientists observe and study brain abnormalities. These methods have been utilised in the analysis of neurological diseases and in the tracking of their development (58).
- AI: AI has the potential to revolutionise the ethnopharmacology drug research and development process. The algorithms used in machine learning can examine massive datasets and spot patterns that humans might miss. In the treatment of neurological illnesses, this method has been utilised to discover promising novel chemicals.
- CRISPR/Cas9 Genome Editing: CRISPR/Cas9 genome editing enables precise alteration of genes in vivo. This method has been utilised in the investigation of genetic causes of neurological illnesses and the creation of treatments that aim to correct these underlying problems (59).

In general, we now know much more about brain problems and have more effective treatments for them thanks to the combination of cutting-edge technology and conventional ethnopharmacology methods.

16.3.4 Development of Evidence-Based Medicine

The development of evidence-based medicine in ethnopharmacology for brain diseases involves the integration of traditional medicine practices and modern scientific research methods to identify effective treatments for brain diseases. Ethnopharmacology is the study of the medicinal use of plants and other natural substances in different cultures (60). One of the challenges in developing evidence-based medicine in ethnopharmacology for brain diseases is the lack of rigorous scientific studies to validate the safety and efficacy of traditional medicine practices. To address this challenge, researchers have been using modern scientific methods, such as randomised controlled trials and systematic reviews, to evaluate the effectiveness of traditional medicine practices. For example, some studies have evaluated the use of traditional Chinese medicine for treating Alzheimer's disease. These studies have found that some traditional Chinese medicines, such as ginkgo biloba and huperzine A, may have beneficial effects on cognitive function in patients with Alzheimer's disease (61). Similarly, some studies have evaluated the use of traditional Indian medicine, such as Ayurveda, for treating depression. These studies have found that some Ayurvedic treatments, such as ashwagandha and Brahmi, may have antidepressant effects.

Overall, the development of evidence-based medicine in ethnopharmacology for brain diseases involves rigorous scientific research to validate the safety and efficacy of traditional medicine practices. This approach can lead to the development of new treatments for brain diseases and the integration of traditional medicine practices into modern healthcare systems.

16.3.5 Development of New Drug Delivery Systems

Integrating ancient medical practises with current scientific research methodologies, as is happening in the field of ethnopharmacology, has led to the identification of effective treatments for brain illnesses and the establishment of evidence-based medicine (62). Ethnopharmacology is the scientific study of traditional medical practises from around the world. However, there is a dearth of high-quality scientific investigations validating the safety and efficacy of traditional medicine practises, which presents a significant barrier to the development of evidence-based medicine in ethnopharmacology for brain illnesses. To overcome this difficulty, scientists have begun conducting randomised controlled trials and thorough evaluations of conventional medical practises (63). Some research has looked at how well traditional Chinese medicine works in treating Alzheimer's, for instance. Some traditional Chinese medicines, like ginkgo biloba and huperzine A, may improve cognitive function in people with Alzheimer's disease, according to these studies (64).

Similarly, Ayurveda and other forms of traditional Indian medicine have been the subject of research into their efficacy in treating depression. Some Ayurvedic remedies, such as ashwagandha and Brahmi, may have antidepressant benefits, according to the results of these investigations. Research into the safety and effectiveness of traditional medicine practises is essential to the growth of evidence-based medicine in ethnopharmacology for brain illnesses (65). New treatments for brain illnesses and the incorporation of old medical practises into contemporary healthcare systems are both possible with this strategy.

16.3.6 Exploration of New Mechanisms of Action

Exploration of new mechanisms of action in ethnopharmacology for brain disorders involves the study of how natural products derived from plants, animals or minerals interact with biological systems in the brain (66). This includes the identification of new targets and pathways, as well as the elucidation of the underlying mechanisms of action.

For example, recent research has explored the potential mechanisms of action of natural products such as curcumin (67), resveratrol (68), and epigallocatechin gallate (EGCG) (69) in the treatment of brain disorders such as Alzheimer's disease and Parkinson's disease. These studies have investigated their effects on oxidative stress, inflammation, amyloid beta protein accumulation (70), and mitochondrial dysfunction. Advanced technologies such as high-throughput screening, proteomics, metabolomics, and bioinformatics are also being used to identify new mechanisms of action in ethnopharmacology for brain disorders (71). These technologies enable researchers to analyse large datasets and identify potential targets and pathways involved in the pathophysiology of brain disorders, as well as the effects of natural products on these targets and pathways (72).

16.4 CHALLENGES AND CONCLUSION

Ethnopharmacological research has recently uncovered a wealth of natural compounds that show promise as treatments for age-related brain problems. Natural agents like these, included in alternative treatments, have been shown to have multiple pharmacological effects, including those that are neuroprotective and neurorestorative. However, there are a number of obstacles that must be overcome before these results may be used in clinical practise based on preclinical research. An important obstacle is the time and effort required to isolate the therapeutically active molecules. Multi-component formulations are the norm in traditional medicine, making it difficult to isolate the bioactive ingredients. The effectiveness and safety of natural products may also be impacted by interactions between their various components. The lack of uniformity in the manufacture and dosage of traditional medicines presents another difficulty. It is difficult to acquire consistent treatment results with traditional medicines since they are often prepared and administered differently by different practitioners. The potential for toxicity and other unintended consequences is increased by this lack of standardisation. In addition, there is a lack of a solid regulatory framework for the research, development, and clinical application of natural products. Natural product evaluations can be challenging because there are few standards by which to judge their safety and effectiveness. Lastly, there are substantial impediments to the greater use of natural goods in clinical practise, including the cost and availability of such products. The manufacture of some of these goods may be restricted to certain regions, making them more expensive. This can prevent patients who could benefit from their therapeutic effects from having access to them.

In conclusion, new findings in ethnopharmacology point to potential avenues for treating age-related brain diseases. For generations, people have turned to traditional remedies, which have been shown to be effective in treating the wide range of symptoms that accompany these diseases. The ability to discover and isolate new compounds for use in the creation of innovative medicinal agents has greatly increased with the advent of current technologies. But we shouldn't ignore the difficulties of combining traditional and modern medicine, or the necessity for more study into the mechanisms of action of these drugs. The ultimate key to realising the potential of ethnopharmacology in controlling age-related brain problems will be a multidisciplinary strategy combining collaboration between traditional healers, researchers, and healthcare providers.

REFERENCES

1. Adams M, Gmünder F, Hamburger M. Plants traditionally used in age related brain disorders—A survey of ethnobotanical literature. Journal of ethnopharmacology. 2007;113(3):363–381.
2. Azam S, Haque ME, Balakrishnan R, Kim IS, Choi DK. The ageing brain: Molecular and cellular basis of neurodegeneration. Frontiers in cell and developmental biology. 2021;9:683459.
3. Hou Y, Dan X, Babbar M, Wei Y, Hasselbalch SG, Croteau DL, Bohr VA. Ageing as a risk factor for neurodegenerative disease. Nature reviews neurology. 2019;15(10):565–581.
4. Hensel A, Ibach B, Muller U, Kruggel F, Kiefer M, Gertz HJ. Does the pattern of atrophy of the corpus callosum differ between patients with frontotemporal dementia and patients with Alzheimer's disease? Dementia and geriatric cognitive disorders. 2004;18(1):44–49.
5. Patwardhan B. Ethnopharmacology and drug discovery. Journal of ethnopharmacology. 2005;100(1–2): 50–2.
6. Heinrich M, Jäger AK, editors. Ethnopharmacology. John Wiley & Sons; 2015.

7. Touwaide A, Appetiti E. Knowledge of Eastern materia medica (Indian and Chinese) in pre-modern mediterranean medical traditions: A study in comparative historical ethnopharmacology. Journal of ethnopharmacology. 2013;148(2):361–78.
8. Scott G, Hewett ML. Pioneers in ethnopharmacology: The Dutch East India company (VOC) at The cape from 1650 to 1800. Journal of ethnopharmacology. 2008;115(3):339–360.
9. Yang XQ, Liu L, Ming SP, Fang J, Wu DN. Tian wang bu xin dan for insomnia: A systematic review of efficacy and safety. Evidence-based complementary and alternative medicine. 2019;2019.
10. Hung I, Chung CJ, Hu WL, Liao YN, Hsu CY, Chiang JH, Hung YC. Chinese Herbal medicine as an adjunctive therapy improves the survival rate of patients with ischemic heart disease: A nationwide population-based cohort study. Evidence-based complementary and alternative medicine. 2022;2022.
11. Kamer AR, Craig RG, Niederman R, Fortea J, de Leon MJ. Periodontal disease as a possible cause for Alzheimer's disease. Periodontology 2000. 2020;83(1):242–271.
12. Ganapathy AA, Haripriya VM, Acharya N, Somappa SB, Kumaran A. Ethnobotanical significance of medicinal plants: Beta-amyloid and tau aggregation inhibitors against Alzheimer's disease. Journal of biochemical and molecular toxicology. 2023:e23339.
13. Robinson B. Biological activities of the alkaloids of the Calabar bean. In The Calabar bean and its alkaloids: From magic, via miracle, to memory 2023 (pp. 169–224). Dordrecht: Springer Netherlands.
14. Liu Y, Yin Y, Lu QL, Dan Y, Xu MS, Song G, Li C. Vinpocetine in the treatment of poststroke cognitive dysfunction: A protocol for systematic review and meta-analysis. Medicine. 2019;98(6).
15. Vyas P, Sethy S, Sinha J, Garg A. Insights into emerging natural products for treating Alzheimer's disease. Journal of pharmaceutical negative results. 2022:4896–906.
16. Abeysinghe AA, Deshapriya RD, Udawatte C. Alzheimer's disease: A review of the pathophysiological basis and therapeutic interventions. Life sciences. 2020;256:117996.
17. Atanasova M, Dimitrov I, Ivanov S, Georgiev B, Berkov S, Zheleva-Dimitrova D, Doytchinova I. Virtual screening and hit selection of natural compounds as acetylcholinesterase inhibitors. Molecules. 2022;27(10):3139.
18. Almalki WH, Alosaim ZA, Al-Sulami MM, Alamri FS, Alzahrani B, Alkabi MB. Cannabinoids as a potential therapeutic agent: A comprehensive review. NVEO-NATURAL VOLATILES & ESSENTIAL OILS journal| NVEO. 2021:3532–3549.
19. Lokollo L, Salamor YB, Ubwarin E. Kebijakan Formulasi Undang-Undang Narkotika Dalam Legalisasi Penggunaan Ganja Sebagai Bahan Pengobatan di Indonesia. Jurnal belo. 2020;5(2):1–20.
20. Talbott SM, Talbott JA, Stephens BJ, Oddou MP. Effect of coordinated probiotic/prebiotic/phytobiotic supplementation on microbiome balance and psychological mood state in healthy stressed adults. Functional foods in health and disease. 2019;9(4):265–75.
21. Lee YK, Choi IS, Ban JO, Lee HJ, Lee US, Han SB, Jung JK, Kim YH, Kim KH, Oh KW, Hong JT 4-o-methylhonokiol attenuated β-amyloid-induced memory impairment through reduction of oxidative damages via inactivation of p38 MAP kinase. The journal of nutritional biochemistry. 2011;22(5):476–86.
22. Coelho F, Birks J. Physostigmine for Alzheimer's disease. The cochrane database of systematic reviews. 2001;1(2):CD001499.
23. De B, Debnath M, Das S, Bhowmick S, Karak S, Saha A. Anti-Alzheimer's potential of different varieties of *Piper betle* leaves and molecular docking analyses of metabolites. Free radicals and antioxidants. 2021;11(1):13–8.
24. Pandey K, Sinha A, Perween Z. Important medicinal plants with their medicinal uses from Jharkhand state. International journal of research in engineering, science and management. 2020;3(8):532–42.
25. Amato A, Terzo S, Mulè F. Natural compounds as beneficial antioxidant agents in neurodegenerative disorders: A focus on Alzheimer's disease. Antioxidants. 2019;8(12):608.
26. Schapira AH, Jenner P. Etiology and pathogenesis of Parkinson's disease. Movement disorders. 2011;26(6): 1049–1055.
27. Raja Sankar S, Manivasagam T, Surendran S. Ashwagandha leaf extract: A potential agent in treating oxidative damage and physiological abnormalities seen in A mouse model of Parkinson's disease. Neuroscience letters. 2009;454(1):11–15.
28. Cheng MM, Tang G, Kuo SH. Harmaline-induced tremor in mice: Videotape documentation and open questions about the model. Tremor and other hyperkinetic movements. 2013;3.
29. Fisher R, Lincoln L, Jackson MJ, Abbate V, Jenner P, Hider R, Lees A, Rose S. The effect of *Banisteriopsis caapi* (*B. caapi*) on the motor deficits in the MPTP-treated common marmoset model of Parkinson's disease. Phytotherapy research. 2018;32(4):678–87.
30. Xian YF, Lin ZX, Mao QQ, Hu Z, Zhao M, Che CT, Ip SP. Bioassay-guided isolation of neuroprotective compounds from Uncaria rhynchophylla against beta-amyloid-induced neurotoxicity. Evidence-based complementary and alternative medicine. 2012;2012.

31. Wei X, Jiang LP, Guo Y, Khan A, Liu YP, Yu HF, Wang B, Ding CF, Zhu PF, Chen YY, Zhao YL. Indole alkaloids inhibiting neural stem cell from uncaria rhynchophylla. Natural products and bioprospecting. 2017;7:413–419.
32. Wang YH, Samoylenko V, Tekwani BL, Khan IA, Miller LS, Chaurasiya ND, Rahman MM, Tripathi LM, Khan SI, Joshi VC, Wigger FT. Composition, standardization and chemical profiling of *Banisteriopsis caapi*, a plant for the treatment of neurodegenerative disorders relevant to Parkinson's disease. Journal of ethnopharmacology. 2010;128(3):662–671.
33. Jia L, Quan M, Fu Y, Zhao T, Li Y, Wei C, Tang Y, Qin Q, Wang F, Qiao Y, Shi S. Dementia in China: Epidemiology, clinical management, and research advances. The lancet neurology. 2020;19(1):81–92.
34. Weinmann S, Roll S, Schwarzbach C, Vauth C, Willich SN. Effects of Ginkgo biloba in dementia: Systematic review and meta-analysis. BMC geriatrics. 2010;10(1):1–2.
35. Mohamed NE, Abd El-Moneim AE. Ginkgo biloba extract alleviates oxidative stress and some neurotransmitters changes induced by aluminum chloride in rats. Nutrition. 2017;35:93–9.
36. Saini N, Singh D, Sandhir R. *Bacopa monnieri* prevents colchicine-induced dementia by anti-inflammatory action. Metabolic brain disease. 2019;34:505–18.
37. Sathyanarayanan V, Thomas T, Einöther SJ, Dobriyal R, Joshi MK, Krishnamachari S. rahmi for the better? New findings challenging cognition and anti-anxiety effects of Brahmi (Bacopa monniera) in healthy adults. Psychopharmacology. 2013;227:299–306.
38. Majdi A, van Boekholdt L, Sadigh-Eteghad S, Mc Laughlin M. A systematic review and meta-analysis of transcranial direct-current stimulation effects on cognitive function in patients with Alzheimer's disease. Molecular psychiatry. 2022;27(4):2000–9.
39. Howes MJ, Perry E. The role of phytochemicals in the treatment and prevention of dementia. Drugs & aging. 2011;28:439–68.
40. Petersen RC. Mild cognitive impairment. CONTINUUM: Lifelong learning in neurology. 2016; 22(2 Dementia):404.
41. Farhan M, Rizvi A. The pharmacological properties of red grape polyphenol resveratrol: Clinical trials and obstacles in drug development. *Nutrients*. 2023;15(20):4486. doi: 10.3390/nu15204486.
42. Knopman DS, Petersen RC. Mild cognitive impairment and mild dementia: a clinical perspective. InMayo Clinic Proceedings 2014 (Vol. 89, No. 10, pp. 1452–1459). Elsevier.
43. Huang HS, Lin YE, Panyod S, Chen RA, Lin YC, Chai LM, Hsu CC, Wu WK, Lu KH, Huang YJ, Sheen LY. Anti-depressive-like and cognitive impairment alleviation effects of Gastrodia elata Blume water extract is related to gut microbiome remodeling in ApoE–/– mice exposed to unpredictable chronic mild stress. Journal of ethnopharmacology. 2023;302:115872.
44. Iwasaki Y. Creutzfeldt-Jakob disease. Neuropathology. 2017;37(2):174–88.
45. Lin CF, Yu KH, Jheng CP, Chung R, Lee CI. Curcumin reduces amyloid fibrillation of prion protein and decreases reactive oxidative stress. *Pathogens*. 2013;2(3):506–19. doi: 10.3390/pathogens2030506.
46. Kartika B, Muralidharan P, Rahman H. Herbal treatment of parkinsonism: A review. International journal of pharmaceutical sciences review and research. 2010;5(3):185–91.
47. Singh J, Singh A, Sharma V, Singh TD, Singh M, Garg R, Sharma R, Gautam DN. Ethnopharmacology, phytochemistry, And pharmacology of ashtanga ghrita: An ayurvedic polyherbal formulation for neurological disorders. Current pharmacology reports. 2022;8(5):376–407.
48. Li S, Wu Z, Le W. Traditional Chinese medicine for dementia. Alzheimer's & dementia. 2021;17(6): 1066–71.
49. Lan B, Ge JW, Cheng SW, Zheng XL, Liao J, He C, Rao ZQ, Wang GZ. Extract of naotaifang, a compound Chinese herbal medicine, protects neuron ferroptosis induced by acute cerebral ischemia in rats. Journal of integrative medicine. 2020;18(4):344–50.
50. Calfio C, Gonzalez A, Singh SK, Rojo LE, Maccioni RB. The emerging role of nutraceuticals and phytochemicals in the prevention and treatment of Alzheimer's disease. Journal of Alzheimer's disease. 2020;77(1):33–51.
51. Dixit H, Kumar CS, Dasgupta D, Gadewal N. Molecular docking analysis of hyperphosphorylated tau protein with compounds derived from *Bacopa monnieri* and *Withania somnifera*. Bioinformation. 2021;17(9):798.
52. Witter S, Witter R, Vilu R, Samoson A. Medical plants and nutraceuticals for amyloid-β fibrillation inhibition. Journal of Alzheimer's disease reports. 2018;2(1):239–52.
53. Cascella M, Bimonte S, Muzio MR, Schiavone V, Cuomo A. The efficacy of epigallocatechin-3-gallate (green tea) in the treatment of Alzheimer's disease: An overview of pre-clinical studies and translational perspectives in clinical practice. Infectious agents and cancer. 2017;12(1):1–7.

54. Jung I, Kim H, Moon S, Lee H, Kim B. Overview of Salvia miltiorrhiza as a potential therapeutic agent for various diseases: An update on efficacy And mechanisms of action. Antioxidants. 2020;9(9):857.
55. Elsaid S, Le Foll B. The complexity of pharmacology of cannabidiol (CBD) and its implications in the treatment of brain disorders. Neuropsychopharmacology. 2020;45(1):229.
56. Khalil M, Teunissen CE, Otto M, Piehl F, Sormani MP, Gattringer T, Barro C, Kappos L, Comabella M, Fazekas F, Petzold A. Neurofilaments as biomarkers in neurological disorders. Nature reviews neurology. 2018;14(10):577–89.
57. Bühler M, Mann K. Alcohol and the human brain: A systematic review of different neuroimaging methods. Alcoholism: Clinical and experimental research. 2011;35(10):1771–93.
58. Alshdoukhi I. Review of application of artificial intelligence in natural products research. Journal of health informatics in developing countries. 2022;16(2).
59. Giau VV, Lee H, Shim KH, Bagyinszky E, An SS. Genome-editing applications of CRISPR–Cas9 to promote in vitro studies of Alzheimer's disease. Clinical interventions in aging. 2018:221–33.
60. Akhlaq S, Ara SA, Fazil M, Ahmad B, Akram U, Haque M, Khan AA. Ethno pharmacology, phytochemical analysis, safety profile, prophylactic aspects, and therapeutic potential of *Asarum europaeum* L. in Unani medicine: An evidence-based appraisal. Phytomedicine plus. 2022:100226.
61. Baskys A, Cheng JX. Pharmacological prevention and treatment of vascular dementia: Approaches and perspectives. Experimental gerontology. 2012;47(11):887–91.
62. Fokunang CN, Ndikum V, Tabi OY, Jiofack RB, Ngameni B, Guedje NM, Tembe-Fokunang EA, Tomkins P, Barkwan S, Kechia F, Asongalem E. Traditional medicine: Past, present and future research and development prospects and integration in the national health system of Cameroon. African journal of traditional, complementary and alternative medicines. 2011;8(3).
63. Dong H, Ren JX, Wang JJ, Ding LS, Zhao JJ, Liu SY, Gao HM. Chinese Medicinal leech: Ethnopharmacology, phytochemistry, and pharmacological activities. Evidence-based complementary and alternative medicine. 2016 Oct;2016.
64. Singhal AK, Naithani V, Bangar OP. Medicinal plants with a potential to treat Alzheimer and associated symptoms. International journal of nutrition, pharmacology, neurological diseases. 2012;2(2):84–91.
65. Shankar PS, Preeti B, Santanu B, Gajanan D, Brahmi RD (*Bacopa monnieri*) as functional food ingredient in food processing industry. Journal of pharmacognosy and phytochemistry. 2018;7(3):189–194.
66. Luo Y, Wang Q, Zhang Y. A systems pharmacology approach to decipher the mechanism of danggui-shaoyao-san decoction for the treatment of neurodegenerative diseases. Journal of ethnopharmacology. 2016;178:66–81.
67. Tang M, Taghibiglou C. The mechanisms of action of curcumin in Alzheimer's disease. Journal of Alzheimer's disease. 2017;58(4):1003–1016.
68. Xu Q, Si LY. Resveratrol role in cardiovascular and metabolic health and potential mechanisms of action. Nutrition research. 2012;32(9):648–658.
69. Aggarwal V, Tuli HS, Tania M, Srivastava S, Ritzer EE, Pandey A, Aggarwal D, Barwal TS, Jain A, Kaur G, Sak K. Molecular mechanisms of action of epigallocatechin gallate in cancer: Recent trends and advancement. In Seminars in Cancer Biology 2022 (Vol. 80, pp. 256–275). Academic Press.
70. Syed AA, Reza MI, Singh P, Thombre GK, Gayen JR. *Withania somnifera* in neurological disorders: Ethnopharmacological evidence, mechanism of action and its progress in delivery systems. Current drug metabolism. 2021;22(7):561–571.
71. Lao Y, Wang X, Xu N, Zhang H, Xu H. Application of proteomics to determine the mechanism of action of traditional Chinese medicine remedies. Journal of ethnopharmacology. 2014;155(1):1–8.
72. da Rocha, M. D., Viegas, F. P., Campos, H. C., Nicastro, P. C., Fossaluzza, P. C., Fraga, C. A., Barreiro, E. J., & Viegas, C., Jr (2011). The role of natural products in the discovery of new drug candidates for the treatment of neurodegenerative disorders II: Alzheimer's disease. CNS & neurological disorders drug targets, 10(2), 251–270. https://doi.org/10.2174/187152711794480429

17 Anthocyanins and Its Derivatives

Systemic Approach in Drug Delivery

Veni Subramanyam[1], Ramakrishnan Ganapathy[1], Agiesh Kumar Balakrishna Pillai[1], Varsha Virendra Palol[1], Haema G[2], Manikandan Sivaprakasam[1], and Indrani Biswas[1]

[1]MGM Advanced Research Institute (MGMARI), Sri Balaji Vidyapeeth (SBV) (Deemed to be University) Pillaiyarkuppam, Puducherry, India

[2]Pharmaceutical Chemistry, School of Pharmacy, A Unit of Sri Balaji Vidyapeeth (Deemed to be University) Pillaiyarkuppam, Puducherry, India

17.1 INTRODUCTION

Anthocyanins (ACns) are found in vegetables and fruits, one of the most abundant phytochemicals over 700 polyphenolic, water-soluble natural pigments that plays important functions in human health. Nature is treasured with 640 distinct varieties of ACns found in nature, of which 80% are found in coloured leaves, 69% in fruits and 50% in flowers (Cai et al., 2022). ACn is a phytochemical consisting of flavonoids composed of polyphenols. ACn is a secondary metabolite sensitive to pH, light, oxygen, temperature, presence of metal ions and structure. Change of colour intensity from red to blue represents the change in pH from acidic to alkaline with structural modifications of methoxyl or hydroxyl (–OH) groups on B-ring upon polymerization or cleavage and derivatization making ACn nucleophilic (Bhatt et al., 2022). ACns are less toxic according to Food and Agriculture Organization (FAO)/World Health Organization (WHO) Expert Committee on Food Additives (JECFA); therefore, its usage as food additives is legally accepted in countries like Europe (E163), Japan and United States (Joint, 2019. ACns provide cardioprotection, reduce lipid pre-oxidation, neuroprotective, anti-inflammatory, anti-ageing, anti-obesity, anti-diabetic, anti-allergy, anti-microbial, anti-viral and free radical scavenging (Panche et al., 2016). It is important to note that the effectiveness of ACns can vary based on factors such as the specific type of ACn, its concentration in a particular food source and individual variations in response. ACns are well known and well documented for their therapeutic efficacies *in vitro* and *in vivo*, elucidating from nutrients to pharmaceuticals for many human diseases. But ACn undergoes less or harmful effects in some physiological conditions in humans; this entails disadvantages and the need to enhance their pharmacokinetic, pharmacodynamic and therapeutic properties. The development of novel delivery systems of nanoformulations was found to be a key strategy to enhance their bioaccessibility, bioactivity, absorption, enhanced stability in the gastrointestinal tract, solubility and control of release, biocompatibility, increased circulation time, encapsulation and protection, combination therapies, tailored formulations and optimized dosage regimens (Gonçalves et al., 2018; Hamedi and Koosha, 2020; Hanafy, 2021). This chapter is to help the readers learn about the chemical

DOI: 10.1201/9781003486237-17

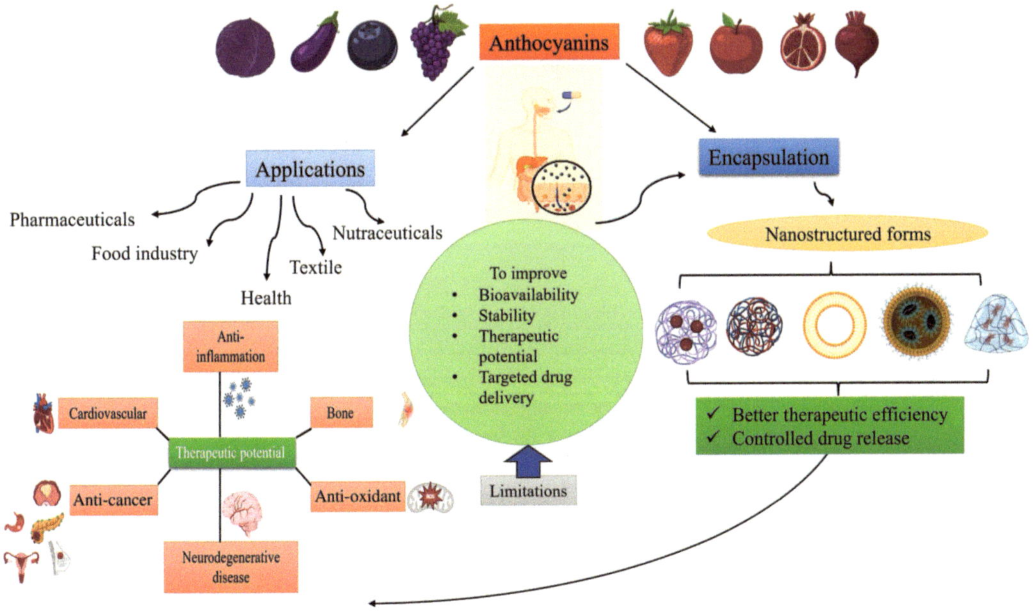

FIGURE 17.1 Graphical abstract

nature, structural modifications benefiting therapeutics with systemic approaches in drug delivery and explore more into ACns and its derivatives applications in drug delivery for human health and diseases (Figure 17.1).

17.2 CHEMISTRY AND PROPERTIES OF ACn

17.2.1 Molecular Structure

ACns are polyhydroxy and polymethoxy 2-phenylbenzopyrylium derivatives constituted by a core called anthocyanidin bound to several glycosidic moieties (glucose, galactose, xylose, arabinose, fructose and rhamnose) attached at the C_3, C_5 or C_7 positions (Prior and Wu 2006). They are characterized by the presence of two benzyl rings (A and B) and a heterocyclic ring (C). The C ring is joined to the A and B rings through a three-carbon bridge. The structural formula of the flavylium cation is common among ACns (Figure 17.2).

There are about 17 anthocyanidins found in nature, whereas only six of them, cyanidin (Cy), delphinidin (Dp), petunidin (Pt), peonidin (Pn), pelargonidin (Pg) and malvidin (Mv), are ubiquitously distributed. The differences in the chemical structure of these six common anthocyanidins occur at the 3′ and 5′ positions of the B-ring (Kong et al., 2003) (Figure 17.3).

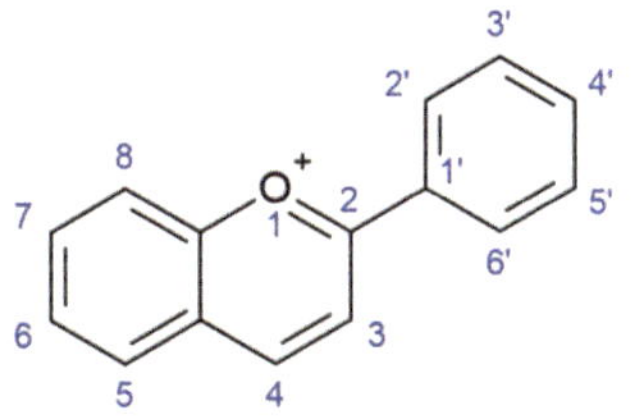

FIGURE 17.2 Chemical structure of flavylium cation.

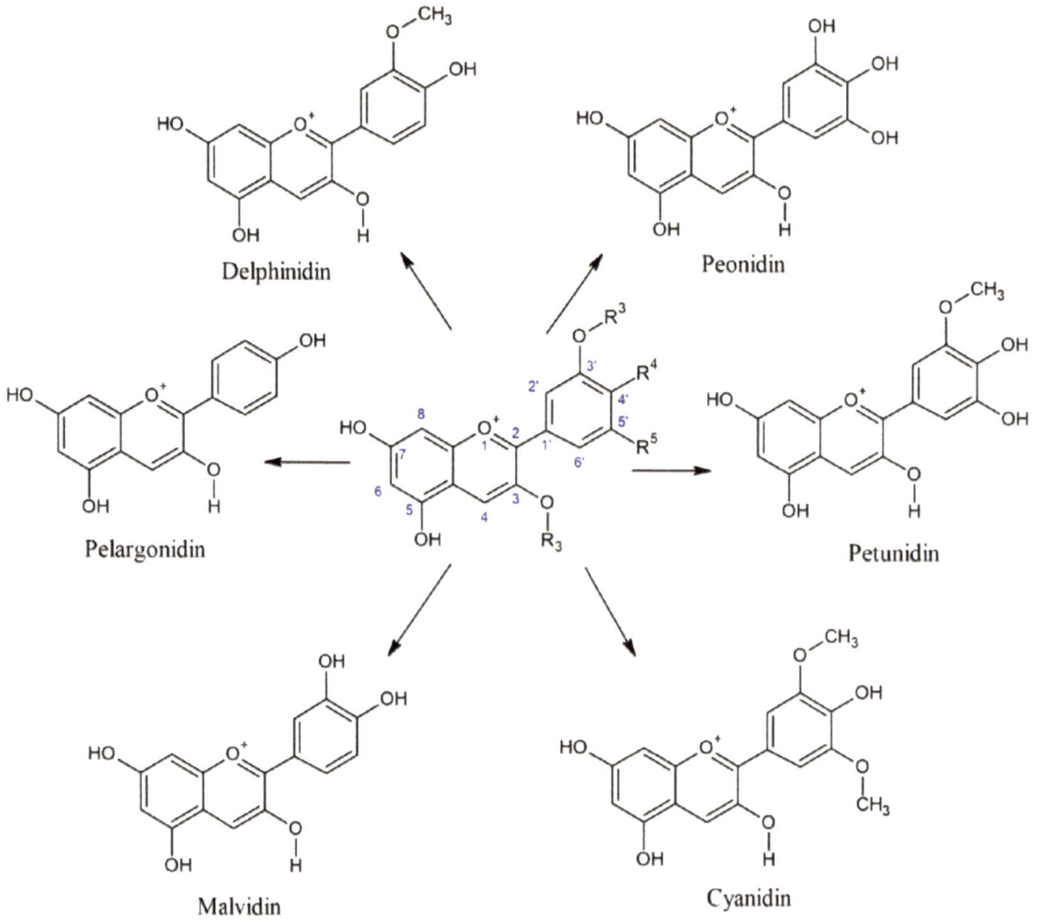

FIGURE 17.3 Chemical structure of commonly used six anthocyanidins.

17.3 PHYSICOCHEMICAL CHARACTERISTICS OF ACn

By harnessing their distinctive physicochemical characteristics, ACns can be tailored for specific applications, facilitating the development of novel products with improved efficiency and health benefits. Based on this study, ACns could be functionalized or encapsulated to enable targeted drug delivery to specific tissue or cells.

17.3.1 pH

The colour of the ACns is strongly influenced by pH of the solution. Since ACns are amphoteric in nature, they function in a manner similar to that of pH indicators. Under an acidic environment, at pH 1, the ACns exist in the form of the flavylium cation predominantly to produce stable red to orange pigment which makes them very soluble in water. There is an abundance of quinoidal blue species at pH values between 2 and 4, and at pH values between 5 and 6, colourless substances called chalcones and carbinol pseudo bases are visible (Kang et al., 2021). Ultimately, the ACns will break down in accordance with the substituent groups at a pH greater than 7. Nonetheless, the four ACn forms (Figure 17.4) can coexist in a pH range of 4–6, with the flavylium cation preserving the equilibrium between these forms.

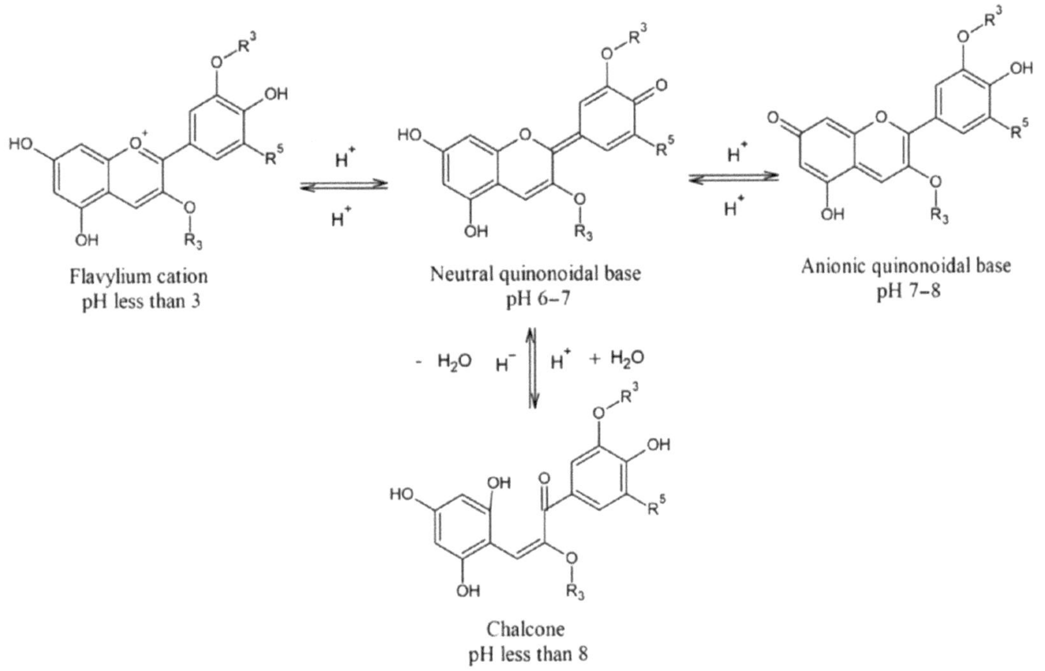

FIGURE 17.4 Four forms of anthocyanin.

ACn encapsulation is beneficial when food is engineered with water-in-oil-in-water (W/O/W). The ACn emulsion prepared in combination with hydrophobic and hydrophilic emulsifiers of polyglycerol polyricinoleate and quillaja saponin, respectively, reduced pH-induced colour changes and retained their functional properties with the changes in water phases (Sharif et al., 2020). The β-lactoglobulin stabilized lipid nanoparticles loaded with ACns demonstrated a higher retention rate percentage when entraining oral pH 6.8, simulated gastric pH 2 and simulated intestinal pH 6.9 by increasing the heat stability and bioavailability of ACn (Salah et al., 2020).

17.3.2 Temperature

Effect of temperature will tend to change the colour and nature of the ACn. When heat is applied between 50 and 180°C, ACns will endure a series of mechanisms such as cleavage, polymerization, glycosylation and nucleophilic attack of water that leads to their deterioration and loss of pigment. Consequently, temperature is another element that influences the stability of the ACns' molecular structure. As a result, the rise in temperature causes the compounds to degrade and the degree to which products turn brown when exposure to oxygen increases (Laleh et al., 2006). It has been demonstrated that the stability of all dietary pigments, including ACns, declines with rising temperatures. Concurrently, recent studies have revealed a strong correlation between these compounds' chemical structure and stability, with the sugar fraction as an important factor. The microencapsulation of ACn extract with soy protein isolate and gum arabic was thermally stable ranging from 80°C to 114°C with enhanced release under simulated gastrointestinal conditions (Mansour et al., 2020).

17.3.3 Solubility

ACns are soluble in both water and most of the organic solvents. In contrast, the apolar organic solvent does not dissolve ACn. Meanwhile, it is not stable in neutral or alkaline solutions. While glycosylated ACn is extremely soluble in water, ACn aglycone is more soluble in alcohol than its glucoside counterpart. ACn's hydrophobic property is generated by its polyphenolic structure, which also makes it soluble in organic solvents like methanol and ethanol. In addition to the flavylium cation, another factor contributing to anthocyanidin's solubility in water may be the 3-OH group in the compound's C-ring, which is always bonded to a sugar moiety and creates a stable ACn in water (Iosub et al., 2014). When intense protonation occurs at lower pH values, ACns become more soluble in water. Alcohol and HCl together improve ACn's solubility. Low methoxyl pectin, sodium alginate and Ticafilm embedded with ACn from wine grape pomace were combined to create an ACn-based polysaccharide film matrix. This film matrix allowed phenolic release under hydrophilic conditions, displaying potential anti-bacterial and antioxidant properties for producing oral fast-dissolving films in pharmaceutical or nutraceutical applications. The regulated release with controlled solubility of ACn from nanoparticles was facilitated by the ACn-encapsulated pectin and chitosan (which dissolve only in diluted acidic conditions), protecting against acrylamide-induced kidney damage in rat cells.

17.3.4 Co-pigmentation

Co-pigmentation of ACn aglycone is referred to as a phenomenon where anthocyanidins are reinforced by metallic ions or flavonoids. Co-pigmentation process is pH sensitive. Since an increase in pH causes a reduction in colour intensity of the ACns. Thermal degradation also has an impact on the stability of co-pigmented compounds. When combined with ACns, the co-pigments exhibit an interaction that causes a hyperchromic effect and a bathochromic shift change in the absorption spectra (Trouillas et al., 2016). Since, some ACns, which are pigment-free, absorb light at a specific wavelength and this value greatly increases upon co-pigmentation. This effect is linked to low electron flavylium cations, which increase the stability. Chitosan polysaccharide when co-pigmented with ACn encapsulated in counter-ionic polyelectrolyte complexes forming dense hydrogen bonding at pH 3.3 was more effective at pH 4.0 and 5.0 enhanced the retention capacity and promoting fabrication of robust delivery system in presence of nucleophilic ascorbic acid (Tan et al., 2018, 2021).

17.3.5 Stability

By virtue of their relative unstable condition, ACns are susceptible to degradation by various factors such as light, heat, oxygen, pH and enzymatic activity. The stability of ACns can be attributed to ring B substituents and the presence of auxiliary hydroxyl or methoxyl groups, which in neutral conditions reduce the stability of aglycones (Fei et al., 2021). In contrast, while aglycones are unstable in a neutral environment, monoglycosides and their derivatives are more stable because sugar molecules prevent unstable intermediates from deterioration. ACns generate van der Waals and hydrogen bonds as well as hydrophobic interactions with nanocarriers to increase their stability (Nazareth et al., 2021). The chemical stability of cyanidin-3-O-glucoside (C3G) was improved under simulated gastrointestinal conditions by C3G-loaded nanogel encapsulation. The clathrin-mediated pathway and macropinocytosis followed the transcytosis of the C3G-loaded nanogel via endosome, endoplasmic reticulum, golgi complex and basolateral plasma membrane route. In comparison to free-C3G, nano-C3G effectively restored the survival of cells and antioxidant enzymes activity in oxidative stress models (Feng et al., 2019).

17.4 HEALTH BENEFITS AND THERAPEUTIC POTENTIAL OF ACn

ACns present in phytochemicals of red wine, red and black grapes, pomegranates, etc., possess several health benefits with therapeutic potential. ACns are often linked to their antioxidant properties, which help combat oxidative stress and inflammation in the body. These biologically active compounds demonstrate significant health benefits, including preventive effects against metabolic diseases, cancers, cardiovascular diseases (CVDs), diabetes, microbial infections, obesity, atherosclerosis and also neurodegenerative diseases as evidenced by numerous *in vitro* and *in vivo* studies (Salehi et al., 2020). In terms of applications, utilized in the food and textile industry to enhance colour as a natural dye and enhance nutritional content. Additionally, ACns have also been applied in various areas of photosensitizers for dye-sensitized solar cells (DSSCs), nutraceuticals, cosmetics and photodynamic therapeutics (PDT) which works on the principle of activation of compounds upon contact with light. As ACn is pigmented, it can absorb red light at physiological pH, which helps it combat biofilms (Oliveira, et al., 2019; Teerakapong et al., 2017). Moreover, the role of ACns in the cosmetic industry is widely explored in the areas of ageing, melanin and skin erythema for their natural colouring and potential skin health benefits (Câmara et al., 2022).

17.4.1 Mechanism of Action of ACNs in Disease Management

The mechanism of action of ACns in disease management are multifaceted and often involve their interactions at the molecular and cellular levels. The mechanisms underlying their biological activity involve pathways such as cyclooxygenase (COX) and mitogen-activated protein kinases (MAPKs), signalling of inflammatory cytokines, free-radical scavenging and alterations in blood biomarkers. The chemical structures of ACns are noted to significantly influence their biological activities, with certain types demonstrating stronger chemoprotective effects. Investigations on the molecular mechanisms reveal that ACns can inhibit signalling pathways associated with tumour growth and apoptosis. Examples include cell cycle arrest induction, caspase-dependent and independent apoptosis and modulation of pathways like MAPK, c-Jun N-terminal kinase (JNK), nuclear factor κB (NF-κB) and the Wnt pathway. Additionally, ACns' potential role in autophagy, a novel therapeutic target for cancer treatment, is explored. There are different anti-cancer mechanisms of ACns, citing their effectiveness against breast, prostate, liver, colorectal, intestinal, blood, cervical, lung, fibrosarcoma and metastatic melanoma cancers (Li et al., 2017).

ACns have shown neuroprotective effects by mitigating reactive oxygen species (ROS)-induced damage in the brain sourced from Korean black beans to suppress neuroinflammation and neurodegeneration (Fathallah et al., 2023). The primary anti-inflammatory mechanism of ACns revolves around inhibiting the activation of pro-inflammatory molecules and enzymes. These include cytokines such as tumour necrosis factor (TNF)-α, interleukins (IL)-1β, IL-6, IL-8 and caspase-1, as well as regulatory proteins like kappa B (NF-κB), COX-1 and COX2, nitric oxide synthase (iNOS) and MAPK (Vendrame, & Klimis-Zacas, 2015). In cellular models, ACns inhibit the induced secretion of various molecules associated with inflammatory modulation, particularly vascular endothelial growth factor and intracellular adhesion molecule-1. The intake of ACns has reduced the risk of CVDs by targeting the lipids and downregulation of proinflammatory genes such as ALOX5AP, CX3CL1 and TNFRSF14 (Mauray et al., 2010) (Figure 17.5).

ACns are considered safe for consumption and have been classified as Generally Recognized as Safe (GRAS) by regulatory authorities. This designation is based on their long history of use in the human diet without reported adverse effects. They are known for their low toxicity and typical dietary intake poses no significant health risks (Forman et al., 2014). It is crucial to comprehend the bioavailability of ACns since, upon ingestion, their constituents experience several modifications in the digestive system (digestion, absorption, metabolism and excretion), all of which have a substantial influence on their advantageous and health-promoting characteristics. The low bioavailability of ACns in humans is only about 1–2% due to their hydrophilic nature which gets affected by pH, temperature and chemical structures while restricting its bioactivity (Jiang et al., 2019; Passamonti, 2019). However, several ongoing research disproves the low bioavailability of ACns showing that microbially produced metabolites of ACns are significantly absorbed and have increased bioavailability. As a result, they are frequently regarded as biocompatible and safe for ingestion in a variety of applications (Ayvaz et al., 2022).

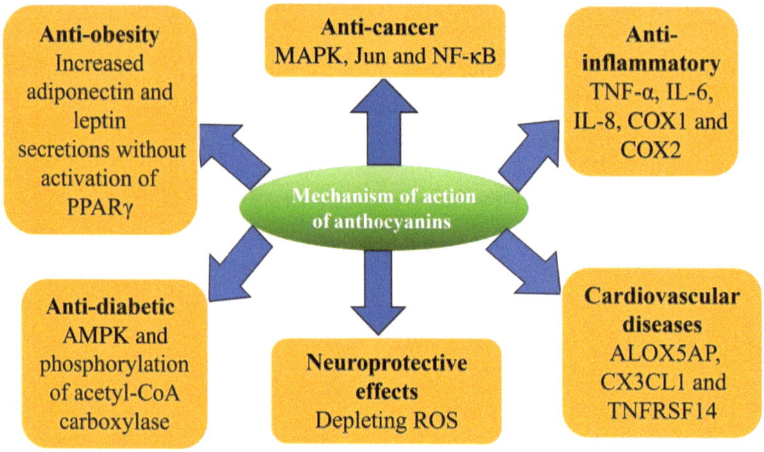

FIGURE 17.5 Overview on the different therapeutic mechanism of action of ACns' safety and biocompatibility.

17.4.2 Derivatization of Anthocyanins for Drug Delivery

Derivatization of ACns for drug delivery offers numerous opportunities to enhance their pharmaceutical properties and enable their use as effective carriers for delivering therapeutic agents to target sites in the body. These strategies can be tailored based on the specific requirements of the drug delivery system and the desired therapeutic outcomes.

17.5 CHEMICAL MODIFICATION STRATEGIES

Chemical modification strategies of ACns involve altering their chemical structure through reactions such as acylation, esterification, glycosylation, methylation and more. These modifications can be performed to enhance the functional properties or stability of ACns for various applications, including food colouring, pharmaceuticals, cosmetics and materials science.

17.5.1 Glycosylation

Glycosylation is a crucial step in the biosynthesis and function of ACns. The process entails affixing one or more glycosides (sugar molecules) to the hydroxyl groups present on the anthocyanidin aglycone, which is the primary component of ACns (Brouillard, 1982). Glycosylation of ACns is catalysed by glycosyltransferase enzymes, which transfer active sugar moieties from donor molecules (such as Uridine diphosphate glucose or Uridine diphosphate-galactose) to specific hydroxyl groups on the anthocyanidin molecule. Naturally occurring ACns are always glycosylated at the C_3–OH position or C_3– and C_5–OH positions (Borkowski et al., 2005). Furthermore, glycosylation of anthocyanidins can occur at the C_7–, $C_{3'}$– or $C_{5'}$–OHs, and less frequently at the C_4–OH (Prior and Wu, 2006).

In ACn glycosylation, only two types of glycosidic bonds, i.e., O- and C-glycosidic, are found. Often the sugar moieties usually connect to the anthocyanidins through O-glycosidic linkage. Nevertheless, C-glycosidic linkages were also identified in several ACns (Tatsuzawa et al., 2004). The sugars glycosylating ACns mainly include mono-, di- and trisaccharides. The most common monosaccharide is glucose, followed by rhamnose, galactose, arabinose, xylose and fructose; glucuronic acid is the rarest monosaccharide; and the disaccharides may be homogeneous or heterogeneous, e.g., rutinose (6-O-α-L-rhamnosyl-D-glucopyranose), sophorose (2-O-β-D-glucopyranosyl-D-glucopyranose) and gentiobiose (6-O-β-D-glucopyranosyl-D-glucopyranose). The trisaccharide mainly includes glycosyl rutinose (Andersen & Markham, 2005).

ACn glycosylation can generally increase its stability to pH changes, temperature and can lend hydrophilicity or polarity, to the anthocyanidins because the –OHs of the exterior can easily form H-bonds with the surrounding water molecules. Meanwhile, anthocyanidin glycosylation enlarges the size of the corresponding ACns and produces steric hindrance to diminish the coplanarity of the B-ring with the rest of the anthocyanidin or ACn (Kren, 2001) (Figure 17.2). Glycosylated ACns are generally biocompatible and biodegradable, making them suitable carriers for drug delivery applications. These conjugates are less likely to induce cytotoxicity or adverse immune responses, enhancing their safety profile for therapeutic use.

17.5.2 Acylation

Acylation is another crucial modification that ACns undergo, further influencing their properties and functions when compared to non-acylated analogues. It refers to the process of attaching an acyl group (derived from an organic acid) to one or more hydroxyl groups (OH) present in the ACn molecule. This might take place at various regions, including the anthocyanidin aglycone or the sugar moieties attached to it (glycosyl acylation) (Luo et al., 2018). Acylated ACns are ACn derivatives with intricate glycosylation and acylation patterns found in almost every part of the plant usually formed by enzymatic or chemical modification pathways. Acyltransferase enzyme aids the enzymatic acylation of ACns using appropriate aliphatic or aromatic acyl groups. Acylation usually occurs at C_3 sugar residue using a specific acyl donor. Acylation of sugar residues can be done by using aliphatic acids like malonic, acetic, malic, succinic or oxalic acids, or by aromatic acids like p-coumaric, caffeic, ferulic, sinapic, gallic or p-hydroxybenzoic acids (Rinaldo et al., 2015). Chemical acylation of ACns involves the reaction of ACn molecules with acylating agents under appropriate conditions. The acylation reaction can occur at various positions on the ACn molecule, typically involving a nucleophilic attack by the oxygen atom on the ACn's flavylium cationic structure. In contrast with chemical acylation, enzymatic acylation of ACns offers greater region-selectivity and yield, and under certain circumstances, it can be utilized to generate certain acylated ACns with a high degree of stability and fat solubility (Chebil et al., 2006). Thus, the free radical scavenging ability of ACn could be enhanced through acylation of ACn whose antioxidant nature was decreased on prior glycosylation. It also exhibits improved lipophilicity, pH resistivity, thermostability and improved permeability across cell membranes.

17.5.3 Esterification

Esterification of ACns involves the formation of ester by the reaction between an ACn molecule and an organic acid. This process also includes the attachment of an acyl group (acylation) or a methyl group (methylation) to one of the hydroxyl groups present in an ACn. Certain enzymes called O-methyltransferases catalyze the enzymatic O-methylation of ACns. These enzymes create methylated ACn products by transferring a methyl group from a methyl donor molecule, typically S-adenosyl methionine (SAM), to the hydroxyl group of the ACn substrate. Chemical reagents are used in chemical O-methylation, which adds methyl groups to the ACn molecule. Methyl iodide, methyl sulphates and diazomethane are examples of common methylating agents. Under the right circumstances, these substances react with the hydroxyl groups of ACns to create methylated ACn derivatives.

17.6 NANOFORMULATIONS OF ANTHOCYANIN DERIVATIVES

Nanoformulations in ACns represent an innovative approach to optimize the delivery of bioactive compounds, enhancing their potential for various applications in health and industry. Encapsulating ACns in nanostructures ensures structural stability throughout the digestive process, facilitating absorption, targeted delivery to tissues/organs and enhancing their biological utilization (Rosales et al., 2023).

17.6.1 Nanoparticles

The co-encapsulation of ACn and cinnamaldehyde within nanoparticle-filled carrageenan films for potential applications in active packaging aiming to enhance their functional properties such as antioxidant and anti-microbial capabilities. These nanoparticles (NPs) were prepared by anti-solvent precipitation and solvent casting method (Zhou et al., 2024). An ACn-rich extract, primarily containing C3G, was obtained from fully ripened haskap (*Lonicera caerulea* L.) berries (HB). This extract was then encapsulated in three distinct polymeric nanocarrier systems: polyethylene glycol-poly(lactide-co-glycolide), maltodextrin and carboxymethyl chitosan (CMC) for targeted delivery system for ACns directly into pulmonary tissues using encapsulated nanocarriers which shows therapeutic potential to treat lung diseases including cancer (Amararathna et al., 2022). Nano-encapsulation through amylopectin nanoparticles (APNPs) serves to stabilize ACns, ensuring their efficient delivery to tissues and improving their physiological functions by enhancing protection against oxidative damage in vivo (Tong et al., 2023). The gold nanoparticles (Au NPs) synthesized from ACn-rich berry extracts haveattributed to their notable colloidal dispersibility and salt tolerance, emphasizing their potential suitability for various biomedical uses (Demirbas et al., 2019). ACn-based nanoparticle with polyethylene glycol (PEG-2000) and poly (lactide-co-glycolide) (PLGA)-mitigated neurotoxicity in SH-SY5Y cell line by scavenging free radicals increased cellular absorption and bioactivity. The neuroprotective properties of ACn-loaded nanoparticles showed notable antioxidant, anti-apoptotic and anti-inflammatory properties which has the ability to prevent Alzheimer's disease. Nanophytosomes of ACns derived from elderberries (*Sambucus nigra*) was combined with membrane polar lipids enhancing complexes I and II of mitochondrial respiratory chain in the presence of rotenone and protected against rotenone- and glutamate-induced neurotoxicity facilitating targeted mitochondrial therapy in the treatment of neurodegenerative diseases (Amin et al., 2017; Deng & Zhao 2011; Mendes et al., 2021; Zhao et al., 2020).

17.6.2 Liposomes

Liposomes are spherical, closed structures made of bilayer lipids that spontaneously develop in aquatic settings and partially enclose the solvent in which they are found. This property leads to the formation of distinct liposomal properties, such as self-sealing, and has garnered significant interest as advantageous carrier systems in a variety of sectors, including tissue engineering, medicines, cosmetics and food. The liposome formulation of blueberry ACn was achieved through response surface methodology, with high entrapment efficiency. Subsequently, chitosan was incorporated to enhance stability and functionality. This combination enhanced antioxidant properties of ACns, with significant application potential in the food, dietary supplement and pharmaceutical industries (Wang et al., 2021). The feasibility of using nanoliposomes composed of rapeseed lecithin for encapsulating ACns from *Berberis vulgaris* enhanced cellular metabolic activity, replication which attributed to their antioxidant and energy generation activities. Therefore, encapsulation of ACn in nanoliposomes not only enhanced stability but also enabled targeted delivery to specific sites of interest through optimized formulation (Homayoonfal et al., 2021).

17.6.3 Micelles

Micelles represent spherical amphiphilic structures featuring a hydrophobic core and a hydrophilic shell. This hydrophilic shell imparts water solubility to the micelle, enabling intravenous delivery, while the hydrophobic core serves as a carrier for therapeutic drug payloads (Aguilar, 2013). The acid-responsive *Lonicera edulis pomace* ACn micelles prepared were proven to show effective toxicity to liver cancer cells. This amphiphilic CMC derivative significantly improved the stability of ACns, making it a potential nanomicelle carrier for ACns and other active food ingredients (Shan et al., 2024). A study on the impact of casein micelle aggregation during *in vitro* digestion upon release of loaded ACns was performed to improve the utilization of serum Ca^{2+}. They used micellar casein as a wall material in an effort to provide a simple and efficient way to regulate micelle structure, which would enable different ways for loaded ACns to be released in the gut (Ren et al., 2024). A study on the development of casein micelle hydrogels involved the use of transglutaminase as a crosslinking agent to encapsulate ACns to provide controlled delivery of jaboticaba fruit (*Myrciaria cauliflora*) (Nascimento et al., 2020). In a related study, stable reverse micelles were developed of crude ACns extracted from grape skin leading to enhanced colour intensity and a bathochromic shift which were used to study these in apolar mediums (Kim et al., 2003).

17.6.4 Encapsulation Techniques

Encapsulation, defined as the process of covering a bioactive agent with a polymeric material or introducing it into a matrix, serves to protect the agent from environmental factors. This process allows for the control of parameters such as the location and timing of delivery for optimal efficiency, producing vesicles ranging in size from 1 to 1000 μm. In recent years, numerous techniques for ACn encapsulation have been developed and optimized in both micro- and nanocarriers. Common methods include spray-drying, freeze-drying and electro-spinning/spraying (Enaru et al., 2021; Sharif et al., 2020). Spray-drying stands out as the oldest and most commonly employed method for microencapsulating ACns. The process involves atomizing a fluid feed into a drying chamber, creating a solution, emulsion or suspension. This method is characterized by its speed, versatility, cost-effectiveness and scalability, offering high encapsulation efficiency and relatively robust storage stability. However, it has limitations, as it produces non-uniform particles of varied sizes and the size and morphology of the carriers cannot be precisely controlled (Mohammadalinejhad and Kurek, 2021). Freeze-drying, also known as lyophilization, presents an alternative to spray-drying by employing low temperatures for dehydration. This method is particularly suitable for encapsulating temperature-sensitive compounds like ACns. The process involves freezing, followed by sublimation (primary drying), desorption (secondary drying) and eventual storage of the resulting dry material (Ozkan et al., 2019). New research directions aim to increase their effectiveness by including ACns into specific delivery methods; thus, the compound's load is projected by the external environment, which may have an impact on their stability. As a matter of fact, ACn encapsulation and formulation are acknowledged as useful tactics for enhancing ACn release control and getting around the restrictions on ACn bioavailability (Martínez-Ballesta et al., 2018). Black soybean skin isolated ACn-loaded with whey protein was microencapsulated to serve as an important sources of probiotics with improved stability, retention in both gastric and intestinal phases and a higher antioxidative potential than free ACn and C3G. With an expanded colony of *Faecalibacterium* and *Akkermansia* boosting the formation of short-chain fatty acids, particularly acetic and propionic acids, the ACn-loaded microcapsule modified the intestinal gut microbiota (Li et al., 2024). Co-encapsulation of ACn of blackberry juice and *Lactobacillus acidophilus* by gum arabic–maltodextrin are super protective for ACn encapsulation for being effective probiotic bacteria for improving gut microbiome responsible for human health conditions (Neuenfeldt et al., 2022).

17.7 ANTHOCYANINS AS DRUG DELIVERY CARRIERS

In the case of acai berry, ACn loaded in water/oil nano-emulsion-based drug delivery enhanced the stability and bioavailability of ACns under both *in vitro* and *in vivo* experimental conditions, as had exhibited highest retention, or highest stability for 30 days. A balance of hydrophilic and hydrophobic/lipophilic ACns' nanoemulsion maintains the highest release rate in the buffer to roughly 40%. Likewise, the microencapsulated ACns' rich black extract being compounded with gelatin arabic gum and chitosan carboxymethyl cellulose exerted a highest shelf life along with highest temperature resistance (Cheng et al., 2023). A nanoemulsion of ACn thyme oil upon mixed with chitosan arabic gum film matrix exerted multiple functional biological properties like enhancement of extension associated with increased ultraviolet (UV) light barrier protection. Also, it had elevated the shelf life of milk via reduced proliferation and growth of bacteria, affecting rancidity of milk (Cheng et al., 2023). Similarly, it was found in another study wherein, liposome of ACns derived from blueberries encoated with pectin depicted release of this ACns in a very controlled manner in gastric fluid, whereas a faster rate of release was observed within intestinal fluid mediated via enzymatic deformation of these liposomal ACns. Cyclodextrins encapsulated ACns also enhanced the bioavailable conditions accompanying stability with concurrent reduction in side effects. Above all, the overall stability of these ACns is affected by low solubility under both organic and polar solvents. This feature necessitates the conversion of intact ACns to other active metabolites via different modes, i.e., either by chemical, enzyme or chemo-enzymatic methods.

17.7.1 Pharmacokinetics and Pharmacodynamics of Anthocyanin-Based Drug Delivery Systems

It is now well accepted that bioactivity of ACns is related to the stability and the encapsulation material. ACns are basically composed of aglycone unit (polyhydroxy and polymethoxy derivatives of 2-phenylbenzopyrylium or flavylium salts) along with glycone and a sugar moiety. It was observed upon administration of ACns in rats was rapidly absorbed via bilitranslocase, which was evident in plasma in a very short duration. In comparison to stomach, there was rapid absorption of ACns in small intestine and it was found in degraded form in urine and bile in the form of glycosides or methylated glucuronidated derivatives. Most of the ACns are availed in the form of metabolites in blood, since it is quite unstable at different physiological conditions. These pigmented metabolites exerted diverse pathophysiological roles like antioxidant and anti-inflammatory. A detailed pharmacokinetic response with respect to antioxidant activity was interpreted as per the profiling of tart cherry ACns in time and dose-dependent manner found in plasma and urine samples. Furthermore, the pharmacodynamics of the same study depicted that an alleviated response of antioxidant could affect consequential synthesis and thereby interfering with re-absorption of metabolites within the systemic circulation (Seymour et al., 2014). Apart from the aforementioned features, encapsulated ACns or derivatives of ACns exhibited neuroprotective effects in cerebral ischaemic rat models, wherein, orally administered ACns had depicted decrease in cerebral infarction. Also, these derivatives significantly reduced the infarction when injected with low doses of stable isotope-labelled C3G in a typical cerebral infarction model (Salehi et al., 2020). ^{13}C-labelled ACn in the urine and breath showed the relative bioavailability of C3G of 12.4% $\pm$ 1.4% ensuring the extent of absorption and metabolism. Metabolites of ACn benefits cardiovascular system upon ACn consumption (Cassidy et al., 2011; 2013; Erdman et al., 2007; Mink et al., 2007) are likely due to the circulation for the longer period, at higher concentrations than the parent ACns, given the degradation, short t½, and low Cmax of the parent ACns.

17.8 APPLICATIONS OF MODIFIED ANTHOCYANINS IN DISEASE TREATMENT

17.8.1 Exploring the Therapeutic Potential of Anthocyanins against Viral Infections

Several natural substances have already been shown to have anti-viral properties in the prevention and they are being studied for therapeutic applications. These natural chemicals work as anti-influenza agents (Mohammadi Pour et al., 2019). Additionally, ACns demonstrated bacteriostatic properties (by upregulating inflammatory cytokines and mediators such as IL-1, IL-6, IL-10, TNF-a, NF-kB, inducible iNOS and COX-2) (Sodagari et al., 2015). The pharmacological properties of ACns encompass their noteworthy anti-viral activity, underscoring their significance in promoting health and hygiene. Additionally, there have been documented reports highlighting the anti-viral effects of plants and fruits enriched with ACns. ACns have been shown to inhibit the replication of a list of viruses *in vitro* as depicted in Figure 17.6. ACns harbour a rich source of antioxidants and thus can help ward off

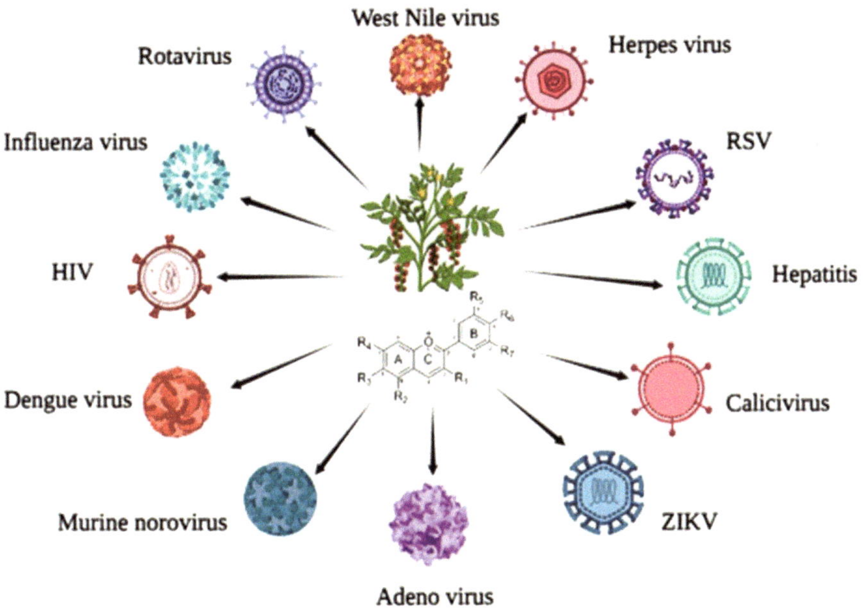

FIGURE 17.6 Anthocyanin's anti-viral efficacy against various viral diseases.

diabetes, obesity and now esteemed as a potent cure for viral diseases. The capacity of ACns to exhibit anti-viral effects by attaching to host cells, disrupting the viral life cycle or enhancing host immunity reinforces the notion that ACns could serve as a crucial element and a promising therapeutic agent in the search for innovative anti-viral lead compounds (Table 17.1).

17.9 CHALLENGES AND FUTURE PERSPECTIVES

The classical methods of microencapsulation yield positive results, but generally, they fall short in delivering bioaccessible and bioavailable ACns to their absorption sites. Additionally, these methods often require substantial amounts of reagents, leading to increased waste generation. While emerging microencapsulation methods show promise, they are not yet economically viable for the food industry due to low yields and high costs. The potential hazard of encapsulation particles crashing into the shell exists. The high temperatures during processing can lead to the denaturation of wall materials like proteins or reactions with carbonyl compounds, resulting in the formation of harmful products such as Maillard reactions, caramelization products and acrylamide (Lin et al., 2023). Consequently, it is crucial not only to draw attention to harmful residues introduced by modification but also to conduct further studies on the stabilization and processing of ACns. Throughout the progression of ACn conservation science, efforts have been made to reduce process costs and enhance environmental protection simultaneously. The genuine challenge in ACn encapsulation lies in integrating these three aspects into a unified process. A promising alternative lies in the combination of methods, as it has the potential to address delivery and environmental concerns as well as overcome technical issues like solubility, particle size and flow rate. However, despite these advantages, further research is needed to optimize the balance between achieving optimal protection, environmental sustainability, high yield and cost-effectiveness (Tarone et al., 2020). A nanotechnology-based drug delivery system offering ACn nanoparticle encapsulation enhanced pharmacokinetics, pharmacodynamics and controlled-release profiles in addition to blood–brain barrier permeability targets intracellular signalling mitochondrial apoptosis, cytotoxicity and ROS scavenging capabilities of ACns for human diseases. The bioavailability, stability and metabolite products upon modification become an effective measure of drug discovery in the world of scientific

TABLE 17.1
Exploring the Therapeutic Applications of ACns and Their Nanodelivery Systems for Various Diseases

ACns and Their Various Nanoforms' Role in Disease	Anthocyanins-Sources	Reference
Anti-Viral Activity		
InfV A and B; stop the internalization or attachment step and deal with the ROS-mediated cell damage brought on by the InfV infection	*Morus alba*	Kim and Chung (2018)
HSV-1; reduce the expression of viral proteins, inhibit HSV-1 replication and lower NOX4 expression	*Solanum melongena*	Di Sotto et al. (2018)
Adenovirus 36; inhibits viral replication	*Morus alba*	Na et al. (2014)
HAV, FCV-F9 and MNV-1; lessen the impact on viral multiplication and prevent viral adsorption	*Vitis vinifera*	Joshi et al. (2015)
Neurodegenerative Diseases		
Polyethylene glycol-gold nanoparticles (PEG-AuNPs) reduced neuroinflammatory	-gold NP	Kim et al. (2017), Ali et al. (2017)
ACn-loaded PEG-gold nanoparticles reduced neuroinflammatory and neuroapoptotic markers via the p-JNK/NF-κB/p-GSK3β signalling pathways	gold NP	Kim et al. (2017)
PEG-PLGA NPs scavenged ROS via p38/JNK pathway against Aβ1–42-induced oxidative stress	-NP	Amin et al. (2017)
Niosomes decreased cell viability of neuroblastoma neuro 2A cells	Black carrots, dendrimers	Fidan-Yardimci et al. (2019), Yesil-Celiktas et al. (2017)
Niosomes non-toxic to normal HT22 hippocampal neuronal cells and internalized in BV2 microglial cells	Niosomes	Kim et al. (2017)
Nanophytosomes formulation	*Sambucus nigra*	Mendes et al. (2021)
Cancer		
Exosomes isolated from mature bovine milk	Bilberries	Mudd, et al. (2020), Munagala et al. (2017)
Exosomes loaded are non-toxic to human aortic endothelial cells, reversing the NO levels and reducing angiogenesis	Delphinidin	Barkallah et al. (2021)
Exosomes isolated from mature bovine milk	Bilberries	Aqil et al. (2017)
Biopolymers chitosan and chondroitin sulphate nanoparticles	Black rice	Liang et al. (2019)
Polysaccharide-based chondroitin sulphate	Black soybean	Jeong et al. (2016)
Exosomes – ACn arrest the growth of A2780, A2780/CP70, OVCA432 and OVCA433 ovarian cancer cells by decreasing P-glycoprotein expression and inhibiting cisplatin-resistant ovarian cancer cells. ExoPAC (paclitaxel)- and ExoAnthos-enhanced oral bioavailability for the management of ovarian cancer in xenograft	Berries	Aqil et al. (2017)
Exosomes are anti-proliferative in colon cancer cell lines HCT-116 and HT-29 human lung cells A549 and H1299, breast cells MDA-MB-231 and MCF7, pancreatic cells PANC1 and Mia PaCa2, prostate cells PC3 and DU145, colon cells HCT-116 and ovarian cells OVCA432	Blueberries	Munagala et al. (2017)
Nanopolysaccharide chitosan and chondroitin sulphate-HCT-116 by decreasing mitochondrial response and apoptosis	Black rice	Liang et al. (2019)
Hyaluronic acid, a polysaccharide-enhanced CD44+ apoptosis in colon cancer cells HT29	Cornflower	Xiong et al. (2021)
Hyperlipidaemia		
Encapsulated anthocyanin loaded-chitosan nanoparticles	Black carrot	Sreerekha et al. (2021)

(Continued)

TABLE 17.1 ***(Continued)***
Exploring the Therapeutic Applications of ACns and Their Nanodelivery Systems for Various Diseases

ACns and Their Various Nanoforms' Role in Disease	Anthocyanins-Sources	Reference
Vision		
Improving visual function by improving the blood flow of retin and decreasing eye fatigue	Anthocyanins-Natural source	T.Tsuda (2012)
Nutraceutical Supplement		
Nanoencapsulation using low-molecular-weight chitosan	Black carrot	Chatterjee et al. (2021)
Cyanidin-3-O-glucoside (C3G)-loaded nanoparticles	Crabapple	Sun et al. (2020)
Pectin-based nanoparticles	Blackberry	Rosales et al. (2023)
Obesity		
Non-ionic niosomes	*Vaccinium meridionale*	Colorado et al. (2020)
Cardiac		
Starch-based hydrogel NPs	Anthocyanins- NP	Hanafy (2021)
Chitosan nanoparticles	Black carrot	Sreerekha et al. (2021)
Silver Ag@ Gold AuNPs	Blackberry	Hussein et al. (2020)
Pain		
Poly (methyl methacrylate) nanoparticles	Anthocyanins-NP	Widyadharma et al. (2021)
Diabetic		
Carboxymethyl chitosan and alginate nanocapsules	Purple sweet potato extract	I'tishom et al. (2021)
Poly (D, L-lactide-co-glycolide) (PLGA) nanoparticles	Pelargonidin	Roy et al. (2017)
Liposomal forms	Delphinidin and cyanidin hydrochloride	Gharib et al. (2013)
Nanoencapsulation poly-lactide-co-glycolic-acid	Pelargonidin	Samadder et al. (2017)
Bone		
Bone metabolism	Bilberry extract	Shimizu et al. (2018)
Anti-inflammatory and Antioxidant		
Nanoparticles based on chitosan derivatives	Cherry extracts	Beconcini et al. (2020)
Pulmonary Disease		
Nanocarrier-polyethylene glycol-poly(lactide-*co*-glycolide), maltodextrin and carboxymethyl chitosan (CMC) human normal lung epithelial BEAS-2B cells-antioxidant	Cyanidin-3-*O*-glucoside (C3G)-rich, haskap (*Lonicera caerulea* L.) berries	Amararathna et al. (2022)

valuation for their safety. Additionally, while there is promising research, further studies are needed to establish specific recommendations and mechanisms of action for different health conditions. Targeted delivery systems aim to deliver ACns specifically to the intended site of action. This can minimize off-target effects, reduce side effects and enhance the therapeutic efficacy of ACns. ACn drug delivery systems may help mitigate potential side effects by minimizing exposure to non-target tissues or organs. Targeted delivery and controlled release contribute to reducing the adverse effects associated with ACn administration.

17.10 CONCLUSION

In summary, ACn drug delivery systems are designed to address challenges associated with pH change, temperature, derivatization, co-pigmentation and co-administration and maximize their therapeutic potential with an increase in circulation, stability, solubility and thermostability in a targeted and controlled manner. These systems play a crucial role in advancing the application of ACns in various fields, including pharmaceuticals and nutraceuticals, for the benefit of human health and diseases. Further research should focus on identifying more potent and new plant-based lead compounds, emphasizing those with synergistic effects to improve treatment outcomes. This includes bolstering immunity, lowering costs and toxicity, minimizing resistance, pinpointing their specific targets and understanding their pharmacological modes of action. Above all, more *ex vivo* and *in vivo* studies should be undertaken to ascertain the therapeutic role of ACn drug delivery systems from various plant sources as a strong therapeutic agent.

REFERENCES

Aguilar, Zoraida P. 2013. "Chapter 2 – Types of Nanomaterials and Corresponding Methods of Synthesis." In Nanomaterials for Medical Applications, edited by Zoraida P. Aguilar, 33–82. Elsevier. https://doi.org/10.1016/B978-0-12-385089-8.00002-9

Ali, Tahir, Min Ju Kim, Shafiq Ur Rehman, Ashfaq Ahmad, and Myeong Ok Kim. 2017. "Anthocyanin-Loaded PEG-Gold Nanoparticles Enhanced the Neuroprotection of Anthocyanins in an Aβ1-42 Mouse Model of Alzheimer's Disease." Molecular Neurobiology 54 (8): 6490–506. https://doi.org/10.1007/s12035-016-0136-4

Amararathna, Madumani, David W. Hoskin, and H. P. Vasantha Rupasinghe. 2022. "Anthocyanin Encapsulated Nanoparticles as a Pulmonary Delivery System." Oxidative Medicine and Cellular Longevity 2022 (September): 1422929. https://doi.org/10.1155/2022/1422929

Amin, Faiz Ul, Shahid Ali Shah, Haroon Badshah, Mehtab Khan, and Myeong Ok Kim. 2017. "Anthocyanins Encapsulated by PLGA@PEG Nanoparticles Potentially Improved Its Free Radical Scavenging Capabilities via P38/JNK Pathway against Aβ1–42-Induced Oxidative Stress." Journal of Nanobiotechnology 15 (1): 12. https://doi.org/10.1186/s12951-016-0227-4

Andersen, Oyvind M., and Kenneth R. Markham, eds. 2005. Flavonoids: Chemistry, Biochemistry and Applications. CRC Press.

Aqil, Farrukh, Jeyaprakash Jeyabalan, Ashish K. Agrawal, Al-Hassan Kyakulaga, Radha Munagala, Lynn Parker, and Ramesh C. Gupta. 2017. "Exosomal Delivery of Berry Anthocyanidins for the Management of Ovarian Cancer." Food & Function 8 (11): 4100–4107.

Ayvaz, H., Cabaroglu, T., Akyildiz, A., Pala, C. U., Temizkan, R., Ağçam, E., Ayvaz, Z., Durazzo, A., Lucarini, M., Direito, R., & Diaconeasa, Z. (2022). Anthocyanins: Metabolic Digestion, Bioavailability, Therapeutic Effects, Current Pharmaceutical/Industrial Use, and Innovation Potential. Antioxidants (Basel, Switzerland), 12(1), 48. https://doi.org/10.3390/antiox12010048

Barkallah, Merwan, Judith Nzoughet-Kouassi, Gilles Simard, Loric Thoulouze, Sébastien Marze, Marie-Hélène Ropers, and Ramaroson Andriantsitohaina. 2021. "Enhancement of the Anti-Angiogenic Effects of Delphinidin When Encapsulated Within Small Extracellular Vesicles." Nutrients 13 (12). Multidisciplinary Digital Publishing Institute: 4378. https://doi.org/10.3390/nu13124378

Beconcini, Denise, Francesca Felice, Angela Fabiano, Bruno Sarmento, Ylenia Zambito, and Rossella Di Stefano. 2020. "Antioxidant and Anti-Inflammatory Properties of Cherry Extract: Nanosystems-Based Strategies to Improve Endothelial Function and Intestinal Absorption." Foods 9 (2): 207. https://doi.org/10.3390/foods9020207

Bhatt, U., Shah, G. and Soni, V., 2022. Therapeutic, protective and industrial significances of anthocyanins: A review. Avicenna J. Med. Biochem, 10, pp.82-93. doi: 10.34172/ajmb.2338

Borkowski, T., H. Szymusiak, A. Gliszczyńska-Swigło, and B. Tyrakowska. 2005. "The Effect of 3-O-β-Glucosylation on Structural Transformations of Anthocyanidins." Food Research International, Third International Congress on Pigments in Food 38 (8): 1031–37. https://doi.org/10.1016/j.foodres.2005.02.020

Brouillard, Raymond. 1982. "Chemical Structure of Anthocyanins." In Anthocyanins as Food Colors. Elsevier, 1–40. https://doi.org/10.1016/b978-0-12-472550-8.50005-6

Cai, Dongbao, Xusheng Li, Jiali Chen, Xinwei Jiang, Xiaoqiang Ma, Jianxia Sun, and Lingmin Tian, et al. 2022. "A Comprehensive Review on Innovative and Advanced Stabilization Approaches of Anthocyanin by Modifying Structure and Controlling Environmental Factors." Food Chemistry 366 (January): 130611. https://doi.org/10.1016/j.foodchem.2021.130611

Câmara, José S., Monica Locatelli, Jorge A. M. Pereira, Hélder Oliveira, Marco Arlorio, Iva Fernandes, Rosa Perestrelo, Victor Freitas, and Matteo Bordiga. 2022. "Behind the Scenes of Anthocyanins—From the Health Benefits to Potential Applications in Food, Pharmaceutical and Cosmetic Fields." Nutrients 14 (23): 5133. https://doi.org/10.3390/nu14235133

Cassidy, Aedín, Kenneth J. Mukamal, Lydia Liu, Mary Franz, A. Heather Eliassen, and Eric B. Rimm. 2013. "High Anthocyanin Intake Is Associated with a Reduced Risk of Myocardial Infarction in Young and Middle-Aged Women." Circulation 127 (2): 188–96. https://doi.org/10.1161/CIRCULATIONAHA.112.122408

Cassidy, Aedín, Éilis J. O'Reilly, Colin Kay, Laura Sampson, Mary Franz, J. P. Forman, Gary Curhan, and Eric B. Rimm. 2011. "Habitual Intake of Flavonoid Subclasses and Incident Hypertension in Adults." The American Journal of Clinical Nutrition 93 (2): 338–47. https://doi.org/10.3945/ajcn.110.006783

Chatterjee, Niladri Sekhar, Pavan Kumar Dara, Sreerekha Perumcherry Raman, Divya K. Vijayan, Jayashree Sadasivam, Suseela Mathew, Chandragiri Nagarajarao Ravishankar, and Rangasamy Anandan. 2021. "Nanoencapsulation in Low-Molecular-Weight Chitosan Improves In Vivo Antioxidant Potential of Black Carrot Anthocyanin." Journal of the Science of Food and Agriculture 101 (12): 5264–71. https://doi.org/10.1002/jsfa.11175

Chebil, Latifa, Catherine Humeau, Aude Falcimaigne, Jean-Marc Engasser, and Mohamed Ghoul. 2006. "Enzymatic Acylation of Flavonoids." Process Biochemistry 41 (11): 2237–51. https://doi.org/10.1016/j.procbio.2006.05.027

Cheng, Yingying, Jiayi Liu, Ling Li, Jiali Ren, Jun Lu, and Feijun Luo. 2023. "Advances in Embedding Techniques of Anthocyanins: Improving Stability, Bioactivity and Bioavailability." Food Chemistry: X 20 (December): 100983. https://doi.org/10.1016/j.fochx.2023.100983

Colorado, Diana, Maritza Fernandez, Jahir Orozco, Yasmin Lopera, Diana Lorena Muñoz, Sergio Acín, and Norman Balcazar. 2020. "Metabolic Activity of Anthocyanin Extracts Loaded into Non-Ionic Niosomes in Diet-Induced Obese Mice." Pharmaceutical Research 37 (8): 152. https://doi.org/10.1007/s11095-020-02883-z

Demirbas, Ayse, Kasım Büyükbezirci, Cagla Celik, Emine Kislakci, Zehra Karaagac, Ersen Gokturk, Ahmet Kati, Behzat Cimen, Vedat Yilmaz, and Ismail Ocsoy. 2019. "Synthesis of Long-Term Stable Gold Nanoparticles Benefiting from Red Raspberry (*Rubus idaeus*), Strawberry (*Fragaria ananassa*), and Blackberry (*Rubus fruticosus*) Extracts–Gold Ion Complexation and Investigation of Reaction Conditions." ACS Omega 4 (20): 18637–44. https://doi.org/10.1021/acsomega.9b02469

Deng, Qian, and Yanyun Zhao. 2011. "Physicochemical, Nutritional, and Antimicrobial Properties of Wine Grape (cv. Merlot) Pomace Extract-Based Films." Journal of Food Science 76 (3). https://doi.org/10.1111/j.1750-3841.2011.02090.x

Di Sotto, Antonella, Silvia Di Giacomo, Donatella Amatore, Marcello Locatelli, Annabella Vitalone, Chiara Toniolo, Giuseppe Leonardo Rotino, et al. 2018. "A Polyphenol Rich Extract from *Solanum melongena* L. DR2 Peel Exhibits Antioxidant Properties and Anti-Herpes Simplex Virus Type 1 Activity In Vitro." Molecules (Basel, Switzerland) 23 (8): 2066. https://doi.org/10.3390/molecules23082066

Enaru, Bianca, Georgiana Drețcanu, Teodora Daria Pop, Andreea Stănilă, and Zorița Diaconeasa. 2021. "Anthocyanins: Factors Affecting Their Stability and Degradation." Antioxidants 10 (12): 1967. https://doi.org/10.3390/antiox10121967

Erdman, John W., Douglas Balentine, Lenore Arab, Gary Beecher, Johanna T. Dwyer, John Folts, James Harnly, et al. 2007. "Flavonoids and Heart Health: Proceedings of the ILSI North America Flavonoids Workshop, May 31–June 1, 2005, Washington, DC." The Journal of Nutrition 137 (3 Suppl 1): 718S–737S. https://doi.org/10.1093/jn/137.3.718S

Fathallah, S., Abdellatif, A., & Saadeldin, M. K. (2023). Unleashing nature's potential and limitations: Exploring molecular targeted pathways and safe alternatives for the treatment of multiple sclerosis (Review). Medicine international, 3(5), 42. https://doi.org/10.3892/mi.2023.102

Fei, Peng, Fansen Zeng, Shiyu Zheng, Qiaoling Chen, Yonghua Hu, and Jie Cai. 2021. "Acylation of Blueberry Anthocyanins with Maleic Acid: Improvement of the Stability and Its Application Potential in Intelligent Color Indicator Packing Materials." Dyes and Pigments 184 (January): 108852. https://doi.org/10.1016/j.dyepig.2020.108852

Feng, Jin, Yinghui Wu, Lixia Zhang, Ying Li, Songbai Liu, Hua Wang, and Chunyang Li. 2019. "Enhanced Chemical Stability, Intestinal Absorption, and Intracellular Antioxidant Activity of Cyanidin-3-O-Glucoside by Composite Nanogel Encapsulation." Journal of Agricultural and Food Chemistry 67 (37): 10432–47. https://doi.org/10.1021/acs.jafc.9b04778

Fidan-Yardimci, Melike, Seref Akay, Fatemeh Sharifi, Canan Sevimli-Gur, Gaye Ongen, and Ozlem Yesil-Celiktas. 2019. "A Novel Niosome Formulation for Encapsulation of Anthocyanins and Modelling Intestinal Transport." Food Chemistry 293 (September): 57–65. https://doi.org/10.1016/j.foodchem.2019.04.086

Forman, Henry Jay, Maret Traber, and Fulvio Ursini. 2014. "Antioxidants: GRABbing New Headlines." Free Radical Biology & Medicine. https://doi.org/10.1016/j.freeradbiomed.2013.10.808

Gharib, Amir, Zohreh Faezizadeh, and Masoud Godarzee. 2013. "Treatment of Diabetes in the Mouse Model by Delphinidin and Cyanidin Hydrochloride in Free and Liposomal Forms." Planta Medica 79 (17): 1599–604. https://doi.org/10.1055/s-0033-1350908

Gonçalves, Ana C., Márcio Rodrigues, Adriana O. Santos, Gilberto Alves, and Luís R. Silva. 2018. "Antioxidant Status, Antidiabetic Properties and Effects on Caco-2 Cells of Colored and Non-Colored Enriched Extracts of Sweet Cherry Fruits." Nutrients 10 (11). Multidisciplinary Digital Publishing Institute: 1688. https://doi.org/10.3390/nu10111688

Hamedi, Sepideh, and Mojtaba Koosha. 2020. "Designing a pH-Responsive Drug Delivery System for the Release of Black-Carrot Anthocyanins Loaded in Halloysite Nanotubes for Cancer Treatment." Applied Clay Science 197 (November): 105770. https://doi.org/10.1016/j.clay.2020.105770

Hanafy, Nemany A N 2021. "Starch Based Hydrogel NPs Loaded by Anthocyanins Might Treat Glycogen Storage at Cardiomyopathy in Animal Fibrotic Model." International Journal of Biological Macromolecules 183 (July): 171–81. https://doi.org/10.1016/j.ijbiomac.2021.04.131

Homayoonfal, Mina, Seyed Mohammad Mousavi, Hossein Kiani, Gholamreza Askari, Stephane Desobry, and Elmira Arab-Tehrany. 2021. "Encapsulation of *Berberis vulgaris* Anthocyanins into Nanoliposome Composed of Rapeseed Lecithin: A Comprehensive Study on Physicochemical Characteristics and Biocompatibility." Foods 10 (3): 492. https://doi.org/10.3390/foods10030492

Hussein, Jihan, Mehrez E. El-Naggar, Moustafa M. G. Fouda, Osama M. Morsy, Jamaan S. Ajarem, Ahmed M. Almalki, Ahmed A. Allam, and Enas Mahmoud Mekawi. 2020. "The Efficiency of Blackberry Loaded AgNPs, AuNPs and Ag@AuNPs Mediated Pectin in the Treatment of Cisplatin-Induced Cardiotoxicity in Experimental Rats." International Journal of Biological Macromolecules 159 (September): 1084–93. https://doi.org/10.1016/j.ijbiomac.2020.05.115

Iosub, I., F. Kajzar, M. Makowska-Janusik, A. Meghea, I. Geana, and I. Rau. 2014. "Electronic Structure, Optical and Electrochemical Properties of Malvidin Molecule Extracted from Grapes." Display and Imaging 1: 175–93.

I'tishom, Reny, Ifan Ali Wafa, David Setyo Budi, and Nando Reza Pratama. 2021. "Oral Delivery of Purple Sweet Potato (*Ipomoea batatas* L.) Extract-Loaded Carboxymethyl Chitosan and Alginate Nanocapsule in Streptozotocin-Induced Diabetic Mice." Indian Journal of Pharmaceutical Education and Research 55 (3): 709–14. https://doi.org/10.5530/ijper.55.3.143

Jeong, Dooyong, Byoung-Chan Bae, Sin-Jung Park, and Kun Na. 2016. "Reactive Oxygen Species Responsive Drug Releasing Nanoparticle Based on Chondroitin Sulfate-Anthocyanin Nanocomplex for Efficient Tumor Therapy." Journal of Controlled Release: Official Journal of the Controlled Release Society 222 (January): 78–85. https://doi.org/10.1016/j.jconrel.2015.12.009

Jiang, Xinwei, Jianxia Sun, and Weibin Bai. 2019. "Anthocyanins in Food." In Handbook of Dietary Phytochemicals, edited by Jianbo Xiao, Satyajit D. Sarker, and Yoshinori Asakawa, 1–52. Springer. https://doi.org/10.1007/978-981-13-1745-3_13-1

Joint FAO/WHO Expert Committee on Food Additives (JECFA),World Health Organization, WHO Expert Committee on Food Additives. 2019. Evaluation of certain food additives: (eighty-seventh report of the Joint FAO/ WHO Expert Committee on Food Additives) WHO Technical Reprt series, No, 1020. https://www.who.int/publications/i/item/9789241210294

Joshi, Snehal S., Xiaowei Su, and Doris H. D'Souza. 2015. "Antiviral Effects of Grape Seed Extract against Feline Calicivirus, Murine Norovirus, and Hepatitis A Virus in Model Food Systems and under Gastric Conditions." Food Microbiology 52 (December): 1–10. https://doi.org/10.1016/j.fm.2015.05.011

Kang, Hye-Ji, Min-Jung Ko, and Myong-Soo Chung. 2021. "Anthocyanin Structure and pH Dependent Extraction Characteristics from Blueberries (*Vaccinium corymbosum*) and Chokeberries (*Aronia melanocarpa*) in Subcritical Water State." Foods 10 (3): 527. https://doi.org/10.3390/foods10030527

Kim, Hyojin, and Mi Sook Chung. 2018. "Antiviral Activities of Mulberry (*Morus alba*) Juice and Seed against Influenza Viruses." Evidence-Based Complementary and Alternative Medicine 2018 (November): e2606583. https://doi.org/10.1155/2018/2606583

Kim, Min Ju, Shafiq Ur Rehman, Faiz Ul Amin, and Myeong Ok Kim. 2017. "Enhanced Neuroprotection of Anthocyanin-Loaded PEG-Gold Nanoparticles Against Aβ1-42-Induced Neuroinflammation and Neurodegeneration via the NF-KB/JNK/GSK3β Signaling Pathway." Nanomedicine: Nanotechnology, Biology, and Medicine 13 (8): 2533–44. https://doi.org/10.1016/j.nano.2017.06.022

Kim, Seok Joong, Young-Hee Cho, Whansoo Park, Daeseok Han, Chang-Hoon Chai, and Jee-Young Imm. 2003. "Solubilization of Water Soluble Anthocyanins in Apolar Medium Using Reverse Micelle." Journal of Agricultural and Food Chemistry 51 (26): 7805–9. https://doi.org/10.1021/jf034843w

Kong, Jin-Ming, Lian-Sai Chia, Ngoh-Khang Goh, Tet-Fatt Chia, and R. Brouillard. 2003. "Analysis and Biological Activities of Anthocyanins." Phytochemistry 64 (5): 923–33. https://doi.org/10.1016/s0031-9422(03)00438-2

Kren, Vladimir. 2001. "Chemical Biology and Biomedicine of Glycosylated Natural Compounds." In Glycoscience: Chemistry and Chemical Biology I–III, edited by Bertram O. Fraser-Reid, Kuniaki Tatsuta, and Joachim Thiem, 2471–529. Springer. https://doi.org/10.1007/978-3-642-56874-9_60

Laleh, G. H., H. Frydoonfar, R. Heidary, R. Jameei, and S. Zare. 2006. "The Effect of Light, Temperature, pH and Species on Stability of Anthocyanin Pigments in Four Berberis Species." Pakistan Journal of Nutrition 5 (1): 90–92. https://doi.org/10.3923/pjn.2006.90.92

Li, Daotong, Pengpu Wang, Yinghua Luo, Mengyao Zhao, and Fang Chen. 2017. "Health Benefits of Anthocyanins and Molecular Mechanisms: Update from Recent Decade." Critical Reviews in Food Science and Nutrition 57 (8): 1729–41. https://doi.org/10.1080/10408398.2015.1030064

Li, Xusheng, Yuxin Wang, Yan Jiang, Chuqi Liu, Wenbao Zhang, Weiwen Chen, Lingmin Tian, Jianxia Sun, Caiyong Lai, and Weibin Bai. 2024. "Microencapsulation with Fructooligosaccharides and Whey Protein Enhances the Antioxidant Activity of Anthocyanins and Their Ability to Modulate Gut Microbiota In Vitro." Food Research International 181 (April): 114082. https://doi.org/10.1016/j.foodres.2024.114082

Liang, Tisong, Zhentao Zhang, and Pu Jing. 2019. "Black Rice Anthocyanins Embedded in Self-Assembled Chitosan/Chondroitin Sulfate Nanoparticles Enhance Apoptosis in HCT-116 Cells." Food Chemistry 301 (December): 125280. https://doi.org/10.1016/j.foodchem.2019.125280

Lin, Yang, Cong Li, Lejuan Shi, and Lixia Wang. 2023. "Anthocyanins: Modified New Technologies and Challenges." Foods 12 (7): 1368. https://doi.org/10.3390/foods12071368

Luo, Chun-Li, Qing Zhou, Zi-Wei Yang, Rui-Dan Wang, and Jiu-Liang Zhang. 2018. "Evaluation of Structure and Bioprotective Activity of Key High Molecular Weight Acylated Anthocyanin Compounds Isolated from the Purple Sweet Potato (*Ipomoea batatas* L. Cultivar Eshu No. 8)." Food Chemistry 241 (February): 23–31. https://doi.org/10.1016/j.foodchem.2017.08.073

Mansour, Mohammed, Mahmoud Salah, and Xiaoyun Xu. 2020. "Effect of Microencapsulation Using Soy Protein Isolate and Gum Arabic as Wall Material on Red Raspberry Anthocyanin Stability, Characterization, and Simulated Gastrointestinal Conditions." Ultrasonics Sonochemistry 63 (May): 104927. https://doi.org/10.1016/j.ultsonch.2019.104927

Martínez-Ballesta, MCarment, Ángel Gil-Izquierdo, Cristina García-Viguera, and Raúl Domínguez-Perles. 2018. "Nanoparticles and Controlled Delivery for Bioactive Compounds: Outlining Challenges for New 'Smart-Foods' for Health." Foods 7 (5): 72. https://doi.org/10.3390/foods7050072

Mauray, Aurelie, Catherine Felgines, Christine Morand, Andrzej Mazur, Augustin Scalbert, and Dragan Milenkovic. 2010. "Nutrigenomic Analysis of the Protective Effects of Bilberry Anthocyanin-Rich Extract in Apo E-Deficient Mice." Genes & Nutrition 5 (4): 343–53. https://doi.org/10.1007/s12263-010-0171-0

Mendes, Daniela, Patrícia Valentão, Maria Manuel Oliveira, Paula Andrade, and Romeu A. Videira. 2021. "A Nanophytosomes Formulation Based on Elderberry Anthocyanins and Codium Lipids to Mitigate Mitochondrial Dysfunctions." Biomedicine & Pharmacotherapy = Biomedecine & Pharmacotherapie 143 (November): 112157. https://doi.org/10.1016/j.biopha.2021.112157

Mink, Pamela J., Carolyn G. Scrafford, Leila M. Barraj, Lisa Harnack, Ching-Ping Hong, Jennifer A. Nettleton, and David R. Jacobs. 2007. "Flavonoid Intake and Cardiovascular Disease Mortality: A Prospective Study in Postmenopausal Women." The American Journal of Clinical Nutrition 85 (3): 895–909. https://doi.org/10.1093/ajcn/85.3.895

Mohammadalinejhad, Samira, and Marcin Andrzej Kurek. 2021. "Microencapsulation of Anthocyanins—Critical Review of Techniques and Wall Materials." Applied Sciences 11 (9): 3936. https://doi.org/10.3390/app11093936

Mohammadi Pour, Pardis, Sajad Fakhri, Sedigheh Asgary, Mohammad Hosein Farzaei, and Javier Echeverría. 2019. "The Signaling Pathways, and Therapeutic Targets of Antiviral Agents: Focusing on the Antiviral Approaches and Clinical Perspectives of Anthocyanins in the Management of Viral Diseases." Frontiers in Pharmacology 10 (November): 1207. https://doi.org/10.3389/fphar.2019.01207

Mudd, Ashley M., Tao Gu, Radha Munagala, Jeyaprakash Jeyabalan, Nejat K. Egilmez, and Ramesh C. Gupta. 2020. "Chemoprevention of Colorectal Cancer by Anthocyanidins and Mitigation of Metabolic Shifts Induced by Dysbiosis of the Gut Microbiome." Cancer Prevention Research (Philadelphia, Pa.) 13 (1): 41–52. https://doi.org/10.1158/1940-6207.CAPR-19-0362

Munagala, Radha, Farrukh Aqil, Jeyaprakash Jeyabalan, Ashish K. Agrawal, Ashley M. Mudd, Al Hassan Kyakulaga, Inder P. Singh, Manicka V. Vadhanam, and Ramesh C. Gupta. 2017. "Exosomal Formulation of Anthocyanidins against Multiple Cancer Types." Cancer Letters 393 (May): 94–102. https://doi.org/10.1016/j.canlet.2017.02.004

Na, Ha-Na, Sooho Park, Hyun-Jung Jeon, Hyun-Bok Kim, and Jae-Hwan Nam. 2014. "Reduction of Adenovirus 36-Induced Obesity and Inflammation by Mulberry Extract." Microbiology and Immunology 58 (5): 303–6. https://doi.org/10.1111/1348-0421.12146

Nascimento, Luis Gustavo Lima, Federico Casanova, Naaman Francisco Nogueira Silva, Álvaro Viana Novaes de Carvalho Teixeira, Paulo Peres de Sá Peixoto Júnior, Márcia Cristina Teixeira Ribeiro Vidigal, Paulo Cesar Stringheta, and Antônio Fernandes de Carvalho. 2020. "Use of a Crosslinked Casein Micelle Hydrogel as a Carrier for Jaboticaba (*Myrciaria cauliflora*) Extract." Food Hydrocolloids 106 (September): 105872. https://doi.org/10.1016/j.foodhyd.2020.105872

Nazareth, Maria Sheeba, S. V. Shreelakshmi, Pooja J. Rao, and Nandini P. Shetty. 2021. "Micro and Nanoemulsions of *Carissa spinarum* Fruit Polyphenols, Enhances Anthocyanin Stability and Anti-Quorum Sensing Activity: Comparison of Degradation Kinetics." Food Chemistry 359 (October): 129876. https://doi.org/10.1016/j.foodchem.2021.129876

Neuenfeldt, Naiara Hennig, Carla Andressa Almeida Farias, Renius de Oliveira Mello, Silvino Sasso Robalo, Juliano Smanioto Barin, Leila Picolli da Silva, Edson Irineu Müller, Erico Marlon Moraes Flores, Milene Teixeira Barcia, and Cristiano Ragagnin de Menezes. 2022. "Effects of Blueberry Extract Co-Microencapsulation on the Survival of *Lactobacillus rhamnosus*." LWT 155 (February): 112886. https://doi.org/10.1016/j.lwt.2021.112886

Oliveira, Hélder, Nuno Basílio, Fernando Pina, Iva Fernandes, Victor de Freitas, and Nuno Mateus. 2019. "Purple-Fleshed Sweet Potato Acylated Anthocyanins: Equilibrium Network and Photophysical Properties." Food Chemistry 288 (August): 386–94. https://doi.org/10.1016/j.foodchem.2019.02.132

Ozkan, Gulay, Paola Franco, Iolanda De Marco, Jianbo Xiao, and Esra Capanoglu. 2019. "A Review of Microencapsulation Methods for Food Antioxidants: Principles, Advantages, Drawbacks and Applications." Food Chemistry 272 (January): 494–506. https://doi.org/10.1016/j.foodchem.2018.07.205

Panche, A. N., A. D. Diwan, and S. R. Chandra. 2016. "Flavonoids: An Overview." Journal of Nutritional Science 5 (December): e47. https://doi.org/10.1017/jns.2016.41

Prior, Ronald L., and Xianli Wu. 2006. "Anthocyanins: Structural Characteristics That Result in Unique Metabolic Patterns and Biological Activities." Free Radical Research 40 (10): 1014–28. https://doi.org/10.1080/10715760600758522

Ren, Jinbo, Minjie Liao, Kaixin Li, Fang Chen, Xiaosong Hu, Lingjun Ma, and Junfu Ji. 2024. "The Aggregation of Casein Micelles Induced by Ca^{2+} During *In Vitro* Digestion: Effects on the Release of Loaded Anthocyanins." Food & Function 15 (2): 503–15. https://doi.org/10.1039/D3FO03684G

Rinaldo, Amy R., Erika Cavallini, Yong Jia, Sarah M. A. Moss, Debra A. J. McDavid, Lauren C. Hooper, Simon P. Robinson, et al. 2015. "A Grapevine Anthocyanin Acyltransferase, Transcriptionally Regulated by VvMYBA, Can Produce Most Acylated Anthocyanins Present in Grape Skins." Plant Physiology 169 (3): 1897–916. https://doi.org/10.1104/pp.15.01255

Rosales, Thiécla Katiane Osvaldt, Lucas de Freitas Pedrosa, Karen Rebouças Nascimento, Alexandre Minami Fioroto, Taíse Toniazzo, Carmen Cecília Tadini, Eduardo Purgatto, Neuza Mariko Aymoto Hassimotto, and João Paulo Fabi. 2023. "Nanoencapsulated Anthocyanins: A New Technological Approach to Increase Physical-Chemical Stability and Bioaccessibility." Food Hydrocolloids 139 (May): 108516. https://doi.org/10.1016/j.foodhyd.2023.108516

Roy, Moumita, Rajat Pal, and Abhay Sankar Chakraborti. 2017. "Pelargonidin-PLGA Nanoparticles: Fabrication, Characterization, and Their Effect on Streptozotocin Induced Diabetic Rats." Indian Journal of Experimental Biology 55:819–30.

Salah, Mahmoud, Mohammed Mansour, Daniel Zogona, and Xiaoyun Xu. 2020. "Nanoencapsulation of Anthocyanins-Loaded β-Lactoglobulin Nanoparticles: Characterization, Stability, and Bioavailability In Vitro." Food Research International 137 (November): 109635. https://doi.org/10.1016/j.foodres.2020.109635

Salehi, Bahare, Javad Sharifi-Rad, Francesca Cappellini, Željko Reiner, Debora Zorzan, Muhammad Imran, Bilge Sener, et al. 2020. "The Therapeutic Potential of Anthocyanins: Current Approaches Based on Their Molecular Mechanism of Action." Frontiers in Pharmacology 11. https://www.frontiersin.org/journals/pharmacology/articles/10.3389/fphar.2020.01300

Samadder, Asmita, Debojyoti Tarafdar, Suresh K. Abraham, Kumaresh Ghosh, and Anisur Rahman Khuda-Bukhsh. 2017. "Nano-Pelargonidin Protects Hyperglycemic-Induced L6 Cells against Mitochondrial Dysfunction." Planta Medica 83 (5): 468–75. https://doi.org/10.1055/s-0043-100017

Seymour, E. Mitchell, Sara M. Warber, Ara Kirakosyan, Kathleen R. Noon, Brenda Gillespie, Virginia E. Uhley, Jenna Wunder, Daniel E. Urcuyo, Peter B. Kaufman, and Steven F. Bolling. 2014. "Anthocyanin Pharmacokinetics and Dose-Dependent Plasma Antioxidant Pharmacodynamics Following Whole Tart Cherry Intake in Healthy Humans." Journal of Functional Foods 11 (November): 509–16. https://doi.org/10.1016/j.jff.2014.08.007

Shan, L. I., Yanlong Zhang, Weimin Zeng, L. I. U. Yang, Dandan Zhao, and J. I. A. Xiangqian. 2024. "Study on Preparation and Stability of Anthocyanin Micelles from *Lonicera edulis* Pomace." Science and Technology of Food Industry 45 (2): 21–29. https://doi.org/10.13386/j.issn1002-0306.2023060191

Sharif, Niloufar, Sara Khoshnoudi-Nia, and Seid Mahdi Jafari. 2020. "Nano/Microencapsulation of Anthocyanins; a Systematic Review and Meta-Analysis." Food Research International (Ottawa, Ont.) 132 (June): 109077. https://doi.org/10.1016/j.foodres.2020.109077

Shimizu, S., Matsushita, H., Morii, Y., Ohyama, Y., Morita, N., Tachibana, R., Watanabe, K., & Wakatsuki, A. (2018). Effect of anthocyanin-rich bilberry extract on bone metabolism in ovariectomized rats. Biomedical reports, 8(2), 198–204. https://doi.org/10.3892/br.2017.1029

Sodagari, Hamid Reza, Mohammad Hosein Farzaei, Roodabeh Bahramsoltani, Amir Hossein Abdolghaffari, Maryam Mahmoudi, and Nima Rezaei. 2015. "Dietary Anthocyanins as a Complementary Medicinal Approach for Management of Inflammatory Bowel Disease." Expert Review of Gastroenterology & Hepatology 9 (6): 807–20. https://doi.org/10.1586/17474124.2015.1002086

Sreerekha, P. R., Pavan Kumar Dara, Divya K. Vijayan, Niladri Sekhar Chatterjee, Mahadevan Raghavankutty, Suseela Mathew, Chandragiri Nagarajarao Ravishankar, and Rangasamy Anandan. 2021. "Dietary Supplementation of Encapsulated Anthocyanin Loaded-Chitosan Nanoparticles Attenuates Hyperlipidemic Aberrations in Male Wistar Rats." Carbohydrate Polymer Technologies and Applications 2 (December): 100051. https://doi.org/10.1016/j.carpta.2021.100051

Sun, Jianxia, Jiali Chen, Zhouxiong Mei, Ziying Luo, Lijun Ding, Xinwei Jiang, and Weibin Bai. 2020. "Synthesis, Structural Characterization, and Evaluation of Cyanidin-3-O-Glucoside-Loaded Chitosan Nanoparticles." Food Chemistry 330 (November): 127239. https://doi.org/10.1016/j.foodchem.2020.127239

Tan, Chen, Giovana B. Celli, Michael Joseph Selig, and Alireza Abbaspourrad. 2018. "Catechin Modulates the Copigmentation and Encapsulation of Anthocyanins in Polyelectrolyte Complexes (PECs) for Natural Colorant Stabilization." Food Chemistry 264 (October): 342–49. https://doi.org/10.1016/j.foodchem.2018.05.018

Tan, Chen, Younas Dadmohammadi, Michelle C. Lee, and Alireza Abbaspourrad. 2021. "Combination of Copigmentation and Encapsulation Strategies for the Synergistic Stabilization of Anthocyanins." Comprehensive Reviews in Food Science and Food Safety 20 (4): 3164–91. https://doi.org/10.1111/1541-4337.12772

Tarone, Adriana Gadioli, Cinthia Baú Betim Cazarin, and Mario Roberto Marostica Junior. 2020. "Anthocyanins: New Techniques and Challenges in Microencapsulation." Food Research International 133 (July): 109092. https://doi.org/10.1016/j.foodres.2020.109092

Tatsuzawa, Fumi, Norio Saito, Kazumitsu Miyoshi, Koichi Shinoda, Atsushi Shigihara, and Toshio Honda. 2004. "Diacylated 8-C-Glucosylcyanidin 3-Glucoside from the Flowers of Tricyrtis Formosana." Chemical and Pharmaceutical Bulletin 52 (5): 631–33. https://doi.org/10.1248/cpb.52.631

Teerakapong, Aroon, Teerasak Damrongrungruang, Sajee Sattayut, Noppawan Phumala Morales, and Surada Tantananugool. 2017. "Efficacy of Erythrosine and Cyanidin-3-Glucoside Mediated Photodynamic Therapy on Porphyromonas Gingivalis Biofilms Using Green Light Laser." Photodiagnosis and Photodynamic Therapy 20 (December): 154–58. https://doi.org/10.1016/j.pdpdt.2017.09.001

Tong, Yuqi, Li Li, and Xianjun Meng. 2023. "Anthocyanins from *Aronia melanocarpa* Bound to Amylopectin Nanoparticles: Tissue Distribution and In Vivo Oxidative Damage Protection." Journal of Agricultural and Food Chemistry 71 (1): 430–42. https://doi.org/10.1021/acs.jafc.2c06115

Trouillas, Patrick, Juan C. Sancho-García, Victor De Freitas, Johannes Gierschner, Michal Otyepka, and Olivier Dangles. 2016. "Stabilizing and Modulating Color by Copigmentation: Insights from Theory and Experiment." Chemical Reviews 116 (9): 4937–82. https://doi.org/10.1021/acs.chemrev.5b00507

Tsuda, Takanori. 2012. "Dietary Anthocyanin-Rich Plants: Biochemical Basis and Recent Progress in Health Benefits Studies." Molecular Nutrition & Food Research 56 (1): 159–70. https://doi.org/10.1002/mnfr.201100526

Vendrame, Stefano, and Dorothy Klimis-Zacas. 2015. "Anti-Inflammatory Effect of Anthocyanins via Modulation of Nuclear Factor-κB and Mitogen-Activated Protein Kinase Signaling Cascades." Nutrition Reviews 73 (6): 348–58. https://doi.org/10.1093/nutrit/nuu066

Wang, Lei, Lulu Wang, Xi Wang, Baojun Lu, and Jing Zhang. 2021. "Preparation of Blueberry Anthocyanin Liposomes and Changes of Vesicle Properties, Physicochemical Properties, In Vitro Release, and Antioxidant Activity before and after Chitosan Modification." Food Science & Nutrition 10 (1): 75–87. https://doi.org/10.1002/fsn3.2649

Widyadharma, I. Putu Eka, Agung Bagus Sista Satyarsa, Feliani Sanjaya, Ni Made Gitari, I. Wayan Niryana, Thomas Eko Purwata, I. Made Jawi, Dewa Ngurah Suprapta, and A. A. Raka Sudewi. 2021. "Potential of Anthocyanin Based Poly (Methyl Methacrylate) Nanoparticles Specific Activated Microglia in Management Inflammatory Pain on Herniated Nucleus Pulposus: A Literature Review." MNJ (Malang Neurology Journal) 7 (1): 40–47. https://doi.org/10.21776/ub.mnj.2021.007.01.9

Xiong, Yongfu, Gang Yang, He Zhou, Weinan Li, Ji Sun, Tang Tao, and Jingdong Li. 2021. "A ROS-Responsive Self-Assembly Driven by Multiple Intermolecular Interaction Enhances Tumor-Targeted Chemotherapy." Journal of Pharmaceutical Sciences 110 (4): 1668–75. https://doi.org/10.1016/j.xphs.2020.09.046

Yesil-Celiktas, Ozlem, Cansu Pala, E. Oyku Cetin-Uyanikgil, and Canan Sevimli-Gur. 2017. "Synthesis of Silica-PAMAM Dendrimer Nanoparticles as Promising Carriers in Neuro Blastoma Cells." Analytical Biochemistry 519 (February): 1–7. https://doi.org/10.1016/j.ab.2016.12.004

Zhao, Xue, Xuedi Zhang, Shanshan Tie, Shuai Hou, Haitao Wang, Yukun Song, Rewa Rai, and Mingqian Tan. 2020. "Facile Synthesis of Nano-Nanocarriers from Chitosan and Pectin with Improved Stability and Biocompatibility for Anthocyanins Delivery: An In Vitro and In Vivo Study." Food Hydrocolloids 109 (December): 106114. https://doi.org/10.1016/j.foodhyd.2020.106114

Zhou, Mingrui, Yahong Han, David Julian McClements, Cheng Cheng, and Shuai Chen. 2024. "Co-Encapsulation of Anthocyanin and Cinnamaldehyde in Nanoparticle-Filled Carrageenan Films: Fabrication, Characterization, and Active Packaging Applications." Food Hydrocolloids 149 (April): 109609. https://doi.org/10.1016/j.foodhyd.2023.109609

18 Toxic Phytochemicals and Their Potential Risks for Human Health

Winkins Santosh S[1], *Ramasamy Vasantharekha*[2],
Usha Rani Balu[2], *and Barathi Seetharaman*[2]
[1]PG & Research Department of Advanced Zoology & Biotechnology, Government Arts College for Men Nandanam, Chennai, India
[2]Endocrine disruption and Reproductive Toxicology Laboratory, Department of Biotechnology, SRM Institute of Science & Technology Kattankulathur, Chennai, India

18.1 INTRODUCTION

Plants and plant products are used in food and medicine for thousands of years. It was thought that natural means safe, but this is not always the case as many plants or plant constituents are toxic or even poisonous especially those containing cardiac, cyanogenic, and thioglycosides and certain types of alkaloids. Toxic natural products may cause hepatotoxicity, neurotoxicity, gastrointestinal toxicity, renal toxicity, and cytotoxicity (1). Diet-derived compounds are generally regarded as safe based on their long history of use in the diet and/or as traditional medicines, it is becoming increasingly apparent that these compounds could have deleterious effects (i) at pharmacological concentrations, (ii) in certain vulnerable populations, and (iii) in certain disease or polypharmaceutical contexts (2). For example, the major polyphenolic compounds in green tea (*Camellia sinensis*, *Theaceae*), the catechins, have been extensively studied for their cancer preventive activity and have shown activity in a number of animal models of carcinogenesis at organ sites including the lung, prostate, oral cavity, intestine, colon, liver, and skin (3) (Figure 18.1).

18.2 BENEFITS OF PHYTOCHEMICALS

Phytochemicals, as plant components with discrete bioactivities toward animal biochemistry and metabolism are being widely examined for their ability to provide health benefits. It is important to establish the scientific rationale to defend their use in foods, as potential nutritionally active ingredients. Phytochemicals could provide health benefits as: (a) substrates for biochemical reactions; (b) cofactors of enzymatic reactions; (c) inhibitors of enzymatic reactions; (d) absorbents/sequestrants that bind to and eliminate undesirable constituents in the intestine; (e) ligands that agonize or antagonize cell surface or intracellular receptors; (f) scavengers of reactive or toxic chemicals; (g) compounds that enhance the absorption and/or stability of essential nutrients; (h) selective growth factors for beneficial gastrointestinal bacteria; (i) fermentation substrates for beneficial oral, gastric, or intestinal bacteria; and (j) selective inhibitors of deleterious intestinal bacteria. Such phytochemicals include terpenoids, phenolics, alkaloids, and fiber (Figure 18.1). Research supporting beneficial roles for phytochemicals against cancers, coronary heart disease, diabetes, high blood pressure, inflammation, microbial, viral and parasitic infections, psychotic

DOI: 10.1201/9781003486237-18

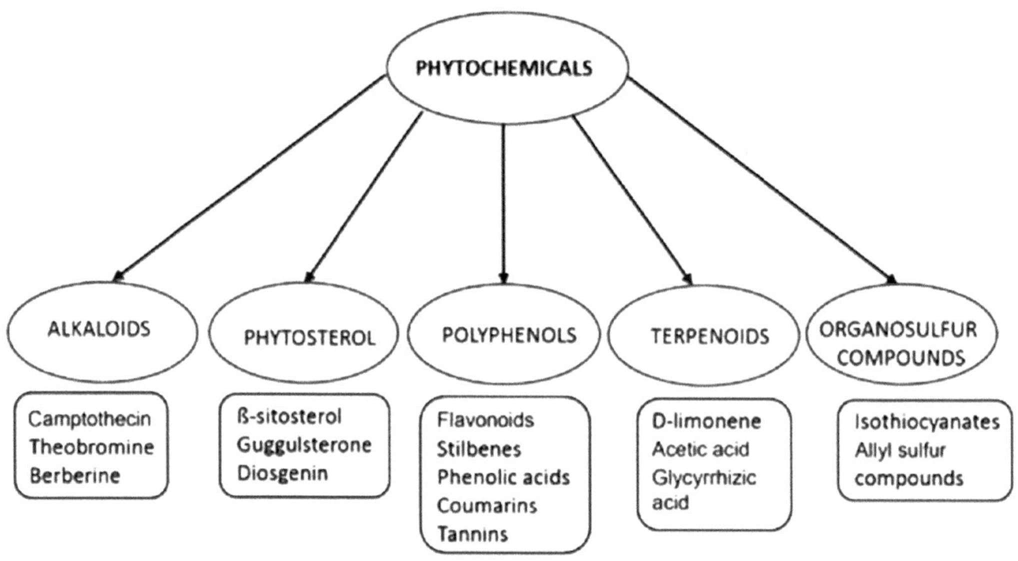

FIGURE 18.1 Diagram showing the basic types or classification of phytochemicals and active compounds.

diseases, spasmodic conditions, ulcers, etc. is based on chemical mechanisms using in vitro and cell culture systems, various disease states in animals, and epidemiology of humans (4).

18.3 NATURALLY OCCURRING PLANT FOOD TOXICANTS (5)

Several plant species grown for food produce a large number of phytotoxins that are different from the primary metabolic intermediates and products (6). Some of the inherent toxicants with adverse health consequences produced as secondary metabolites by different plant foods are tabulated as follows (Table 18.1).

Various novel food-processing methods have been developed in recent years to improve both the palatability and durability of food. Moreover, these methods are widely employed in detoxifying a range of toxicants from food. These toxins can be detoxified using different methods and techniques both at home and in laboratories. The common methods at home may be peeling, grating, boiling, cooking, microwave heating, fermenting, soaking and germination, etc. and in the laboratory, ultrasonic treatment, autoclaving, sonication, solvent extraction, steam explosion, high-pressure processing (HPP), gamma irradiation, and instant controlled pressure drop.

18.4 POTENTIAL TOXICITY OF DIETARY PHENOLS

Although most phenolic compounds are considered safe and with health benefits, some in vitro and in vivo animal studies have demonstrated that dietary polyphenols can exhibit many adverse effects on diverse biological systems (7). For example, several flavonoids inhibit thyroid peroxidase, an enzyme that oxidizes iodide ions to form iodine and thereby interfering with thyroid hormone biosynthesis and ultimately thyroid function. The elevated consumption of soy isoflavones has been associated with fertility reduction and retardation of sexual maturation, while polyphenols exhibit antinutritional properties through the chelation of transition metals such as iron, which can lead to iron deficiency (8). Caffeic acid, a phenolic compound synthesized by all plant species but mostly present in foods such as coffee, wine, and tea, was found to induce forestomach and kidney tumors in rats and mice when administered at 2% (9).

TABLE 18.1
Toxins Secreted by Plants and Their Possible Toxicity

S. No.	Name of the Plant	Toxin Secreted	Toxicity
1.	Cassava, flaxseed, sorghum, apricot fruits, and bamboo	Cyanogenic glycosides (they produce hydrogen cyanide [HCN])	Acute poisoning, tropical ataxic neuropathy, blood glucose and lactic acid levels rise, and the ATP/ADP ratio falls
2.	Potatoes, tomatoes, and eggplants	Glycoalkaloids (α-solanine and α-chaconine)	Inhibits acetylcholinesterase, cell disruption and organ damage, diarrhea, fever, vomiting, gastrointestinal discomfort, gastroenteritis, neurological disorders, a high pulse rate, and low blood pressure
3.	Broccoli, cauliflower, cabbage, horseradish root, and mustard seed	Glucosinolates (β-D-thioglucose and isothiocyanates)	Enlarged thyroid, decreased plasma thyroid hormone levels, organ abnormalities (liver and kidney), impaired growth, and diminished reproductive performance
4.	Mostly flowering plants from *Asteraceae* and *Boraginaceae* families	Pyrrolizidine alkaloids (necic acids and necines)	Veno-occlusive disease, hepatic sinusoidal obstruction syndrome, vomiting, bloody diarrhea, liver enlargement, hemorrhagic necrosis, and hepatomegaly
5.	Legumes, cereals, beans, wheat, quinoa, peas, kidney beans, bananas, lentils, soybeans, mushrooms, and rice	Lectins (phytohemagglutinins)	Inhibit the activity of digestive enzymes, lowering protein in vitro digestibility, Crohn's disease, coeliac disease, colitis, oedema, hyperemia, hemorrhages in lymphatic tissues, and induce autoimmune disorders

Source: Ref. (5).

18.5 POTENTIAL TOXICITY OF ALKALOIDS

The occurrence of alkaloids in the food chain and possible adverse effects of plant-derived alkaloids have been reported in a number of studies. Research has revealed that some alkaloids known to be present in the modern food chain such as piperine, nicotine, theobromine, and theophylline are considered to have low if not negligible risk (11). However, pyrrolizidine alkaloids (PAs) and tropane alkaloids and their derivatives have shown variable levels of toxicity in humans. The toxicity of PA in humans has been established from various incidences of deaths and poisoning following ingestion of PA-containing herbal medicines, teas, and weeds (10), with liver and lung as the main target organs. The acute toxicity of 1,2-unsaturated PAs in humans is characterized mainly by the onset of hepatic veno-occlusive disease (HVOD), associated with high mortality and possibly progressing to liver cirrhosis. Potential health hazards of the ingestion of high doses of tropane alkaloids have been reported in the literature. While tropane alkaloids can be beneficial for medical treatment at low doses, severe unwanted toxic effects have been recorded in case of an over dose. The most common effects reported on humans caused by toxic doses include dizziness, blurred vision, pupil dilation, dry mouth, red skin, vomiting, clouded consciousness, muscle spasms, low body temperature, hallucinations, delirium, tachycardia, and even death (11, 12).

18.6 POTENTIAL TOXICITY OF CAROTENOIDS

According to the United States Food and Drug Administration (FDA), supplements for several carotenoids enriched products including palm oil carotenoids are categorized as generally recognized as safe (GRAS) products (13). Dietary carotenoids are generally considered safe at levels much higher than those observed in normal food patterns. Therefore, the European Food Safety Authority (EFSA) has made recommendations defining safe daily intakes, including in some cases acceptable daily intake (ADI), of certain carotenoids (13).

Carotenoids are generally non-toxic, even when taken in high doses as purified supplements. There appear, however, to be a few exceptions. High doses of canthaxanthin can cause a reversible form of retinopathy (14). Also, high intake of β-carotene (in supplement form, 20–30 mg/d) is contraindicated for smokers due to the increased risk of lung and stomach cancers found in some studies (15). Excessive intake of some of the major carotenoids can also cause carotenemia, a reversible yellowing of the skin.

18.7 POTENTIAL ADVERSE EFFECTS OF PHYTOCHEMICAL ACCUMULATION ALTERATIONS IN PLANT PRODUCTS

Fruits and vegetables are known to possess phytochemical compounds with biological activities that have the ability to prevent and reverse the development of chronic diseases such as diabetes and cancer (16, 17). The secondary metabolites found in different plants parts, including fruits and vegetables, possess many biological activities. Among the biological activities exerted by different phytochemicals are the enzyme inhibitory properties. There are two enzymes that are reported to contribute to the fast postprandial release of glucose from a carbohydrate-rich meal, namely the alpha-amylase and alpha-glucosidase. Inhibitory actions of these enzymes that are inherent in some phytochemicals present in plant products such as fruits and leafy vegetables contribute to the regulation of blood glucose levels as the slow release of glucose from diet sources may afford its proper metabolism, which mitigates against the development of diabetes mellitus (18). In addition, the enzyme inhibitory properties inherent in some phytochemicals may affect kinases that are involved in cell cycle progression, which may contribute to the mitigation against the development of cancer. Cancer is a disease of uncontrolled cell growth, which is propelled through up-normal continuous cell division cycle (19). Cell division cycle progression is regulated through the activity of phosphorylation enzymes known as kinases (20). Cell division cycle progression is also depended on the intactness of the DNA that is safeguarded by the action of tumor suppressor genes. Tumor suppressor genes act by way of effecting cell division cycle arrest in case of DNA damage that affords DNA repair or cell death. One of the major causes of DNA damage is oxidative stress that emanates from the imbalances between reactive oxygen species (ROS), known as free radicals and antioxidants. In the case, where DNA damage results in the malfunctioning of tumor suppressor genes, cell division cycle continues even in situations, where it should not of which the result is uncontrolled cell growth. Oxidative DNA damage may also affect metabolism regulatory genes that may lead to inadequate or impaired insulin. As such, oxidative DNA damage may culminate in the development of diseases such as cancer and diabetes mellitus (20, 21).

18.8 NATURAL TOXINS IN FOOD (NATURAL TOXINS IN FOOD | FDA REPORT – 2023)

Natural toxins are often produced in plants as a defense against predators, insects, or infestation. Below are some examples of vegetables, fruits, plants, and other foods that can sometimes contain a natural toxin.

18.8.1 Ackee Fruit (Hypoglycin A)

Hypoglycin A is a heat stable toxin found in high levels in the rinds and seeds of ackee, a tropical fruit that features prominently in Jamaican cuisine. Reactions to hypoglycin A can range from no symptoms to vomiting, coma, and death. Profound hypoglycemia (blood glucose levels as low as 3 mg/dL) is observed in most cases. Properly harvested and prepared ackee will have low levels of hypoglycin and is safe to eat. Potential risk behaviors for ackee poisoning include (a) selection and cooking of unripe ackee; (b) purchase of tampered, forcibly opened ackee; and (c) reuse of the water in which unripe ackee has been cooked. Toxins such as hypoglycin and their metabolites inhibit beta-oxidation pathways and can produce Reye's syndrome-like presentations and sometimes resulting in death, especially among the malnourished.

18.8.2 Blue-Green Algae (Microcystins)

Microcystins (MCs) are widely distributed cyanotoxins and have often been implicated in accidental human and animal poisonings. They are produced by several genera, including the planktonic *Microcystis*, *Planktothrix*, *Anabaena* species, and the benthic *Oscillatoria*. Nodularins are only produced by the species *Nodularia spumigena*, which occurs in brackish water. Clinical signs of MC or nodularin intoxication in mammals are diarrhea, vomiting, piloerection, weakness, and pallor. Acute episodes of gastroenteritis due to oral contamination by MCs have been observed.

Produced by *Microcystis* and other types of blue-green algae (BGA), MCs are natural toxins that can cause gastrointestinal distress and potentially damage your liver over time.

Microcystis cells can contaminate other BGA used in foods, such as wild-harvested *Aphanizomenon flos-aquae* (AFA). Properly harvested BGA is safe to eat and is used in some dietary supplements and food products.

18.8.3 Honey (Grayanotoxins)

Plants contain numerous compounds that, when beneficial to humans, are categorized as "medicinal" and when harmful they are termed "poisonous". Secondary products derived from plants, such as honey, can contain a number of chemical compounds that, depending on their concentration and application, can also be considered medicinal or poisonous (22). Grayanotoxin containing honey, called "mad honey", can cause dramatic effects when ingested as has already been recorded by the Greek warrior-writer Xenophon in 401 BC in his Anabasis. Bees collect nectar from flowers in order to create honey. If those flowers have a natural toxin, that toxin may end up in the honey from that beehive. One example is grayanotoxins, which plants such as rhododendrons and mountain laurel naturally produce. Eating honey with a high amount of this toxin can lead to "mad honey" poisoning, with symptoms such as nausea, vomiting, or dizziness. This type of poisoning is rare. A number of clinical signs have been associated with mad honey disease. Most often hypotension, cardiac rhythm disorders (first-, second-, and third-degree AV block; asystole; and sinus bradycardia), nausea and/or vomiting, sweating, dizziness, and impaired consciousness have been observed. More rarely, syncope, blurred vision or diplopia, and salivation have been described. In some rare cases, convulsion, atrial fibrillation, asystole, and myocardial infarction were observed (23–26).

18.8.4 Beans (Phytohemagglutinin)

Lectins and phytohemagglutinins (PHAs) are natural toxicants present in many foods, especially in beans and other dietary pulses, which can have toxic effects when consumed without adequate cooking, occasionally leading to an acute gastroenteritis. PHA, as its name implies, can agglutinate many mammalian red blood cells and interfere with cellular metabolism. Moreover, PHA is an antinutrient, which can interfere with the absorption of minerals, particularly calcium, iron, phosphorus, and zinc. But at high levels in raw beans, PHA can lead to nausea, severe vomiting, and diarrhea.

Plant lectins that are not efficiently degraded by digestive enzymes and that have an affinity for the surface of gut epithelial cells, such as those present in the *Leguminosae* family, can be poisonous. Lectins are proteins that bind to carbohydrates and some plants produce them as a natural defense mechanism. In canned and properly cooked kidney beans, the low levels of PHA won't affect you. But at high levels in raw beans, PHA can lead to nausea, severe vomiting, and diarrhea. Soaking the beans for a minimum of 5 hours and then boiling them in fresh water for at least 30 minutes will remove and destroy this toxin.

18.9 MUSHROOMS (MUSHROOM TOXINS)

It's well known that some mushrooms are toxic.

Depending on the type of mushroom you eat, you could experience milder reactions like nausea or diarrhea. More severe reactions include coma and death.

18.9.1 Amygdalin (D-Mandelonitrile 6-O-B-D-Glucosido-B-D-Glucoside)

Stone fruit, such as peaches, apricots and plums having a substance called amygdalin. Amygdalin is a natural cyanogenic glycoside occurring in the seeds of some edible plants, such as bitter almonds and peaches. It is a medically interesting but controversial compound as it has anticancer activity on one hand and can be toxic via enzymatic degradation and production of hydrogen cyanide on the other hand. Despite numerous contributions on cancer cell lines, the clinical evidence for the anticancer activity of amygdalin is not fully confirmed. Moreover, high dose exposures to amygdalin can produce cyanide toxicity. Excessive consumption of seeds may have a negative effect on the body, causing a number of adverse reactions of the following types: diarrhea, vomiting, abdominal pain, and in extreme cases, may lead to death.

Amygdalin, also called laetrile, is a natural chemical found in the seeds of apricots, bitter almonds, apples, peaches, and plums. The chemical is not found in the fruit itself and accidentally eating a seed or pit will not harm you. However, consuming a large amount of the seeds or pits can be problematic because enzymes in your intestines can turn amygdalin into cyanide and cause cyanide poisoning.

18.9.2 Mycotoxins in Foods

Mold or fungus can infect some foods with mycotoxins while the crop is growing or is being stored. Only certain molds and fungi produce mycotoxins that can make a person sick when consumed. Mycotoxins associated with human food include aflatoxins, patulin (PAT), fumonisin, ochratoxin A (OTA), and deoxynivalenol (DON).

18.9.2.1 Aflatoxin

Aflatoxins are a family of toxins produced by certain fungi that are found on agricultural crops such as maize (corn), peanuts, cottonseed, and tree nuts. The main fungi that produce aflatoxins are *Aspergillus flavus* and *Aspergillus parasiticus*, which are abundant in warm and humid regions of the world. Aflatoxin-producing fungi can contaminate crops in the field, at harvest, and during storage. People can be exposed to aflatoxins by eating contaminated plant products (such as peanuts) or by consuming meat or dairy products from animals that ate contaminated feed. Farmers and other agricultural workers may be exposed by inhaling dust generated during the handling and processing of contaminated crops and feeds. Aflatoxin toxicity may result in nausea, vomiting, abdominal pain, convulsions, and other signs of acute liver injury. Long-term exposure also leads to various complications like growth retardation, cirrhosis, and hepatocellular carcinoma (27).

18.9.2.2 Patulin

The mycotoxin PAT (4-hydroxy-4H-furo[3,2c]pyran-2[6H]-one) is a secondary metabolic product of molds such as *Penicillium*, *Aspergillus*, and *Byssochlamys* species. PAT is a common contaminant of fruit- and vegetable-based products, most notably apples. It is particularly associated with apples exhibiting "brown rot" or other rotting characteristics and is now known to occur worldwide in apples and apple products. In vivo toxicity assessment shows damage to vital organs and systems including the liver, kidney, intestinal tissues, and immune system. Several studies have revealed its mutagenicity, teratogenicity, chromosomal aberration, DNA strand damage, and micronuclei formation in mammalian cells (28).

18.9.2.3 Fumonisin

Fumonisins are secondary metabolites produced in cereals by pathogenic fungi, namely *Fusarium verticillioides*, *Fusarium proliferatum*, and related species. Moreover, Aspergillus niger also produces fumonisins in the crop plants of peanut, maize, and grape. The maize and maize-based products are most commonly infected with fumonisins besides their presence in several other grains (rice, wheat, barley, maize, rye, oat, and millet) and grain products (tortillas, corn flask, and chips) which have major influences on health. Further intake of fumonisin-affected diets by pregnant women causes neural tube defects in the developing fetus and a dose above the threshold level may cause fetal death. Similarly, the exposure of fumonisin and its effect on esophageal and liver cancer is rare. While no direct evidence of fumonisin hazard is found, its prolonged exposure may lead to cancer and birth defects in humans (29).

18.9.2.4 Ochratoxin A

OTA is a naturally occurring foodborne mycotoxin found in a wide variety of agricultural commodities worldwide, ranging from cereal grains to dried fruits to wine and coffee. It is produced by several different fungi including *Aspergillus ochraceus*, *Aspergillus carbonarius*, *A. niger*, and *Penicillium verrucosum*. Contamination generally occurs as a result of poor storage of commodities and suboptimal agricultural practices during the drying of foods. OTA causes nephrotoxicity and renal tumors in a variety of animal species; however, human health effects are less well-characterized. Various studies have linked OTA exposure with the human diseases Balkan endemic nephropathy (BEN) and chronic interstitial nephropathy (CIN), as well as other renal diseases (30).

18.9.2.5 Deoxynivalenol

DON is one of several mycotoxins produced by certain *Fusarium* species that frequently infect corn, wheat, oats, barley, rice, and other grains in the field or during storage. The exposure risk to human is directly through foods of plant origin (cereal grains) or indirectly through foods of animal origin (kidney, liver, milk, and eggs). It has been detected in buckwheat, popcorn, sorghum, triticale, and other food products including flour, bread, breakfast cereals, noodles, infant foods, pancakes, malt, and beer. DON affects animal and human health causing acute temporary nausea, vomiting, diarrhea, abdominal pain, headache, dizziness, and fever (31).

18.10 POTENTIAL DIETARY CARCINOGENS

A number of phytochemicals with potential carcinogenic effects are found naturally in many foods and seasonings that have been consumed for centuries by humans as food, supplements to maintain optimal health, or used as treatments for disease (Table 18.2) (32).

Phytochemicals are found in various plants that are frequently included in the human diet and are generally thought to be safe for consumption because they are produced naturally. However, this is not always the case and in fact many natural compounds found in several commonly consumed plants are potential carcinogens or tumor promoters and should be avoided.

18.11 REGULATIONS, PRECAUTIONS, AND SAFETY MEASURES FOR USING MEDICINAL PLANTS, PHYTONUTRIENTS, AND PHYTOCHEMICALS

The use of herbal medicines and phytonutrients or nutraceuticals continues to expand rapidly across the world with many people now resorting to these products for treatment of various health

TABLE 18.2
Common Sources of Dietary Phytochemicals and Their Chemical NAME and CAS Number

S. No.	Dietary Phytochemical/Source	Chemical Name	CAS Number
1.	Capsaicin (chili pepper)	8-Methyl-N-vanillyl-trans-6-nonenamide	404-86-4
2.	Cycasin	(Methyl-ONN-azoxy)methyl b-D-glucopyranoside	14901-08-7
3.	Methylazoxymethanol (MAM) (Cycas seed)	Methyl-ONN-azoxymethanol	590-96-5
4.	Phytoestrogens genistein (soybeans, fava beans, and red clover)	4′,5,7-Trihydroxyisoflavone	446-72-0
5.	Phytoestrogens (lignans matairesinol) (sesame and flax seed)	(3R,4R)-3,4-bis[(4-hydroxy-3-methoxyphenyl) methyl] oxolan-2-one	580-72-3
6.	Ptaquiloside (bracken fern)	(2′R)-7′aβ-(β-D-glucopyranosyloxy)-1′,3′aβ,4′7′a-tetrahydro-4′α-hydroxy-2′β,4′,6′-trimethylspiro[cyclopropane-1,5′[5H]inden]-3′(2′H)-one	87625-62-5
7.	Safrole (sassafras plant)	5-(2-propenyl)-1,3-benzodioxole	94-59-7

challenges in different national healthcare settings (33). This past decade has obviously witnessed a tremendous surge in acceptance and public interest in natural therapies both in developing and developed countries, with these herbal remedies being available not only in drug stores but now also in food stores and supermarkets. It is estimated that up to four billion people (representing 80% of the world's population) living in the developing world rely on herbal medicinal products as a primary source of healthcare and traditional medical practice which involves the use of herbs is viewed as an integral part of the culture in those communities (34, 35).

Taking consumer health into account, WHO collaborated with the FAO of the United Nations to develop the Joint Expert Committee on Food Additives (JECFA), responsible for assessing health risks posed by foods containing natural toxins. Based on JECFA evaluations, Codex Alimentarius established international rules and guidelines to limit the amount of natural poisons in food (36). The EU Council Directive on Flavorings (88:388:EU) regulates intrinsic toxicants from many plant source materials, including plant foods, herbs, and spices, such as coumarin, thujone, safrole, and hydrocyanic acid (originating from cyanogenic glycosides) (37).

Consumption of under processed cassava (tapioca) has been linked to cyanide poisoning and tropical ataxic neuropathy (TAN) disease; hence, cassava should be processed and cooked well before consuming. **Peeling** is the first stage in the processing of cassava roots. Cassava peel has a higher concentration of cyanide than cassava pulp. In the sweet varieties where the peel has been removed, the pulp can be consumed immediately after boiling, while bitter varieties require additional detoxification measures before eating. **Grating** is a size reduction process, and therefore it increases the surface area that allows for greater contact of linamarin and linamarase for an easier detoxification process. **Drying** is one of the most effective methods for removing cyanogenic glycosides from plant foods. There are several drying methods available to reduce cyanogen content in food products, including sun, oven, freeze, and superheated steam. **Cooking and boiling** are among the most efficient procedures for reducing cyanogenic chemicals in plant foods. **Fermentation** is an old method of food preservation that has gained popularity in many cultures due to its nutritional value and diverse sensory qualities. **Germination** is a traditional food processing technique that is effective in lowering the cyanogenic glycoside level.

Some novel food processing techniques effective in detoxifying inherent plant food toxicants are microwave heating, hydrothermal treatment, ultrasonic treatment, solvent extraction, sonication, autoclaving, HPP, ultrasound treatment, steam explosion, gamma irradiation, and instant controlled pressure drop (5).

Twenty common herbal supplements, adverse reactions, and cautions is given in Table 18.3.

TABLE 18.3
Twenty Common Herbal Supplements, Adverse Reactions, and Cautions

S. No.	Herbal Supplement	Adverse Reactions	Caution
1.	Capsicum pepper *Capsicum frutescens* or *Capsicum annuum*	Anaphylaxis, abdominal pain, belching, bloating, anal burning, renal/hepatic damage, local burning, contact dermatitis, edema, erythema, flushing, rhinorrhea, lacrimation, sneezing, and mucous membrane irritation	Avoid if upcoming surgery or hypersensitivity to peppers; consult provider before use with history of bleeding disorders, diabetes mellitus, or hypertension
2.	Chamomile, German *Matricaria recutita*	Anaphylaxis, hypersensitivity reactions, confusion, drowsiness, atopic dermatitis, ecchymosis, contact dermatitis, eczema, and ocular irritation	Avoid if age less than 2 years; pregnant; hypersensitivity to *Asteraceae* and *Compositae* plants, including chrysanthemums, daisies, marigolds, and ragweed; caution with asthma, hormone-sensitive conditions, or upcoming surgery
3.	Cranberry *Vaccinium macrocarpon*	Abdominal pain, nausea, vomiting, diarrhea, nocturia, nephrolithiasis, candidiasis, vaginal pruritus, and altered blood glucose	Avoid with hypersensitivity to aspirin or other salicylates; caution with asthma, atrophic gastritis, hypochlorhydria, or nephrolithiasis
4.	Echinacea *Echinacea purpurea* and related species	Anaphylaxis, angioedema, bronchospasm, hypersensitivity reaction, abdominal pain, diarrhea, nausea, vomiting, constipation, heartburn, hepatitis, headache, dizziness, drowsiness, and alteration of fertility	Avoid if history of autoimmune disorders, including multiple sclerosis, rheumatoid arthritis, or systemic lupus erythematosus; caution with atopic conditions or attempting to conceive a pregnancy
5.	Evening primrose *Oenothera biennis*	Abdominal pain, distention, dyspepsia, flatulence, indigestion, vomiting, taste disturbance, dizziness, headache, acne, skin rash, increased risk of bleeding, and decreased aggregation	Avoid if upcoming surgery; caution with coagulation disorders, schizophrenia, seizure disorder, pregnancy, or lactation
6.	Feverfew *Tanacetum parthenium*	Abdominal pain, nausea, vomiting, constipation, diarrhea, indigestion, flatulence, indigestion, dysgeusia, oral ulcers, gingival bleeding, anxiety, and insomnia	Avoid if pregnant; age less than 18 years; hypersensitivity to *Asteraceae* or *Compositae* plants, including chrysanthemums, daisies, and marigolds
7.	Flax/flaxseed oil *Linum usitatissimum*	Anaphylactic or hypersensitivity reactions, abdominal pain diarrhea, intestinal obstruction, nausea, vomiting, hypotension, and mania	Avoid with GI disorders, obstruction, stricture, and inflammation; breast, uterine or ovarian cancer, uterine fibroids, or endometriosis; hypertriglyceridemia and open wounds; caution with bipolar disorder, bleeding disorders, diabetes mellitus, or hypothyroidism
8.	Garlic *Allium sativum*	Anaphylaxis, angioedema, urticaria, abdominal pain, irritation, burning, belching, nausea, dyspepsia, vomiting, diarrhea, flatulence, hypotension, and dizziness	Avoid with bleeding disorders; recent or upcoming surgery; or hypersensitivity to garlic or members of the *Liliaceae* family, including lily, hyacinth, tulip, onion, leek, and chive; caution if diabetes mellitus, GI infection, peptic ulcer disease, or inflammatory bowel disease
9.	Ginger *Zingiber officinale*	Abdominal pain, nausea, belching, distention, flatulence, diarrhea, dermatitis, hypoglycemia, CNS depression, and arrhythmias	Avoid with history of gallbladder dysfunction, inflammatory bowel disease, or GI obstruction; caution with history of cardiac disease, gastric or duodenal ulcers, diabetes mellitus, HTN, hypotension, breastfeeding, or pregnancy

(Continued)

TABLE 18.3 *(Continued)*
Twenty Common Herbal Supplements, Adverse Reactions, and Cautions

S. No.	Herbal Supplement	Adverse Reactions	Caution
10.	Ginkgo *Ginkgo biloba*	Nausea, vomiting, constipation, diarrhea, xerostomia, arrhythmias, palpitations, transient ischemic attacks, cerebrovascular accidents, and edema Increased risk of bleeding, hematuria, weakness, restlessness, dizziness, drowsiness, headache, skin hypersensitivity reactions, and Stevens-Johnson syndrome	Avoid if attempting to conceive, are pregnant, or having surgery within 2 weeks; caution if history of bleeding or seizure disorders, diabetes mellitus, or G6PD deficiency
11.	Ginseng, American *Panax quinquefolius*	Headache	Avoid if pregnant or having surgery within 2 weeks; caution with bleeding disorders, diabetes mellitus, insomnia, hormone-sensitive conditions, ovarian or uterine cancers, uterine fibroids, endometriosis, or breast cancer
12.	Ginseng, Asian *Panax ginseng*	Anaphylaxis, skin hypersensitivity reactions, headache, euphoria, insomnia, abdominal pain, nausea, decreased appetite, diarrhea, cholestatic hepatitis, hypertension, hypotension, tachycardia, vaginal bleeding, and amenorrhea	Avoid with history of bleeding and clotting disorders, organ transplant recipients, pediatric patients, or pregnancy; caution with a history of cardiovascular disease, arrhythmias, cerebrovascular disease, autoimmune disorders, endometrial cancer or endometriosis, hormone-sensitive conditions, insomnia, or schizophrenia
13.	Goldenseal Hydrastis canadensis	Abdominal pain and distention, flatulence, diarrhea, constipation, vomiting, headache, photosensitivity, skin rash, and hypernatremia	Avoid with breastfeeding, pregnancy, or infants; caution with history of bleeding disorders, diabetes mellitus, hypertension, or hypotension
14.	Green tea *Camellia sinensis*	Nausea, vomiting, flatulence, abdominal distention, diarrhea, constipation, arrhythmias, palpitations, tachycardia, HTN, dizziness, nervousness, anxiety, restlessness, irritability, insomnia, delirium, rash, urticaria, muscle spasm, and seizures	Avoid with breastfeeding, pregnancy, and hepatic impairment or disease; caution with elderly patients and postmenopausal women; caution with history of cardiovascular disease or arrhythmias, anxiety, diabetes mellitus, hypertension, headaches, glaucoma, psychiatric disorders, prostate cancer, renal disease, osteoporosis, and gastric or duodenal ulcers
15.	Hawthorn *Crataegus monogyna*	Nausea, dizziness, agitation, headache, insomnia, palpitations, arrhythmias, circulatory disturbance, and tachycardia	Avoid if known hypersensitivity; caution with elderly patients and concurrent cardiovascular agents, CNS depressants, and vasodilatory agents
16.	Kava *Piper methysticum*	Hepatotoxicity, hepatic failure, hepatitis, cirrhosis, weight loss, xerostomia, mouth numbness, skin hypersensitivity reactions, scaly rash, mydriasis, oculomotor equilibrium and visual accommodation disturbances, headache, motor reflex impairment, and dyskinesia	Avoid if pregnant or breastfeeding, history of or acute hepatitis, long-term ethanol use, chronic lung disease, Parkinson's disease, operating heavy machinery, or taking concurrent CNS depressants or hepatotoxic drugs; caution in patients suffering from depressive disorders
17.	Milk thistle *Silybum marianum*	Anaphylaxis or hypersensitivity reaction, abdominal pain, dyspepsia, nausea, vomiting, constipation, diarrhea, flatulence, laxative effect, loss of appetite, weakness, rash, pruritus, eczema, and dysgeusia	Avoid with breast, ovarian, or uterine cancers, endometriosis, uterine fibroids, hormone-sensitive conditions, hypersensitivity to *Asteraceae* or *Compositae* plants, including chrysanthemums, daisies, or marigolds; caution in patients with diabetes mellitus

(Continued)

TABLE 18.3 *(Continued)*
Twenty Common Herbal Supplements, Adverse Reactions, and Cautions

S. No.	Herbal Supplement	Adverse Reactions	Caution
18.	Saw palmetto *Serenoa repens*	Abdominal pain, heartburn, vomiting, dyspepsia, diarrhea, constipation, duodenal ulcer, diarrhea, constipation, hepatitis, arrhythmias, angina, heart failure, hypertension, hypotension, myocardial infarction, tachycardia, dyspnea, testicular pain, UTI, decreased libido, erectile dysfunction, headache, dizziness, weakness, myalgias, insomnia, and rhabdomyolysis	Avoid if breastfeeding, pregnant, or having surgery within 2 weeks; caution with bleeding disorders or hormone-sensitive conditions
19.	St. John's wort *Hypericum perforatum*	Anorexia, nausea, vomiting, diarrhea, constipation, dyspepsia, hypoglycemia, xerostomia, agitation, dizziness, anxiety, fatigue, hypomania, irritability, lethargy, panic attack, psychosis, restlessness, serotonin-like syndrome, suicidal or homicidal ideation, vivid dreams, tremor, withdrawal symptoms, pruritus, jaundice, muscle pain, joint stiffness, palpitations, hypertension, tachycardia, and sexual dysfunction	Avoid if having difficulty conceiving a pregnancy or having surgery within 2 weeks; caution with dementia of Alzheimer's type, bipolar disorder, depression, schizophrenia, concurrent use of anesthesia, or dose greater than 2 grams per day
20.	Valerian *Valeriana officinalis*	Agitation, excitability, fatigue, insomnia, mental dullness, morning drowsiness, restlessness, sedation, uneasiness, abdominal pain, diarrhea, hepatotoxicity, headache, dizziness, and tachycardia	Avoid if pregnant or having surgery within 2 weeks

Source: Ref. (38).
Abbreviations: CNS, central nervous system; GI, gastrointestinal; HTN, hypertension.

18.12 CONCLUSION

Plants and plant products serve as a vital source of food for most living organisms. Some plants secrete secondary metabolites (toxins) to protect themselves from predators, such as cyanogenic glycosides, glucosinolates, glycoalkaloids, lectins, and PAs, are widely found in the most important plant foods, which include cassava, flaxseed, potato, broccoli, rapeseed, chickpea, soybean, and lentil. Some of these toxic chemicals have health benefits and are used to protect against chronic health complications such as cancer. Inversely, short- and long-term exposure to significant amounts of these phytotoxins may end up with chronic irreversible negative health problems in important organ systems such as the immune system, kidneys, and reproductive system, and in severe cases, they can be carcinogenic and fatal.

WHO, FAO, FDA, and other organizations are strictly regulate the production, processing, and consumption of herbal medicines and other phytonutrients. FDA does not approve herbal medicines. This is because the FDA does not consider herbal medicines drugs. Instead, it considers them dietary supplements for complementary therapy. The NHS (National Health Services, UK) warns that "natural" does not mean safe if a person uses a product without a doctor's prescription. This is because some herbs can have adverse drug interactions with other medications. Some may also cause fatal side effects. A person should always check with a doctor before taking supplements if they have a health condition or are taking prescribed medication. The risks of obtaining fake, substandard, unlicensed, or contaminated herbal medicines or phytonutrients and supplements are increased by buying online or by mail order. Herbal products sold online may also contain banned ingredients and toxic substances. Herbal slimming products and products that claim to improve sexual performance are best avoided. Talk to a doctor before taking any nutritional supplements.

REFERENCES

1. Salem MA, Serag A, El-Seedi HR, Hamdan DI, Ezzat SM, Zayed A. 23 – identification and analysis of toxic phytochemicals. In Andrew G. Mtewa, Chukwuebuka Egbuna (Eds.), Phytochemistry, the Military and Health 2021 (pp. 443–479). Elsevier. https://doi.org/10.1016/B978-0-12-821556-2.00015-3
2. Schilter B, Andersson C, Anton R, Constable A, Kleiner J, O'Brien J, Renwick AG, Korver O, Smit F, Walker R, Guidance for the safety assessment of botanicals and botanical preparations for use in food and food supplements. Food and Chemical Toxicology. 2003;41(12):1625–1649. https://doi.org/10.1016/S0278-6915(03)00221-7
3. Yang, Chung S, Maliakal, Pius, Meng, Xiaofeng. Inhibition of carcinogenesis by tea. Annual Review of Pharmacology and Toxicology. 2002;42:25–54. doi:10.1146/annurev.pharmtox.42.082101.154309
4. Dillard CJ, German JB. Phytochemicals: Nutraceuticals and human health. Journal of the Science of Food and Agriculture. 2000;80(12):1744–1756. https://doi.org/10.1002/1097-0010(20000915)80:12<1744::AID-JSFA725>3.0.CO;2-W
5. Urugo MM, Tringo TT. Naturally occurring plant food toxicants and the role of food processing methods in their detoxification. International Journal of Food Science. 2023 Apr 27;2023:9947841. doi:10.1155/2023/9947841.
6. Hajšlová J, Schulzová V, Botek P, Lojza J. Natural toxins in food crops and their changes during processing. Czech Journal of Food Sciences. 2004;22.
7. Martin KR, Appel CL. Nutrition and dietary supplements polyphenols as dietary supplements: A double-edged sword. Nutrition and Dietary Supplements. 2010;2:1–12. www.dovepress.com
8. Giahi L, Mohammadmoradi S, Javidan A, Sadeghi MR. Nutritional modifications in male infertility: A systematic review covering 2 decades. Nutrition Reviews. 2016;74(2):118–130.
9. Holst B, Williamson G. Nutrients and phytochemicals: From bioavailability to bioefficacy beyond antioxidants. Current Opinion in Biotechnology. 2008;19(2):73–82.
10. Wink M. Quinolizidine and pyrrolizidine alkaloid chemical ecology—A mini-review on their similarities and differences. Journal of Chemical Ecology. 2019;45:109–115.
11. Samtiya M, Aluko REDT. Plant food anti-nutritional factors and their reduction strategies: An overview. Food Production, Processing and Nutrition. 2020;2:1–14.
12. Spina SP, Taddei A. Teenagers with jimson weed (*Datura stramonium*) poisoning. The Journal of Emergency Medicine. 2007;9(6):467–469.
13. Meléndez-Martínez AJ. An overview of carotenoids, apocarotenoids, and vitamin a in agro-food, nutrition, health, and disease. Molecular Nutrition & Food Research. 2019;63(15):1–11.
14. Hueber A, Rosentreter A, Severin M. Canthaxanthin retinopathy: Long-term observations. Ophthalmic Research. 2011;46:103–106.
15. Druesne PN, Latino-Martel P, Norat T, Barrandon E, Bertrais S, Galan P, Hercberg S. Beta-carotene supplementation and cancer risk: A systematic review and meta-analysis of randomized controlled trials. International Journal of Cancer. 2010;127:172–184.
16. Babu PVA, Liu D, Gilbert ER. Recent advances in understanding the anti-diabetic actions of dietary flavonoids. Journal of Nutritional Biochemistry. 2013;24:1777–1789. doi:10.1016/j.jnutbio.2013.06.003
17. Chen AY, Chen YC. A review of the dietary flavonoid, kaempferol on human health and cancer chemoprevention. Food Chemistry. 2013;138:2099–2107. doi:10.1016/j. foodchem.2012.11.139
18. Zengin G, Guler GO, Aktumsek A, Ceylan R, Picot CMN, Mahomoodally MF. Enzyme inhibitory properties, antioxidant activities, and phytochemical profile of three medicinal plants from Turkey. Advances in Pharmacological Sciences. 2015;410675. doi:10.1155/2015/410675
19. Collins K, Jacks T, Pavletich NP. The cell cycle and cancer. Proceedings of the National Academy of Sciences of the United States of America. 1997;94:2776–2778. doi:10.1073/pnas.94.7.2776
20. Alberts B, Johnson A, Lewis J, Raff M, Roberts K, Walter P. Molecular Biology of the Cell. 4th ed. 2002 (p. 1616). Garlan Science. doi:10.1093/aob/mcg023

21. Wright, E., Jr, Scism-Bacon, J. L., & Glass, L. C. (2006). Oxidative stress in type 2 diabetes: the role of fasting and postprandial glycaemia. International journal of clinical practice, 60(3), 308–314. https://doi.org/10.1111/j.1368-5031.2006.00825.x
22. Viuda-Martos, M., Ruiz-Navajas, Y., Fernández-López, J., & Pérez-Alvarez, J. A. (2008). Functional properties of honey, propolis, and royal jelly. Journal of food science, 73(9), R117–R124. https://doi.org/10.1111/j.1750-3841.2008.00966.x
23. Dilber E, Kalyoncu M, Yariş N, Őkten A. A case of mad honey poisoning presenting with convulsion: Intoxication instead of alternative therapy. Turkish Journal of Medical Science. 2002;32:361–362.
24. Bayram NA, Keles T, Durmaz T, Dogan S, Bozkurt E. A rare cause of atrial fibrillation: Mad honey intoxication. Journal of Emergency Medicine. 2011. doi:10.1016/j.jemermed.2011.02.021
25. Gunduz A, Durmus I, Turedi S, Nuhoglu I, Ozturk S. Mad honey poisoning-related asystole. Emergency Medicine Journal. 2007;24:592–593. doi:10.1136/emj.2006.045625
26. Akıncı S, Arslan U, Karakurt K, Çengel A. An unusual presentation of mad honey poisoning: Acute myocardial infarction. International Journal of Cardiology. 2008;129:e56–e58. doi:10.1016/j.ijcard.2007.06.129
27. Dhakal A, Hashmi MF, Sbar E. Aflatoxin toxicity. [Updated 2023 Feb 19]. In StatPearls [Internet] 2024 Jan. StatPearls Publishing. https://www.ncbi.nlm.nih.gov/books/NBK557781
28. Pal S, Singh N, Ansari KM. Toxicological effects of patulin mycotoxin on the mammalian system: An overview. Toxicology Research. 2017;6(6):764–771. doi:10.1039/c7tx00138j
29. Kamle M, Mahato DK, Devi S, Lee KE, Kang SG, Kumar P. Fumonisins: Impact on agriculture, food, and human health and their management strategies. Toxins. 2019;11(6):328. doi:10.3390/toxins11060328
30. Bui-Klimke TR, Wu F. Ochratoxin A and human health risk: A review of the evidence. Critical Reviews in Food Science and Nutrition. 2015;55(13):1860–1869. doi:10.1080/10408398.2012.724480
31. Sobrova P, Adam V, Vasatkova A, Beklova M, Zeman L, Kizek R. Deoxynivalenol and its toxicity. Interdisciplinary Toxicology. 2010;3(3):94–99. doi:10.2478/v10102-010-0019-x
32. Bode AM, Dong Z. Toxic phytochemicals and their potential risks for human cancer. Cancer Prevention Research (Philadelphia, Pa.). 2015;8(1):1–8. doi:10.1158/1940-6207.CAPR-14-0160
33. WHO. WHO Guidelines on Safety Monitoring of Herbal Medicines in Pharmacovigilance Systems 2004. World Health Organization.
34. Chung RT. Detoxification effects of phytonutrients against environmental toxicants and sharing of clinical experience on practical applications. Environmental Science and Pollution Research International. 2017;24(10):8946–8956. doi:10.1007/s11356-015-5263-3.
35. Ekor M. The growing use of herbal medicines: Issues relating to adverse reactions and challenges in monitoring safety. Frontiers in Pharmacology. 2014;4:177. doi:10.3389/fphar.2013.00177
36. Logrieco AF, Miller JD, Eskola M, et al. The mycotox charter: Increasing awareness of, and concerted action for, minimizing mycotoxin exposure worldwide. Toxins. 2018;10(4):149.
37. Essers AJA, Alink GM, Speijers GJA, et al. Food plant toxicants and safety: Risk assessment and regulation of inherent toxicants in plant foods. Environmental Toxicology and Pharmacology. 1998;5(3):155–172.
38. Williams CT. Herbal supplements: Precautions and safe use. Nursing Clinics of North America. 2021;56(1): 1–21. doi:10.1016/j.cnur.2020.10.001.

Index